W9-CRV-831

BEACH PROCESSES AND SEDIMENTATION

Approaching Storm: Beach Near Newport
by the late nineteenth-century American artist Martin Johnson Heade.
(Courtesy of the Museum of Fine Arts, Boston; Karolik Collection)

BEACH PROCESSES AND SEDIMENTATION

PAUL D. KOMAR

School of Oceanography
Oregon State University

PRENTICE-HALL, INC., Englewood Cliffs, New Jersey

Library of Congress Cataloging in Publication Data

Komar, Paul D (date)
Beach processes and sedimentation.

Includes bibliographies and index.
1. Coast changes. 2. Marine sediments. 3. Ocean waves. 4. Tides. I. Title.
GB451.2.K65 551.4′5 75-44005
ISBN 0-13-072595-1

Printed in the United States of America

10 9 8 7 6 5 4 3 2 1

Prentice-Hall International, Inc., *London*
Prentice-Hall of Australia Pty. Limited, *Sydney*
Prentice-Hall of Canada, Ltd., *Toronto*
Prentice-Hall of India Private Limited, *New Delhi*
Prentice-Hall of Japan, Inc., *Tokyo*
Prentice-Hall of Southeast Asia Pte. Ltd., *Singapore*

To Jan and Kristi,
who have seen many beaches
but few mountains

CONTENTS

PREFACE

At Oregon State University I teach a one-term course in beach processes. The first time this course was offered I was confronted with graduate students in physical, geological and biological oceanography, civil engineering, geology, geophysics, and geography. It soon became apparent that lecturing to such a diverse group was not possible. I turned to writing and distributing my lectures to the students, allowing them to proceed at their own natural paces and understanding. This approach had the additional advantage of permitting the students to pursue in depth those subjects of interest to them individually. I assigned outside readings on their individual special interests. Class periods became "bull sessions" in which I answered questions or gave some indication of the direction of current research. Fortunately, my classes were small, so they could be handled in this way, and at the end of the term I gave the students oral exams, each tailored to what I felt they should have learned from the course.

Having originated in this way, the book is designed as much for self-study as it is for classroom use. It is meant only to be a starting point, a launching pad. In a book of this size and scope I could, unfortunately, give the material only a summary coverage. I have attempted to outline what is presently known but have not gone into the details or given derivations when mathematical formulations are called for. I hope the reader, like my class, will become sufficiently stimulated to go into greater depths on some of the subjects. For this purpose I provide an extensive reference section in each chapter. There is a vast literature written by civil engineers, oceanographers, geologists, and geographers. One of the main goals of this book is to attempt to bring this literature together. However, I am certain that I must have overlooked many important papers that deserved discussion. Unfortunately, I simply had to leave some out to keep the size of the book manageable. If, while reading this book, you find that I have overlooked *the* paper that should have been included, I would appreciate hearing from you.

This book is designed for the first- or second-year graduate level, although senior undergraduates, I suspect, could follow the material just as well. It is intended to encompass a wide spectrum of students interested in beach processes. Students entering my class always worry about whether they have sufficient mathematical skills. The book itself is not especially mathematical; mathematics through calculus is considered basic although people with only algebra and trigonometry can follow about 95 percent of the material. This is because derivations are not included. My purpose was to present only the results of the derivations and discuss what they physically mean and how they actually agree with the real world. I know that there will be some objections to not including the derivations, but if I included them, I would have lost some of the audience to which this book is directed, and at least three chapters would have been expanded to three separate texts. The purpose of this book is to meet the needs of a more general audience. The student who wants the full derivations is better served by seeking out the original sources in the references.

As with most books, this one has my own personal view of the subject. I have tried to treat all viewpoints equally, but I may not have always succeeded. In this regard, my own views continue to evolve, so please do not hold me to what I have written here, say, in two years.

Now, as to the content of the book. It begins with an introductory chapter that is the usual pep talk to convince the reader that the subject is relevant and worth knowing about. It also describes the uses of beaches, the various disciplines that study beaches, and pertinent literature sources. Chapter Two deals with coastal geomorphology and presents the general terminology that will be needed throughout the remainder of the book. Because there are other books already available on coastal geomorphology, the length of Chapter Two was kept short on purpose. The next few chapters deal with nearshore processes: water waves (their generation, travel, refraction, and breaking); tides and longer-term sea level changes; longshore current generation and prediction; and sand transport on beaches. The subsequent chapters deal with the response of the beach to these processes: shoreline configuration, including the variety of rhythmic shoreline forms; the beach profile; and the effects of constructing engineering structures in the nearshore and how this alters the equilibrium. The final chapter deals principally with the geological aspects of beach sedimentation, including what governs the composition of the beach, the roundness and shape of its grains, the many sedimentary structures found there, and the organisms found in the nearshore and how they affect the sedimentation. It ends with a summary discussion, including examples, of the identification of ancient beach deposits in the geologic rock record.

I am obligated to all who have assisted me in so many ways. First, there are my colleagues at Oregon State University who read parts of the manuscript and gave me encouragement. They include John V. Byrne, G. Ross Heath, Vern Kulm, John H. Nath, Tj. H. van Andel, Michael K. Gaughan, and Bruce Malfait. Others who have read various chapters include Robert Dean, William Fox, Donn S. Gorsline, G. V. Middleton, and Darwin Spearing. Their efforts are also much appreciated. I would, of course, like to thank the students of my class who gave the book the acid test and quickly demonstrated the weaknesses of some discussions and explanations. Over the four years in the preparation of the book three draftspersons were employed: Ron Hill, Natasa Sotiropolous, and Kathryn Torvik. Thanks also to those who supplied original photographs and diagrams from their publications. Finally, I would like to express my gratitude to Dr. Douglas L. Inman, Scripps Institution of Oceanography, who first stimulated my interest in the study of beaches.

P.D.K.

Corvallis, Oregon

NOTATIONS

a'	Pore-space factor for immersed weight of sand ($\simeq 0.6$)
C	Velocity of wave propagation (phase velocity)
c_f	Frictional drag coefficient for longshore current
D	Sediment grain diameter
d	Horizontal major diameter of the elliptical water particle motions under waves
E	Total wave energy density
f	Wave frequency ($= 1/T$)
F	Fetch or distance over which winds blow in generating waves
g	Acceleration of gravity
H	Wave height
h	Still-water depth
I_l	Total immersed weight sand transport rate along a beach
k	Wave number ($= 2\pi/L$)
L	Wave length

L_e	Length of edge waves
m	Longshore modal number for edge waves
n	Offshore modal number for edge waves
n	Ratio of wave group velocity to wave phase velocity [equation (3-23)]
P	Wave energy flux (power)
P_l	$= (ECn)_b \sin \alpha_b \cos \alpha_b$
p	Pressure under waves
q	Water discharge resulting from the mass transport $\bar{U}$ induced by waves
Q	Water discharge associated with the advance of a solitary wave [equation (3-53)]
S_{xx}	One component of the radiation stress (momentum flux due to the presence of the waves); the flux of x-directed momentum in the x-direction
S_{yy}	One component of the radiation stress (momentum flux due to the presence of the waves); the flux of y-directed momentum in the y-direction
S_{xy}	The longshore component of the radiation stress; the flux of y-directed momentum in the x-direction (onshore)
S_l	Total volume transport rate of sand along a beach
s	Vertical major diameter of the elliptical water particle motions under waves (Chapter 3); the spacing between wave rays during wave refraction
T	Wave period
T_e	Period of edge waves
t	Time
u	Horizontal component of the water particle velocity under waves
U	Wind velocity in the generation of waves
$\bar{U}$	Mass transport induced by wave motions [equations (3-33), (3-34), and (3-36)]
u_m	Maximum value of the horizontal orbital velocity, evaluated at the breaker zone
$\bar{v}_l$	Longshore current velocity
w	Vertical component of the water particle velocity under waves
X_b	Width of surf zone
α	Angle between the wave crest length and a line parallel to the shoreline

β	Angle of beach slope
γ	Ratio of wave height to water depth
η	Elevation of water surface due to waves
$\bar{\eta}$	Wave set-down or set-up—the difference between the still-water and the mean water level due to the presence of waves
λ	Spacing of sediment ripples
ν	Kinematic viscosity of water
ρ	Density of water
ρ_s	Density of sediment grains
σ	Wave radian frequency ($= 2\pi/L$)
ϕ	Velocity potential

Subscripts

∞	Denotes deep-water conditions
s	Denotes shallow-water wave conditions
b	Denotes breaking wave conditions
o	Denotes bottom orbital motions
1/3	Denotes significant wave evaluation of parameters ($H_{1/3}$, $T_{1/3}$)

Chapter 1

INTRODUCTION TO THE STUDY OF BEACHES

The shore is an ancient world, for as long as there has been an earth and sea there has been this place of the meeting of land and water. Yet it is a world that keeps alive the sense of continuing creation and of the relentless drive of life. Each time that I enter it, I gain some new awareness of its beauty and its deeper meanings, sensing that intricate fabric of life by which one creature is linked with another, and each with its surroundings.

Rachel Carson
The Edge of the Sea (1959)

The seacoast has always had great appeal; presently about two-thirds of the world's population lives within a narrow belt directly landward from the ocean edge. In the United States the thirty coastal states contain 75 percent of the total population and twelve of the thirteen largest cities (Soucie, 1973). Beaches and estuaries have therefore been the first to feel the full impact of our growing populations. It has been estimated (Inman and Brush, 1973, p. 26) that if everyone in the world all at once decided to spend some time along the 440,000 kilometers of the world shoreline (including the Arctic and Antarctic), each person would have less than 13 centimeters of shoreline. The population pressure is observable in crowded public beaches and in the proliferation of seaside condominiums, trailer parks, motels, and gas stations, which in many cases destroy the aesthetic value which originally drew people to the coast.

Because of these problems, concern over the condition of coasts and beaches is at an all-time high. Many states in the United States have adopted plans for management of the coastal zone. Britain with its National Trust already has a working system of coastal protection from unrestricted development, as do many other countries.

There are certain inherent dangers in living on the coast, since it is definitely less stable than interior areas. Steep rocky coasts are subject to landslides that can cause homes and sections of highway to fall into the sea (Figures 1-1 and 1-2). Sand beaches are particularly unstable, as the sand is constantly shifted about by waves, currents, and the wind. Hurricanes have destroyed property worth millions of dollars on the Atlantic and Gulf coasts of the United States. The concern with property erosion is demonstrated by the detailed attention given it in the *National Shoreline Survey* issued by the U.S. Army Corps of Engineers (1971). The survey showed that one-fourth of all the U.S. shoreline is experiencing significant erosion; if Alaska's coasts are excluded, the proportion is raised to 43 percent. Because the destruction of homes has an immediate visual impact, much interest and news coverage has been focused on coastal erosion.

Figure 1-1 Houses toppled because of beach erosion during a winter storm on North Carolina's Outer Bank. [*From* Soucie (1973); *photo by* Aycock Brown]

The American naturalist William Beebe wrote that the beach is "the battleground of the shore." Waves and currents constantly shift sand about, removing it from one area and depositing it in another. Sand added to the beach during calm weather is removed to the offshore during storms, only to be returned again with the next calm period (see Chapter 11). The natural beach therefore shows short-term fluctuations within a sort of dynamic equilibrium. Over hundreds of years a particular coastal area may be slowly eroding and retreating, or building outward into the sea. Man's construction often interferes with this natural variability in that once the house or road is built it is necessary to maintain the property upon which it rests.

> The real conflict of the beach is not between sea and shore, for theirs is only a lover's quarrel, but between man and nature. On the beach, nature has achieved a dynamic equilibrium that is alien to man and his static sense of equilibrium. Once a line has been established, whether it be a shoreline or a property line, man unreasonably expects it to stay put. (Soucie, 1973, p. 56)

Figure 1-2 The progressive destruction of the natatorium on Bayocean Spit, Oregon, because of coastal erosion. See the discussion in Chapter 12 for the reasons for the erosion. [*courtesy* Tillamook Pioneer Museum]

To maintain the property line, man has pitted himself against nature, constructing seawalls and piles of rock known as riprap. Not only is this a furtive attempt at battling nature, it is expensive and destroys the aesthetic value of the coast.

One additional factor in the population pressure on the coast is the impact of man's waste discharge into the shallow waters adjacent to the beaches. This is industrial waste as well as the community sewage. Many areas have had to close down their beaches for extensive periods and prohibit swimming because of the resulting danger to the public. Even the famed Waikiki Beach in Hawaii has suffered from this ignoble humiliation, at least temporarily. The requirement for additional electricity has led to the construction of nuclear and conventional power plants on the coast which use the water for cooling. In some cases this has led to thermal and radioactive pollution that affects the plants and animals living in the coastal waters.

USES OF THE BEACH

To anyone who lives near the coast or has been there on a holiday the uses of the beach are obvious: swimming, surfing, sunbathing, beachcombing, walking or jogging, fishing, picnicking (Figure 1-3). Good beaches attract thousands of visitors each year and are often economically important to the adjacent communities for that reason. The particular activities carried out at a specific beach are dependent upon such factors as the temperature of the water, the size of the waves, and the coarseness of the beach sediments. Beaches in Florida and California are used for swimming, whereas only the hardiest individual would venture into the waters off Maine or Washington State. Surfers travel long distances to find the beach with the "right kind of waves."

Groups other than the general public have an interest in the use of beaches. In certain parts of the world valuable minerals are mined from the beach sands, including diamonds in South Africa and gold in Nome, Alaska. Minable quantities of minerals such as magnetite (Japan), ilmenite, zircon, and platinum are also found in some beaches. In the past, beaches have been important sources of sand, gravel, and cobbles, but this use quickly destroys the beach and so is not now generally permitted. Boat users, whether commercial or pleasure, are frequently interested in the construction of boat basins in the nearshore region, often to the detriment of the local beaches, as will be seen in a later chapter. Finally, coastal communities and industries find the coastal waters convenient locations to dump their wastes.

In addition to these direct uses of the beach, there is an indirect use that is not generally recognized: beaches serve as natural buffers between the ocean and the land. The destruction and removal of these buffers leaves the coast exposed to the erosion effects of the waves. Attempts are then made to use riprap, junk cars, and so on to prevent erosion and property destruction, a job done formerly and much more effectively by the beach.

It is thus apparent that many potential uses of the beach and adjacent waters are mutually incompatible. As the beaches become even more intensely utilized because of increasing population, we must establish priorities in the possible uses of the coast and beaches. The beach is one of our finest natural resources. Once destroyed, its repair is difficult if not impossible, and it is always costly.

THE STUDY OF BEACHES AND COASTAL PROCESSES

Workers in several distinct professions or disciplines have participated in the study of beaches and coasts. Often two or more of these groups attack the same problems without recognizing the contributions being made by their counterparts, and generally they have different approaches and sets of prejudices. We shall see in the next section that each group also has its journals in which they publish their findings.

Historically, the first professionals who seriously studied beaches and coasts were the geographers and geomorphologists, and their contribution is continuing today. Their interest lies in recognizing and understanding the evolution of various coastal forms. They have done such things as study changes in particular stretches of coast by comparing existing configurations with old surveys and photographs; they have described various beach structures; and they have attempted to establish classifications of different coastal types and their developmental sequence through time. Their more recent contribution is in coastal planning, drawing up guidelines for the wise utilization of our coastal resources.

Geologists (sedimentologists) have also made an important contribution to the study of beaches, especially to knowledge of the nature of the sediments found there. Aside from their interest in modern sediments in their own right, geologists' study of present-day beach sediments is expected to aid them in recognizing ancient beach deposits in the geologic rock column. In their quest they concentrate on various properties of the sediments, such as the statistical properties of grain-size distribution, grain-size variations across the beach, degree of roundness and the shape of sediment grains, and sedimentary structures found in beach deposits. These aspects of the study of beaches are considered mainly in Chapter 13, although the geologists have also made important contributions in their study of beach profiles, in exploring the utilization of heavy minerals as natural tracers to determine sediment source, and in determining the origin and developmental history of barrier islands and related environments—subjects considered in other chapters.

The coastal engineer has the unenviable task of protecting our coasts from erosion, maintaining our beaches, and at the same time building jetties, boat harbors, and other similar structures on the coast. In general these tasks are not compatible with one another. Commonly the construction of such structures has pronounced deleterious effects on the adjacent beaches and may even affect beaches some distance away. The coastal engineer must anticipate this and try to prevent such destruction. To do this requires an understanding of the basic processes involved in the nearshore region. This interest has led to some fundamental studies by coastal engineers and resulted in important contributions to our understanding of that environment.

Figure 1-3 Recreational uses of the beach.

[*From* Co Rentmeester, *Life*, © Time Inc.]

[Astronaut John Glenn *courtesy* NASA]

World War II found us with insufficient knowledge to conduct the landing operations required in the Pacific war and in Europe such as occurred in the D-Day Normandy invasion. If such operations were to avoid disaster, we needed to know much more about the generation and propagation of waves on the ocean and their breaking characteristics in shallow water, about the character of longshore currents in the nearshore zone, and about factors affecting the slope of the beach face. There resulted a crash program that led to a quantum jump in our knowledge of the nearshore zone. Although much of this effort was conducted by coastal engineers, there sprang from this program a group of university scientists interested in the problems of the nearshore environment. Today the latter are concentrated mainly in the oceanography departments and special coastal studies groups of the universities. There they are able to examine problems of a fundamental nature which are not of obvious immediate applicability, and they are able to participate in studies of tides and waves which require the use of ships and other facilities that are more readily available at such institutions.

LITERATURE SOURCES

One of the principal difficulties in the study of beaches and coasts is sorting through the large number of journals, technical reports, memos, and other publications which contain pertinent information. This proliferation results chiefly from different groups working in the same environment, each with their distinct manner of publication. Table 1-1 contains a brief list and approximate classification of some of the more important sources. (This list is by no means exhaustive; its length could easily be doubled or tripled.) The group classification in Table 1-1 is oversimplified in that there is some overlap, however small, of publication sources of the several disciplines. There is some problem with a lack of communication, with each group in part ignoring the literature of the other groups and duplicating work. A more unified attack on the problems of the nearshore environment is required.

Several books pertinent to the study of nearshore processes are listed in the references at the end of this chapter. The geographers and geomorphologists have been the most prolific. The classic treatment in this area is the book by Johnson (1919); even though many of its concepts are outdated, much of it is still useful and provides a good accounting of the early literature. The books by Bird (1969), Davies (1973), Guilcher (1958), Ottmann (1965), Steers (1948, 1962), and Zenkovich (1967) are more recent examples of the geomorphological literature. The text by King (1972) is widely used and includes more discussion of processes on beaches than any of the above books. Shepard and Wanless (1971) attempted a comprehensive survey of the shoreline of the United States, including Alaska and Hawaii, documenting coastal changes that have occurred during historical times; it is a very useful starting point for any study of a specific area of coastline in the United States. The books by coastal engineers—Ippen (1966), Muir Wood (1969), and Wiegel (1964)—present the more technical side of the subject, with emphasis on engineering applications. The "bible" of the coastal engineer is the *Shore Protection Manual* (Coastal Engineering Research

TABLE 1-1 Literature Sources in Coastal Studies

GEOGRAPHY-GEOMORPHOLOGY	GEOLOGY
Geography	*Bulletin of the American Association of Petroleum Geologists*
Geographical Journal	*Geological Society of America Bulletin*
Journal of Geomorphology	*Journal of Geology*
Zeitschrift für Geomorphologie	*Journal of Sedimentary Petrology*
	Sedimentology
COASTAL ENGINEERING	OCEANOGRAPHY
Bulletins of the Coastal Engineering Research Center	*Journal of Geophysical Research*
Coastal Engineering in Japan	*Journal of Fluid Mechanics*
Dock and Harbor Authority	*Journal of Marine Research*
Journal of the Institute of Civil Engineers	*Proceedings of the Royal Society of London*
Proceedings of the Conferences on Coastal Engineering	
Proceedings of the American Society of Civil Engineers (Journal of Waterways, Harbors, and Coastal Engineering)	

Center, 1973), a compendium of tables and graphs. There are also several books that concentrate on wind waves; these are included among the references to Chapter 3. Finally, the books by Manley (1968) and Bascom (1964) provide somewhat lighter reading on the subject. Manley presents the "lives, legends, and lore" of beaches, and Bascom gives a very readable account of waves and beaches on an elementary level.

REFERENCES

BASCOM, W. (1964). *Waves and beaches.* Doubleday, Garden City, N.Y., 267 pp.

BIRD, E. C. F. (1969). *Coasts.* M.I.T. Press, Cambridge, 246 pp.

Coastal Engineering Research Center (1973). *Shore protection manual.* U.S. Army Corps of Engineers, Washington, D.C., 3 volumes.

DAVIES, J. L. (1973). *Geographical variation in coastal development.* Hafner, New York, 204 pp.

DOLAN, R., and J. MCCLOY (1965). *Selected bibliography on beach features and related near-shore processes.* Louisiana State Univ. Press, Baton Rouge, 59 pp.

GUILCHER, A. (1958). *Coastal and submarine morphology.* Methuen, London, 274 pp.

INMAN, D. L., and B. M. BRUSH (1973). The coastal challenge. *Science,* 181: 20–32.

IPPEN, A. T. (1966). *Estuary and coastline hydrodynamics.* McGraw-Hill, New York, 744 pp.

JOHNSON, D. W. (1919). *Shore processes and shoreline development.* Wiley, New York, 584 pp. Facsimile edition: Hafner, New York. (1965).

KING, C. A. M. (1972). *Beaches and coasts.* 2nd ed. St. Martin's Press, New York, 570 pp.

MANLEY, S., and R. MANLEY (1968). *Beaches; their lives, legends, and lore.* Chilton, Philadelphia, 383 pp.

MUIR WOOD, A. M. (1969). *Coastal hydraulics.* Macmillan, London, 187 pp.

OTTMANN, F. (1965). *Introduction à la géologie marine et littorale.* Masson, Paris, 259 pp.

SHEPARD, F. P., and H. R. WANLESS (1971). *Our changing coastlines.* McGraw-Hill, New York, 579 pp.

SOUCIE, G. (1973). Where beaches have been going: into the ocean. *Smithsonian*, 4, no. 3: 55–61.

STEERS, J. A. (1948). *The coastline of England and Wales.* Cambridge Univ. Press, Cambridge, 644 pp.

——— (1962). *The sea coast.* 3rd ed. Collins, London, 292 pp.

U.S. Army Corps of Engineers (1971). National shoreline study: shore management guidelines. Department of the Army, Corps of Engrs., Washington, D.C., 56 pp.

WIEGEL, R. L. (1964). *Oceanographical engineering.* Prentice-Hall, Englewood Cliffs, N. J., 532 pp.

ZENKOVICH, V. P. (1967). *Processes of coastal development.* Translated by D. G. Fry and edited by J. A. Steers. Oliver and Boyd, Edinburgh, 738 pp.

Chapter 2

COASTAL GEOMORPHOLOGY

If the great geological labours of the oceans, such as the erosion of cliffs, the demolition of promontories, and the construction of new shores, astonish the mind of man by their grandeur, on the other hand, the thousand details of the strands and beaches charm by their infinite grace and marvelous variety.

Elisee Reclus
The Ocean, Atmosphere, and Life (1873)

Several entire books have been written on coastal geomorphology (see Chapter 1); for this reason the subject will only be outlined here. The main purpose of this chapter is to present the basic terminology that will be needed in all future discussions. We shall also examine such topics as the mechanisms of coastal erosion and ideas concerning the origin of barrier islands. Other topics generally included in coastal geomorphology will be considered in later chapters: shoreline configuration, beach profile morphology, and so on.

GENERAL BEACH NOMENCLATURE

It is necessary to begin with the introduction of some basic terminology which describes the beach profile and the overlying nearshore waters. Here an attempt is made to conform as closely as possible to the definitions given in Coastal Engineering Research Center (1973, appendix) and to the established usage of coastal geomorphologists. Some departure is necessary where the customary nomenclature is inadequate.

The *beach* is an accumulation of unconsolidated sediment (sand, shingle, cobbles, and so forth) extending shoreward from the mean low-tide line to some physiographic change such as a sea cliff or dune field, or to the point where permanent vegetation is established. This term may satisfy the sunbather, but it is unsatisfactory from our standpoint since the beach under this definition does not include any portion that is permanently underwater; thus the beach includes only that portion which is exposed to the atmosphere at some time. We require a more inclusive term, one that will encompass this underwater portion of the environment, since that is where the more important processes occur which are responsible for the beach formation. We shall therefore use the term *littoral* to denote this entire environment. The ecologists use the term *littoral* to include only the intertidal zone between mean high- and mean low-tide levels. The geologists, however, frequently employ the term *littoral* to extend across the beach and into the water to a depth at which the sediment is less actively transported by surface waves. This depth varies, of course, but is generally considered to be some 10 to 20 meters. We shall use the term in this more inclusive sense. In actual practice the term *beach* is used more loosely than indicated above, and commonly is almost synonymous with our definition of the *littoral* zone. The term *coastal* is even more inclusive in that the coastal region extends further inland to include the sea cliff, any marine terraces, dune fields, and so on; the shoreward limit is very indefinite. The *nearshore zone* extends seaward from the shoreline to just beyond the region in which the waves break, so that this term is particularly useful when discussing waves and currents within this environment.

Figure 2-1 illustrates the terminology used to describe the beach profile, while Figure 2-2 diagrams the terms for the wave action in the nearshore zone. An alphabetical list of definitions for the terms follows:

Backshore: The zone of the beach profile extending landward from the sloping foreshore to the point of development of vegetation or change in physiography (sea cliff, dune field, and so on).

Beach face: The sloping section of the beach profile below the berm which is normally exposed to the action of the wave swash.

Beach scarp: An almost vertical escarpment notched into the beach profile by wave erosion. Its height is commonly less than a meter, although higher examples are found.

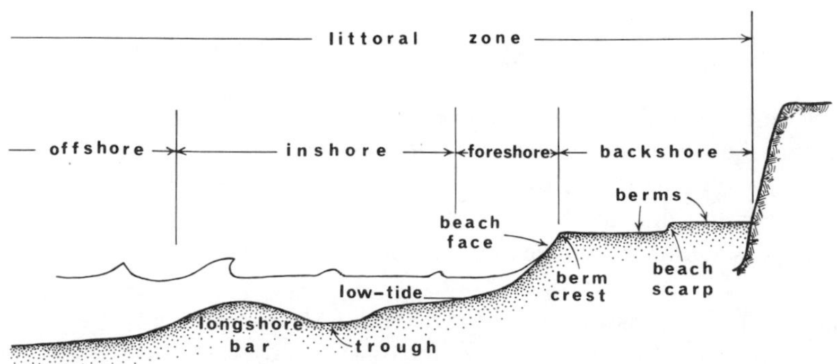

Figure 2-1 The terminology used to describe the beach profile. See text for definitions.

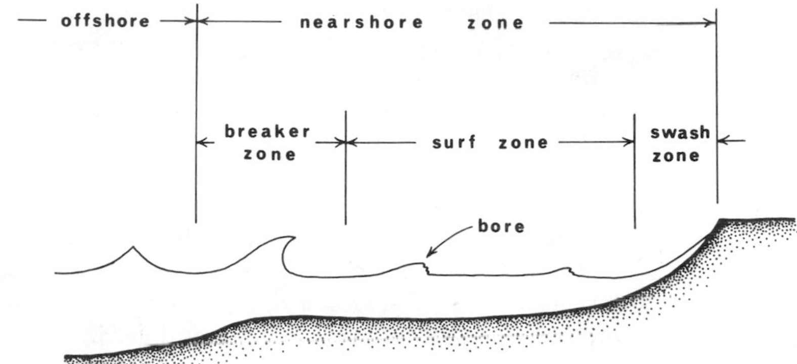

Figure 2-2 The terminology used to describe the wave action and currents in the nearshore region. See text for definitions.

Berm (beach berm): A nearly horizontal portion of the beach or backshore formed by the deposition of sediment by the receding waves. Some beaches have more than one berm, while others have none.

Berm crest (berm edge): The seaward limit of a berm.

Breaker zone: The portion of the nearshore region in which the waves arriving from offshore reach instability and break. With very simple uniform waves, such as may be generated in a laboratory wave tank, the zone may be reduced to a breaker line. On a wide, flat beach secondary breaker zones may occur in which reformed waves break for a second time.

Foreshore: The sloping portion of the beach profile lying between a berm crest (or in the absence of a berm crest, the upper limit of wave swash at high tide) and the low-water mark of the backrush of the wave swash at low tide. This term is often nearly synonymous with the beach face but is commonly more inclusive, containing also some of the flat portion of the beach profile below the beach face.

Inshore: The zone of the beach profile extending seaward from the foreshore to just beyond the breaker zone.

Longshore bar: A ridge of sand running roughly parallel to the shoreline. It may become exposed at low tide. At times there may be a series of such ridges parallel to one another but at different water depths.

Longshore trough: An elongated depression extending parallel to the shoreline and any longshore bars that are present. There may be a series at different water depths.

Offshore: The comparatively flat portion of the beach profile extending seaward from beyond the breaker zone (the inshore) to the edge of the continental shelf. This term is also used to refer to the water and waves seaward of the nearshore zone.

Shore: The strip of ground bordering any body of water, whether the ground is rock or loose sediment. If it is unconsolidated sediment, then *shore* becomes synonymous with *beach* used in its restricted sense.

Shoreline: The line of demarcation between the water and the exposed beach.

Surf zone: The portion of the nearshore region in which borelike translation waves occur following wave breaking. This portion extends from the inner breakers shoreward to the swash zone.

Swash zone: The portion of the nearshore region in which the beach face is alternately covered by the uprush of the wave swash and exposed by the backwash.

In Chapter 7 we shall see that longshore currents and currents associated with nearshore circulation cells are generated in the surf zone. This contrasts with the swash zone where true longshore currents do not develop and instead the water motion consists of a series of surges, a zigzag or sawtooth longshore movement. Because of this the processes of sediment movement differ in these two zones (Chapter 8). The distinction between surf and swash zones is therefore real and not artificial. Schiffman (1965) defines a transition zone between the surf and swash zones (not included in Figure 2-2) in which the return flow of the swash collides with the incoming surf bores, creating high turbulence, a broad energy spectrum, and a bimodal sand-size distribution.

The presence and width of the surf zone is primarily a function of the beach slope and the tidal stage. Beaches of low slope, those normally composed of fine sand, are characterized by wide surf zones. In contrast, the steeply sloping shingle and cobble beaches seldom possess a surf zone; the waves break close to shore and develop directly into an intense swash surging up and down the beach face. Moderately sloping beaches commonly lack a surf zone at high tide, when the waves break close to shore over the steeper beach face, but develop a surf zone at low tide, when the wave action is over the flatter portions of the beach profile.

With a large tidal range there may be both a "high-tide beach" and a "low-tide beach" separated by a stretch of tidal flat of low slope (Figure 11-6). Usually the "high-tide beach" is composed of material that is coarser than found in the remainder of the profile; it could consist of cobbles, shingle, or gravel, while the tidal flat and "low-tide beach" is of sand. The "low-tide beach" is generally not much coarser than the tidal flat sediments but is usually better sorted.

EROSIONAL COASTAL FORMS

When waves strike a sea cliff, they cause erosion and a general retreat of the cliff. This retreat may be relatively rapid, the changes wrought being readily apparent within one's lifetime (Figure 2-3). Homes and other structures unwisely built too close to the cliff may be undermined and destroyed (Figure 2-4). In England entire medieval villages are known to have disappeared, their former locations now being out to sea (Figure 2-5).

Associated with the retreating rocky coast is a series of typical erosional features, diagramed in Figure 2-6. The erosion of the cliff leaves a gently sloping rock platform called the *wave-cut platform*, leveled off near the water surface. The waves may find local weaknesses in the sea cliff, which they excavate to form *sea caves*. Commonly observed are the *sea stacks*, where a more resistant portion of the rock remains in the

Figure 2-3 Retreat of the sea cliffs at Sunset Cliffs, San Diego, California, between 1946 (upper photo) and 1968 (lower photo). The sea stacks have been destroyed by the wave action; also note the slide that has reduced the cliff in the lower lefthand corner of the 1968 photo. The erosion has left a flat wave-cut platform exposed at low tide. [*photographs from* F. P. Shepard]

surf, separated from the retreating cliff (Figures 2-7 and 2-8). More spectacular but rarer is the natural *sea arch* (Figure 2-9), where the wave erosion has hollowed out a line of weakness in an otherwise resistant promontory. When the sea cliff retreat is faster than a stream is able to cut down its bed, the stream is left hanging and a waterfall results (Figure 2-10), plunging into the sea.

Wave erosion of sea cliffs is achieved principally through the hydraulic pressure exerted by the wave impact, which may reach immense values, and by the abrasive action of sand and rock fragments hurled at the cliff by the waves. Wave erosion of the cliffs occurs chiefly during storms. The waves undercut the sea cliff at its base

Figure 2-4 German World War II bunkers on the northwest coast of Jutland, Denmark, undermined by the retreating sea cliffs.

which, left unsupported, may collapse and slide down to the shoreline (Figure 2-11). The slide debris is then worn down by the wave action and is either retained as a beach or is carried out to sea. A well-known example is the Portuguese Bend landslide, Palos Verdes Hills, California, studied by Merriam (1960). The slide covers some 4×10^6 m^2 and is moving seaward at rates of 1 to 3 cm/day, carrying homes and roads with it.

Little actual quantitative work has been conducted on the processes of cliff erosion by waves. Shepard and Grant (1947) and others have shown qualitatively that rock type, jointing and bedding patterns, and wave exposure are important factors. The series of studies by Horikawa and Sunamura (1967, 1968, 1970) and Sunamura and Horikawa (1969, 1971) provide the best attempts at quantifying sea cliff erosion. They conducted field studies on the coast of Japan and attempted to correlate cliff erosion rates to wave height and period and to rock properties (compression, penetration strength, and so on). They also conducted laboratory wave tank studies of erosion of artificial cliffs composed of mixtures of cement and fine sand. The wave action in the tank cut notches into the artificial sea cliffs, and Horikawa and Sunamura related the rates of notch erosion again to the wave conditions and properties of the artificial stone. They also studied the effectiveness of engineering protection measures in decreasing the erosion rate. Many more studies of this sort are required if we are better to understand sea cliff erosion.

In addition to wave action, there are various weathering processes, ice wedging, rain wash, and so on that act to reduce the sea cliffs just as they do inland slopes. Chemical solution of the rock by seawater and the activities of organisms which burrow into the rock or scrape vegetation from its surface may in certain instances be important in eroding the sea cliff, although in general these factors are not as significant as wave action. Burrowing clams such as *Pholadidae* (Figure 2-12) may considerably weaken an otherwise resistant massive rock so that it more quickly succumbs to wave attack. Jehu (1918) reported localities on the coast of England where pholads

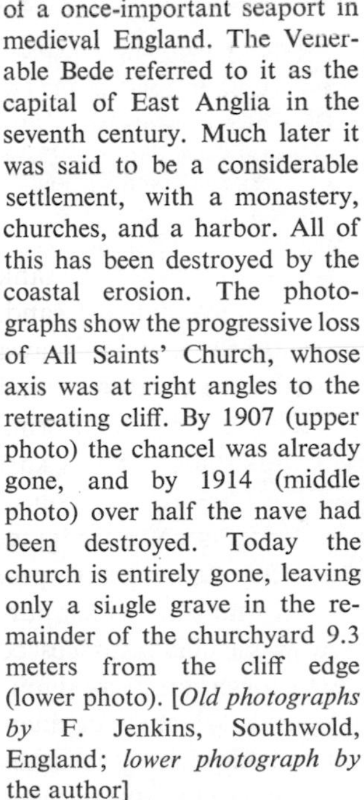

Figure 2.5 Dunwich, the site of a once-important seaport in medieval England. The Venerable Bede referred to it as the capital of East Anglia in the seventh century. Much later it was said to be a considerable settlement, with a monastery, churches, and a harbor. All of this has been destroyed by the coastal erosion. The photographs show the progressive loss of All Saints' Church, whose axis was at right angles to the retreating cliff. By 1907 (upper photo) the chancel was already gone, and by 1914 (middle photo) over half the nave had been destroyed. Today the church is entirely gone, leaving only a single grave in the remainder of the churchyard 9.3 meters from the cliff edge (lower photo). [*Old photographs by* F. Jenkins, Southwold, England; *lower photograph by* the author]

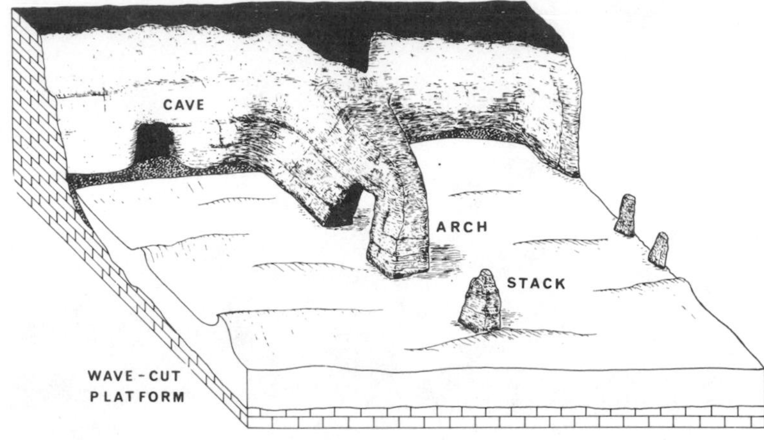

Figure 2-6 Erosional features associated with the retreating rocky coast.

Figure 2-7 Stacan Gobhlach on the north coast of the Isle of Skye, Scotland. These stacks and sea cliffs are formed of a resistant columnar jointed sill. The term *stack* is derived from the Gaelic *stacan.*

locally lowered the chalk at rates of from 1.8 to 3.3 cm per year, averaging 2.3 cm. Vita-Finzi and Cornelius (1973) found cliff-sapping rates of 0.25 cm/yr in Oman, caused by the burrowing date-stone mollusk *Lithophaga.* North (1954), by studying the ingestion rate and stomach content of periwinkles, a small intertidal snail, estimated that with their scraping radulae one hundred snails could remove 86 cm^3 of rock material yearly, or almost one liter in a decade. Since the rock in question is a rather friable sandstone, it would appear that their contribution is generally relatively small compared to erosion by waves. In addition to mollusks, other organisms may aid in rock erosion: marine worms, boring barnacles (Ahr and Stanton, 1973), boring sponges, and sea urchins. Normal seawater is saturated with calcium carbonate and so would not dissolve limestone or sediments cemented together by $CaCO_3$. However, Emery (1946) has demonstrated that at night organisms locally increase the acidity of the water and may cause some solution, producing high-tide rock basins. The small pocket marks sometimes observed near the water's edge, commonly housing

Figure 2-8 The Needles, Isle of Wight, England, a series of stacks in the massive Cretaceous chalk. [*Etching from* T. H. Huxley, *Physiography* (London: Macmillan, 1880)]

Figure 2-9 Durdle Door, a natural sea arch on the Dorset coast of England. This arch is just to the west of Lulworth Cove (Figures 2-14 and 2-15) and is composed of the resistant Jurassic limestone.

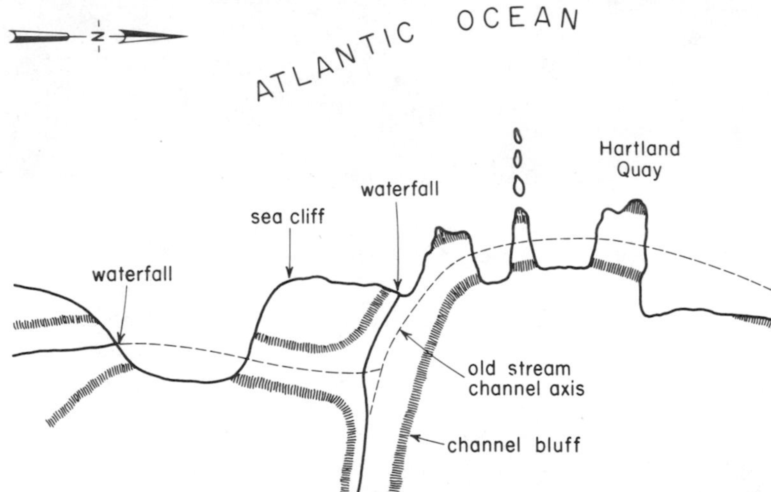

Figure 2-10 (Upper photo) Waterfall of Wargery Water on the northern Devon coast at Hartland Quay, England, formed by the retreat of the coast intersecting the stream channel. (Lower) Field sketch of the cliff intersection with the old stream course (shown dashed) producing waterfalls. For more information on the waterfalls of north Devon, consult E. A. N. Arber, *The Coast Scenery of North Devon* (London, Cambridge Univ. Press, 1911).

periwinkles, chitons, and so on, may also be attributed to biochemical leaching in this way. As well as providing food for periwinkles to scrape off the rocks, blue-green algae may also cause some direct dissolution of calcium carbonate in the rock.

Marine erosion produces a general retreat of the coast, but the retreat is usually not uniform because of the variations in rock resistance. The more resistant coastal rock formations retreat slowly and remain as headlands, stacks, and offshore islands, while the weaker formations are cut back to form embayments between the headlands. Solid and massive igneous rocks, most metamorphic rocks, and some limestones are very resistant to wave attack and hence are found in headlands. In contrast, friable

Figure 2-11 Land slump which has destroyed some 800,000 square meters of land on Cascade Head, Oregon. Note the cliff cut into the slide debris by the wave action. [*From* North and Byrne (1965)]

sandstones, shales, and rocks which have many bedding planes or are weakened by closely spaced joints or faults are easily eroded by waves. The most rapid erosion occurs in unconsolidated sediments such as glacial deposits. The most famous example of sea cliff erosion by wave action is the Holderness, England, where the glacial deposits are retreating at rates up to 1.75 m/yr, destroying farms and the ancient town of Dunwich (Figure 2-5) (Valentin, 1954).

The orientation of rock formations is an important factor in the resulting configuration of the eroding coast. When the beds trend normal to the coast and stand nearly vertical, the result is pronounced headlands where the resistant strata are found, separated by bays where the weaker formations occur (Figure 2-13). Because of wave refraction (Chapter 4) the energy of the waves is concentrated on the resistant headlands and weakened in the bays. This and simple protection of the bays from wave attack from many directions prevents a continued higher erosion rate within the bays, so that there exists a maximum permitted shore relief to which the coast is likely to erode.

When the rock strata trend parallel to the coast, the erosion proceeds somewhat differently. The classic example of this situation is the Dorset coast of Purbeck, southern England. Several stages in the development of coast erosion can be seen there (Figure 2-14). Stair Hole [Figures 2-14 and 2-15(upper)] demonstrates the earliest phase. The waves have cut out a series of caves and arches into the resistant Jurassic limestone, one cave having collapsed, leaving a gap. The waves are now beginning to scour out the less resistant Cretaceous sands and clays behind the resistant limestone. The next stage of advancing erosion is Lulworth Cove [Figures 2-14 and 2-15(lower)], close by Stair Hole. At Lulworth Cove the gap in the Jurassic is wider, and the weak Cretaceous beds have been carved out into an almost circular bay. Erosion further landward is being hampered by the high ridge of resistant chalk backing Lulworth

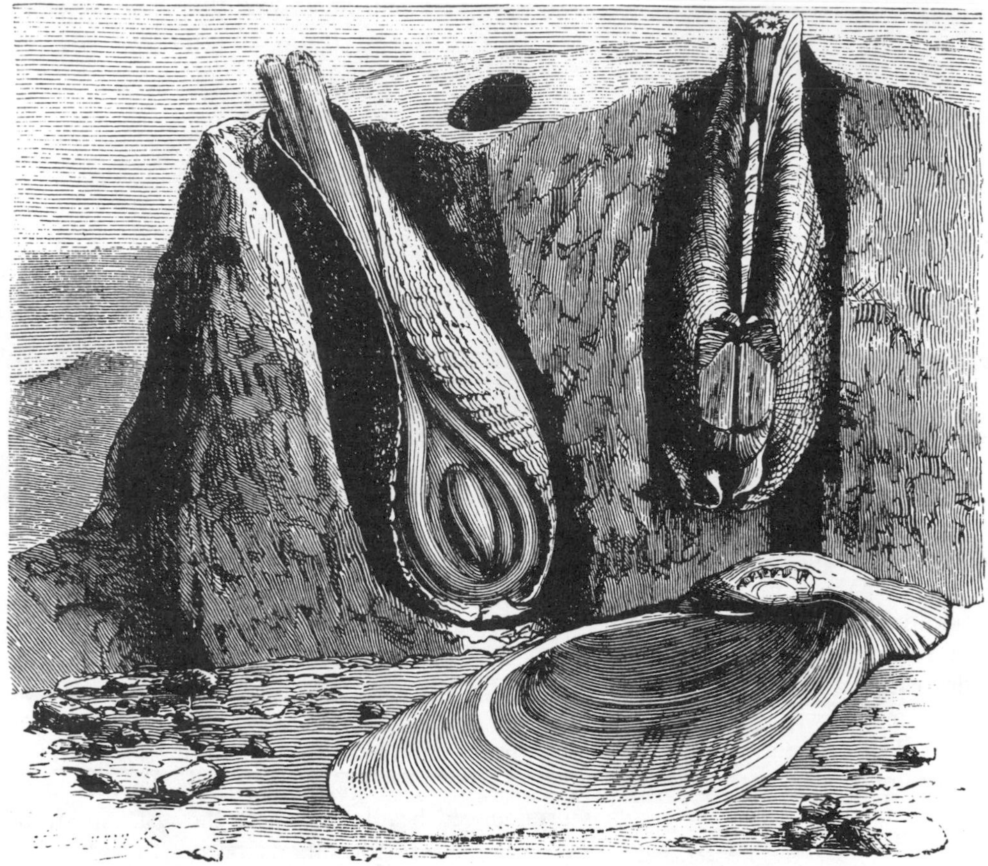

Figure 2-12 The rock-burrowing clam *Pholadidae*. "Their ceaseless, and as yet inexplicable, toil enables them to pierce deeply into the most compact stones. We are astonished, in splitting marble, to find living shells in the midst of blocks which the chisel of the sculptor cuts with difficulty." [*Etching and quotation from* Pouchet, *The Universe* (Edinburgh, Blackie and Sons, 1884)]

Cove. An even more advanced stage of the coast erosion is demonstrated by Worbarrow Bay further to the east. A broad embayment has developed, and the waves are now concentrating on eroding the Cretaceous chalk ridge.

Where the rock strata trend nearly parallel to the coast and dip seaward (Figure 2-16), landsliding and rock falls are an especially important process, since following wave undercutting the rock readily slides down the bedding-plane surface into the sea. This leaves the uncovered bedding planes as the coastal slope. North of Yaquina Bay, Oregon, such sliding is produced in the Quaternary terrace deposits which rest unconformably on the seaward-dipping Miocene sandstones and shales (Byrne, 1963). On the south coast of England, coastal rock slides are common where the permeable

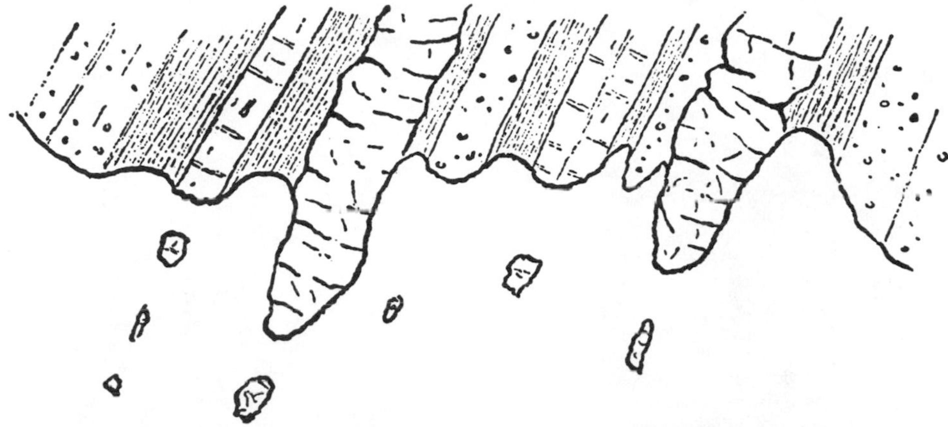

Figure 2-13 Rock strata trending normal to the coastline, the more resistant beds forming headlands and sea stacks and the weaker beds forming embayments. [*From* James Geikie, *Outlines of Geology* (London, Edward Stanford, 1896)]

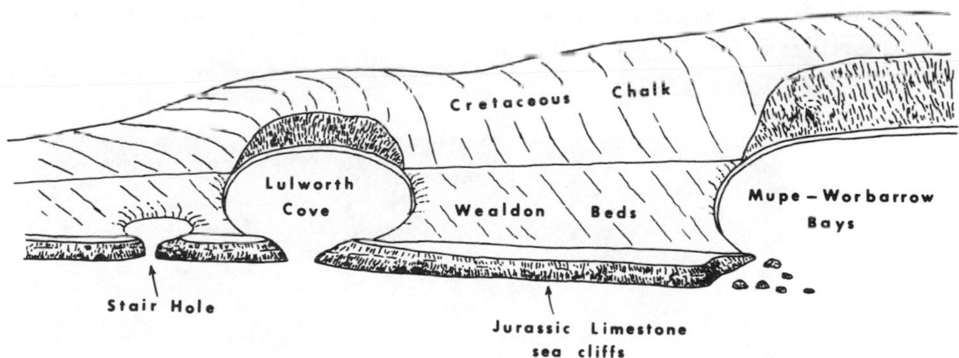

Figure 2-14 The erosion of the Purbeck coast of Dorset, England. The bays are formed by the waves eroding the weak Wealdon Beds (shales and friable sandstones) between the very resistant Jurassic limestone and the ridge of Cretaceous chalk.

Chalk and Upper Greensand formations dip seaward, resting on the impermeable Gault Clay (Bird, 1969, p. 53). The giant slide on the Palos Verdes Hills, California, also occurs in seaward-dipping deposits (Merriam, 1960).

Horizontally stratified formations and rocks that are of a massive uniform nature lead to the development of uniform sea cliffs and coastlines. The chalk cliffs of Dover and the Seven Sisters, Sussex, in the same deposits, are examples of this common condition.

Figure 2-15 (upper) Stair Hole, Dorset, the initial stage of bay formation by breaching the Jurassic limestone and removal of the weak Wealdon Beds. (lower) Lulworth Cove, Dorset, a more advanced stage of erosion where the entire width of the Wealdon Beds has been eroded away. [Aerofilms Ltd.]

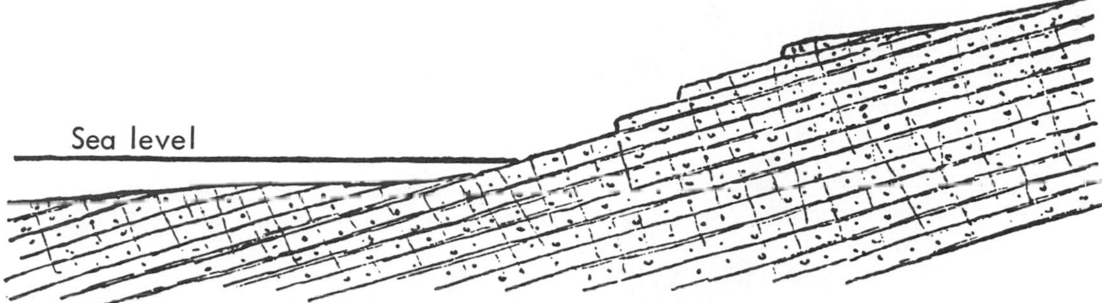

Figure 2-16 When the rock strata dip seaward, landslides and rock falls are important and leave the bedding-plane surface as the coastal slope. [*From* Geikie, *Outlines of Geology*]

DEPOSITIONAL COASTAL FORMS

Once sediment deposition on the coast becomes important and the sea cliffs are fronted by beaches, a new set of coastal formations prevail. Besides the beach itself, one of the more common depositional features is the *spit* (Figures 2-17 and 2-18), a beach that is tied to the coast at one end and free at the other. A spit grows in the direction of predominant longshore sediment drift (Figure 2-19), and is often a continuation of the beach that is adjacent to the coast; at other times the spit departs from the trend of the coast and aligns itself nearly at right angles to the prevailing wave direction. The free end terminates in a hook or recurve (Figures 2-20 and 2-21), formed either by wave refraction around the terminal end (Evans, 1942) or by the interplay of wave trains arriving from different directions (King and McCullagh, 1971). Often older hooks can be seen trailing landward from the spit, with almost the same configuration as the active hook. Spits are most common on irregular coasts, where they grow across bay mouths, sometimes completely closing them. The growth of spits

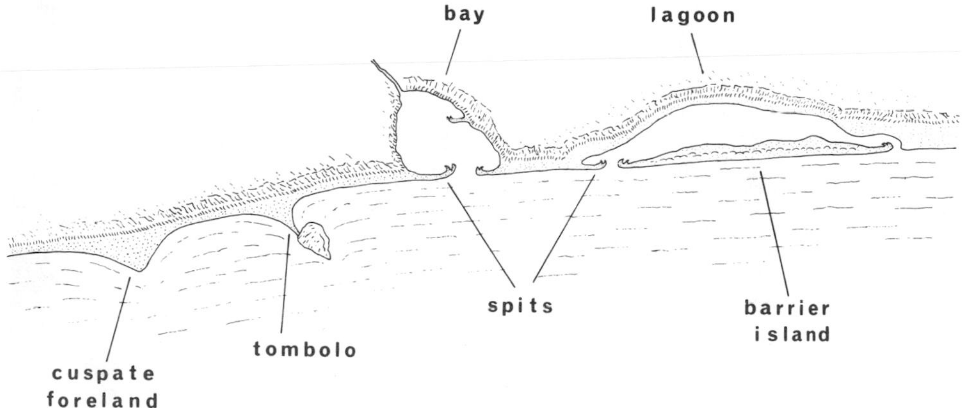

Figure 2-17 Features associated with the depositional coast.

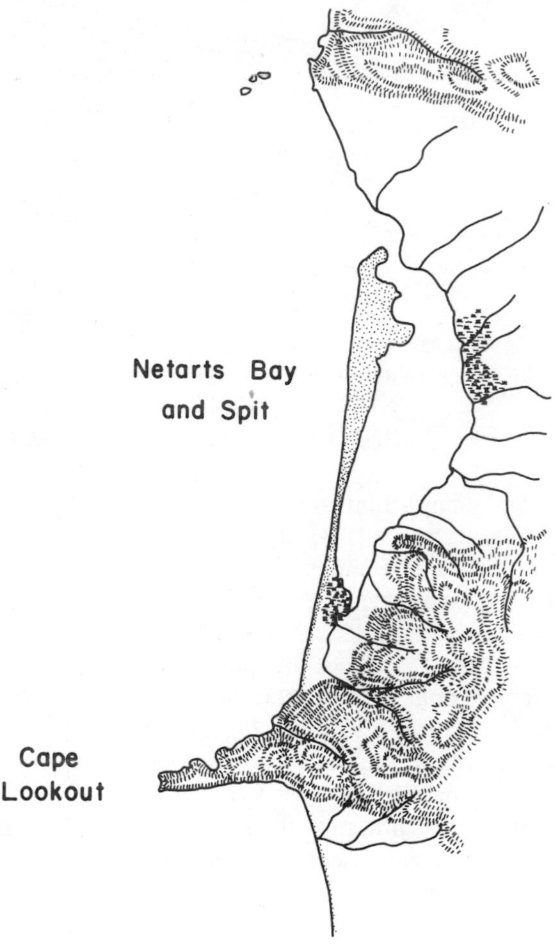

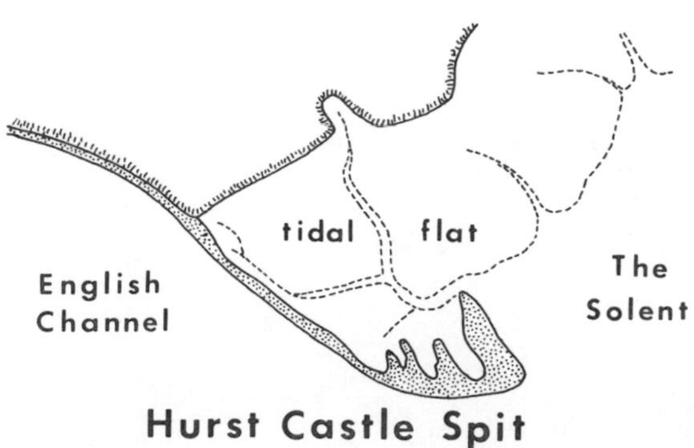

Figure 2-18 A variety of spits from throughout the the world: Netarts Spit, Oregon; Hurst Castle Spit, England; and Presque Isle, Lake Erie [*From* Johnson (1919)]

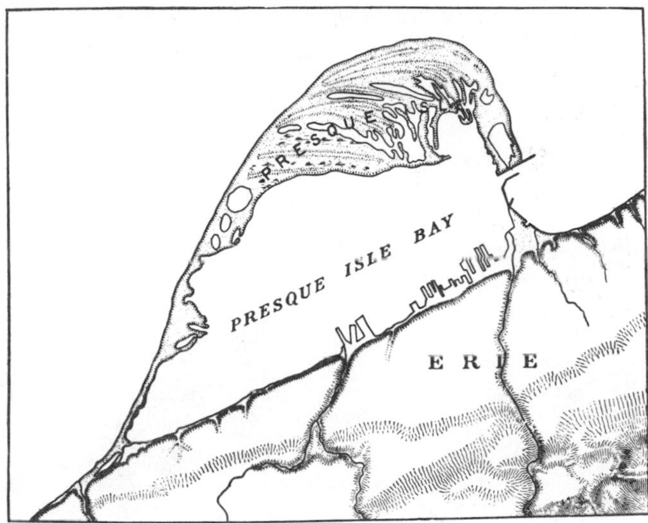

Figure 2-18 (continued)

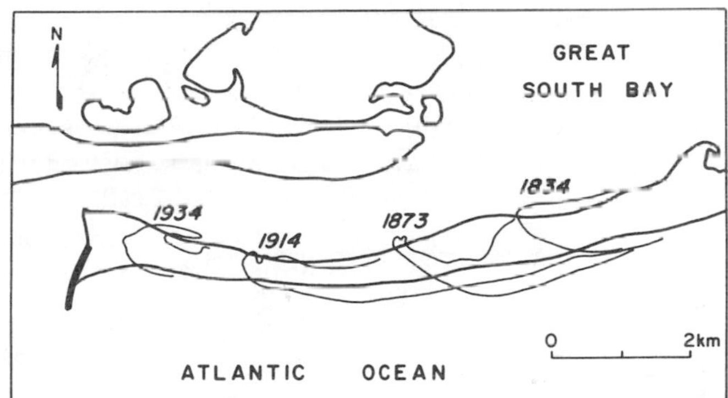

Figure 2-19 The longshore growth of Fire Island, a sand spit on the south shore of Long Island, New York. [*After* Saville (1961)]

Figure 2-20 Recurved sand spit, Duck Point, Grand Traverse Bay, Lake Michigan. The spit has been built out from a point of land on the left, just out of view. [*From* I. C. Russell, *Lakes of North America* (Boston, Ginn and Company, 1895)]

Figure 2-21 The recurved terminal end of a 15-meter-long spit in Loch Rannoch, Scotland. The maximum height of the unbroken waves was 4 cm and period 0.8 sec. The length of the spit was oriented parallel to the crests of the waves. Of interest, this spit showed longshore variations in the grain size much like observed at Chesil Beach (Chapter 13) grading from fine sand to granules (2—4 mm), with the coarsest material at the terminal end shown in the photograph.

in this way is an important factor in smoothing an initially irregular coastline (Johnson, 1919, p. 301).

A growing spit often deflects the mouth of a river or the entrance to a bay, prolonging it in the direction of longshore sediment drift. The mouth of the River Alde, Suffolk, England (Figure 2-22), has been deflected for about 18 km by a shingle spit, after having come within 45 meters of the sea (Steers, 1953, p. 152). Records indicate that at the time Orford Castle was built in 1165, it stood at the very mouth

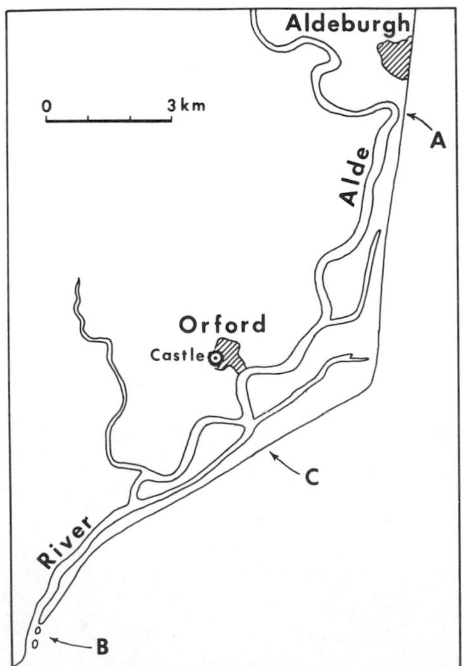

Figure 2-22 The River Alde, Suffolk, England, whose mouth has been deflected for about 18 km by the growth of the shingle spit of Orford Ness. (A) The river comes within 45 meters of the sea. (B) The present river mouth (C) Approximate position of the river mouth in 1165, at the time of the building of Orford Castle. [*After* Steers (1953)]

of the river. Maps dating back to the sixteenth century show the spit still ending opposite Orford early in the century but growing rapidly southward at the time (Steers, 1953). Since the early sixteenth century the spit has lengthened itself by 6.5 km to the south. Old maps and written records are very useful in this way for examining the very-long-term changes in the coast. This kind of information, however, is generally limited to western Europe and parts of North America. Comparisons of dated aerial photographs are now being used to trace the shorter-term evolution of spits and other coastal features (Shepard and Wanless, 1971).

When a spit links the mainland to an offshore island it becomes a *tombolo* (Figures 2-17 and 2-23). A tombolo may also develop from a cusp-shaped body growing in the lee and eventually connecting with an island. The term comes from Italy, where these features are particularly well developed. Johnson (1919) has described examples from New England, and Bird (1969) provides examples from Australia.

A variety of cuspate shoreline features may be observed (Chapter 10). The largest, the *cuspate forelands*, are sometimes associated with offshore islands or shoals, the shape of the cusp being governed by wave refraction around the island (Russell, 1958). Other cuspate forelands are clearly independent of such a mode of formation. Dungeness, on the south coast of England (Figure 2-24), is thought to have built out at a point of convergence of the longshore sediment drift, sediment building up from two directions. Waves arriving diagonally at that portion of coast are stronger and more frequent than those coming directly offshore because of the positioning of

Figure 2-23 A double tombolo, two spits connecting with an offshore island on the west coast of Italy. [*From* Johnson (1919)]

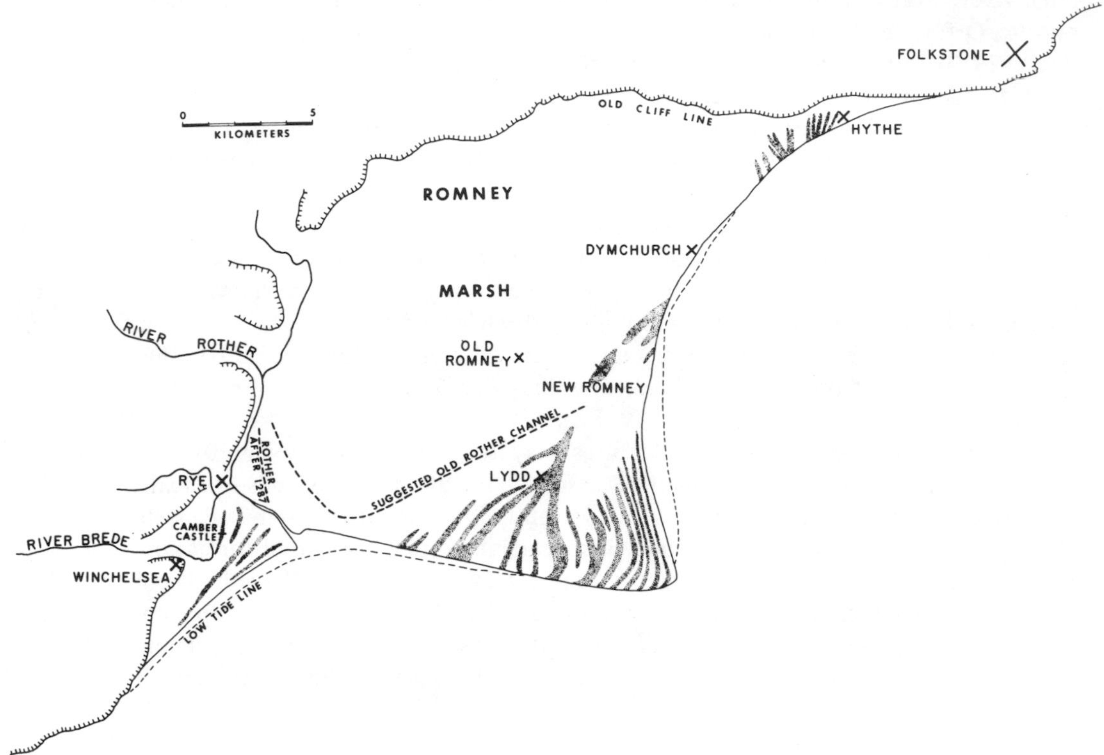

Figure 2-24 Dungeness, a type of cuspate foreland on the Sussex coast of England. The shaded areas represent old beach ridges. [*After* Lewis (1932)]

France across the English Channel. The development of the cuspate forelands on the southeast coast of the United States (Capes Hatteras, Lookout, Fear, Canaveral) has received considerable attention. Suggestions of origin have included eddies associated with the Gulf Stream, structural uplift, wave refraction by offshore shoals (Russell, 1958), and the reworking of river delta sediments (Hoyt and Henry, 1971). There are also a variety of cuspate shoreline features on a smaller scale than the cuspate forelands: sand waves, giant cusps, and beach cusps. The origins of all these features will be considered in Chapter 10.

Along most of the lowland coasts of the world the beaches occur on *barrier islands* which run parallel to the mainland coast but are separated from the mainland by lagoons and bays (Figure 2-17). Barriers are best developed where the tidal range is relatively low and where the wave energy is low. In the United States barrier islands extend along the east coast from Long Island to Florida, and almost all the beaches of the Gulf coast are of this type. They are essentially absent on the west coast, except for Alaska. Outside the United States chains of barrier islands are found on the North Sea coasts of the Netherlands, Germany, and Denmark, in the Mediterranean along the south of France and on the Nile Delta, on the southeast coasts of Africa, Brazil, India, and Ceylon, and on the southeast coast of Australia.

There has been a continuing debate over the origin of barrier islands. De Beaumont (1845) contended that they develop by the erosion of the bottom under the wave breaker zone, the sediment being piled up into a ridge shoreward of the breakers. Gilbert (1885, p. 87) attributed their formation to the extension of longshore spits parallel to the coast. Johnson (1919, pp. 348–92) suggested that most barriers result from the emergence of a flat, gently sloping continental shelf which permits the waves to break at a considerable distance from the shore. It is doubtful that Johnson is correct in this, however, as all the present barrier islands grew during a period marked by a rapid rise in sea level brought about by the melting of glaciers ending the last ice age (Chapter 6); an emergent coast is not required for barrier island development.

Steers (1953) has suggested that Chesil Beach, on the south coast of England (Figure 13-6), developed during the earlier stages of the ocean transgression following the last ice age, when the rising sea collected gravels that had been previously strewn across the continental shelf, and swept them shoreward to their present position. Shepard (1960, 1963) has invoked a similar sweeping action of the water transgression to explain the formation of the Gulf coast barrier islands. LeBourdiec (1958) reached a similar conclusion for barrier islands on the Ivory Coast, and Zenkovich (1967) invoked the process to explain barriers of sand and shingle that develop in front of uncliffed mainland coasts in the Soviet Union.

The most recent episode of this debate on the origin of barrier islands involves the papers of Hoyt (1967) and Otvos (1970) and discussions of them (Cooke, 1968; Fisher, 1968; Hoyt, 1970). Hoyt puts forward the hypothesis that barrier islands are initiated by the building of a ridge immediately landward of the shoreline, formed from wind- or water-deposited sediments, followed by a slow submergence that floods the area landward of the ridge, forming the lagoon and transforming the ridge into a barrier island. Otvos indicates that drilling results in the Gulf of Mexico suggest that those barrier islands were formed by the upward aggradation of submerged shoal areas. Otvos also stresses that subsequent barrier migration may be extensive, obscuring the conditions of formation.

It appears that we have not heard the last word in this long debate. In all probability, barrier islands have originated in a variety of ways, and this is the reason why there is supporting evidence for and objections to all the suggested hypotheses.

One of the earliest and most extensive studies of barrier islands was Project 51 of the American Petroleum Institute (Shepard, Phleger, and van Andel, 1960). This study investigated the organisms as well as the sediments of the Gulf coast barrier islands, the lagoons and bays, and the adjacent continental shelf. Since that time there has been a proliferation of literature on barrier islands, mostly on their geologic history and sedimentation. Allen (1965) studied the barrier islands on the Nigerian coast, and the Gulf coast barriers have also been investigated by Fisk (1959), LeBlanc and Hodgson (1959), and Conatser (1971), and those on the east coast of the United States by Fisher (1973), Ruzyla (1973), Godfrey and Godfrey (1973), Dolan (1973), and

Figure 2-25 Beach ridges and the modern shoreline on the coast of Nayarit, Mexico. [*From* Curray and Moore (1964)]

Kraft, Biggs, and Halsey (1973). This last series of papers is contained in *Coastal Geomorphology*, edited by D. R. Coates; this book has other relevant articles as well.

The term *beach ridge* is used in the literature with two different meanings. Gravel and shingle beaches sometimes have a beach ridge on their shoreward limits where storm waves have piled up the coarse material considerably above the normal high-tide level. A series of ridges on a prograding beach may result from successive storms, each of which forms a ridge parallel to the shoreline (Lewis and Balchin, 1940). In some areas this type of beach ridge or storm ridge consists of shells. They are almost never formed of sand, since the finer sediments are swept into deeper water by the storm waves rather than being built into ridges.

The more common use of the term *beach ridge* is in reference to a series of long parallel ridges which are typically spaced from 25 to 500 meters apart. The relief of the individual ridges is small and can be best viewed from the air. Each ridge was formed individually as a shoreline deposit, the present beach often representing the most recent ridge of the series. They constitute lines of growth of the shoreline and when preserved enable one to decipher the history of development. The mechanism most often proposed for their formation is similar to the de Beaumont (1845) suggestion for the origin of barrier islands: each ridge started as a submerged longshore bar in front of the then-existing beach. Such a bar could build up to the surface at high tide and then become exposed at low tide. If low-water conditions persisted through the low tide and several tidal cycles thereafter, the longshore bar could be built up still further, permanently isolating the former beach. Most of the observed series of beach ridges were built since sea level reached its present position some 3,000 to 5,000 years ago (Chapter 6). Curray and Moore (1964), who studied a series of some 250 such beach ridges on the Costa de Nayarit, Mexico, calculated that an average of 12 to 20 years were required for the formation of each ridge, assuming a regular and uniform rate of formation.

REFERENCES

AHR, W. M., and R. J. STANTON (1973). The sedimentologic and paleoecologic significance of *Lithotyra*, a rock-boring barnacle. *J. Sediment. Petrol.* 43: 20–23.

ALLEN, J. R. L. (1965). Coastal geomorphology of eastern Nigeria: beachridge barrier islands and vegetated tidal flats. *Géologie en Mijnbouw*, 44: 1–21.

BIRD, E. C. F. (1969). *Coasts.* M.I.T. Press, Cambridge, 246 pp.

BYRNE, J. V. (1963). Coastal erosion, northern Oregon. In *Essays in marine geology in honor of K. O. Emery*, ed. T. CLEMENTS, pp. 11–33. Univ. of Southern California Press, Los Angeles.

Coastal Engineering Research Center (1973). *Shore protection manual.* U.S. Army, Corps of Engineers, Washington, D.C., 3 volumes.

COATES, D. R. (1973), ed. Coastal geomorphology. Publications in geomorphology, State Univ. of N.Y., Binghamton, 403 pp.

CONATSER, W. E. (1971). Grand Isle: a barrier island in the Gulf of Mexico. *Geol. Soc. Am. Bull.*, 82: 3049–68.

COOKE, C. W. (1968). Barrier island formation: discussion. *Geol. Soc. Am. Bull.*, 79: 945.

CURRAY, J. R., and D. G. MOORE (1964). Holocene regressive littoral sand, Costa de Nayarit, Mexico. In *Deltaic and shallow marine deposits*, ed. L. M. J. U. van Straaten, pp. 76–82. Elsevier, Amsterdam.

DE BEAUMONT, E. (1845). *Leçons de géologie pratique.* Paris.

DOLAN, R. (1973). Barrier islands: natural and controlled. In *Coastal geomorphology*, ed. D. R. Coates, pp. 263–78.

EMERY, K. O. (1946). Marine solution basins. *J. Geol.*, 54: 209–28.

EVANS, O. F. (1942). The origin of spits, bars, and related structures. *J. Geol.*, 50: 846–63.

FISHER, J. J. (1968). Barrier island formation: discussion. *Geol. Soc. Am. Bull.*, 79: 1421–26.

——— (1973). Bathymetric projected profiles and the origin of barrier islands: Johnson's shoreline of emergence, revisited. In *Coastal geomorphology*, ed. D. R. Coates, pp. 161–79.

FISK, H. N. (1959). Padre Island and Laguna Madre flats, coastal south Texas. *Proc. 2nd Coastal Geog. Conf.*, Louisiana State Univ., pp. 103–51.

GILBERT, G. K. (1885). The topographic features of lake shores. *U.S. Geol. Survey 5th Ann. Rpt.*, pp. 69–123.

GODFREY, P. J., and M. M. GODFREY (1973). Comparison of ecological and geomorphic interactions between altered and unaltered barrier island systems in North Carolina. In *Coastal geomorphology*, ed. D. R. Coates, pp. 239–58.

HORIKAWA, K., and T. SUNAMURA (1967). A study of erosion of coastal cliffs by using aerial photography. *Coastal Eng. in Japan*, 10: 67–83.

——— (1968). An experimental study of erosion of coastal cliffs due to wave action. *Coast. Eng. in Japan*, 11: 131–47.

——— (1970). A study of coastal cliffs and of submarine bedrocks. *Coast. Eng. in Japan*, 13: 127–39.

HOYT, J. H. (1967). Barrier island formation. *Geol. Soc. Am. Bull.*, 78: 1125–36.

——— (1970). Development and migration of barrier islands, northern Gulf of Mexico: discussion. *Geol. Soc. Am. Bull.*, 81: 3779–82.

HOYT, J. H., and V. J. HENRY (1971). Origin of capes and shoals along the southeastern coast of the United States. *Geol. Soc. Am. Bull.*, 82: 59–66.

JEHU, T. J. (1918). Rock boring organisms as agents in coast erosion. *Scot. Geog. Mag.*, 34: 1–11.

JOHNSON, D. W. (1919). *Shore processes and shoreline development.* Wiley, New York, 584 pp., [Facsimile edition: Hafner, New York (1965)].

KING, C. A. M., and M. J. MCCULLAGH (1971). A simulation model of a complex recurved spit. *J. Geol.*, 79: 22–37.

KRAFT, J. C., R. B. BIGGS, and S. D. HALSEY (1973). Morphology and vertical sedimentary sequence models in Holocene transgressive barrier systems. In *Coastal geomorphology*, ed. D. R. Coates, pp. 321–54.

LEBLANC, R. J., and W. D. HODGSON (1959). Origin and development of the Texas shorelines. *Gulf Coast Assoc. Geol. Soc. Trans.*, 9: 197–220.

LEBOURDIEC, P. (1958). Aspects de la morphogenese plio-quaternaire en basse Côte d'Ivoire. *Rev. Geomorph. Dyn.*, 9: 33–42.

LEWIS, W. V. (1932). The formation of Dungeness foreland. *Geog. J.*, 80: 309–24.

LEWIS, W. V., and W. G. V. BALCHIN (1940). Past sea levels at Dungeness: *Geog. J.*, 96: 258–85.

MERRIAM, R. (1960). Portuguese Bend landslide, Palos Verdes Hills, California. *J. Geol.*, 68: 140–53.

NORTH, W. B., and J. V. BYRNE (1965). Coastal landslides of northern Oregon. *Ore Bin*, 27, no. 11: 217–41.

NORTH, W. J. (1954). Size distribution, erosion activities, and gross metabolic efficiency of the marine intertidal snails, *Littorina planaxis* and *L. scutulata*. *Biol. Bull.*, 106: 185–97.

OTVOS, E. G. (1970). Development and migration of barrier islands, northern Gulf of Mexico. *Geol. Soc. Am. Bull.*, 81: 241–46.

RUSSELL, R. J. (1958). Long straight beaches. *Ecol. Geol. Helv.*, 51: 591–98.

RUZYLA, K. (1973). Effects of erosion on barrier-island morphology, Fire Island, New York. In *Coastal geomorphology*, ed. D. R. Coates, pp. 219–37.

SAVILLE, T. (1961). Sand transfer, beach control, and inlet improvements, Fire Island inlet to Jones Beach, New York. *Proc. 7th Conf. Coastal Eng.*, pp. 785–807.

SCHIFFMAN, A. (1965). Energy measurements in the swash-surf zone. *Limnol. and Oceanog.*, 10: 255–60.

SHEPARD, F. P. (1952). Revised nomenclature for depositional coastal features. *Bull. Am. Assoc. Petrol. Geol.*, 36: 1902–12.

——— (1960). Gulf coast barriers. In Shepard, Phleger, and van Andel (1960), pp. 197–220.

——— (1963). *Submarine geology*. 2nd ed. Harper & Row, New York, 557 pp.

SHEPARD, F. P., and U. S. GRANT IV (1947). Wave erosion along the southern California coast. *Geol. Soc. Am. Bull.*, 58: 919–26.

SHEPARD, F. P., F. B. PHLEGER, and TJ. H. VAN ANDEL, eds. (1960). *Recent sediments, northwest Gulf of Mexico*. American Association of Petroleum Geologists, Tulsa, Okla. 394 pp.

SHEPARD, F. P., and H. R. WANLESS (1971). *Our changing coastlines.* McGraw-Hill, New York, 579 pp.

STEERS, J. A. (1953). *The sea coast.* Collins, London, 292 pp. [The page numbers in text refer to the 3rd edition, 1962.]

SUNAMURA, T., and K. HORIKAWA (1969). A study on erosion of coastal cliffs by using aerial photographs. Report No. 2. *Coast. Eng. in Japan*, 12: 99–120.

——— (1971). A quantitative study on the effect of beach deposits upon cliff erosion. *Coast. Eng. in Japan*, 14: 97–106.

VALENTIN, H. (1954). Der landverlust in Holderness, Ostengland, von 1852 bis 1952. *Die Erde*, 3, no. 4: 296–315.

VITA-FINZI, C., and P. F. S. CORNELIUS (1973). Cliff sapping by molluscs in Oman. *J. Sediment. Petrol,.* 43: 31–32.

ZENKOVICH, V. P. (1967). *Processes of coastal development.* Translated by D. G. Fry and edited by J. A. Steers. Oliver and Boyd, Edinburgh, 738 pp.

Chapter 3

THEORIES OF WAVE MOTIONS

. . . And, believe me, if I were again beginning my studies, I would follow the advice of Plato and start with mathematics.

Galileo
Dialogue Concerning the Two World Systems (1630)

Wind-generated surface waves are the principal source of input energy into the littoral zone. They are responsible for the erosion of the coast and for the formation of depositional beach features. In subsequent chapters we shall see that the waves generate currents in the nearshore region, currents that are in turn responsible for the littoral drift of sediments along the beach. An understanding of wave action is therefore fundamental to an understanding of the processes that take place on beaches.

In Chapter 4 we shall concern ourselves with the generation of ocean waves by storms and with the travel of the waves from their source area to the beach. Ocean waves are irregular and complex, and it is best if we first consider regularized, idealized, theoretical waves. This chapter will examine the various theories of water waves that yield simplex mathematical formulas that may be applied to waves on the sea. We shall look at the assumptions involved in the derivations of these formulas and the resulting limitations of application, and review the laboratory investigations which test the theoretical wave theories.

The wave theories assume trains of long, smooth, uniform waves such as may be approximately generated in a laboratory wave tank. *Swell waves*, distant from the area of generation, often approximate these conditions so that one may apply the theories to describe the wave motions and energy transfer. At other times the real waves are short-crested and more random and irregular than depicted by the theories. At the end of this chapter we shall examine how these irregular waves are measured and analyzed.

PERIODIC WAVES

Wave motion is periodic; that is, it is repetitive through fixed periods of time. At some stationary position—say, the pile in Figure 3-1—a succession of wave crests (or troughs) pass at fixed intervals of time T, the *wave period.* If L is the *wavelength,*

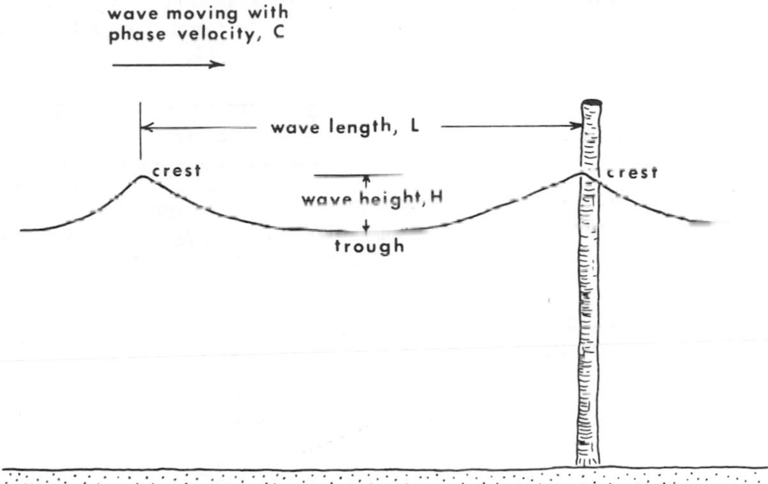

Figure 3-1 The parameters that, along with the wave period T, describe simple oscillatory wave motion.

the horizontal distance between successive crests (or troughs), then the velocity of propagation of the wave form, C, is simply

$$C = \frac{L}{T} \tag{3-1}$$

This relationship holds for any periodic waves, since it is based solely on the geometry. We shall see in a moment that there is a direct relationship in many instances between the wave length and the period of the waves. When the waves are in shallow water, the water depth also becomes an important parameter.

The other parameter required to describe this simple periodic wave motion is the *wave height* H, the vertical distance from trough to crest (Figure 3-1). The wave height is independent of the other wave parameters, and to a first approximation the reverse is true as well. To higher approximations the wave height causes a slight increase in the wave velocity C: the higher the wave, the greater the increase in C over that of a wave of negligible height. This effect is small, however.

If one follows the motion of a cork floating on the water surface as waves pass, it will be observed that the cork rises and falls and at the same time moves back and forth, describing a circular motion whose diameter is the wave height H and whose period is T. The cork makes no net advance in the direction of wave motion. Waves may therefore transfer energy and momentum across the water surface, often for thousands of kilometers, with negligible net drift of the water itself; the water does not move along with the wave form.

WAVE THEORIES

There are five wave theories that are commonly applied to describe wave motion on the sea. The general form, application, and founding references for the five are outlined in Table 3-1. Airy wave theory represents the simplest formulation, but because of the assumptions involved in the derivation its application should be limited

TABLE 3-1 Outline of Water Wave Theories

AIRY WAVES (sinusoidal)

Application: Waves of small amplitude in deep water.

References: Laplace (1776), Airy (1845).

STOKES AND GERSTNER WAVES (trochoidal)

Application: Waves of finite amplitude in deep, intermediate, and shallow water.

References: Gerstner (1802), Stokes (1847), Froude (1862), Rankine (1863), Rayleigh (1877).

CNOIDAL WAVES

Application: Waves of finite amplitude in intermediate to shallow water.

References: Korteweg and DeVries (1895), Keller (1949).

SOLITARY WAVES

Application: Solitary or isolated crests of finite amplitude moving in shallow water.

References: Scott-Russell (1844), Boussinesq (1871), Rayleigh (1876), McCowan (1891).

to waves of small amplitude. The theory is actually strictly true only for the limiting case of waves of zero height. Where waves of appreciable height are involved, one of the other, more complex wave theories may be required.

Airy Wave Theory

Laplace (1776) presented the first satisfactory treatment for waves of small amplitude in water of arbitrary depth. Airy (1845) developed a theory for irrotational waves traveling over a horizontal bottom in any depth of water. In the derivation of this theory the equations are linearized, and for this reason the theory is often referred to as "linear wave theory."

If the viscosity of the water is ignored, and if one considers waves outside the area of generation, then there will be no energy loss or gain. In addition, as shown by Lord Kelvin, the water motion will be *irrotational*; that is, the individual particles of which the fluid is composed will not spin but will rather retain their original orientations in space throughout the movement. Stated another way, without internal friction (viscosity), particles that have no angular momentum must remain without angular momentum. Although one's initial feeling is that these assumptions are wrong, water wave motion is in large part (except near the bottom) reasonably regarded as nonviscous and irrotational.

If the motion is irrotational, then the horizontal velocity u and the vertical velocity w of the water particles can be derived from a velocity potential $\phi(x, z, t)$ such that

$$u(x, z, t) = \frac{\partial \phi(x, z, t)}{\partial x}$$
$$w(x, z, t) = \frac{\partial \phi(x, z, t)}{\partial z} \qquad (3\text{-}2)$$

where t is time and the coordinates are diagramed in Figure 3-2. This is a considerable simplification, since one need only consider the single parameter ϕ throughout the derivation, rather than u and w separately.

Water can be considered to be homogeneous and incompressible (the density $\rho =$ constant) so that there must at all times be a continuity of the water mass; water cannot be miraculously created or destroyed. This property of continuity can be expressed as the equation

$$\frac{\partial u}{\partial x} + \frac{\partial w}{\partial z} = 0 \qquad (3\text{-}3)$$

or, in terms of the velocity potential,

$$\frac{\partial^2 \phi}{\partial x^2} + \frac{\partial^2 \phi}{\partial z^2} = 0 \qquad (3\text{-}4)$$

which is known as the *Laplace equation of continuity*. Any solution for the water wave motion must obey this continuity relationship.

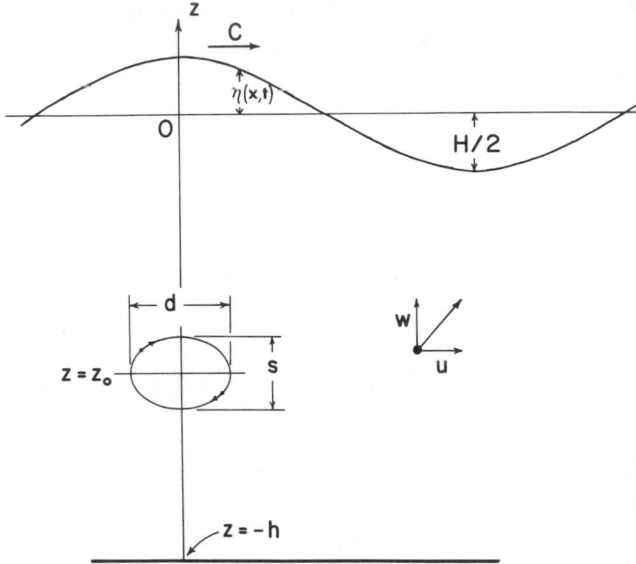

Figure 3-2 Notation associated with the Airy wave theory. Each water particle moves with a horizontal velocity u and vertical velocity w in an elliptical orbit of major diameter d and minor diameter s.

To represent the wave motions requires that the solution satisfy certain boundary conditions. These are:

1. No fluid must pass through the sea floor:

$$w(x, z = -h, t) = 0$$

2. The pressure p at the sea surface must equal the atmospheric pressure.

The pressure requirement at the surface is applied by way of the Bernoulli relationship

$$\frac{p}{\rho} = \frac{\partial \phi}{\partial t} - \frac{1}{2}(u^2 + w^2) - gz \tag{3-5}$$

Assuming that the wave slope is small ($H/L \ll 1$) and that the water depth h is much greater than the wave height ($h/H \gg 1$), a solution is obtained in which the elevation η of the water surface is

$$\eta(x, t) = \frac{H}{2} \cos \left(\frac{2\pi}{L}x - \frac{2\pi}{T}t\right) \tag{3-6}$$

The resulting wave profile is therefore found to be sinusoidal with a period T, wavelength L, and height H (Figure 3-2). The factors $2\pi/L$ and $2\pi/T$ occur repeatedly and so are denoted by k and σ, respectively; k is commonly called the *wave number*. The inverse of the wave period, $f = 1/T$, is the *wave frequency*, so that σ is the radian frequency. Using this notation, equation (3-6) can then be abbreviated

$$\eta(x, t) = \frac{H}{2} \cos (kx - \sigma t) \tag{3-7}$$

Figure 3-3 compares the theoretical sinusoidal profile with the surface time history of a laboratory wave. Table 3-2 lists the formulas for the velocity potential, the orbital

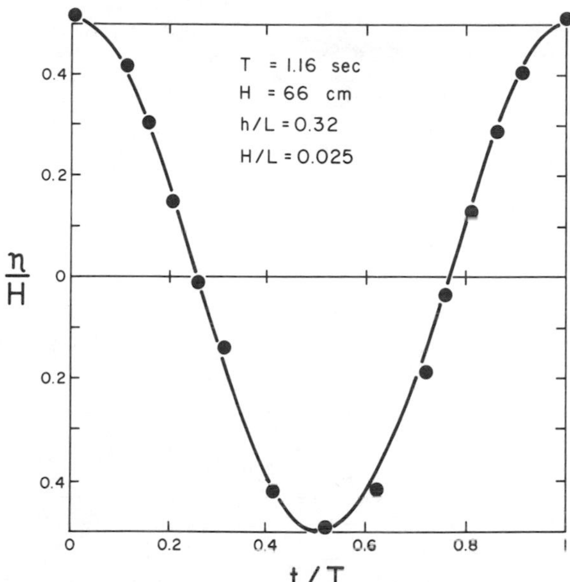

Figure 3-3 Comparison of water surface time history of laboratory waves with the theoretical sinusoidal profile of the Airy wave. [*After* Wiegel (1964)]

dimensions and velocities, and the pressure changes with depth which are obtained in the solution. Figure 3-2 illustrates the notation used. It should be noted that the coordinates are such that the x-axis is positive in the direction of wave motion and the z-axis is positive upward; the origin of the coordinates is on the still-water surface. This means that if you wish to compute the orbital velocity at 200 cm below the surface, for example, you must set $z = -200$ cm in the appropriate equation.

One fundamental result obtained in the solution is the relationship

$$\sigma^2 = gk \tanh (kh) \tag{3-8}$$

or, substituting the identities for k and σ,

$$L = \frac{g}{2\pi} T^2 \tanh \left(\frac{2\pi h}{L}\right) \tag{3-9}$$

There is obviously some difficulty is applying this equation, since it contains L on both sides, L being tucked under the hyperbolic tangent on the right side. We shall see in a moment that tables have been developed which give us the desired solutions in the general case, but there are limiting cases in which the relationship simplifies that should be examined first.

Figure 3-4 contains a graph of the function tanh (r) for the r-range of interest. It is apparent from this graph that as $r = kh = 2\pi h/L$ becomes large, we have

$$\tanh \left(\frac{2\pi h}{L}\right) \simeq 1 \tag{3-10}$$

This approximation has application to deep water where h is large in comparison to the wave length L. If we substitute this approximation into equation (3-9), we obtain

$$L_\infty = \frac{g}{2\pi} T^2 \tag{3-11}$$

TABLE 3-2 Linear Airy Wave Equations

PARAMETER	GENERAL EXPRESSION	DEEP WATER ($h/L_\infty > \frac{1}{4}$)	SHALLOW WATER ($h/L_\infty < \frac{1}{20}$)
surface elevation $\eta(x, t)$	$\eta = \frac{H}{2}\cos(kx - \sigma t)$		
phase velocity C	$C = \frac{gT}{2\pi}\tanh\frac{2\pi h}{L}$	$C_\infty = \frac{gT}{2\pi} = \frac{g}{\sigma}$	$C_s = \sqrt{gh}$
wave length L	$L = \frac{gT^2}{2\pi}\tanh\frac{2\pi h}{L}$	$L_\infty = \frac{gT^2}{2\pi}$	$L_s = T\sqrt{gh}$
velocity potential ϕ	$\frac{HC}{2}\frac{\cosh[k(z+h)]}{\sinh(kh)}\sin(kx - \sigma t)$	$\frac{HC_\infty}{2}e^{kz}\sin(kx - \sigma t)$	$\frac{HgT}{4\pi}\sin(kx - \sigma t)$
horizontal orbital diameter d	$d = H\frac{\cosh[k(z_o+h)]}{\sinh(kh)}$	$d = He^{kz_o}$	$d = \frac{HL}{2\pi h} = \frac{HT}{2\pi}\sqrt{\frac{g}{h}}$
vertical orbital diameter s	$s = H\frac{\sinh[k(z_o+h)]}{\sinh(kh)}$	$s = He^{kz_o}$	$s = 0$
horizontal orbital velocity u	$u = \frac{\pi H}{T}\frac{\cosh[k(z+h)]}{\sinh(kh)}\cos(kx - \sigma t)$	$u = \frac{\pi H}{T}e^{kz}\cos(kx - \sigma t)$	$u = \frac{H}{2}\sqrt{\frac{g}{h}}\cos(kx - \sigma t)$
vertical orbital velocity w	$w = \frac{\pi H}{T}\frac{\sinh[k(z+h)]}{\sinh(kh)}\sin(kx - \sigma t)$	$w = \frac{\pi H}{T}e^{kz}\sin(kx - \sigma t)$	$w = 0$
pressure deviation $\Delta p(x, z, t)$	$\frac{\rho gH}{2}\frac{\cosh[k(z+h)]}{\cosh(kh)}\cos(kx - \sigma t)$	$\frac{\rho gH}{2}e^{kz}\cos(kx - \sigma t)$	$\frac{\rho gH}{2}\cos(kx - \sigma t)$

for the wave length L_∞ in deep water. We shall use the subscript "∞" to denote this specific case of waves in deep water. The wave phase velocity in deep water is then simply

$$C_\infty = \frac{L_\infty}{T} = \frac{g}{2\pi}T \tag{3-12}$$

We see that for deep-water conditions the wave length and phase velocity are dependent on the wave period, both being greater for the longer-period waves.

We must define exactly what is meant by "deep water." This will be governed by the range of the ratio h/L within which the above approximation in equation (3-10) will yield results of acceptable accuracy. The selection is therefore somewhat arbitrary. A commonly accepted limit of the deep-water approximation is placed at $h > L_\infty/4$, a water depth greater than one-fourth the deep-water wave length. This limit gives a 5% error, which is acceptable for most applications. Geologists sometimes refer to "deep water" as depths that exceed one-half the deep-water wave length ($h > L_\infty/2$). This limit of application is much too stringent for most practical cases, as the error is only 0.37%.

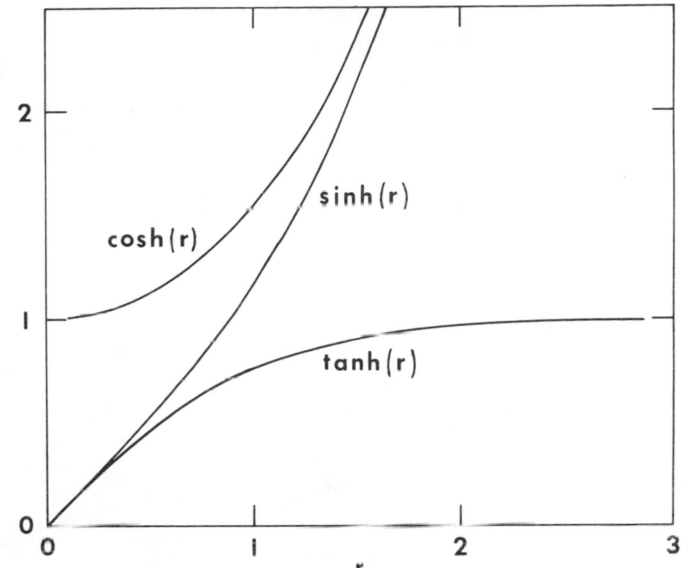

Figure 3-4 Values of the hyperbolic functions sinh (r), cosh (r), and tanh(r).

Now let us examine the other extreme, where $r = kh = 2\pi h/L$ in tanh (r) becomes small. This will represent the shallow-water conditions, where h is small compared to the wave length. Referring again to the graph of Figure 3-4, we see that for this extreme we have as an approximation

$$\tanh\left(\frac{2\pi h}{L}\right) \simeq \frac{2\pi h}{L} \tag{3-13}$$

Substituting this approximation in equation (3-9) we obtain

$$L_s = T\sqrt{gh} \tag{3-14}$$

for the shallow-water wave length. The phase velocity then becomes

$$C_s = \sqrt{gh} \tag{3-15}$$

for waves in shallow water. The subscript "s" will be used to denote this shallow-water approximation. In shallow water the speed of advance of the wave is dependent on the water depth and is no longer a function of the wave period as it is in deep water. If, as before, a 5% error is acceptable, then the limit of application of these shallow-water equations is approximately $h < L_\infty/20$, water depths less than one-twentieth of the deep-water wave length.

Summarizing the regions of application of the approximations:

Deep water	$\frac{h}{L_\infty} > \frac{1}{4}$
Intermediate water (general equations)	$\frac{1}{4} > \frac{h}{L_\infty} > \frac{1}{20}$
Shallow water	$\frac{h}{L_\infty} < \frac{1}{20}$

In the intermediate-depth range, where we cannot use these approximations if the errors are not to exceed 5%, it is necessary to employ the general equation (3-9) directly. Because, as we have already seen, there is a problem with its solution, Wiegel has employed a computer to obtain solutions and has tabulated the results in terms of the ratios h/L and h/L_∞. These tables may be found in Wiegel (1954, 1964) and in the appendices of Coastal Engineering Research Center (1973). The variations of L/L_∞ and C/C_∞ with h/L_∞, obtained from these tables, are graphed in Figure 3-5. The other

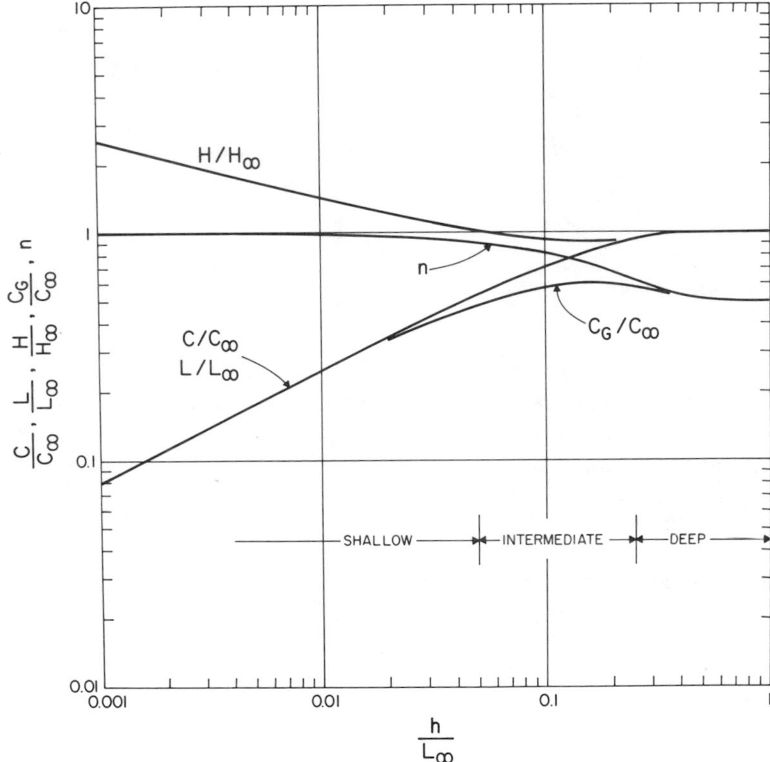

Figure 3-5 The shoaling transformations for Airy waves as functions of the ratio of the water depth, h, to the deep-water wave length L_∞.

curves in the graph will be discussed later in this chapter and in Chapter 4. If an equation is desired rather than a set of tables or a graph, an approximation devised by Eckart (1952) that is often useful is

$$L = L_\infty\left[\tanh\left(\frac{2\pi h}{L_\infty}\right)\right]^{1/2} \tag{3-16}$$

Little error is introduced by using this relationship in the intermediate-depth range.

If one follows the motions of a floating cork as waves pass, it will be observed that the cork moves roughly in a circular path with a diameter approximately equal to the wave height H. After each wave passes, the cork returns to nearly its original position. A particle suspended some distance below the surface also follows a circular (elliptical) motion, but the diameter is smaller than at the surface. At still greater depths, the water motion from the surface waves may be barely perceptible. Let us examine the equations obtained for the wave orbital motion (Table 3-2) to see how they conform to this experience. The equations for the wave orbits appear very complicated in the general solution, but they are simply those of an ellipse of horizontal major diameter d and vertical minor diameter s (Figure 3-2). The elliptical particle motion becomes more circle like as the surface is approached and becomes flatter with depth. At the bottom itself, the ellipse is reduced to a completely to-and-fro horizontal motion ($s = 0$). This results from the boundary requirement that the water not pass through the bottom.

In deep water, both sinh (r) and cosh (r) reduce to $e^r/2$, so that the equations for the orbital diameter reduce to

$$d = s = He^{kz_o} \tag{3-17}$$

where z_o is the depth to the center of the orbit. It is seen that in deep water the orbits become true circles, equal to the wave height at the surface ($z_o = 0$) and decreasing in diameter exponentially with depth. The rate of the exponential decrease with depth is dependent on the value of

$$k = \frac{2\pi}{L_\infty} = \frac{2\pi}{(g/2\pi)T^2}$$

the larger the value of k the more rapid the decrease with depth. It is seen that the decrease is more marked for short-period waves than for those with long periods.

In shallow water, the orbital diameter reduces to

$$d = \frac{HT}{2\pi}\sqrt{\frac{g}{h}} = \frac{H}{kh}$$
$$s = 0 \tag{3-18}$$

The ellipses have flattened out and become horizontal straight lines through the entire water column. In shallow water kh is always less than 0.6, so that the orbital diameter is greater than the wave height. There is no dependence on z, so that the orbital diameter is constant from the surface to the bottom and the orbital velocity u is likewise independent of z.

Although there is no net movement of the water under the wave action, the particles returning to their original positions, the motion of the wave itself constitutes a transfer of energy over the sea surface. It is especially in this regard that waves are important to the beaches. The displacement of the wave surface away from the flat, still-water condition gives the wave form a potential energy. At the same time, the orbital motion of the water under the waves constitutes a kinetic energy for the wave. If E_p and E_k denote the potential and kinetic energies, then the total energy becomes

$$E = E_p + E_k \tag{3-19a}$$

$$= \frac{1}{L}\int_0^L \int_0^\eta \rho g z \, dz \, dx + \frac{1}{L}\int_0^L \int_0^{-h} \frac{1}{2}\rho(u^2 + w^2)\, dz \, dx \tag{3-19b}$$

$$= \tfrac{1}{16}\rho g H^2 + \tfrac{1}{16}\rho g H^2 \tag{3-19c}$$

$$= \tfrac{1}{8}\rho g H^2 \tag{3-20}$$

For Airy waves the potential energy is equal to the kinetic energy. Our energies are averaged over the wave length and so become energies per unit area. Alternatively, energies are commonly expressed as wave energy per unit wave crest length, which is equivalent to our EL.

The *energy flux*, the rate at which the wave energy is transmitted in the direction of wave propagation, is given by (Reynolds, 1877; Rayleigh, 1877)

$$P = \frac{1}{T}\int_0^T \int_0^{-h} [\Delta p(x, z, t)]u \, dz \, dt \tag{3-21}$$

where $\Delta p(x, z, t)$ is the pressure deviation from the static pressure $p_o = \rho g z$ and is given in Table 3-2. Integration of equation (3-21) yields

$$P = \frac{1}{8}\rho g H^2 C \frac{1}{2}\left[1 + \frac{2kh}{\sinh(2kh)}\right] \tag{3-22}$$

Substituting

$$n = \frac{1}{2}\left[1 + \frac{2kh}{\sinh(2kh)}\right] \tag{3-23}$$

the relationship for the energy flux becomes

$$P = ECn \tag{3-24}$$

Energy flux has the units of power, and for that reason it is denoted by P; it is commonly referred to as the *wave power*. Snyder, Wiegel, and Bermel (1958) have shown that equation (3-24) is essentially correct by generating waves at one end of a channel and measuring the transmitted power at the other end.

In deep water $n = \frac{1}{2}$ but increases in value as the waves travel into water of intermediate depth, becoming $n = 1$ in shallow water (Figure 3-5). This means that in deep water, individual waves are advancing with the phase velocity C, which is twice the rate of transmission of the wave energy $C_g = Cn$, often called the *group velocity*. If several waves are generated in a long wave tank, and if the crest of one particular wave is followed, you will observe that the wave progresses toward the front, eventually becoming the lead wave of the group, at which point the crest diminishes in size and eventually disappears. This is because that particular wave is advancing faster than its energy. At the same time new waves are formed in the tail of the moving wave group, so that the total energy flux of the group is preserved. Similarly, in the ocean, where wave trains of limited length exist, a particular wave may "outrun" the wave group and its energy.

Conservation of the energy flux will be used in Chapter 4 to examine the wave height variations in shoaling waves and to relate the height of breaking waves to the deep-water wave conditions. We shall also see in Chapter 8 that the rate of sand transport along beaches is commonly correlated with the "longshore component of the energy flux."

Also associated with the wave advance is a flux or transmission of momentum. Longuet-Higgins and Stewart (1960) (see physical discussion in Longuet-Higgins and Stewart, 1964) define the *radiation stress* as "the excess flow of momentum due to the presence of the waves." The radiation stress is the "excess" momentum flux in that the dynamic pressure is used, the hydrostatic pressure being subtracted from the absolute pressure. The details of the development should be studied in Longuet-Higgins and Stewart (1964). If the x-axis is placed in the direction of wave advance and the y-axis parallel to the wave crests, then there are two nonzero components to the radiation stress: the x- and y-fluxes of x-momentum and y-momentum. The radiation stress (momentum flux) across the plane $x =$ constant (parallel to shore) in the direction of wave advance (the x-direction) is found to be given by

$$S_{xx} = E\left[\frac{2kh}{\sinh(2kh)} + \frac{1}{2}\right] = E\left(2n - \frac{1}{2}\right) \tag{3-25}$$

Despite the orbital velocity component $v = 0$ parallel to the wave crest, there is also a flux of momentum in the y-direction because the pressure departs from the hydrostatic when the waves are present. This flux of y-momentum across the plane $y =$ constant is shown to be

$$S_{yy} = E\left[\frac{kh}{\sinh(2kh)}\right] = E\left(n - \frac{1}{2}\right) \tag{3-26}$$

In deep water $n = \frac{1}{2}$, so that $S_{xx} = E/2$ and $S_{yy} = 0$; in shallow water $S_{xx} = 3E/2$ and $S_{yy} = E/2$, since $n = 1$.

Radiation stress has proved to be a very powerful tool in the study of a variety of oceanographic phenomena. In the context of littoral processes, it has been used to predict changes in the mean water level (set-down and set-up) in the nearshore region and to analyze the generation of longshore currents (Chapter 7). Other applications have been to the generation of surf beat, the interaction of waves with steady currents, and the steepening of short gravity waves on the crests of longer waves.

Stokes Waves of Finite Height

Airy wave theory, presented in the preceding section, omits terms involving the wave height to the second (H^2) and higher orders. If the wave height is large, such an approximation is not adequate, and it is necessary to retain these higher-order terms if the theory is to represent wave motions. Many applications involve considerations that make this necessary.

The theoretical development of waves of finite height is basically the same as that for Airy waves with the exception that the higher-order terms are considered important and so retained. Stokes (1847, 1880) considered waves of small but finite height progressing over still water of finite depth and presented a second-order theory. The solutions he obtained are approximate and in the form of series with coefficients that require considerable detailed calculations. The series were proved to be convergent for infinitely deep water by Levi-Civita (1925) and for water of finite depth by Struik (1926). The method of Stokes has since been extended to higher orders of approximation by Borgman and Chappelear (1958), Skjelbreia (1959) [third order], and Skjelbreia and Hendrickson (1961) [fifth order]. All of the equations are difficult to apply directly. Skjelbreia (1959) and Skjelbreia and Hendrickson (1962) present tabulated solutions to third and fifth orders that are very useful in applications.

To the second-order approximation, the wave profile of the Stokes wave is given by

$$\eta = \frac{H}{2}\cos(kx - \sigma t) + \frac{\pi}{8}\frac{H^2}{L}\frac{\cosh(kh)[2 + \cosh(2kh)]}{[\sinh(kh)]^3}\cos[2(kx - \sigma t)] \tag{3-27}$$

which in deep water ($h/L_\infty > \frac{1}{4}$) reduces to

$$\eta_\infty = \frac{H_\infty}{3}\cos 2\pi\left(\frac{x}{L_\infty} - \frac{t}{T}\right) + \frac{\pi H_\infty^2}{4L_\infty}\cos 4\pi\left(\frac{x}{L_\infty} - \frac{t}{T}\right) \tag{3-28}$$

It is seen that if H/L is small, the profile reduces to the sinusoidal form given by the Airy wave theory. For finite height waves there is an additional term added onto the basic sinusoidal shape. The effect of this term is demonstrated in Figure 3-6, where a finite wave form is compared to a sine wave with the same height. The added term enhances the crest amplitude and detracts from the trough amplitude, so that the Stokes wave profile has steeper crests separated by flatter troughs than does the sinusoidal Airy wave. The waves are symmetrical about vertical planes through crests and troughs. The shallower the water, the more pronounced the peaking of the wave and the flatter the troughs. The general shape of the Stokes wave more closely conforms with the profile of swell waves as they enter shallow water and begin to transform.

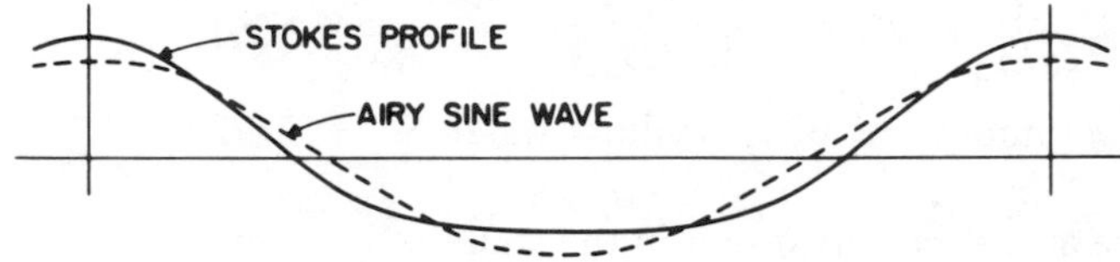

Figure 3-6 Comparison of the theoretical profile of the Stokes wave with a sinusoidal Airy wave of the same height and length.

To the second order of approximation, the equations of wave length and velocity are the same as those for the linear Airy wave theory. To the third order of approximation, the wave velocity is dependent on the wave height as well as on the period and water depth, being given by

$$C = \frac{gT}{2\pi}\tanh\left(\frac{2\pi h}{L}\right)\left[1 + \left(\frac{\pi H}{L}\right)^2 \frac{7 + 2\cosh^2(4\pi h/L)}{8\sinh^4(2\pi h/L)}\right] \tag{3-29}$$

This relationship corresponds to that developed by Hunt (1953); somewhat different forms may be found in the literature. It is seen that if $(H/L)^2$ is small, the relationship reduces to equation (3-9) for Airy waves. In deep water the equation approximates to

$$C_\infty = \frac{g}{2\pi}T\left[1 + \left(\pi\frac{H_\infty}{L_\infty}\right)^2\right] \tag{3-30}$$

A finite wave slope H_∞/L_∞ will cause a slight increase in C_∞ over that calculated with Airy wave theory. Figure 3-7 compares equation (3-30) with measured values of the wave velocity. The maximum value that the deep-water steepness H_∞/L_∞ can achieve is approximately $\frac{1}{7}$; at greater values the wave is unstable and breaks (equation 3-37).

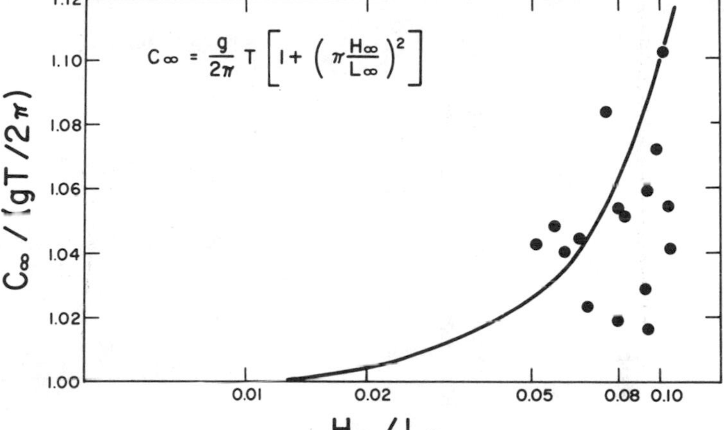

Figure 3-7 Comparison of measured and theoretical wave velocity for the deep-water Stokes wave showing the effect of the wave steepness. Although the comparison is not very convincing, the data does show an increase in wave phase velocity at high steepness values. [*Modified from* Morison (1951)]

For this limiting H_∞/L_∞ we have

$$\left[1 + \left(\pi \frac{H_\infty}{L_\infty}\right)^2\right] = 1.20$$

a 20% increase due to the finite wave height. At more normal wave steepness values the finite height produces only a slight increase in C_∞.

The components of the water particle velocities at any point (x, z) are given by

$$u = \frac{\pi H}{T} \frac{\cosh [k(z + h)]}{\sinh (kh)} \cos (kx - \sigma t) + \frac{3}{4}\left(\frac{\pi H}{L}\right)^2 C \frac{\cosh [2k(z + h)]}{[\sinh (kh)]^4} \cos [2(kx - \sigma t)] \tag{3-31}$$

$$w = \frac{\pi H}{T} \frac{\sinh [k(z + h)]}{\sinh (kh)} \sin (kx - \sigma t) + \frac{3}{4}\left(\frac{\pi H}{L}\right)^2 C \frac{\sinh [2k(z + h)]}{[\sinh (kh)]^4} \sin [2(kx - \sigma t)] \tag{3-32}$$

The first terms in the relationships for u and w are the same as the equations obtained in Airy wave theory (Table 3-2). The second term is positive under the wave crest and trough and negative $\frac{1}{4}$ and $\frac{3}{4}$ wave lengths from the crest. The effect of this second term is to increase the magnitude but shorten the duration of the velocity under the crest, and decrease the magnitude but lengthen the duration of the velocity under the trough (Figure 3-8). This effect is observed in waves in shallow water (Inman and Nasu, 1956).

An interesting departure of the Stokes wave from the Airy wave is that the particle orbits are not closed. This leads to a nonperiodic drift or mass transport in the direction of wave advance, the associated velocity being given by

$$\bar{U} = \frac{1}{2}\left(\frac{\pi H}{L}\right)^2 C \frac{\cosh [2k(z + h)]}{[\sinh (kh)]^2} \tag{3-33}$$

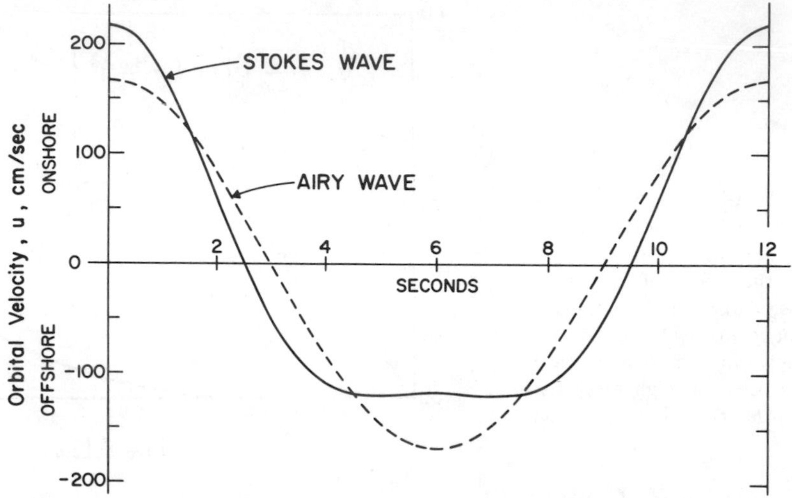

Figure 3-8 Comparison of the theoretical bottom orbital velocity under the Stokes wave with that of the Airy wave of the same height and length. ($H = 4$ m, $h = 15$ m, $T = 12$ sec).

which in deep water reduces to

$$\bar{U}_\infty = \left(\frac{\pi H_\infty}{L_\infty}\right)^2 C_\infty e^{2kz} \tag{3-34}$$

Integration of equation (3-34) with depth yields (Problem 3-13)

$$q = \int_{-h}^{0} \bar{U}_\infty \, dz = \frac{\pi}{4} \frac{H_\infty^2}{T} \tag{3-35}$$

for the *discharge q*, the volume transported forward per unit wave crest length per unit time (cm³/cm · sec).

The above results for the Stokes mass transport were derived for a channel of infinite length and constant depth, with no consideration of viscosity. Longuet-Higgins (1953) has formulated the same problem for a channel of finite length and constant depth, with a real viscous liquid. Because of the finite length of the channel, continuity must be considered and the mass transport of water in the direction of wave advance must be balanced by a return discharge. Longuet-Higgins's solutions yield velocity distributions such as that diagramed in Figure 3-9: a net flow in the direction of wave advance near the water surface and near the bottom, balanced by a net flow in the opposite direction at mid-depths. Measurements by Russell and Osorio (1958) of the vertical distribution of mass transport velocity in a laboratory wave tank found reasonable agreement with the theory of Longuet-Higgins in the range $0.7 < kh < 1.5$. In deep water where $kh > 3$, fairly good agreement was found (except near the bottom) with the Stokes theory modified to produce a curve balanced about the vertical axis (no net discharge).

It is not known how important such wave-induced currents are in the ocean.

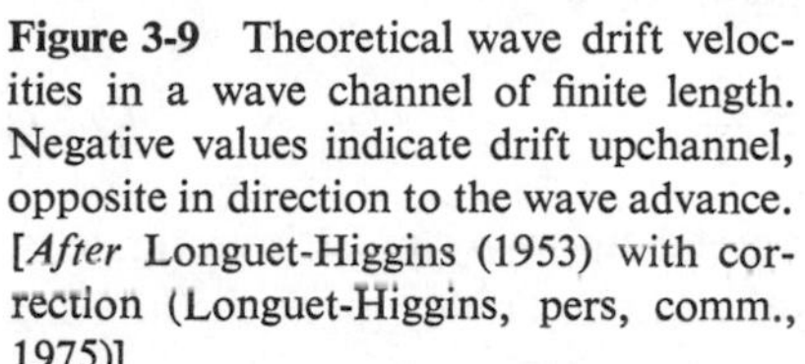

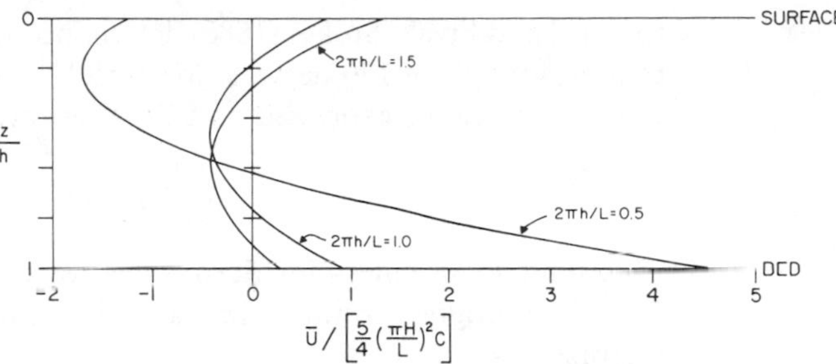

Figure 3-9 Theoretical wave drift velocities in a wave channel of finite length. Negative values indicate drift upchannel, opposite in direction to the wave advance. [*After* Longuet-Higgins (1953) with correction (Longuet-Higgins, pers. comm., 1975)]

The discharge of equation (3-35) is very low (Problem 3-13). Russell and Osorio (1958) note that when the width of the channel is large compared with the water depth, there is a tendency to develop a horizontal circulation with a vertical axis. In the ocean such a circulation may develop so that the velocity distribution of Figure 3-9 need not apply, as continuity would no longer be maintained in two dimensions; the shoreward discharge need not be balanced by a return flow at mid-depth.

The net shoreward velocity near the bottom, given by Longuet-Higgins (1953) as

$$\bar{U}_0 = \frac{5}{4}\left(\frac{\pi H}{L}\right)^2 C \frac{1}{[\sinh (kh)]^2} \tag{3-36}$$

and compared to the data of Russell and Osorio (1958) in Figure 3-10, may be important in producing a slow transport of sediment toward the beach. Such a sediment drift is known to occur commonly, but it is uncertain whether this is the cause.

For any given water depth and wave period, there is an upper limit to the wave height of the Stokes wave at which the wave becomes unstable and breaks. The Stokes criterion for wave breaking is that the water particle velocity at the crest be just equal to the wave propagation velocity C. It is apparent that if the waves were any larger so that the crest water-particle velocity exceeded C, the waves would topple

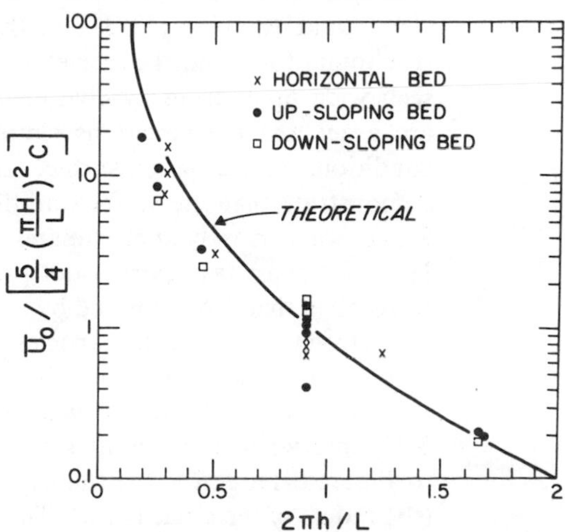

Figure 3-10 Theoretical shoreward bottom drift velocity in a wave channel of finite length. [*After* Russell and Osorio (1958)]

forward and break. Stokes (1880) determined that this breaking condition corresponds to a crest angle of 120 degrees. Mitchell (1893) found that for deep water this condition could also be expressed as a limiting wave steepness

$$\left(\frac{H_\infty}{L_\infty}\right)_{\max} = 0.142 \simeq \frac{1}{7} \tag{3-37}$$

This limit in steepness for deep-water waves is agreed upon by most investigators.

For progressive waves in water of finite depth, Miche (1944) gives the limiting steepness as

$$\left(\frac{H}{L}\right)_{\max} = \left(\frac{H_\infty}{L_\infty}\right)_{\max} \tanh\left(\frac{2\pi h}{L}\right) = 0.142 \tanh(kh) \tag{3-38}$$

The data of Danel (1952) demonstrates that, from an application standpoint, this relationship is satisfactory.

Hunt (1953) has pointed out that Stokes wave solutions in deep water should be limited to waves of steepness less than $\frac{1}{200}$ ($H/L < 0.005$), although with certain refinements the limit can be extended to a wave steepness of $\frac{1}{98}$. In shallow water, the restrictions on wave steepness become more severe because the series in the solution converge much more slowly.

De (1955) has demonstrated that the Stokes theory (to the fifth order) should not be used for h/L values less than about 0.125, the minimum value depending upon the value of H/L; the greater the value of H/L, the greater the value of h/L at which the Stokes wave theory becomes unreliable.

Gerstner Trochoidal Waves

The first solution for periodic waves of finite height was developed by Gerstner (1802). His solution is limited to waves in water of infinite depth. From the developed equations Gerstner concluded that the surface profile is trochoidal in form. Froude (1862) and Rankine (1863), on the other hand, started with the assumption of a trochoidal form and then developed their equations from this curve. This theory has seen wide application by civil engineers and naval architects because the solutions are exact and the equations simple to use. The solutions do satisfy the pressure conditions at the water surface and continuity. In addition, experimental studies demonstrate that the surface profile of the trochoidal wave (as well as that of the Stokes wave) closely approximates the actual profile of waves on a horizontal bottom. However, mass transport is not predicted and the velocity field is rotational. It will be recalled that waves formed by conservative forces must be irrotational. Even worse, in the trochoidal wave the particles rotate in a sense opposite to the rotation that would be present in a wave generated by a wind stress on the water surface.

The trochoid curve is generated by the motion of a point (point A of Figure 3-11) interior to a circle, as the circle rolls along the underside of a line. Referring to Figure 3-11, if R is the radius of the circle, then the wave length will be $L = 2\pi R$ (the radius of the circle is $1/k$). The wave height will be $H = 2r_o$, where r_o is the radial

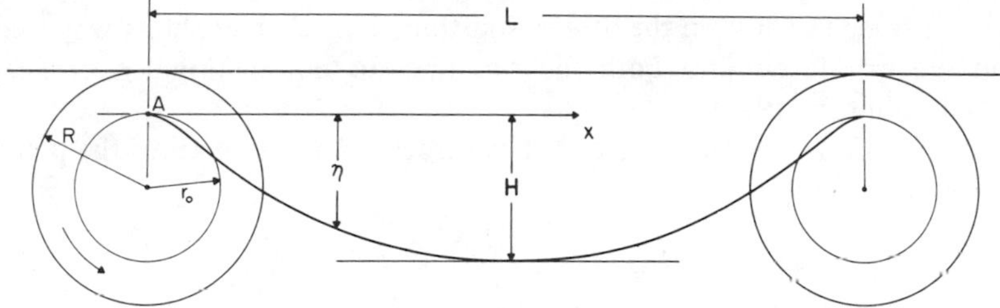

Figure 3-11 The form of the trochoidal curve generated by the motion of point A interior to a circle, as the circle rolls along the underside of a straight line.

distance from the circle center to the point A. For an angle of rotation θ, the surface depression below crest level is

$$\eta = \frac{H}{2}(1 - \cos\theta) \tag{3-39}$$

while the horizontal distance of the surface from the origin at the crest is

$$x = L\left(\frac{\theta}{2\pi} + \frac{H}{2L}\sin\theta\right) \tag{3-40}$$

As H/L becomes small (as point A approaches the center), the surface profile becomes nearly that of the Stokes wave, and with H nearly zero the form tends toward a sine wave. Hence in the limit, the wave corresponds to a deep-water Airy wave.

The positions of the crest and trough relative to the still-water level are

$$\text{Height of crest} = \frac{H}{2} + \frac{\pi H^2}{4L} \tag{3-41}$$

$$\text{Depth of trough} = \frac{H}{2} - \frac{\pi H^2}{4L} \tag{3-42}$$

Thus it is seen that the crest is more than half the wave height above the still-water level and the trough is less than half below.

The particle orbits of the trochoidal wave are

$$d = s = He^{kz} \tag{3-43}$$

the same as deep-water Airy waves, circles whose diameters decrease exponentially with depth. The wave length and phase velocity expressions are also the same as those for a deep-water Airy wave. The energy is given by

$$E = \frac{1}{8}\rho g H^2\left[1 - \frac{1}{2}\left(\frac{\pi H}{L}\right)^2\right] \tag{3-44}$$

Experiments by Wiegel (1950) demonstrate that the surface profile of the trochoid wave closely approximates the actual profile of waves on a horizontal bottom. Experiments performed by the Beach Erosion Board (1941) verify the positions of the crests and troughs relative to the still-water level (equations 3-41 and 3-42). However, both sets of experiments verify the Stokes wave theory equally as well. If the equations for the trochoid are expanded into a series, it is found that the first three terms are the same as those in the Stokes solution. This then explains why the experiments on the profile confirm both theories and do not distinguish between them (Wiegel and Johnson, 1951).

Stokes (1847, 1880) determined that the trochoid wave requires the preexistence of a horizontal velocity

$$\bar{U}_\infty = -\left(\frac{\pi H_\infty}{L_\infty}\right)^2 C_\infty e^{2kz} \tag{3-45}$$

in the direction opposite to that of the wave advance. This is seen to be identical with the drift velocity of the Stokes wave [equation (3-34)] but in the opposite direction. The trochoid wave itself predicts no drift of water (the orbits close). The trochoid wave is rotational (the water particles rotate) and requires a horizontal current opposed to the direction of wave advance; the Stokes wave generates a water drift in the direction of wave advance, currents which actually have been observed. The Stokes wave is therefore on much firmer ground theoretically and agrees with observations better than the Gerstner trochoid wave theory. All of the differences involve small quantities, however, and since the trochoid theory is exact and simpler to apply, it is often found suitable for engineering purposes.

The trochoid theory as developed by Gerstner (1802) is limited to water of infinite depth. Gaillard (1935) has attempted to extend the theory to water of finite depth. The equations of wave velocity and orbital velocities and shapes are the same as those for the general solution for Airy waves, and the other equations are almost identical. This extended theory does not satisfy the conditions of either continuity or dynamic stability except at the troughs and crests, so that the theory is not sound (Wiegel and Johnson, 1951). In spite of this, the theory has seen wide application by engineers.

Solitary Wave Theory

The *solitary wave*, as its name suggests, is a progressive wave consisting of a single crest; it is not oscillatory like the previous waves we have examined. There is therefore no wave period or wave length associated with the solitary wave. It would appear, then, that the solitary wave would not be particularly useful in describing the periodic wind waves we deal with in the ocean. However, when ocean waves enter shallow water their crests peak up and are separated by wide flat troughs and so appear much like a series of individual solitary waves. It is this similarity that first suggested such an application of the solitary wave (Bagnold, 1947; Munk, 1949). We have already seen in the preceding wave theories that in shallow water it is not the wave period that is significant, but rather the water depth. Therefore, we are not particularly oriented toward the periodicity of the waves in shallow water, so that consideration of a solitary wave seems reasonable. Because of its similarity to real waves and because of its simplicity, the solitary wave has seen wide application to nearshore studies.

The character of the solitary wave was first described by J. Scott Russell (1844), who produced such waves in a laboratory wave tank by suddenly releasing a mass of water at one end of the tank. The first theoretical consideration was that of Boussinesq (1872). Rayleigh (1876) and McCowan (1891) developed the theory further and obtained higher approximations. The mathematics of these higher approximations is very tedious, so it is fortunate that, except for the orbital particle velocities, the simpler Boussinesq solution fits the experimental data best. For the orbital velocities, the McCowan solution must be used.

The profile and notation of the solitary wave is shown in Figure 3-12, and its equation is (relative to the moving crest)

$$\eta = H \operatorname{sech}^2\left(\sqrt{\frac{3}{4}\frac{H}{h}}\,\frac{x}{h}\right) \tag{3-46}$$

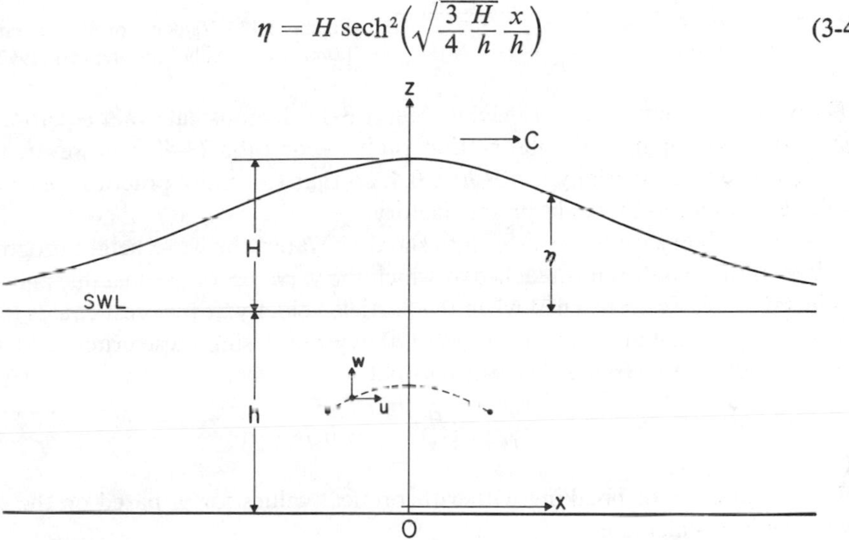

Figure 3-12 Notation associated with the solitary wave; SWL represents the still-water level.

where η is the vertical coordinate above the still-water line at a horizontal distance x from the crest; H is the wave height, and h the water depth below the still-water level. Because of the repeated appearance of the ratio H/h, it is commonly seen abbreviated as $\gamma = H/h$. The wave velocity C is given by Laitone (1959), to higher order, as

$$C = \sqrt{gh}\left[1 + \frac{1}{2}\frac{H}{h} - \frac{3}{20}\left(\frac{H}{h}\right)^2 + \cdots\right] \tag{3-47}$$

which is seen to be greater than the velocity of the shallow-water Airy wave; the solitary wave takes into consideration the finite height of the wave. Equation (3-47) is nearly equal to

$$C = \sqrt{gh\left(1 + \frac{H}{h}\right)} = \sqrt{g(h + H)} \tag{3-48}$$

the equation determined empirically by Russell (1844) and obtained as first approximations by Boussinesq (1872), Rayleigh (1876), and McCowan (1891). The two relationships depart somewhat at higher values of H/h. The laboratory measurements

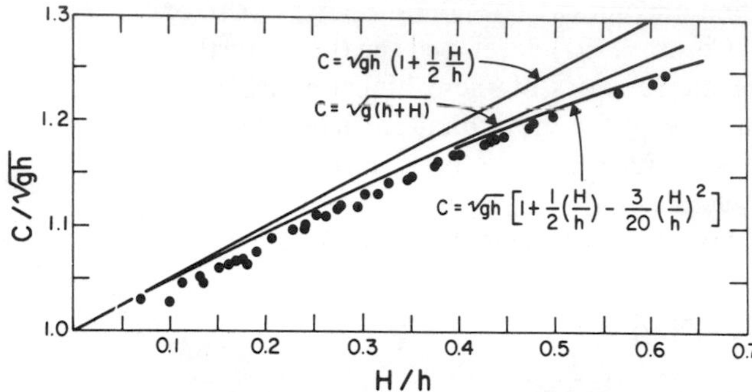

Figure 3-13 Velocity of the solitary wave. [*Data from* Daily and Stephen (1952)]

of Daily and Stephen (1953*a*, 1953*b*) in Figure 3-13 demonstrate that equation (3-47) is in better agreement with observation than is equation (3-48); however, the differences are small, especially for $H/h < 0.4$, so that for many practical purposes the simpler equation (3-48) would be satisfactory.

As the solitary wave advances into shoaling water, the wave height progressively increases until a condition is reached at which the wave becomes unstable and breaks. This instability is again reached when the particle velocity at the crest equals the wave velocity C. Also, the angle at the crest is 120 degrees. Using these criteria, McCowan (1894) further demonstrated theoretically that

$$\gamma_b = \left(\frac{H}{h}\right)_{\max} = 0.78 \tag{3-49}$$

at the critical point of breaking. Other theoretical values for γ_b based on the solitary wave have been calculated:

Source	γ_b
McCowan (1894)	0.78
Boussinesq (1871)	0.73
Rayleigh (1876)	1.0
Gwyther (1900)	0.83
Davies (1951)	0.83
Packham (1952)	1.03
Yamada (1957)	0.83
Laitone (1959)	0.73
Lenau (1966)	0.83

Field measurements on ocean beaches with very low gradients, reported in Sverdrup and Munk (1946), agreed with the $\gamma_b = 0.78$ value of McCowan. For this reason, this value has been the most widely accepted. Ippen and Kulin (1955) conducted laboratory experiments with solitary waves which determined the effects of a sloping bottom on the γ_b value. With a slope of only 0.023 the critical value was about $\gamma_b = 1.2$. For the steepest slope studied, 0.065, values of γ_b reached as high as 2.8, the value depending on the initial wave height. Relatively small bottom slopes can therefore produce pronounced departures from theory. Kishi and Saeki (1967) supported the conclusions of Ippen and Kulin (1955) and, based on data extrapolation, indicated that γ_b would not reduce to 0.78 until slopes became gentler than 0.007.

The total energy of the solitary wave is the sum of approximately equal potential and kinetic energies. The total energy per unit crest length is given by

$$E_{sol} = \frac{8}{3\sqrt{3}} \rho g \left(\frac{H}{h}\right)^{3/2} h^3 \tag{3-50}$$

According to the studies of Daily and Stephen (1952, 1953), the solutions of McCowan (1891) must be used for the particle velocities, as they are most reliable. In this solution, the horizontal and vertical velocities are respectively

$$u = NC \frac{1 + \cos(Mz/h) \cosh(Mx/h)}{[\cos(Mz/h) + \cosh(Mx/h)]^2} \tag{3-51}$$

$$w = NC \frac{\sin(Mz/h) \sinh(Mx/h)}{[\cos(Mz/h) + \cosh(Mx/h)]^2} \tag{3-52}$$

where values of M and N have been calculated by Munk (1949) and are given in Figure 3-14 as functions of H/h.

The solitary wave is a *wave of translation*; that is, the water particles move only in the direction of wave advance—there is no return flow (Figure 3-15). As the crest approaches, the water particles are essentially at rest until approximately $x = 10h$, a distance 10 times the water depth from the crest. The particles then move forward and upward, attaining their maximum velocity and upward displacement at the instant the crest passes. As the crest passes, the particles slow down and move down-

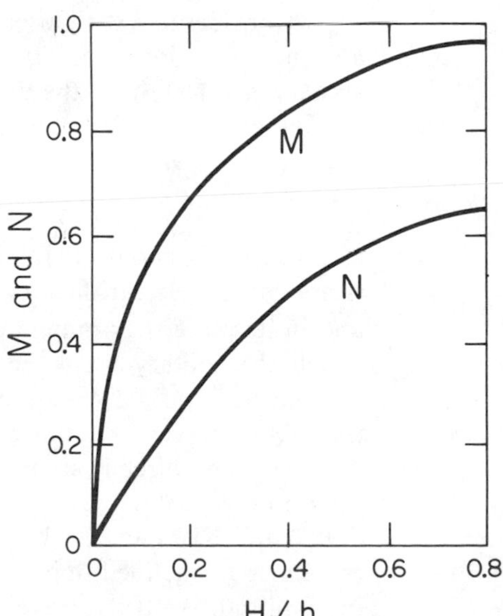

Figure 3-14 The quantities M and N as functions of H/h for the solitary wave. [*After* Munk (1949)]

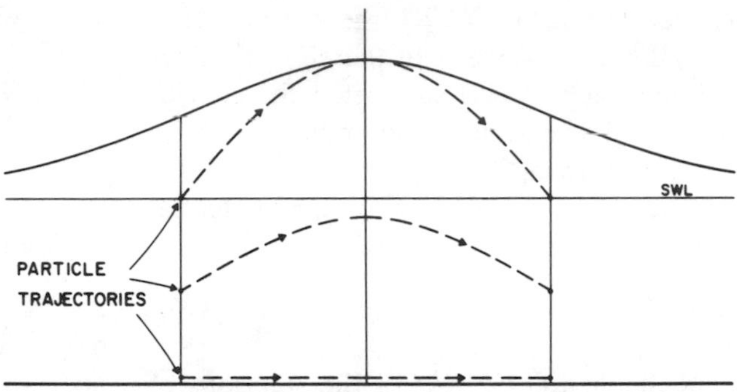

Figure 3-15 Water particle trajectories produced by the passage of a solitary wave.

ward, finally reaching the depth they occupied before the crest passage. As shown in Figure 3-15, they have been displaced in the direction of wave advance in the process. There is, therefore, a net transport of water in the direction of wave advance. This discharge, per unit wave crest length, is equal to the volume contained in the solitary wave above the still-water level and is therefore given by

$$Q = \int_{-\infty}^{\infty} \eta \, dx = 4h^2\left(\frac{1}{3}\frac{H}{h}\right)^{1/2} \tag{3-53}$$

Nearly all this volume is found very near the crest peak. For the wave $H/h = 0.40$, 90% of the volume is found in the range $x = \pm 2.7h$. For the same wave, 90% of the total energy is found in the range $x = \pm 1.7h$. Because of this concentration of volume and energy near the wave crest, it seemed reasonable to apply it to waves near the shore (Munk, 1949). The limbs of the solitary wave are relatively unimportant, so that one can conceive of a series of solitary waves passing without adjacent waves influencing one another appreciably.

Consider just such a series of solitary waves. The assumption of the waves being a sequence of single solitary waves is fulfilled to a sufficient degree of accuracy if the actual wave length of the waves exceeds an effective wave length $L_{\text{eff}} = 2\pi/M$ or if the actual wave period T exceeds an effective wave period

$$T_{\text{eff}} = \frac{2\pi}{M}\sqrt{\frac{h}{g}} \tag{3-54}$$

as given by Bagnold (1947). Housley and Taylor (1957) found that equation (3-54) is not completely satisfactory for defining the region of application of solitary waves, and in its place they present a graph of H/h versus $T\sqrt{g/h}$ that defines the region in which the solitary wave theory can be applied and the region in which the more complicated cnoidal wave theory must be used. We shall see in the next section that the solitary wave is a limiting case of the more general oscillatory cnoidal wave theory.

We have already seen that small bottom slopes produce profound departures of the observed critical γ_b from the theoretical values. Studies such as Ippen and Kulin (1955) and Kishi and Saeki (1967) have also demonstrated that when a solitary wave travels up an inclined slope, as it would in approaching a beach, the observed changes in amplitude, velocity, wave profile, and so on also deviate markedly from the theoretical solitary wave. This, plus the usual difficulties and doubts in applying solitary wave theory to periodic oscillatory waves, make questionable the applicability of the solitary wave to nearshore studies.

Cnoidal Waves

The theory of cnoidal waves was first developed by Korteweg and de Vries (1895). Contributions to the theory have also been made by Keulegan and Patterson (1940) and by Keller (1948). Littman (1957) has demonstrated the existence of periodic waves of this type.

The *cnoidal wave* is a periodic wave that may have widely spaced sharp crests separated by wide troughs and so would be applicable to wave description just outside the breaker zone. The cnoidal wave has the advantage over the solitary wave in that it is periodic. We shall see that in fact the solitary wave is but a limiting case of the cnoidal wave when the wave period becomes infinite. In the other direction, the opposite limiting case of the cnoidal wave is the Airy wave. It is apparent then that we could simply apply the cnoidal wave theory and ignore the other theories. Unfortunately, the mathematics of the cnoidal wave is difficult, so that in practice the cnoidal wave is applied to as limited range as possible. Because of their comparative simplicity the other theories are used instead if they offer suitable approximations to the real waves. For example, if the wave period is adequately long, then the solitary wave theory is applied rather than the cnoidal wave theory.

The wave profile of the cnoidal wave is given by

$$\eta = H\,\mathrm{cn}^2\left[2K(\kappa)\left(\frac{x}{L} - \frac{t}{T}\right), \kappa\right] \tag{3-55}$$

in which $K(\kappa)$ is the complete elliptic integral of the first kind of modulus κ, and η is the vertical coordinate of the water surface above the trough level at the horizontal coordinate x. The term $\mathrm{cn}(r)$ is the Jacobian elliptic function of r and accounts for the name "cnoidal" (analogous to "sinusoidal"). Figure 3-16 illustrates one profile

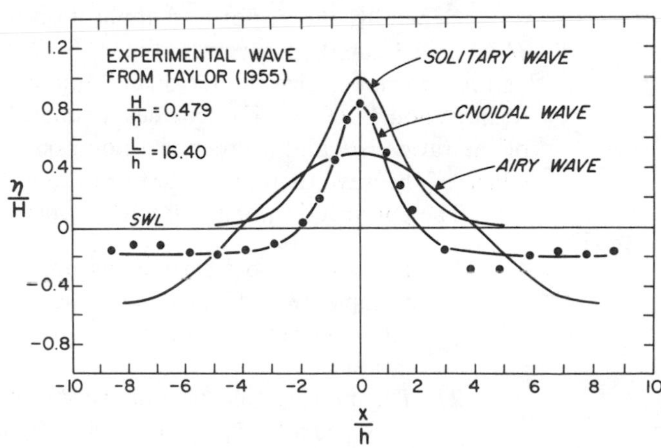

Figure 3-16 Comparison of an experimental wave from Taylor (1955) with the theoretical forms of the solitary wave, the cnoidal wave, and the Airy wave. [*After* Wiegel (1960)]

obtained from equation (3-55) along with the experimental observations of Taylor (1955). The solitary wave and Airy wave profiles are included for comparison. Masch and Wiegel (1961) developed charts of the principal properties of the cnoidal wave for given values of h, H, and L. These graphs are used in lieu of equations whenever the cnoidal wave must be applied.

The limiting cases of the cnoidal wave have already been mentioned. When the modulus κ becomes zero, $\text{cn}(r, \kappa) = \cos(r)$ and $K(\kappa) = \pi/2$ and the cnoidal wave reduces to the linear Airy wave. When $\kappa = 1$, its other extreme value, the period and wave length become infinite and $\text{cn}(r, \kappa) = \text{sech}(r)$: the cnoidal wave becomes equivalent to the solitary wave. At intermediate values of the modulus κ between zero and unity, the wave profile from equation (3-55) demonstrates crests separated by flattened troughs.

As already indicated, due to the complexity of the cnoidal wave, its range of application is limited and preference is given to one of the other simpler wave theories. In deep water, the Airy, Stokes, or Gerstner wave theories are used. At shallower depths, the cnoidal wave theory should be used. If the wave period is sufficiently long and the crests are isolated by wide troughs, then its limiting case, the solitary wave, with its simpler mathematical expressions, can be used.

RANGES OF APPLICATION OF WAVE THEORIES

Having examined five theories for water waves, there must at this point be some confusion as to exactly what are their ranges of application. How does one decide which of the wave theories is applicable to his particular problem?

In our discussions of the individual wave theories some indication was given as to the assumptions involved in obtaining the solutions and how these assumptions affect the range of application. In other cases it was simply pointed out that the theoretical wave closely resembles the real waves in certain depth ranges. In this section, we shall quantify this and put numbers on the limits of application of the theories.

Generally, the regions of application are defined in terms of the ratios H/h, H/L, or h/L, or their inverses. Figure 3-17 has been prepared using the ratios H/h and h/L to summarize the regions of application. Other interpretations are possible; Muir Wood (1969, p. 50) provides a somewhat different development. Calculations of the ratios H/h and h/L define a point on the graph of Figure 3-17 which indicates which of the wave theories should be utilized to describe the wave motions.

The construction of Figure 3-17 is based on the following:

1. The widest possible regions are given to the simpler wave theories. For example, the difficult cnoidal theory is given only a restricted region of application within its actual much greater field; preference is given to the simpler theories.
2. The limiting steepness above which waves are unstable and break is taken to be given by equation (3-38), derived by Miche (1944). In the solitary wave region of application, the breaking criterion $\gamma_b = H_b/h_b = 0.78$ is used.

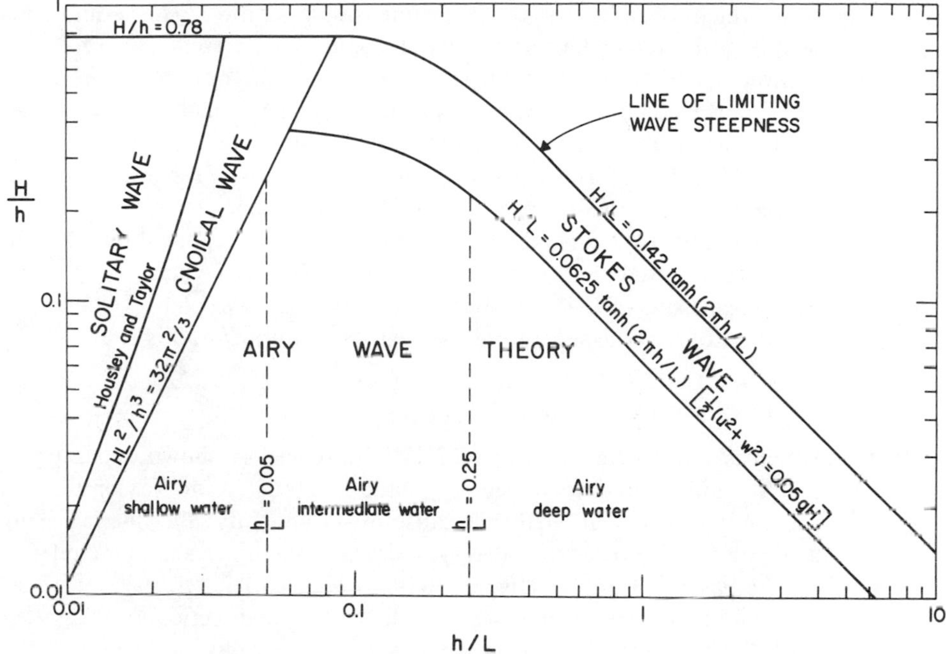

Figure 3-17 The areas of application of the several wave theories as a function of the ratios H/h and h/L.

3. To obtain the linear Airy wave solution, it was assumed that the term $(u^2 + w^2)/2$ in the Bernoulli relationship (equation 3-5) is small and can be neglected. This places a limit on the application of the Airy wave theory, which we shall place at

$$\tfrac{1}{2}(u^2 + w^2) < 0.05gH \tag{3-56}$$

(i.e., 5% of gH). Muir Wood (1969) has shown that this is equivalent to

$$\frac{H}{L} < \frac{1}{16} \tanh\left(\frac{2\pi h}{L}\right) \tag{3-57}$$

The curve of this relationship is shown in Figure 3-17, separating the regions of Airy wave and Stokes wave application.

4. We saw in the section on cnoidal waves that the value of the modulus κ is important in determining the wave form. For $\kappa = 0$ the wave corresponds to the Airy wave, while for $\kappa = 1$ the cnoidal wave reduces to the solitary wave. For intermediate values of κ the wave form is that given by the cnoidal theory in its general form. This suggests that the value of κ might be a good criterion for deciding which wave theory is most applicable. In the cnoidal wave solution, Keller (1948) and Littman (1957) both obtain the relationship

$$\frac{L^2 H}{h^3} = \frac{16}{3}[\kappa K(\kappa)] \tag{3-58}$$

from which it can be seen that the dimensionless ratio L^2H/h^3 could equally well be utilized to define the limits of application. Ursell (1953) has shown theoretically that the linear Airy wave theory is valid only if $L^2H/h^3 \ll 1$, the well-known condition that $H/L \ll 1$ not being sufficient. Longuet-Higgins (1956) later expressed the limitation for linear wave theory as $L^2H/h^3 < 32\pi^2/3$. This limit forms the straight line in Figure 3-17 separating the region of the Airy wave theory from the cnoidal wave theory.

5. Housley and Taylor (1957) defined the regions of application of solitary waves versus oscillatory wave theory in a graph of H/h versus $T\sqrt{g/h}$, based on comparisons of theoretical and experimental wave phase velocities. The equation of the line separating the regions of application is

$$\frac{H}{h} = \frac{1{,}600}{(T\sqrt{g/h})^{2.5}} \tag{3-59}$$

which is constructed in Figure 3-17. The curve is shown separating the solitary and cnoidal wave theories. Actually, Housley and Taylor eliminated the cnoidal theory altogether because of its difficulty: the line is meant to separate solitary from Airy theory. A cnoidal theory region is included here on the basis of Longuet-Higgins's (1956) limit to Airy theory and to show how cnoidal theory becomes solitary theory in one direction and Airy theory in the other. In actual application, you may want to carry the Airy (and Stokes) region over into that shown as cnoidal.

THE NUMERICAL STREAM FUNCTION

Thus far in this chapter we have examined the various analytical wave theories. It will be recalled that these analytical solutions must obey the Laplace equation of continuity (equation 3-4) and the necessary boundary conditions. The Bernoulli relationship (equation 3-5) is used, but to obtain solutions certain approximations are made. The nature of the approximations governs which wave theory is obtained.

With high-speed computers, it is possible to find solutions which obey these same equations and boundary conditions and should therefore yield a more accurate representation of the water motions than do any of the analytical solutions. In a numerical solution, one need not make approximations such as are required in the analytical solutions. The numerical solutions can be worked out to any desired degree of approximation.

The disadvantage of the numerical approach is that it of course yields no simple (or difficult) formulas to represent the wave motion and that a computer run must therefore be carried out for each particular wave condition. However, the numerical approach does yield much more accurate results and, in the case of cnoidal and high-order Stokes wave theories, is actually easier to apply. With respect to expediency of application, it is possible to prepare results in tabular form for the expected range of wave parameters.

Different versions of numerical wave theories have been developed by Chappelear (1961), Dean (1965, 1970), and Dean and Asce (1965). Dean terms his approach *stream function wave theory*. Tables of results will be published by the Coastal Engineering Research Center which should expedite application. These references should be consulted for the details of the numerical approach.

Dean (1970) compared his numerical stream function solutions and the various analytical wave theories to determine their relative validity. This was done by comparing the fits of the solutions to the free surface boundary conditions. The results for the analytical theories alone are shown in Figure 3-18, where the validity is dependent upon providing the best fit to the dynamic free-surface boundary condition which requires that the motions of the water particles at the surface be in accord with the motion of the free surface (i.e., that water particles on the free surface remain on the free surface). The results in Figure 3-18 are somewhat surprising in that the Airy wave theory provides the least amount of error in comparison to the other wave theories over an extended range of h/T^2 values, especially where the waves are steep and approaching the breaking condition. This is in direct contrast with Figure 3-17, which is based on the approximations made in developing the theory. At higher h/T^2 values in Figure 3-18, the Stokes fifth-order theory provides the least error, and at lower h/T^2 the cnoidal first-order theory is best.

As pointed out by Dean (1970), it must be realized that best agreement with the free-surface condition, the basis of Figure 3-18, does not necessarily imply the best overall theory. For example, although the Airy theory provides a better fit than the Stokes theory near breaking in Figure 3-18, we know from observations that the profile, orbital motions, and so forth associated with the Stokes theory are more

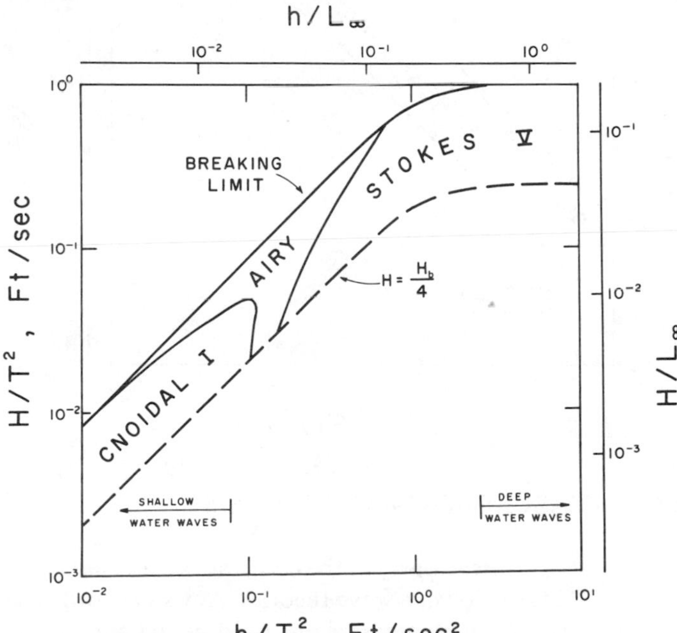

Figure 3-18 Periodic wave analytical theories providing the best fit to the dynamic free-surface boundary condition. The stream function theory is not included in the comparison. [*After* Dean (1970)]

realistic than those of the Airy theory. A far different result would be obtained if the criterion of validity was based on the fit to the profile rather than on the dynamic free-surface condition. This fact is strikingly shown in sample calculations by Dean (1970) of total drag forces. At an intermediate water depth, the Airy theory showed the greatest amount of error of the theories tested. In shallow water, all the analytical theories yielded very large errors compared to the stream function results, demonstrating the decreasing applicability of the analytical approach as the water becomes shallow.

In Figure 3-19, the fifth-order stream function numerical theory is included with the analytical wave theories. With respect to the dynamic free-surface fit, it is seen that the stream function theory yields the least error over nearly the entire range of h/T^2 values. This range would be increased if still higher-order results were used. As before, somewhat different results would be obtained if the criterion for validity was something other than the fit to the dynamic free-surface condition, but it can be expected that the stream function theory would tend to show the best fit whatever the criterion.

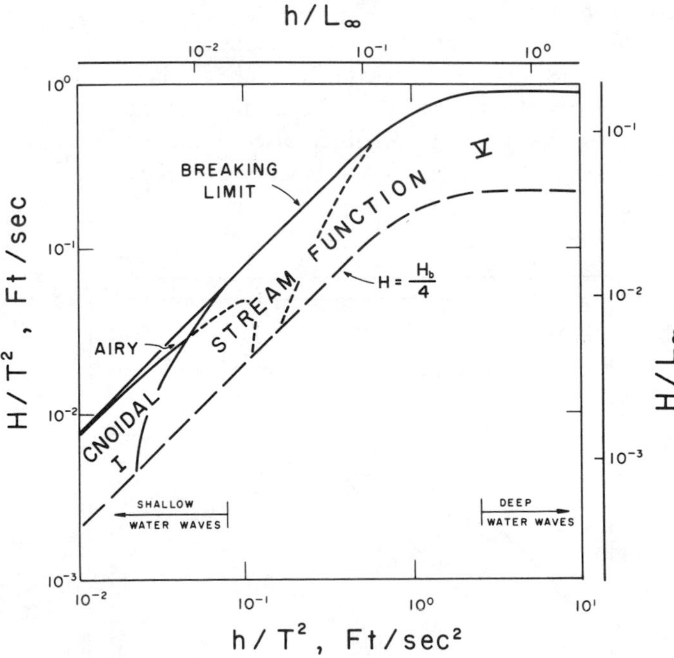

Figure 3-19 Periodic wave theories, including the fifth-order stream function numerical theory, providing the best fit to the dynamic free-surface boundary condition. [*After* Dean (1970)]

WAVE MEASUREMENT AND ANALYSIS

Real waves in the ocean are seldom as simple as we have pretended in the preceding sections on wave theories. Any storm will generate a whole spectrum of waves and not a simple train of one fixed period and height. In addition, more than one major storm may be present, so that waves may be arriving simultaneously at the beach from different directions with different heights and periods. As shown in Figure 3-20, addition of even simple sinusoidal wave trains results in an irregular wave pattern of varying height with little apparent periodicity. Measurement and analysis is then more complex than simply obtaining the wave period with a stopwatch and the height with a meterstick.

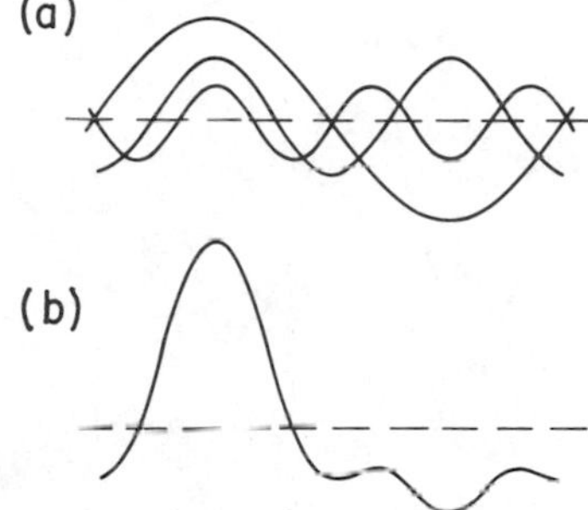

Figure 3-20 The addition of three sinusoidal waves (a) of differing amplitudes and lengths to produce an irregular wave (b) with little apparent periodicity.

To properly analyze the waves we must first obtain a record of their variations. Basically, what is required is a device that can rapidly monitor the changing water level with time and a second device to record these changes. Wave staffs have been designed with a series of bare contacts at equal intervals along the staff length. The staff plus the water makes a complete electrical circuit, the resistance and therefore the current of the circuit thus varying with the water level and the number of staff contacts under water. Resistors in the circuit are so selected that the current flow is proportional to the submerged length of the staff. The fluctuating water level is therefore translated into a fluctuating electrical current which may be recorded. Other staffs operate by electrically scanning in sequence the equally spaced metal contacts (Koontz and Inman, 1967). As each contact is scanned, an electrical pulse is transmitted if the contact is under water; no pulse occurs if the contact is above water. Still another approach is to use a pair of vertical wires, one bare and the other insulated, the capacitance between them depending on the lengths of the wires under water; this method has been developed for recording small waves on lakes and in laboratory wave tanks. Besides the wave staff approach, pressure-type gauges have been devised which convert the pressure fluctuations associated with the wave motion into an electrical current or voltage signal. The pressure–electric current conversion is accomplished through the use of strain gauges, thermocouples, or coils in a magnetic field. Pressure sensors are generally placed on the bottom some short distance outside the breaker zone, the depth depending on the waves to be measured. The equation for the pressure fluctuations associated with the waves (Table 3-2) shows that this pressure decreases with depth, the rate of decrease being greater for the shorter-period waves. Because of this the shorter-period waves are preferentially filtered out by the depth. In some respects this is an advantage in that the small chop from local winds is eliminated, smoothing the record. This pressure decrease with depth restricts the depth range of application of such pressure sensors. In all cases the record obtained by the sensor must be corrected for the water depth to obtain the proper wave heights and energy at the surface.

The electrical current fluctuations emitted by the monitoring sensor may be recorded continuously in graphical form to obtain an analog record such as that of Figure 3-21. With suitable calibration, this record yields a picture of the varying water

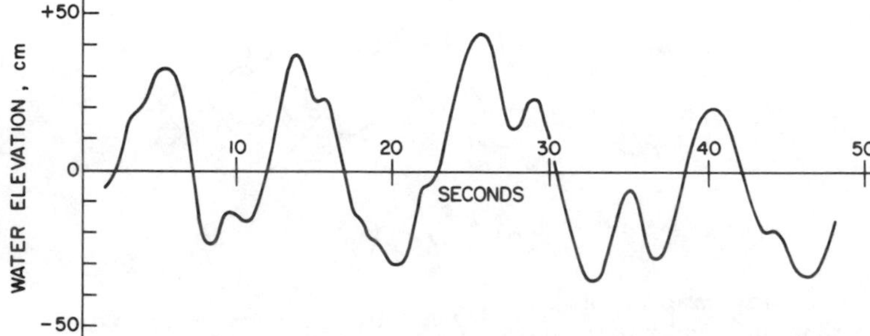

Figure 3-21 An example of an analog record of irregular water waves.

level. Another approach is to digitize the data—that is record the sensor output on magnetic tape at increments of some short period of time, say every one-tenth second. Such a system is described by Koontz and Inman (1967). The advantage of digitized data is that it is already set for analysis on a computer.

We have seen that measured water fluctuations may be very complex because of additions of individual wave trains. A method is needed by which the separate wave trains can be sorted out. Basically, the problem is to work backwards from Figure 3-20(b) to obtain the three sinusoidal curves of Figure 3-20(a). The procedure by which this is done is known as *harmonic* or *spectral analysis*, based on the mathematics of Fourier, who as long ago as 1807 showed that any curve can theoretically be broken down into a series of simple harmonic components (sine waves). The computation techniques were first developed in communications (Blackman and Tukey, 1959) but have been found to be applicable to the analysis of water waves as well (Munk, Snodgrass, and Tucker, 1959). The computations are long and tedious and must be performed on the computer. An example of the results is shown in Figure 3-22, obtained with the system described by Koontz and Inman (1967) from a pair of wave staffs (K and L) placed just offshore from the breaker zone, parallel to shore. The analysis presents the energy density (the energy per unit frequency interval) for each frequency or period. In this example it is seen that there are two pronounced energy peaks at the periods 3.6 seconds and 6.6 seconds, indicating the presence of two dominant wave trains. The total energy in each individual wave train can be obtained by summing the energy densities under its peak. The phase relationship shown below the spectra is obtained by cross-correlating the pair of staffs and can be used to determine the direction of approach of the wave trains. The *coherence* is a measure of the reliability of the spectra and the phase measure—the higher the coherence the better the results.

Figure 3-22 indicates that there is some energy at high frequencies (low periods)

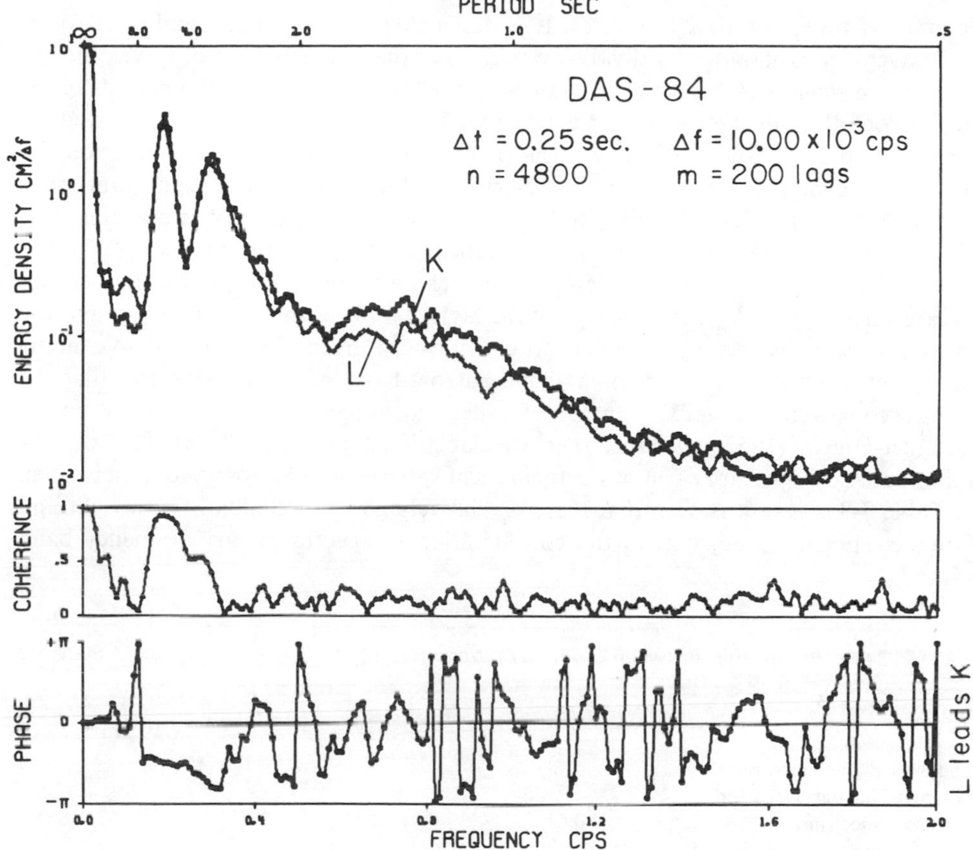

Figure 3-22 Wave spectra obtained from a pair of wave staffs *L* and *K* placed just seaward of the breaker zone and parallel to the shoreline. Pronounced energy peaks appear at the periods 3.6 and 6.6 sec, corresponding to the two wave trains present.

but one or two orders of magnitude lower than in the energy peaks. In the area of wave generation a whole spectrum of periods is produced, so much of this high-frequency energy is associated with the 3.6-second peak generated by local onshore winds. The further from the source of generation the narrower the spectrum; thus the narrowness of the longer period 6.6-second peak. Matters are further complicated in that near the beach the waves become asymmetrical, so that harmonics of the principal frequencies are also introduced.

One commonly meets the term *significant wave height*. This concept developed before the days of harmonic analysis in an attempt to obtain statistics from a record such as that of Figure 3-21. Because small waves are difficult to observe and measure in the presence of large waves, and because the large waves are more important, the significant wave was formulated. The *significant wave height* is defined as the average height of the highest one-third of the waves measured over a stated interval of time, usually 20 minutes. It is designated by $H_{1/3}$. The number of waves to be averaged is obtained by dividing one-third of the time duration of wave observation by the *significant wave period*, in turn defined as the average of the periods of the highest one-third of the waves and determined by averaging the individual periods of large, well-defined groups of waves. The significant wave height corresponds roughly to the wave height that is visually observed and estimated. The significant wave period has less physical meaning and can lead to appreciable error if an attempt is made to use it in calculations with the theoretical wave equations.

Other possible wave statistics are the average wave height $\bar{H}$; the root-mean-square wave height H_{rms}; the average of the highest 10% of the waves, $H_{1/10}$; and the maximum wave height H_{max}, which occurs during a given time interval. Assuming that the wave spectrum contains a single narrow band of frequencies and that the energy comes from a large number of different sources whose phase is random, Longuet-Higgins (1952) developed theoretical relationships for the ratios of the various wave statistics. A comparison of his theoretical values with the observed data is given in Table 3-3, where it is seen that there is relatively good agreement. Cartwright and Longuet-Higgins (1956) extend the consideration to spectra of any frequency band width.

TABLE 3-3 Relationship Between Significant Wave Height and Other Wave Height Statistics

	$\bar{H}/H_{1/3}$	$H_{rms}/H_{1/3}$	$H_{1/10}/H_{1/3}$	$H_{max}/H_{1/3}$
Theoretical prediction by Longuet-Higgins (1952) for a narrow spectrum	0.64	0.71	1.27	1.53–1.85*
Analysis of 25 wave pressure records by Putz (1952)	0.62	—	1.29	1.87
Observations of Goodknight and Russell (1963) of broad-spectrum hurricane waves	0.60	0.69	1.25	1.57

*The value depends on the wave period and on the length of the record.

FURTHER READING

There is a wide choice of books on water waves that can be an aid to additional study. The book by Tricker (1964) is on the layman's level, with interesting discussions of bores, tidal waves, and ship wakes as well as wind waves, and it contains many illustrative photographs; there is essentially no mathematics. The more mathematical side of water waves is presented by Kinsman (1965), a readable book in spite of the mathematics; it is often used as a college-level text. The book by Barber (1969) presents the intuitive side of the physics of wave motion. Even professionals will find it difficult to follow at times, and so one should turn to it only after studying one of the other sources; at that point, the book can lead to a deeper understanding of the basics of wave motions. The first two chapters of the book edited by Ippen (1966) present wave theory with some discussion of applications. Wiegel (1964) presents the engineering interest in waves with numerous diagrams and graphs illustrating the various properties of the several wave types in use. Depending on your particular interests and background, any of these sources are recommended for further study to supplement the material covered in this chapter and the next.

PROBLEMS

3-1 From the surface expression for the form of the Airy wave, equation (3-7), demonstrates that the phase velocity is $C = \sigma/k$.

3-2 Show that the maximum value of the slope $\partial\eta/\partial x$ of the water surface of an Airy wave is $\pi H/L$.

3-3 Show that the sea floor condition

$$w = \left(\frac{\partial\phi}{\partial z}\right)_{z=-h} = 0$$

is true for the general expression for ϕ given in Table 3-2.

3-4 Show that the general expression for ϕ in Table 3-2 obeys the Laplace continuity relationship (equation 3-4).

3-5 Deduce the general expressions for d_o and u_o, the orbital diameter and velocity at the bottom ($z = -h$) for an Airy wave. Confirm that s and w are zero at the bottom.

3-6 Work out the cgs units for gT, $T\sqrt{gh}$, L^2H/h^3, ρgH^2, $HT\sqrt{g/h}$, and ECn.

3-7 If the force of gravity were stronger than at present, how would this effect the motions of water waves?

3-8 Calculate and compare the deep-water Airy wave orbital diameters at water depths $z = -200$ cm, -500 cm, and $-1{,}000$ cm for the following wave trains:

Wave #1	Wave #2	Wave #3
$T = 5$ sec $H = 100$ cm	$T = 10$ sec $H = 100$ cm	$T = 10$ sec $H = 200$ cm

3-9 You calculate that waves from a storm will arrive at a beach in 5 days. They actually do not arrive until 10 days have passed. What error did you probably make?

3-10 Waves are generated by a storm. They range in periods from $T = 5$ to 15 sec. Assuming deep-water conditions for the entire travel distance, apply Airy wave theory to construct a rough graph of travel time versus T for beaches 1,000 km and 3,000 km from the storm area. How much later does the 5-sec wave arrive at the 3,000-km beach than the 15-sec wave? Compare this with their difference in arrival times at the 1,000-km beach.

3-11 Waves arriving at a beach from a distant storm progressively decrease in wave period from $T = 10$ sec to $T = 5$ sec. The 5-sec waves arrive approximately 10 hours after the arrival of the 10-sec waves. Assuming deep-water conditions for the entire travel distance, how far away was the generating storm?

3-12 Using Airy wave theory, determine the wave length L and the phase velocity C for waves whose period is $T = 10$ sec traveling in a water depth $h = 25$ m.

3-13 From the Stokes drift velocity $\bar{U}_\infty$ given by equation (3-34), derive equation (3-35) for the associated discharge. Compute q for the wave $H = 100$ cm, $T = 10$ secs.

3-14 With the identity $\cosh(2r) = \sinh^2(r) + \cosh^2(r)$ and through the use of Figure 3-3, demonstrate that equation (3-30) is the deep-water approximation of equation (3-29).

3-15 Calculate the maximum bottom orbital velocity under the crest and the return velocity at the mid-trough for the Stokes wave with $H = 4$ m, $h = 15$ m, and $T = 12$ sec. The results should compare with Figure 3-8. Calculate the bottom drift velocity $\bar{U}_0$ for this wave, with both equations (3-33) and (3-36).

3-16 Deduce the equation for u_o, the orbital velocity at the bottom, for the Stokes wave. From this relationship derive the acceleration $\partial u_o/\partial t$. Where within the wave motion does the acceleration reach its maximum and minimum values?

3-17 A deep-water wave has a height $H = 100$ cm and length $L = 1{,}000$ cm. Compute the wave energy with both the Airy wave theory and the Gerstner trochoidal wave theory, and compare the results. The maximum possible wave steepness is $H/L = 0.142$ (equation 3-38). For this steepness what is the ratio of the energy as predicted by the Gerstner equation to that of the Airy wave?

3-18 A wave breaking at the beach has a height $H_b = 200$ cm. Using the breaking wave criterion of McCowan for solitary waves, what is the depth of water at which the wave breaks? Compute the phase velocity C_b for the breaker from solitary wave theory.

3-19 An oscillatory wave of period $T = 10$ sec and height $H = 60$ cm travels in water of depth $h = 100$ cm. Is this a shallow-water wave? Compare T with T_{eff} of equation (3-54) to determine whether solitary wave theory can be applied according to the criteria of Bagnold (1947). Now use Figure 3-17 to determine which theory is most applicable.

3-20 We make measurements of wave phase velocity for short-period waves in deep water. Our results depart from Airy wave theory when T drops below about 0.5 sec as follows:

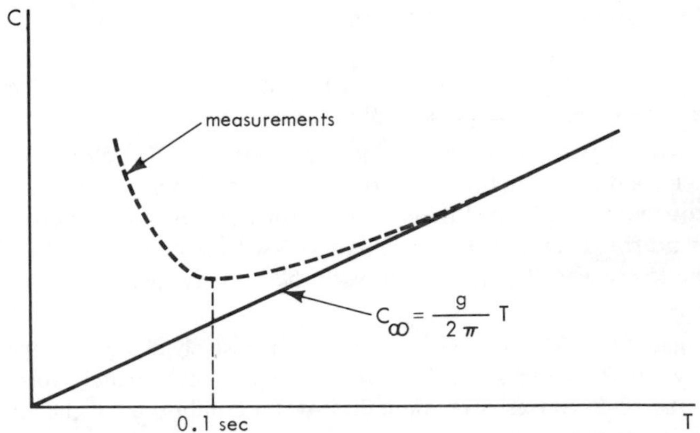

What could account for this systematic departure from Airy theory?

REFERENCES

Airy, G. B. (1845). Tides and waves. *Encyc. Metrop., Art.*, 192: 241–396.

Bagnold, R. A. (1947). Sand movement by waves: some small-scale experiments with sand at very low density. *J. Inst. Civ. Engrs.*, 27, no. 4: 447–69.

Barber, N. F. (1969). *Water waves.* Wykeham, London, 142 pp.

Beach Erosion Board (1941). *A study of progressive oscillatory waves in water.* U.S. Army Corps of Engrs., Washington, D.C., Tech. Report no. 1.

Blackman, R. B., and J. W. Tukey (1959). *The measurement of power spectra from the point of view of communications engineering.* Dover, New York, 190 pp.

Borgman, L. E., and J. E. Chappelear (1958). The use of the Stokes-Struik approximation for waves of finite height. *Proc. 6th Conf. Coast. Eng.*, pp. 252–80.

Boussinesq, J. (1872). Théorie des ondes et de remous qui se propagent le long d'un canal rectangulaire horizontal, en communiquant au liquide contenu dans ce canal des vitesses sensiblement paralleles de la surface au fond. *J. Math. Pures et Appliquées* (Lionvilles, France), 17: 55–108.

Cartwright, D. E., and M. S. Longuet-Higgins (1956). The statistical distribution of the maxima of a random function. *Proc. Roy. Soc.* (London), series A, 237: 212–32.

Chappelear, J. E. (1961). Direct numerical calculation of wave properties. *J. Geophys. Res.*, 62, no. 2: 501–8.

Coastal Engineering Research Center (1973). *Shore protection manual.* U. S. Army, Corps of Engineers, Washington, D.C., 3 volumes.

Daily, J. W., and S. C. Stephen (1953*a*). The solitary wave. *Proc. 3rd Conf. Coast. Eng.*, pp. 13–30.

——— (1953*b*), Characteristics of the solitary wave. *Trans. Am. Soc. Civ. Engrs.*, 118: 575–87.

Danel, P. (1952). On the limiting clapotis. In *Gravity waves*, National Bureau of Standards Circular no. 521, pp. 35–38.

Davies, T. V. (1951). Symmetrical, finite amplitude gravity waves. In *Gravity waves*, National Bureau of Standards Circular no. 521, pp. 55–60.

De, S. C. (1955). Contributions to the theory of Stokes' waves. *Proc. Cambridge Phil. Soc.*, 51: 713–36.

Dean, R. G. (1965). Stream function representation of nonlinear ocean waves. *J. Geophys. Res.*, 70 no. 18: 4561–72.

——— (1970). Relative validities of water wave theories. *Proc. Am. Soc. Civ. Engrs.*, 96, WW1, 105–19.

Dean, R. G., and A. M. Asce (1965). Stream function wave theory: validity and application. *Coast. Eng.*, Santa Barbara Specialty Conf., ASCE, pp. 269–97.

Eckart, C. (1952). The propagation of waves from deep to shallow water. In *Gravity waves*, National Bureau of Standards Circular no. 521, pp. 165–73.

Froude, W. (1862). On the rolling of ships. *Trans. Inst. Nav. Archs.*, 3: 45–62.

Gaillard, D. D. (1935). *Wave action in relation to engineering structures.* Engineering School, Fort Belvoir, Virginia.

GERSTNER, F. (1802). *Theorie der Wellen.* [In English as Tech. Report, series 3, issue 339, Univ. of Calif. Inst. Eng. Res., Waves Research Laboratory, 1952.]

GOODKNIGHT, R. C., and T. L. RUSSELL (1963). Investigation of the statistics of wave heights. *J. Waterways and Harbors Div., Am. Soc. Civ. Engrs.* 89, paper 3524, pp. 29–54.

GWYTHER, R. F. (1900). The classes of long progressive waves. *Phil. Mag.*, 50, no. 5: 213.

HOUSLEY, J. G., and D. C. TAYLOR (1957). Application of the solitary wave theory to shoaling oscillatory waves. *Trans. Am. Geophys. Union*, 38: 56–61.

HUNT, J. N. (1953). A note on gravity waves of finite amplitude. *Quart. J. Mech. Appl. Math.*, 6: 336–43.

INMAN, D. L. (1963). Ocean waves and associated currents. In *Submarine geology*, 2nd ed., F. P. Shepard, pp. 49–81. Harper & Row, New York.

INMAN, D. L., and N. NASU (1956). *Orbital velocity associated with wave action near the breaker zone.* U.S. Army Corps of Engrs., Beach Erosion Board Tech. Memo. no. 79, 43 pp.

IPPEN, A. T. (1966). *Estuary and coastline hydrodynamics.* McGraw-Hill, New York, 744 pp.

IPPEN, A. T., and G. KULIN (1955). The shoaling and breaking of the solitary wave. *Proc. 5th Conf. Coast. Eng.*, pp. 27–49.

KELLER, J. B. (1948). The solitary wave and periodic waves in shallow water. *Comm. Appl. Math.*, 1, no. 4: 323–29.

KEULEGAN, G. H. (1950). Wave motion. In *Engineering hydraulics*, ed. H. ROUSE, Chap. 11. Wiley, New York.

KEULEGAN, G. H., and G. W. PATTERSON (1940). Mathematical theory of irrotational translation waves. *J. Res. Natl. Bur. Std.*, 24, RP 1272: 47.

KINSMAN, B. (1965). *Wind waves.* Prentice-Hall, Englewood Cliffs, N. J., 676 pp.

KISHI, T., and H. SAEKI (1967). The shoaling, breaking, and runup of the solitary wave on impermeable rough slopes. *Proc. 10th Conf. Coast. Eng.*, pp. 322–48.

KOONTZ, W. A., and D. L. INMAN (1967). *A multi-purpose data acquisition system for instrumentation of the nearshore environment.* U.S. Army Coast. Eng. Res. Center Tech. Memo. no. 21, 38 pp.

KORTEWEG, D. J., and G. de VRIES (1895). On the change of form of long waves advancing in a rectangular canal, and on a new type of long stationary waves. *Phil. Mag.*, series 5, 39: 422–43.

LAITONE, E. V. (1959). *Water waves, IV; shallow water waves.* Univ. of California, Berkeley, Inst. Eng. Res. Tech. Report no. 82–11.

LAPLACE, P. S. (1776). *Recherches sur quelques points du système du monde.* Mem. de l'Acad. Roy. des Sciences, *Oeuvres Completes*, 9, no. 88, 187 pp.

LENAU, C. W. (1966). The solitary wave of maximum amplitude. *J. Fluid Mech.*, 26, part 2: 309–20.

LEVI-CIVITA, T. (1925). Determination rigoureuse de ondes d'ampleur finie. *Math. Ann.*, 93: 264–314.

LITTMAN, W. (1957). On the existence of periodic waves near critical speed. *Comm. Pure App. Math.*, 10: 241–69.

LONGUET-HIGGINS, M. S. (1952). On the statistical distribution of the height of sea waves. *J. Mar. Res.*, 11, no. 3: 245–66.

——— (1953). Mass transport in water waves. *Phil. Trans. Roy. Soc.* (London), series A, 245, no. 903; 535–81.

——— (1956). The refraction of sea waves in shallow water. *J. Fluid Mech.*, 1, 163–76.

Longuet-Higgins, M. S., and R. W. Stewart (1960). Changes in the form of short gravity waves on long waves and tidal currents. *J. Fluid Mech.*, 8: 565–83.

——— (1964). Radiation stresses in water waves: a physical discussion, with applications. *Deep-Sea Res.*, 11: 529–62.

Masch, F. D., and R. L. Wiegel (1961). *Cnoidal waves: tables of functions.* Engineering Foundation Council on Wave Research, Berkeley, California.

McCowan, J. (1891). On the solitary wave. *Phil. Mag.*, series 5, 32: 45–58.

——— (1894). On the highest wave of permanent type. *Phil. Mag.*, series 5, 38; 351–57.

Miche, R. (1944). Undulatory movements of the sea in constant and decreasing depth. *Ann. de Ponts et Chaussees*, May-June, July-August, pp. 25–78, 131–64, 270–92, 369–406.

Mitchell, J. H. (1893). On the highest waves in water. *Phil. Mag.*, series 5, 36: 430–37.

Morison, J. R. (1951). The effect of wave steepness on wave velocity. *Trans. Am. Geophys. Union*, 32, no. 2: 201–6.

Morison, J. R., and R. C. Crooke (1953). *The mechanics of deep-water, shallow-water, and breaking waves.* U.S. Army Corps of Engrs., Beach Erosion Board Tech. Memo. no. 40.

Muir Wood, A. M. (1969). *Coastal hydraulics.* Macmillan, London, 187 pp.

Munk, W. H. (1949). The solitary wave theory and its applications to surf problems. *N.Y. Acad. Sci. Ann.*, 51: 376–424.

Munk, W. H., F. E. Snodgrass, and M. J. Tucker (1959). Spectra of low frequency ocean waves. *Bull. Scripps Inst. Oceanog.*, 7, no. 4: 283–362.

Packham, B. A. (1952). The theory of symmetrical gravity waves of finite amplitude, II: the solitary wave. *Proc. Roy. Soc.* (London), series A, 213: 238–49.

Putz, R. R. (1952). Statistical distributions for ocean waves. *Trans. Am. Geophys. Union*, 33: 685–92.

Rankine, W. J. M. (1863). On the exact form of waves near, the surface of deep water. *Phil. Trans. Roy. Soc.* (London), pp. 127–38.

Rayleigh, L. (1876). On waves. *Phil. Mag.*, series 5, 1: 257–79.

——— (1877). On progressive waves. *Proc. London Math. Soc.*, 9: 21–26.

Reynolds, O. (1877). On the rate of progression of groups of waves and the rate at which energy is transmitted by waves. *Nature*, 36: 343–44.

Russell, J. S. (1844). Report on waves: *14th Meeting Brit. Assoc. Advanc. Sci.*, pp. 311–90.

Russell, R. C. H., and J. D. C. Osorio (1958). An experimental investigation of drift profiles in a closed channel. *Proc. 6th Conf. Coast. Eng.*, pp. 171–83.

Skjelbreia, L. (1959). *Gravity waves, Stokes third order approximations, tables of functions.* Council on Wave Research, Eng. Foundation, Univ. of California, Berkeley.

Skjelbreia, L., and J. A. Hendrickson (1961). Fifth order gravity wave theory. *Proc. 7th Conf. Coast. Eng.*, pp. 184–96.

——— (1962). *Fifth order gravity wave theory and tables of functions.* National Engineering Science Co.

Snyder, C. M., R. L. Wiegel, and C. J. Bermel (1958). Laboratory facilities for studying water gravity wave phenomena. *Proc. 6th Conf. Coast. Eng.*, pp. 231–51.

Stokes, G. G. (1847). On the theory of oscillatory waves. *Trans. Cambridge Phil. Soc.*, 8: 441.

[Also in *Mathematical and physical papers*, Cambridge Univ. Press, London, 1880, 1: 197–229.]

——— (1880). On the theory of oscillatory waves: In *Mathematical and physical papers*, Cambridge Univ. Press, London, 1: 314–26.

STRUICK, D. J. (1926). Détermination rigoureuse des ondes irrotationelles périodiques dan un canal a profoundeur finie. *Math. Ann.*, 95: 595–634.

SVERDRUP, H. U., and W. H. MUNK (1946). Theoretical and empirical relations in forecasting breakers and surf. *Trans. Am. Geophys. Union*, 27: 828–36.

TAYLOR, D. C. (1955). *An experimental study of the transition between oscillatory and solitary waves.* Master's thesis, M.I.T., Cambridge.

TRICKER, R. A. R. (1964). *Bores, breakers, waves, and wakes; an introduction to the study of waves on water.* American Elsevier, New York, 250 pp.

URSELL, F. (1953). The long-wave paradox in the theory of gravity waves. *Proc. Cambridge Phil. Soc.*, 49, no. 4: 685–94.

WIEGEL, R. L. (1950). Experimental study of surface waves in shoaling water. *Trans. Am. Geophys. Union*, 31, no. 3: 377–85.

——— (1954). *Gravity wave tables of functions.* Council on Wave Research, Engineering Foundation, Univ. of California, Berkeley.

——— (1960). A presentation of cnoidal wave theory for practical application. *J. Fluid Mech.*, 7: 273–86.

——— (1964). *Oceanographical engineering*. Prentice-Hall, Englewood Cliffs, N. J., 532 pp.

WIEGEL, R. L., and J. W. JOHNSON (1951). Elements of wave theory. *Proc. 1st Conf. Coast. Eng.*, pp. 5–21.

YAMADA, H. (1957). On the highest solitary wave. *Rpt. Res. Inst. Appl. Mech.*, 5, no. 18: 53–155.

Chapter 4

WAVE GENERATION, TRAVEL, AND BREAKING

Dumont d'Urville even asserts that he has seen waves above 108 feet high, to the depths of which the ship descended as into a valley, and M. Fleuriot de Langle attests to the truth of this assertion.

Elisee Reclus
The Ocean, Atmosphere, and Life (1873)

In this chapter we shall examine the *generation of waves*, the mechanisms by which winds blowing over the water surface transfer energy to the water and produce the waves. This process is not fully understood, and we must still utilize semiempirical curves to forecast or predict waves formed by a particular storm.

The term *sea* is used for waves in the process of generation (Figure 4-1). Sea waves are complex and confused, individual waves being shaped like mountains with sharp angular crests. There are small waves superimposed on larger waves and so on down to small ripples. A particular wave crest can be followed for only a short distance before it disappears, for most large crests are produced by the chance summation of several smaller waves and quickly vanish as soon as the latter again move out of phase. Once the waves leave the area of formation and travel across the wide expanse of ocean, they sort themselves out by period and thereby become more regular. The crests become rounded—the sharp peaks are gone. Such regular waves are known as *swell* (Figure 4-1). These wave crests are long, and successive waves are of nearly the same height.

The fundamental difference between sea and swell is determined by the range of periods covered in the spectrum and the energy distribution within that range (Figure 4-1). A sea contains a wide spectrum of periods from 0.1-second ripples to

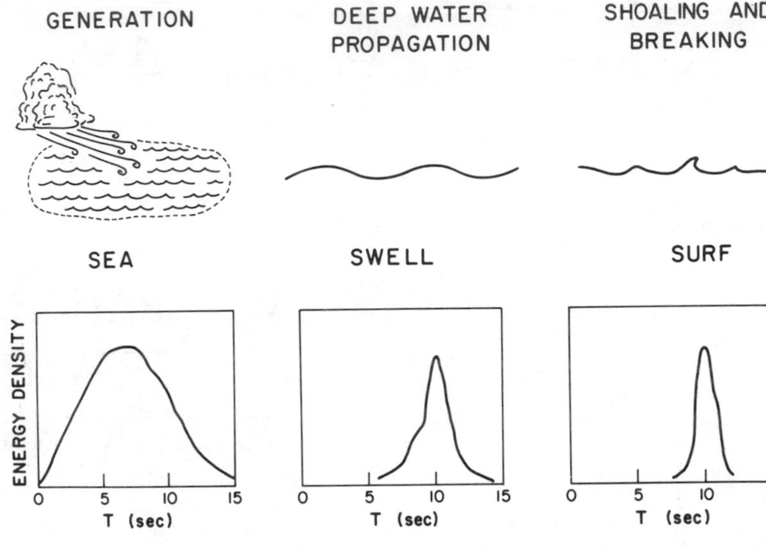

Figure 4-1 Spectrum changes as the waves move from the area of generation, across the deep ocean as swell, and enter shallow water and finally break.

waves of periods from 15 to 20 seconds, there being significant energy within almost the entire range. Swell waves, on the other hand, typically have a very narrow range of periods, the wave form itself being close to a simple sine wave.

In this chapter we shall also examine wave-forecasting procedures and the modifications produced in the waves by the travel stage: dispersion by wave period, attenuation, and the resulting spectrum changes.

When waves enter shallow water they slow down as the depth decreases and undergo a transformation in which the crests peak and increase in height until they become unstable and break at the beach. If the waves approach the coast at an angle, they will also refract or bend, becoming more nearly parallel to the shore. These shallow-water transformations will also be examined.

WAVE GENERATION

We have all seen at some time a breeze blowing over an initially quiet, smooth-surfaced pond. Gentle winds at first disturb the calm water from time to time, producing small capillary waves known as *cat's-paws* wherever a gust brushes the water surface. These local disturbances quickly die away, and the water again becomes glassy. If the winds increase in speed, reaching approximately 1.1 m/sec, ripples develop on the surface, covering the entire pond. If the wind continues, genuine waves may develop. The longer and harder the wind blows, the larger the waves that are produced.

The heights and periods of the waves generated are governed then by the wind velocity U and the duration or time that the wind blows, D. The third important factor is the *fetch*, the distance F over which the wind blows (Figure 4-2). The fetch distance restricts the time during which individual waves can be moving under the action of

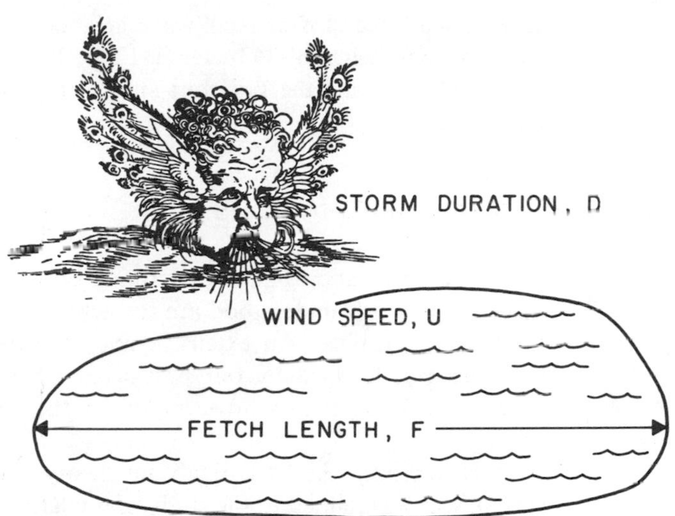

Figure 4-2 Factors that are important in the generation of wind waves.

the wind, and therefore governs the time over which energy can be transferred from wind to waves.

The energy flux or power of a wave train of uniform waves is proportional to the wave period and to the square of the wave height (Chapter 3). Because of this, it takes a greater transfer of power from wind to waves to double the height of a long-period wave than to double the height of a short-period wave. Therefore, longer-period waves can be developed only under conditions of strong winds blowing over large fetch areas with a long storm duration. The fetch is particularly important in limiting the wave periods and heights generated. Long-period waves occur only where the fetch is large. In our pond we might expect to find wave periods no greater than 2 to 3 seconds, while in the ocean waves of periods 20 seconds and greater may be generated, although they most commonly range from 10 to 15 seconds. The wave heights are correspondingly limited, since short-period waves become unstable and break at lower wave heights than do longer-period waves.

Storm waves on the open sea can be an awesome sight. From the bridge of the liner *Majestic*, Cornish (1934) observed waves in the North Atlantic which he estimated at 23 meters in height. Rudnick and Hasse (1971) reported on waves observed from *Flip*, a manned buoy, during exceptional storm conditions north of Hawaii. They observed a maximum wave height of 24 meters, the significant wave height being 15 meters. Offshore oil rigs equipped with automatic wave-recording instruments have reported the measurement of remarkable wave conditions during intense storms. Off the coast of Oregon, waves up to 18 meters, and one 29-meter wave, were reported by Watts and Faulkner (1968), and Rogers (1966) reports seas with waves of 15 meters occurring under winds gusting up to 70 m/sec (150 mph). The highest wave reliably measured on the open ocean was seen from the U.S.S. *Ramapo*, a 146-meter-long naval tanker, while traveling from Manila to San Diego in February 1933. A wind of about 30 m/sec (60 knots) blew steadily from astern. The watch officer noted a sequence of increasing wave heights of 24, 27, 30, and 33 meters, and finally the maximum height of 34 meters (112 feet). The officer lined up the wave crests with the crow's nest and the horizon; simple trigonometry enabled him to calculate the wave heights.

WAVE GENERATION—MECHANISMS

The observation of waves forming on a pond is so familiar that we are apt to forget how poorly understood are the actual processes by which wind imparts energy to the water surface. An extensive line of suggestions and theoretical development has been put forward by our best scientific minds in an attempt to explain wave generation. This remains, however, one of the principal unsolved problems in oceanography.

Historically, the first significant theory is that of Kelvin (1887) and Helmholtz (1888), which pertained to the study of oscillations set up at the interface of two media of differing densities and flowing with different velocities. A complete summary of their approach can be found in Lamb (1932). Applied to the air-sea interface, it is found theoretically that if the wind velocity U exceeds $28C$—28 times the phase velocity of the waves on the surface—an instability develops which should produce a continuous transfer of energy to the waves which goes into increasing their height and velocity. It can be shown that the smallest velocity that a *ripple* (capillary wave) can have is 23.2 cm/sec, which corresponds to a wave length of 1.7 cm and a period of 0.073 sec. At lower periods the surface tension acts to increase the wave velocity, while at higher periods gravity produces a higher velocity (Problem 3-20). According to the Kelvin-Helmholtz theory, conditions will be unstable and the waves will grow if $U > 6.5$ m/sec (28×23.2 cm/sec). This, then, should be the critical wind speed required for gravity wave generation. However, observations indicate that a speed of only 1.1 m/sec will produce ripples. This suggests that waves can be built up by another process at much lower wind speeds. The 6.5-m/sec velocity does approximately correspond to the speed at which the irregularities and the turbulence of the sea suddenly become much more marked (Munk, 1947). This might, however, be a coincidence.

Another early attempt to explain wave generation is the *sheltering theory* of Jeffreys (1925, 1926). According to this theory, the leeward side of the wave form is sheltered by the crests, so that low velocities and pressures are found there. It is envisioned that the boundary layer will separate at the crest with an eddy in the crest lee (Figure 4-3). The normal pressure at the water surface therefore differs between the windward and leeward faces of any wave crest. According to Jeffreys, the wind therefore presses more strongly on the windward slope and aids in the downward motion of the water surface following the passage of the crest (the pressure acts in support of the downward motion of the wave orbital movement). The weakened air pressure in the lee of the crest similarly aids in enhancing the upward motion of the water surface preceding the crest.

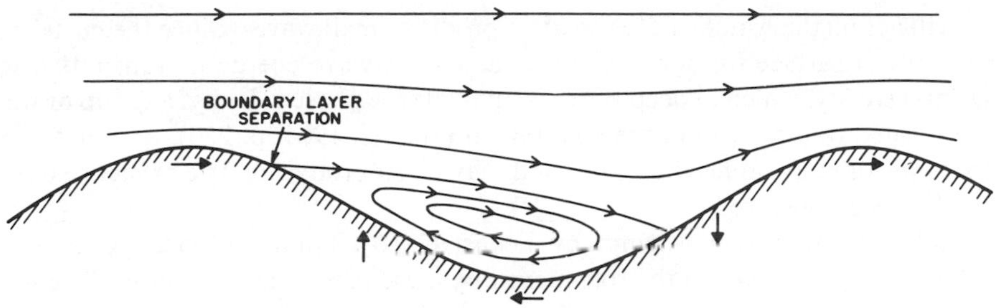

Figure 4-3 The growth of waves due to the sheltering effects of the wave crests according to Jeffreys (1925, 1926).

A series of laboratory studies have been conducted to determine the degree of boundary layer separation and sheltering (Stanton, Marshall, and Houghton, 1932; Mostzfeld, 1937; Bonchkovskaya, 1955). Wave forms constructed of wood or metal covered with plaster of paris and varnish were used rather than real water waves. The studies demonstrated that the sharp-crested wave forms lead to greater separation and sheltering than do the smooth sine waves. Flow over nonsymmetrical model waves also lead to greater separation than flow over symmetrical models. In addition, the degree of sheltering, and hence wave growth, was shown to be strongly dependent upon the wave height and wind speed.

Once waves have attained appreciable steepness and become sharp-crested, it can be expected that flow separation and sheltering might play an important part in the transfer of energy from wind to waves. For waves of low steepness the mechanism would not be effective. The Jeffreys model therefore cannot predict the initial formation and growth of the waves.

Jeffreys also considered the energy transfer from wind to waves through *skin friction*, the drag of the blowing wind on the water surface. It was found that the lowest wind speed needed to transfer energy to the water at a faster rate than the energy is dissipated by water viscosity is 4.8 m/sec. This again is much higher than the value at which the first ripples are formed. However, the mechanism could be important in energy transfer when wind speeds exceed 4.8 m/sec.

Sverdrup and Munk (1947) have also invoked the tangential stresses of the blowing wind on the water surface. Waves acquire energy from the tangential stress of the wind when the water particles in their orbital motion move in the same direction as the wind; energy is lost when the particles move against the wind stress. There would be a net gain in energy because of the mass transport of water in the direction of the blowing wind. They assume that all the energy communicated to the water by the stresses appears in the form of waves and that none goes into generating currents. This is doubtful; in addition, Miles (1957) has shown that tangential wind stress could be only a small part of the total energy input.

Suppose that a turbulent stream of air begins to blow over the water surface, initially at rest. Within the wind will be turbulent velocity eddies which we feel as the "gustiness" of the wind. Associated with these eddies are fluctuations in air pressure which will act on the water surface so as to produce small waves. Since the gusts move over the water surface for some distance, and since wave energy is transmitted with the group velocity, which in deep water is one-half the phase velocity, a group of waves will be formed trailing behind the gust region (Wiegel, 1964, p. 220). A group of five to ten waves of uniform height and period may be so produced. The existence of such groups has been observed.

Following an earlier attempt by Eckart (1953), Phillips (1957, 1958*a*, 1958*b*, 1958*c*) extensively developed the role played by these pressure fluctuations. The waves develop by means of a resonance mechanism which occurs when a component of the surface pressure distribution moves at the same speed as the free-surface waves with the same wave number. Roschke (1954) found that gusts may move for some distance at ground level and that the gusts have periods in the range of 10 to 25 seconds, with some shorter and longer periods as well. Therefore, the gusts are "periodic," similar to ocean waves. The periodic rise and fall in pressure could resonate with water waves of the same length and thereby transfer energy to the waves. For the resonance to exist, the gust "period" must be $2\pi/g$ times the speed of the gust.

Originally it was believed that the model of Phillips correctly predicts wave heights. Measurements of Longuet-Higgins, Cartwright, and Smith (1961) have since shown, however, that the pressure fluctuations assumed by Phillips are 100 times too high.

Miles (1957, 1959) has considered another mechanism of energy transfer from wind to waves. He considers a simple logarithmic velocity profile shear flow of air over a water suface on which there is already present a small sinusoidal wave displacement. He then examines the pressure variations on the water surface that result from the perturbation of the airflow. These produce an air pressure distribution which is greatest over the troughs and least over the crests. This in turn causes the airflow over the crest to turn back, since it is flowing toward the higher pressure in the next trough. The rate at which energy is transferred from the air shear flow to the water waves is proportional to the curvature of the air velocity profile at the elevation where the air velocity is equivalent to the phase velocity of the waves. In some respects the theory resembles the sheltering theory of Jeffreys (1925). However, in Miles's theory the degree of sheltering is calculated from the physical model, whereas Jeffreys had to assume a sheltering coefficient a priori. More recently, Miles (1960) has combined the theories of wave generation by turbulent pressure fluctuations (Phillips, 1957) and by shear flow instability (Miles, 1957).

There can be little doubt that both the pressure fluctuation mechanism of Phillips (1957) and the shear flow mechanism of Miles (1957) are operative to some degree. It seems plausible to suppose that the initiation and early development of waves is a consequence of the pressure fluctuations upon the surface. The shear flow mechanism of Miles (1957) would then become significant in the continued growth of the waves. The Phillips mechanism predicts a linear increase in the wave energy with time or fetch, while the Miles theory predicts an exponential growth. Gilchrist (1966) has qualitatively observed such a switch from linear to exponential growth.

There is relatively little quantitative field information with which to test the theories. Limited field experiments on wave growth have been conducted by Snyder and Cox (1966) and by Barnett and Wilkerson (1967). These demonstrated that the theoretical mechanisms of Miles and Phillips are not adequate, the predicted rates of wave growth being an order of magnitude lower than the observed rates. Experiments have been conducted in the North Sea and elsewhere which hopefully will remedy the lack of satisfactory field data and will lead to a better understanding of wave growth (Barnett, 1970).

Since the above theories were demonstrated to be deficient, Stewart (1967) and Longuet-Higgins (1969*a*, 1969*b*) proposed still other mechanisms to account for the formation and growth of waves. In his review paper Stewart (1967) pointed out that a saturated wave field has about the same energy and momentum as the mean airflow for a height of about one wave length above the water surface. Since the air turbulence has only about 1 percent of the energy of the mean airflow, Stewart argued that the energy for wave generation probably does not come from the turbulent energy itself. The turbulence, however, could in some way enhance the transfer of momentum and energy from the mean wind to the water waves.

Stewart (1967) suggested two additional possible mechanisms for energy transfer which do not rely on normal pressure variations. Each of these was developed in detail by Longuet-Higgins (1969*a*, 1969*b*). The first mechanism results from variations of the wind stress on the water surface in the direction of wave motion. The boundary layer at the surface induced by the stresses is thickest on the upwind slop of the wave; this provides an energy source to the waves. The other mechanism suggested by Stewart (1967) and developed by Longuet-Higgins (1969*b*) is the transfer of momentum from short to long waves as the short waves steepen or break on the crests of the long waves. Longuet-Higgins refers to this as the *maser mechanism*, since it involves the sweeping-up by long waves of momentum from shorter waves. Longer waves passing through shorter waves cause the shorter waves to steepen as the crests of the long waves pass. In doing so, the shorter waves transfer some of their momentum to the long waves. This momentum transfer is especially great if the short waves actually oversteepen and break. This would provide a mechanism for the transfer of momentum and energy from high-frequency waves to low-frequency waves (those with the higher periods and longer wave lengths). The high-frequency waves would be continuously regenerated by the winds through the above mechanisms. Calculations suggest that this mechanism can account for the energy input and wave growth at the observed rate. Field and laboratory studies are required to evaluate the proposed mechanism.

WAVE PREDICTION

In the absence of a satisfactory understanding of the mechanisms of wave generation, semiempirical approaches have been developed for wave prediction. They are semitheoretical in that theory is usually involved in their foundation, semiempirical in that the actual relationships require basic data for the evaluation of various constants and coefficients. Development of the "state of the art" is largely a matter of collecting data on winds and the waves they generate. As new data accumulates, the relationships must be modified and so continue to evolve. The approaches can be classified, on the basis of their results, as (1) the significant wave method, and (2) wave spectrum methods.

The Significant Wave Method

The significant wave approach was originally introduced by Sverdrup and Munk (1947) and later revised with more data by Bretschneider (1952, 1958). The forecasting relationships have acquired the abbreviated name *S-M-B method*, after Sverdrup, Munk, and Bretschneider.

The approach yields predictions of significant wave height $H_{1/3}$ and significant wave period $T_{1/3}$ from the known storm conditions: wind velocity U, fetch distance F, and storm duration D. Bretschneider (1959) and Wiegel (1961) present empirical graphs of all the available data in terms of the dimensionless ratios gF/U^2, gD/U, $gH_{1/3}/U^2$, and $gT_{1/3}/U$; Wiegel's correlation curves are shown in Figure 4-4. From his dimensionless curves, Bretschneider (1959) constructed a graph like that in Figure 4-5; the graph shown is an updated version presented in Coastal Engineering Research Center (1973). With this graph one may estimate the significant wave height and period at the end of a fetch length, knowing the wind speed, duration, and fetch. To use the graph, the U-line is followed from the *left* side of the graph across to its intersection with the fetch length line or the duration line, whichever comes first. This is governed by whether the wave generation is duration-limited or fetch-limited. In fetch-limited waves, for example, the fetch area is too restricted for waves to reach their maximum energy for the given wind speed and duration. The *fully developed sea* requires a storm duration and fetch both long enough so that energy is being dissipated internally and radiated at the same rate as it is being transferred from the wind to the water in the form of waves; a steady state of maximum wave development is thus achieved.

Coastal Engineering Research Center (1973) contains a complete summary and the required graphs for application of the S-M-B significant wave approach to wave prediction. Their summary includes the steps necessary to go from a synoptic surface weather chart to the generated winds and finally to the generated waves. Bretschneider (1957*a*) considers the special case of waves produced by high wind speeds over short fetches and obtains approximating formulas that are simple to apply. Wilson (1955) considers waves generated by hurricane conditions.

The depth of water affects the growth of waves and must be considered in wave prediction in shallow water. Initially this is because the depth modifies the wave group and phase velocities in shallow water. Later the water depth causes increased energy dissipation from bottom drag, and ultimately by virtue of limiting the maximum wave height through wave breaking. Less information is available on wind waves in shallow water than for deep water. Thijsse and Schijf (1949) present dimensionless curves relating the generated waves to the wind conditions, much like those of Figure 4-4 from deep water except that the dimensionless ratio gh/U^2 becomes an additional parameter, where h is the water depth. Bretschneider (1954) takes bottom friction and

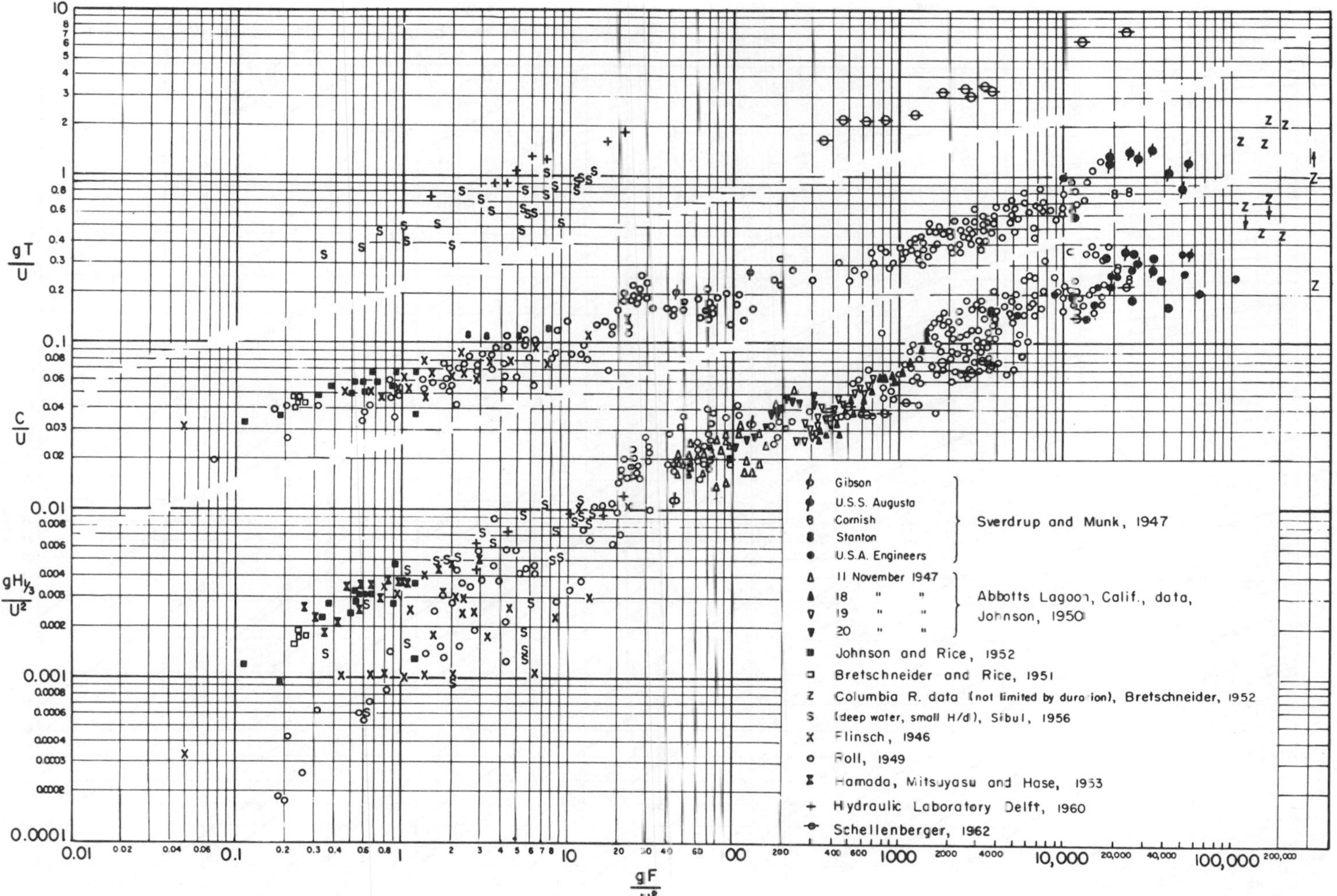

Figure 4-4 Empirical relationships between the wind velocity, fetch, and duration, and the resulting wave height, period, and phase velocity. [*After* Wiegel (1964)]

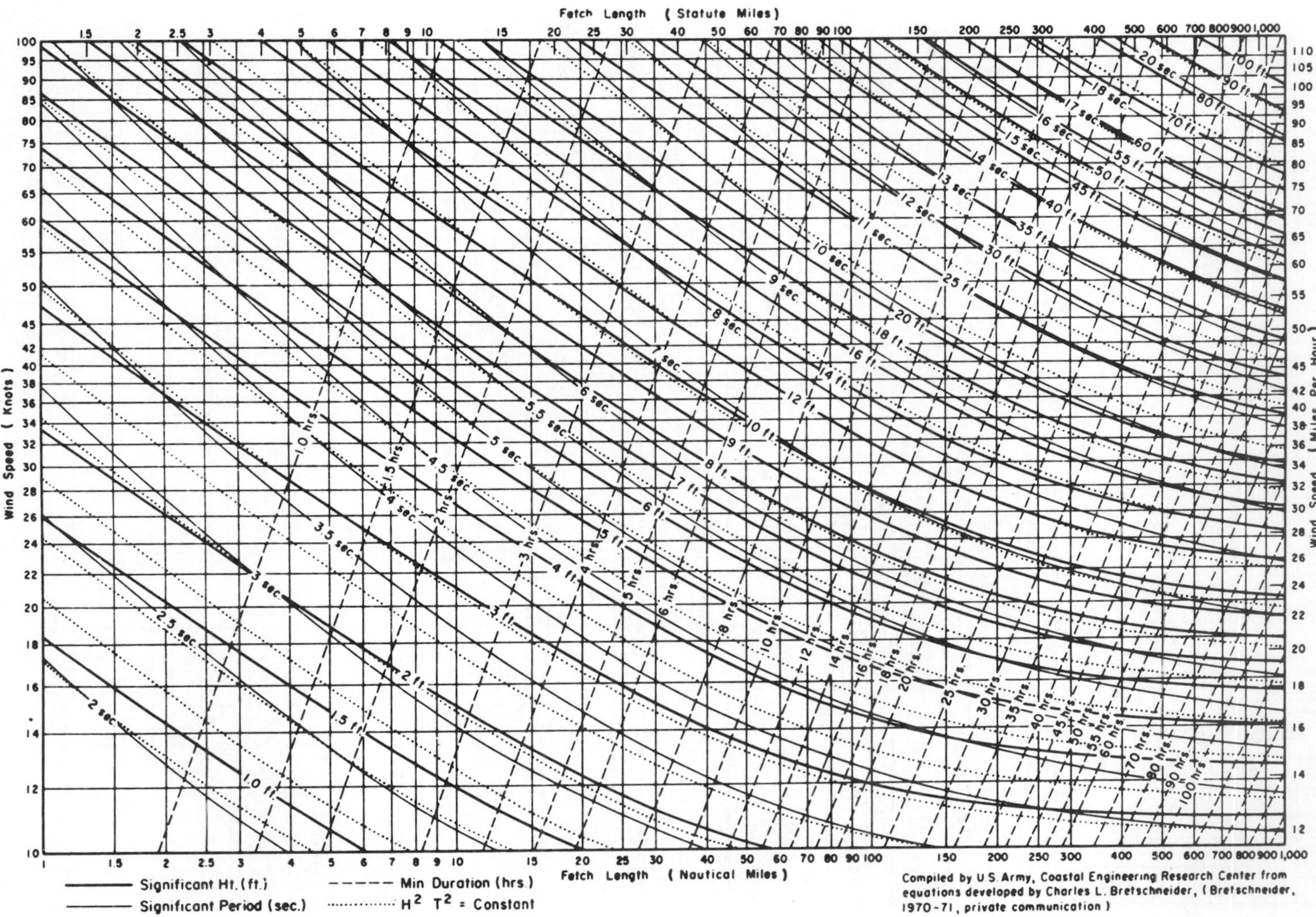
Fetch Length (Statute Miles)
Wind Speed (Knots)
Wind Speed (Miles Per Hour)
Fetch Length (Nautical Miles)
Significant Ht. (ft.)
Significant Period (sec.)
Min Duration (hrs.)
H^2 T^2 = Constant
Compiled by U.S. Army, Coastal Engineering Research Center from equations developed by Charles L. Bretschneider, (Bretschneider, 1970-71, private communication)

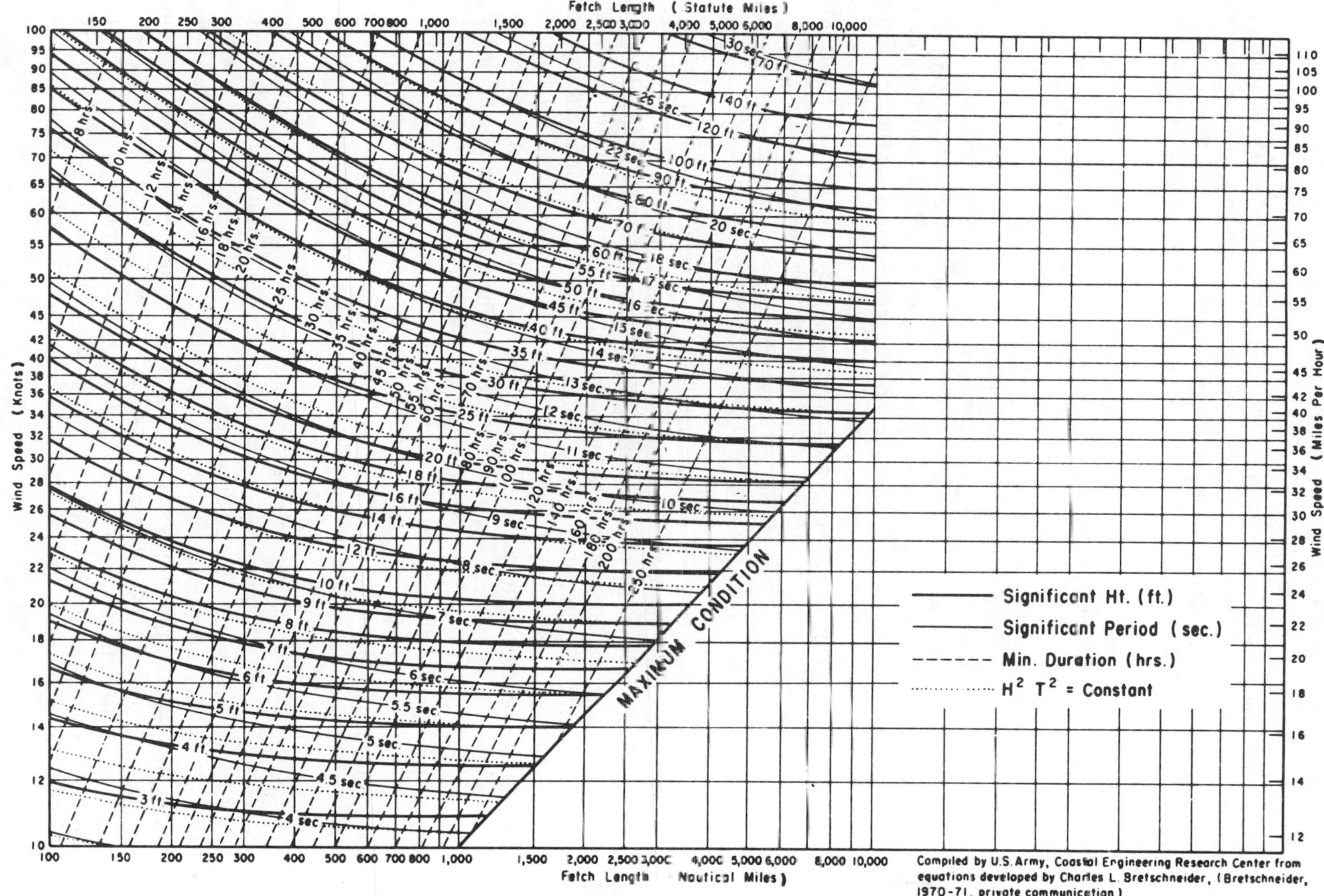

Figure 4-5 Deep-water wave-forecasting curves, based on Bretschneider (1959), relating the storm conditions to the generated significant wave height and period. [*From* Coastal Engineering Research Center (1973)]

percolation into account. Wave energy is added due to wind stress and subtracted due to bottom friction and percolation, this being done numerically in successive approximations. Coastal Engineering Research Center (1973, figures 3-21 through 3-30) presents graphs for the application of Bretschneider's approach as modified by Ijima and Tang (1967).

In any of the above wave prediction methods that yield $H_{1/3}$ and $T_{1/3}$, the findings of Longuet-Higgins (1952) can be applied to obtain the other wave statistics such as $H_{1/10}$, $\bar{H}$, and H_{max}, using the ratios in Table 3-3. Thus it is possible to obtain the complete statistical distribution of wave heights.

The Sverdrup-Munk significant wave forecast method came into being during World War II, wave prediction being a necessity. At the time it was developed, spectral analysis techniques had not been applied to sea waves, and for that reason the significant wave parameters were chosen to describe the sea wave motions. With the present development of spectral analysis techniques and the realization of the theoretical and practical importance of wave spectra, these significant wave methods should be abandoned unless only a crude estimate is required. They have been superseded by the wave spectrum forecast techniques described in the next section.

The Wave Spectrum Methods

The wave spectrum forecast methods describe the generated storm waves in terms of the complete spectrum of periods and energies, not just in terms of a single significant wave height and period. This represents a considerable improvement in wave-forecasting techniques and in the understanding of the nature of wave generation. However, the spectrum methods are more complex in application, and for that reason many engineers still choose to utilize the S-M-B significant wave method. The spectrum methods are preferred, since they arrive at a far closer realization to the wave conditions found in nature, especially with regard to the period structure of the wave motions. For this reason, they should be used for design problems, especially those entailing resonance, where the period becomes critical.

Pierson, Neumann, and James (1955) present a wave spectrum method that is commonly referred to as the *P-N-J method*. The approach is based on the earlier work of Neumann (1953) on thoretical spectra supported by an abundance of wave observations.

Each wind velocity produces a certain range of wave periods with a well-defined maximum, something like the spectrum shown in Figure 3-22—except that example is complicated by the addition of a long-period swell peak to the local wind waves. The total range of periods increases with the wind velocity, as does the energy within the total spectrum. It is found that the period of the maximum energy density is empirically related to the wind velocity by $T_{peak} = 0.405U$, where U is given in knots ($T_{peak} = 0.208U$ with $U =$ m/sec). The total energy in a fully developed sea is found to be proportional to the fifth power of the wind speed. Therefore, since the energy density is proportional to H^2, for unlimited fetch and duration the significant wave height derived from the spectrum is equal to some constant times the wind speed to the $5/2 = 2.5$ power.

As with the S-M-B method, the P-N-J spectrum approach is calibrated by the use of actual wave data and is therefore partly empirical. The P-N-J method can be used to predict the spectrum of the waves from which one can obtain the significant wave height as well as the complete distribution of wave heights ($H_{1/10}$, $\bar{H}$, and so on), the same product given in the S-M-B method. The P-N-J method makes use of the theoretical distribution function given by Longuet-Higgins (1952), while the S-M-B approach uses the semiempirical values of Putz (1950). We saw in Table 3-3 that there is good agreement between Longuet-Higgins's theory and Putz's observations. As a result, if one finds that the P-N-J and S-M-B methods give the same significant wave height, then one obtains almost the exact same wave height distribution. Regarding wave period, the S-M-B method again makes use of Putz's distribution function of wave period variability, whereas the P-N-J method directly predicts the period range and period of the energy peak in the spectrum. Since both methods are based on actual wave observations, their predicted values of wave height should agree fairly closely. One reason for disagreement is probably that not all the same observational data were used by the two methods to establish forecasting relationships (Bretschneider, 1957*b*). Techniques of analysis also differ; and perhaps most important, there is a basic fundamental difference in the proposed proportionality to the mean wind speed, Neumann giving $U^{2.5}$ and the S-M-B method using U^2. This latter difference will be discussed again

The total energy accumulated in the wave spectrum is obtained by integrating the spectrum over the entire possible wave frequency range ($f = 0$ to ∞). If this integration is begun at $f = \infty$ and progresses toward lower frequencies (higher periods), the resulting integral or energy sum is a cumulative curve called the *cocumulative power spectrum* (CCS). One example from Neumann (1953) is shown in Figure 4-6; other curves are given for other mean wind speeds. Because these are cumulative curves, most of the spectral energy is concentrated at the maximum-slope portions of the CCS curve. Theoretically, all wave periods are possible in a fully developed sea, but as can be seen in Figure 4-6 very little energy is added at low and high periods, the cumulative curves leveling off and becoming horizontal. These ends can therefore be neglected without materially affecting the results, and as a rule 5% of the total energy E_f value is cut off the upper end and 3% off the lower end. For example, with a 30-knot (15-m/sec) wind, $E_f = 54{,}350\ \text{cm}^2$, so that

$$\text{Upper cutoff} = 95\%\ \text{of}\ E_f = 51{,}630\ \text{cm}^2$$

$$\text{Lower cutoff} = 3\%\ \text{of}\ E_f = 1{,}630\ \text{cm}^2$$

From the 30-knot curve the corresponding period range for these cutoffs is from 16.7 sec to 4.7 sec ($f = 0.06$ to $0.213\ \text{sec}^{-1}$). This defines the important range of periods that contains most of the wave energy in a fully developed sea generated by a 30-knot wind. Through just such a procedure Table 4-1 was developed for a series of wind speeds. It is seen that as the velocity increases, the range of periods that contain the

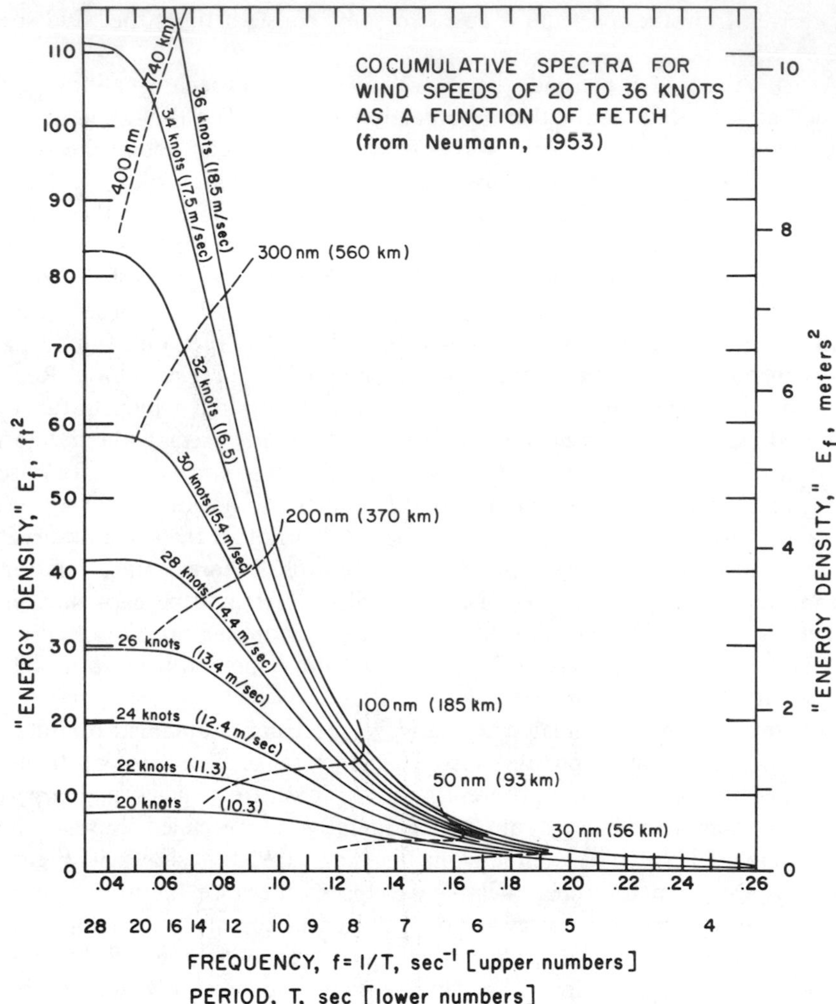

Figure 4-6 The cocumulative power spectra for ocean waves at wind velocities between 20 and 36 knots. The short-dashed curves give the cutoff points of the spectra if the fetch distance is limiting. Neumann (1953) and Pierson, Neumann, and James (1955) give comparable curves for other wind speeds and for conditions where the storm duration is limiting.

dominant amount of wave energy shifts upward toward longer periods as well as broadens to include a greater total span. Table 4-1 also gives T_{peak}, the period of the maximum in the wave spectrum, and $\bar{\bar{T}}$, which is the average "period," the time interval between actual wave crests in the complex sea. The quantity $\bar{\bar{T}}$ is the "period" one would measure with a stopwatch while looking at successive crests. For a fully developed sea it is found that $\bar{\bar{T}} = 0.146U$, where U is in m/sec. Also given in Table 4-1 are the total energies of the spectra, obtained by multiplying E_f from the CCS curve

TABLE 4-1 Characteristics of Fully Developed Sea (*After* Pierson, Neumann, and James, 1955)

WIND VELOCITY U		DOMINANT PERIOD RANGE		PEAK PERIOD	AVERAGE "PERIOD"	TOTAL WAVE ENERGY E	SIGNIFICANT WAVE HEIGHT	AVERAGE WAVE HEIGHT
Knots	*m/sec*	T_{lower}	T_{upper}	T_{peak}	$\bar{\bar{T}}$	*ergs/cm*^{2}	$H_{1/3}(m)$	$\bar{H}(m)$
10	5.1	1.0	6.0	4.0	2.8	0.55×10^5	0.43	0.27
12	6.2	1.0	7.0	4.8	3.4	1.4×10^5	0.68	0.42
14	7.2	1.5	7.8	5.6	4.0	3.0×10^5	0.99	0.62
16	8.2	2.0	8.8	6.5	4.6	6.0×10^5	1.39	0.87
18	9.3	2.5	10.0	7.2	5.1	10.4×10^5	1.86	1.16
20	10.3	3.0	11.1	8.1	5.7	17.6×10^5	2.44	1.52
22	11.3	3.4	12.2	8.9	6.3	28.3×10^5	3.11	1.95
24	12.3	3.7	13.5	9.7	6.8	43.7×10^5	3.81	2.41
26	13.4	4.1	14.5	10.5	7.4	6.5×10^6	4.69	2.93
28	14.4	4.5	15.5	11.3	8.0	9.5×10^6	5.54	3.44
30	15.4	4.7	16.7	12.1	8.5	13.4×10^6	6.58	4.11
32	16.4	5.0	17.5	12.9	9.1	18.4×10^6	7.86	4.91
34	17.5	5.5	18.5	13.6	9.7	25.0×10^6	9.08	5.67
36	18.5	5.8	19.7	14.5	10.2	33.2×10^6	10.5	6.52
38	19.5	6.2	20.8	15.4	10.8	43.5×10^6	12.0	7.47
40	20.6	6.5	21.7	16.1	11.4	56.3×10^6	13.8	8.50
42	21.6	6.8	23.0	17.0	12.0	7.2×10^7	15.3	9.60
44	22.6	7.0	24.0	17.7	12.5	9.1×10^7	17.6	11.0
46	23.6	7.2	25.0	18.6	13.1	11.3×10^7	19.4	12.1
48	24.7	7.5	26.0	19.4	13.7	14.0×10^7	21.6	13.5
50	25.7	7.7	27.0	20.2	14.2	17.2×10^7	23.8	14.9
52	26.7	8.0	28.5	21.0	14.8	20.9×10^7	26.5	16.5
54	27.8	8.2	29.5	21.8	15.4	25.2×10^7	29.0	18.0
56	28.8	8.5	31.0	22.6	15.9	30.3×10^7	31.4	19.5

by $\rho g/8$, and the significant wave heights and average wave heights that correspond to the spectra total energies. With increasing wind speed, T_{peak} also increases, as do the wave heights and therefore the total wave energies within the spectra.

The above discussion and example pertains only to the fully developed sea. Neumann (1953) and Pierson, Neumann, and James (1955) also consider the effects of limited fetch and wind duration. Figure 4-7 gives the minimum fetch and minimum duration required for a fully developed sea for a range of wind speeds. A limitation in either fetch or duration causes a truncation of the wave spectrum toward the larger wave periods, since the higher periods cannot develop under these limited conditions. The spectrum of the nonfully developed sea can be obtained from the spectrum of the fully developed sea by discarding that part of the curve above a certain maximum wave period. In Figure 4-6 the short dashed curves indicate the trucation point for limited fetch. The intersection of the CCS curve for the appropriate velocity with the fetch distance curve defines the total E_f within the spectrum. The

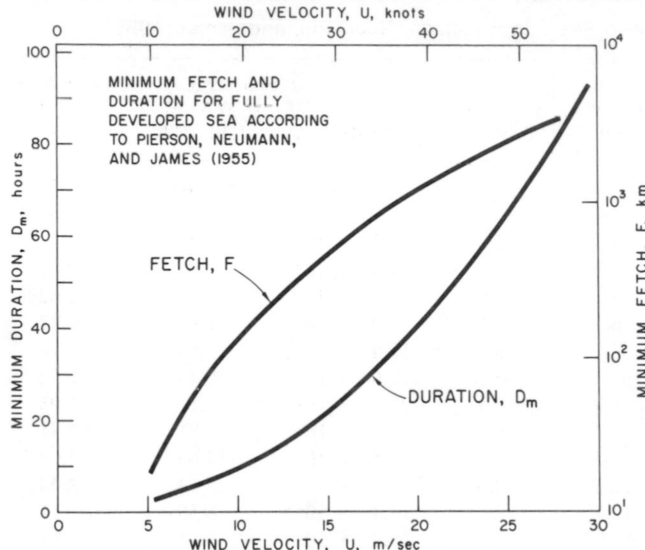

Figure 4-7 The minimum storm duration and fetch required for fully developed sea for a range of wind speeds. [*Based on values given in* Pierson, Neumann, and James (1955, table 2.2)]

wave spectrum is then cut off abruptly at that given maximum period without considering possible wave components of higher period that may be just in the beginning stage of development. These wave components at higher periods than the cutoff probably have a small energy content and so may be neglected in most practical cases in wave forecasting. Neumann (1953) does give empirical procedures by which in certain circumstances this energy can be added into the spectrum.

Neumann (1953) assumed that H^2 is proportional to the energy per unit range of periods and thus deduced that the spectral energy density $E'(f)$ at high frequencies is proportional to f^{-6}; that is, the energy density decreases with increasing wave frequency (decreasing wave period) according to a -6-power law. Roll and Fisher (1956) pointed out that this assumption made by Neumann is not logical, and correcting this they obtained $E'(f) \propto f^{-5}$ at high frequencies. This agrees with the conclusions of Phillips (1958*b*) based on dimensional arguments. Experimental evidence on the whole also supports the -5 exponent rather than -6 (Figure 4-8).

The spectrum of a developing sea may be thought of as divided into two portions. The range of periods less than the peak period, where the decrease in energy is governed by the f^{-5} law, is known as the *equilibrium range* (Figure 4-8). This range of the spectrum is already fully developed, and energy added to this range is either transferred to lower frequencies (higher periods) on the other side of the peak or is lost through wave-breaking turbulence. Periods greater than the optimum peak period comprise the *growth range*, where energy is still being added to the spectrum. Due to the shift of the peak period during wave growth, the growth range also progressively shifts to higher periods, the equilibrium range expanding to occupy periods that previously were in the growth range. The cluster of lines on the left of Figure 4-8 are representative of the growth range for which equilibrium has not been achieved.

Besides the fully developed spectrum form of Neumann (1953), forms have been

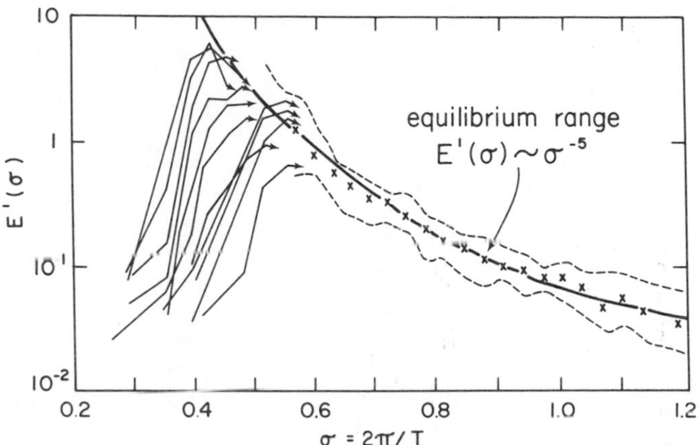

Figure 4-8 Cluster of spectra of wind-generated waves. On the right the curves merge in the range of periods (frequencies) known as the equilibrium range, where conditions are already fully developed. The broken lines indicate the extreme measured value of the energy density at each frequency, and the crosses represent the mean observed value. The heavy line depicts an f^{-5} decrease with increasing frequency. On the left is a cluster of lines representing the spectra at low frequencies in the growth range where equilibrium has not been attained. [*After* Phillips (1958*b*)]

proposed by Darbyshire (1959), Bretschneider (1963), and Pierson and Moskowitz (1964). Fetch-dependent forms have been proposed by Bretschneider (1959) and more recently by Liu (1971). Based on wave data from Lake Michigan, from laboratory studies, and from the ocean, and applying similarity analysis to wind and wave data, Liu obtained the empirical spectral equation

$$E'(\sigma) = \frac{0.4g^2}{F_0^{1/4}\sigma^5} \exp\left[-5.5 \times 10^3 \left(\frac{g}{U_* F_0^{1/3} \sigma}\right)^4\right] \tag{4-1}$$

for fetch-limited deep-water waves. Here U_* is the friction wind velocity, a measure of the stress exerted by the winds; $F_0 = gF/U_*^2$ is the dimensionless fetch parameter; and $\sigma = 2\pi/T$ is the radian frequency. The value of U_* is calculated from the wind velocity U_{10} measured 10 m above the sea surface using the empirical relationship

$$U_* = U_{10}\left(\frac{U_{10}^2}{gF}\right)^{1/3} \tag{4-2}$$

Equation (4-1) is quite similar to the form given by Pierson and Moskowitz (1964), approaching that form as the fetch distance becomes large; and, like all the other newer spectral forms, it gives $E'(\sigma) \propto f^{-5}$. Given the wind velocity U_{10} and the fetch distance F, Liu's equation will yield the predicted fetch-limited spectrum. Figure 4-9 shows the spectra for a wind speed of 20 m/sec at various fetch lengths. The spectra are quite wide at early stages of development with no pronounced peak, becoming narrower and the front face steeper as the wind speeds or fetch distances increase. The spectral energy density $E'(\sigma)$ approaches an equilibrium state at high frequencies as the fetch increases, but the low-frequency waves in the growth range continue to grow, however slowly, not as yet having reached a fully developed sea state. All of these behaviors of the Liu spectrum form agree with observed changes in wind wave spectra with wave growth.

Darbyshire (1952, 1955, 1956, 1959, 1963) has also contributed to our knowledge of wave spectra and wave generation, obtaining results for a large number of storms in the North Atlantic, the Irish Sea, and in the short-fetch Lough Neagh, Northern Ireland. The records were analyzed for the significant wave parameters

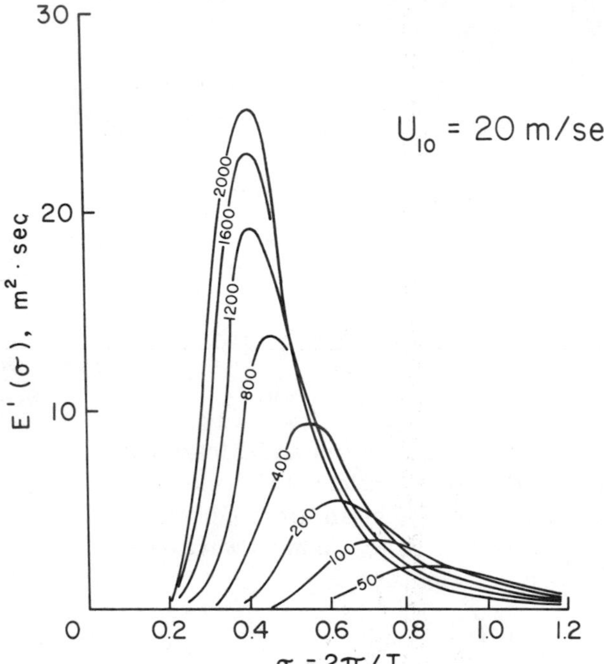

Figure 4-9 The fetch-limited spectra of Liu (1971) given by equation (4-1) for a wind speed of 20 m/sec.

as well as by spectral analysis techniques. It was found, for example, that the wave period T_{peak} of the maximum peak in the spectrum is related to the significant wave period by $T_{peak} = 1.14T_{1/3}$. This would suggest that the significant wave period is meaningful and can be corrected to T_{peak} to be used in calculations of wave orbital velocities and so on.

The results of Darbyshire's investigations provide a third alternative to the S-M-B and P-N-J methods for wave prediction. His approach to the problem is basically different than the others. Darbyshire takes many measured wave spectra and tries to fit purely empirical formulas to them. Darbyshire (1959) plots $E'(f)/E$ against $f - f_0$, where E is the total energy in the spectrum and $E'(f)$ is the energy at the frequency f; f_0 is the frequency maximum of the spectrum. All the spectra fit reasonably well onto a single curve. He also finds empirically that $E \propto U^4$: the total energy is proportional to the fourth power of the wind speed rather than to the fifth power as found by Neumann (1953).

Because his measurements are either from the deep ocean or from shallow coastal waters (the Irish and North seas), Darbyshire gives two sets of empirical results for the two conditions. One set of empirical spectra are required for wave prediction in the deep ocean, for depths greater than 180 meters, and a second set for shallower coastal regions where the depths are between 30 and 45 meters. Winds generate higher waves in the open ocean than in shallow water, and there is also a marked difference in wave period. No research was carried out in areas of intermediate depth, but presumably the results would fall between the values for the open sea and coastal waters.

A convenient set of graphs for wave forecasting have been prepared by Darbyshire and Draper (1963), comparable to that given in Figure 4-5 based on the results of Bretschneider. Again, two sets of curves are prepared, one for the open ocean and a second set for coastal waters.

Comparisons of Forecasting Methods

The state of wave forecasting is still sufficiently primitive that it is not presently possible to choose definitely between the various methods on the basis of predicted wave heights. Arguments for the preference of one method over another are still based mainly on theoretical reasons, with some comparisons between the spectra formulas. With regards to prediction, Table 4-2 compares results of the approaches for a range of wind speeds and fetch distances. It is seen that there are no pronounced differences in the predicted values. This agrees with the conclusion of Neumann and Pierson (1957) that comparisons between predicted wave heights using the different methods and the observed waves will not in general disclose a preference for one method over another. Although Table 4-2 compares significant wave parameters, it should again be recognized that the P-N-J, Liu, and Darbyshire methods all yield the complete spectra, which is potentially closer to nature and more useful than the S-M-B approach, which gives only the significant wave parameters.

One last word about the S-M-B significant wave approach. It has been found that it generally gives a significant wave height that is close to that given by the P-N-J method when the spectrum is converted to an equivalent significant wave. At high wind speeds the S-M-B method gives a slightly lower significant wave height, whereas with low winds it yields a greater height. The differences are small, however. Of greater importance in distinguishing predictions by the two methods is what happens to the

TABLE 4-2 Wave Predictions by Various Methods for a Range of Wind Velocities and Fetches

WIND SPEED (m/sec)	FETCH DISTANCE (km)	SIGNIFICANT WAVE HEIGHT (m)				SIGNIFICANT WAVE PERIOD (sec)			
		S-M-B	*P-N-J*	*Liu*	*Darbyshire**	*S-M-B*	*P-N-J*	*Liu*	*Darbyshire*
10	200	2.1	2.4	2.1	1.2–1.4	7	8	7	7
10	1,000	2.7	2.4	3.4	1.2–1.5	11	8	10	7
20	200	5.2	4.3	5.8	4.3–5.2	10	8.5	10	10
20	1,000	8.9	11	9.2	4.6–5.6	15	16	15	11
30	200	8.2	7.9	11	9.7–12	12	10	12	15
30	1,000	15	15	17	11–13	19	15	18	16

*Darbyshire and Draper (1963) give H_{max}. The theoretical H_{max}/H from Longuet-Higgins (1952) deepnds on the wave period and on the length of the record (Table 3-3). Thus a range of possible H values are given for the Darbyshire approach.

waves once they travel from the storm area to the particular beach of interest. It is in this phase that the S-M-B approach displays its greatest conceptual weakness. Without correctly applying wave spectrum concepts which can account for wave dispersive effects by period, the S-M-B method leads to incorrect results. There is no way to forecast, for example, how long the swell arriving from a distant storm will last at a given point. It cannot answer questions such as what will be the character of the swell a day after the first high swell arrives from the storm. Because of insufficient evaluation of the wave periods, the S-M-B method does poorly at predicting the initial arrival time of the waves. Walden (1954, 1957) compared the P-N-J method with various significant wave prediction approaches for swell arrival times. The predicted values from different methods varied considerably, so that this provided a good test of the approaches. For a swell observed on the coast of Angola in January 1955, the P-N-J method gave a swell period prediction of 16 seconds, while the S-M-B method, as modified by Bretschneider (1959), gave 23.7 seconds. Since the deep-water propagation velocity is proportional to the period, the P-N-J method predicted an arrival time considerably later than that of the S-M-B method. The observed swell arrived with a period of 16 seconds and at the time predicted by the P-N-J approach.

Rattray and Burt (1956) compared the S-M-B and P-N-J methods for wave height prediction for an intense storm in the North Pacific. The wind in the generation area was between 45 and 55 knots (23 to 28 m/sec) for 33 hours; the fetch was 500 nautical miles (925 km). The waves had a significant wave height of 14.6 meters as observed by trained U.S. Weather Bureau personnel. Both methods gave predicted heights of 14.6 meters, in agreement with the observed height. Neumann and Pierson (1957, p. 136) used this example in a comparison with the Darbyshire method and found that the latter's approach gave a significant wave of only 11 meters, considerably smaller than observed. Darbyshire (1957, p. 185) argued, however, that the error in the observed estimate could have been 20 to 30% and that a reduction of this order would bring the observed value into agreement with the predicted value from his method.

The papers of Neumann and Pierson (1957) and Darbyshire (1957) provide an interesting debate on the relative merits and effectiveness of the two methods. Neumann and Pierson compare the theoretical wave spectra of Darbyshire (1952, 1955, 1956), Neumann (1953), and the modification of Neumann's spectra by Roll and Fisher (1949). The comparison is made with one very carefully computed observed spectrum. It is shown that the Neumann spectrum agrees best with the observed spectrum, that it correctly predicts the significant wave height, and that it is consistent with the average period considerations, sea surface slope considerations, and the properties of the sea surface when photographed. Data from a number of different investigations are utilized in the comparisons. The Roll-Fisher modification is found to lead to discrepancies with the observed wave characteristics, and Neumann and Pierson present theoretical arguments against the suggested modifications. The Darbyshire method predicted a total energy much below that observed, and the peak period was on the order of 8 seconds rather than the observed 6 seconds. Neumann and Pierson convincingly attribute this discrepancy to Darbyshire's use of a pressure transducer without proper correction of attenuation with water depth. Darbyshire (1957) presents a rebuttal to the conclusions of Neumann and Pierson (1957) and presents several additional examples of comparisons with waves in the North Atlantic. In each of these examples the Darbyshire approach agreed more closely with the observed waves than did the P-N-J prediction.

In conclusion, as pointed out by Munk (1957), the spectrum methods provide a "conceptual advance" over the significant wave approach, and thus in his opinion the S-M-B method should be retired. On the whole, the P-N-J method has a firmer basis than does the Darbyshire approach, the Darbyshire spectrum being entirely empirical. In general, the P-N-J method appears to give more reasonable predictions than does the Darbyshire approach, especially at high wind speeds and large fetch distances. The Darbyshire method would be preferred in the North Atlantic, the area to which it is empirically "fitted." The fetch-limited spectrum proposed by Liu (1971) has not been adequately tested and compared to make any judgment as to its accuracy.

More studies are obviously required of wave generation and forecasting if the differences between the methods are to be resolved and prediction improved. There may be effects that are important that have not been adequately accounted for in the previous investigations. Cartwright and Darbyshire (1957), for example, point out the importance of the difference in temperature between sea and air, and Fleagle (1956) has shown that an 11°F decrease in air temperature relative to sea temperature causes a doubling of the wave height for the same wind speed. Such factors could account for some of the considerable disagreement between predicted and recorded values.

DEEP-WATER WAVE PROPAGATION

Once waves leave the area of generation and are no longer under the influence of the wind, they begin to sort themselves out by a process known as *wave dispersion* and thereby become more regular. It will be recalled from Chapter 3 that in deep water the wave group velocity, the rate at which the wave energy and the wave group as a whole travels, is given by

$$C_g = \frac{1}{2}C = \frac{1}{2}\left(\frac{g}{2\pi}T\right) \tag{4-3}$$

The velocity of propagation depends on the wave period, the longer-period waves traveling faster than the short-period waves. The important factor is that although the entire spectrum of waves is complex, the individual groups of sine waves that make up the spectrum travel according to equation (4-3). It is apparent that, starting together in the generation area, the longer-period waves will outrun and leave behind the shorter-period waves. This sorting or *wave dispersion* by the wave period is the main factor in producing the more uniform swell waves from the initially complex sea in the area of generation. Dispersion therefore produces a narrowing of the energy spectrum, with a strong energy peak corresponding to the wave period of the swell, and the greater the distance of travel from the area of generation the narrower the spectrum becomes.

Each wave period in the spectrum has a group of waves associated with it, moving outward from the generation area, and to a first approximation, moving independently of different-period waves. The different groups with different periods have nonidentical travel speeds; therefore, although they start out together in the generation area, they soon become separated as travel proceeds. Within each group, waves continuously arise in the rear of the group, travel forward through it, and then die away as they move to the front. The wave group has a permanence that is not shared by individual wave crests. Individual waves can be followed for only a short time, while the progress of wave groups can be traced across entire oceans.

Let us follow the progress of one separate wave group associated with the period T as it moves from the generation area to the coastline shown in Figure 4-10. The

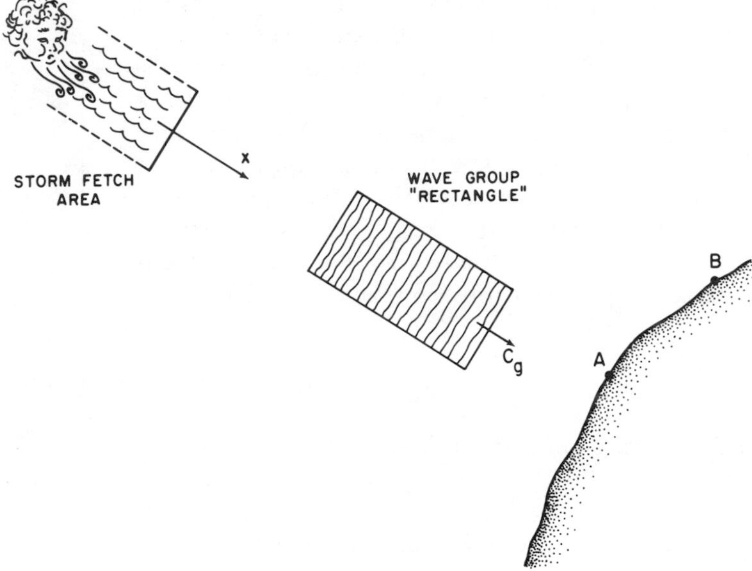

Figure 4-10 The wave group "rectangle" associated with one wave period in the complete spectrum as it moves from the generation area to the forecast point A.

group of waves will occupy a rectangular area whose width is the width of the storm front and whose length is governed by the storm duration and total fetch length. Let the edge of the conceptual rectangle be at $x = 0$ at the time $t = 0$ of first wave formation. The wave group outlined by the rectangle will move downwind in the positive x-direction at a speed given by equation (4-3). If the storm duration is 30 hours, then the windward end of the rectangle would pass the line $x = 0$ approximately 30 hours later than the front of the rectangle, and its total length would be $C_g \times 30$ hours. The edges of the rectangle would not actually remain sharp, as some of the wave disturbance would leak out into the surrounding still water. A few very low waves would run out ahead of the leading edge, and wave energy would likewise diffuse sideways, making the conceptual edge of the rectangle rounded rather than sharp. These effects would be small, however, and waves would not reach point B in Figure 4-10 for this reason. We shall see later that waves might actually reach point B from the storm area, but because of the angular spreading of the waves, not all the waves generated are initially directed in the mean fetch direction (there are slight deviations in direction from the x-axis direction of Figure 4-10). Even with this angular spread the waves are directed down a rather narrow corridor; much of the ocean will not be affected by the storm waves. Waves generated by a storm are not analogous to a pebble thrown into a pond, where the waves spread in every direction and the energy per unit crest length of wave quickly decreases. Because the spread of ocean waves is limited, the wave group will still retain much of its energy and wave height when it reaches the distant shore. There is some ordinary viscous dissipation; as we shall see, however, this is small except for the lowest wave periods.

Going back to our single conceptual wave group of rectangular shape with period T: if R is the distance from the leading edge of the storm fetch to our point A on the coast in Figure 4-10, then the time t_{ob} of first observation of arrival of the waves is

$$t_{ob} = \frac{R}{C_g} = 4\pi\frac{R}{gT} \tag{4-4}$$

It is seen that there is an inverse linear relationship between t_{ob} and T. Therefore, if 10-second-period waves arrive in 2 days from the storm area, 5-second waves will arrive in 4 days; that is, if t_{ob} is plotted against T for a given storm, there is a straight-line relationship (see Figure 4-12, for example). For a line source of waves, the final arrival time of waves of period T will be determined by the storm duration D, being $t_{ob} + D$. There would be some error in this simple estimate if the fetch length were large, so that one could not consider a simple line source. This can be corrected for with relative ease.

Since there is an infinite number of wave periods produced in the generation area—that is, an entire spectrum—there would likewise be an infinite number of conceptual rectangles such as that discussed above, one for each period. In fact, there is of course a continuum, not a number of discrete wave periods and rectangles as we have conceptualized. However, we can picture the wave dispersive process as many different rectangles moving at different velocities depending on the associated wave period, and therefore a succession of rectangles arriving at our coastal point A in Figure 4-10. The first rectangle to arrive would have the longest wave period, followed by rectangles of progressively shorter wave period.

This process of wave dispersion by wave period was first clearly demonstrated in the study of Barber and Ursell (1948). Figure 4-11 contains a series of wave spectra they obtained at 2-hour intervals for 2.5 days from a wave recorder at Pendeen, Cornwall, England. Swell from a tropical storm off the east coast of the United States was first detected in the spectra as an energy peak at 19 seconds at 1900 hours on 30 June 1945. In subsequent spectra it is seen that the mean peak period progressively decreased on account of dispersion, reaching approximately 13 seconds at the end of 2 July. Figure 4-12 illustrates this progressive shift of the maximum and minimum periods which limit the width of the spectrum peak associated with the storm. The irregularities in the points about the straight lines were demonstrated to result from the effects of the passage of the waves through tidal currents near the wave recorder.

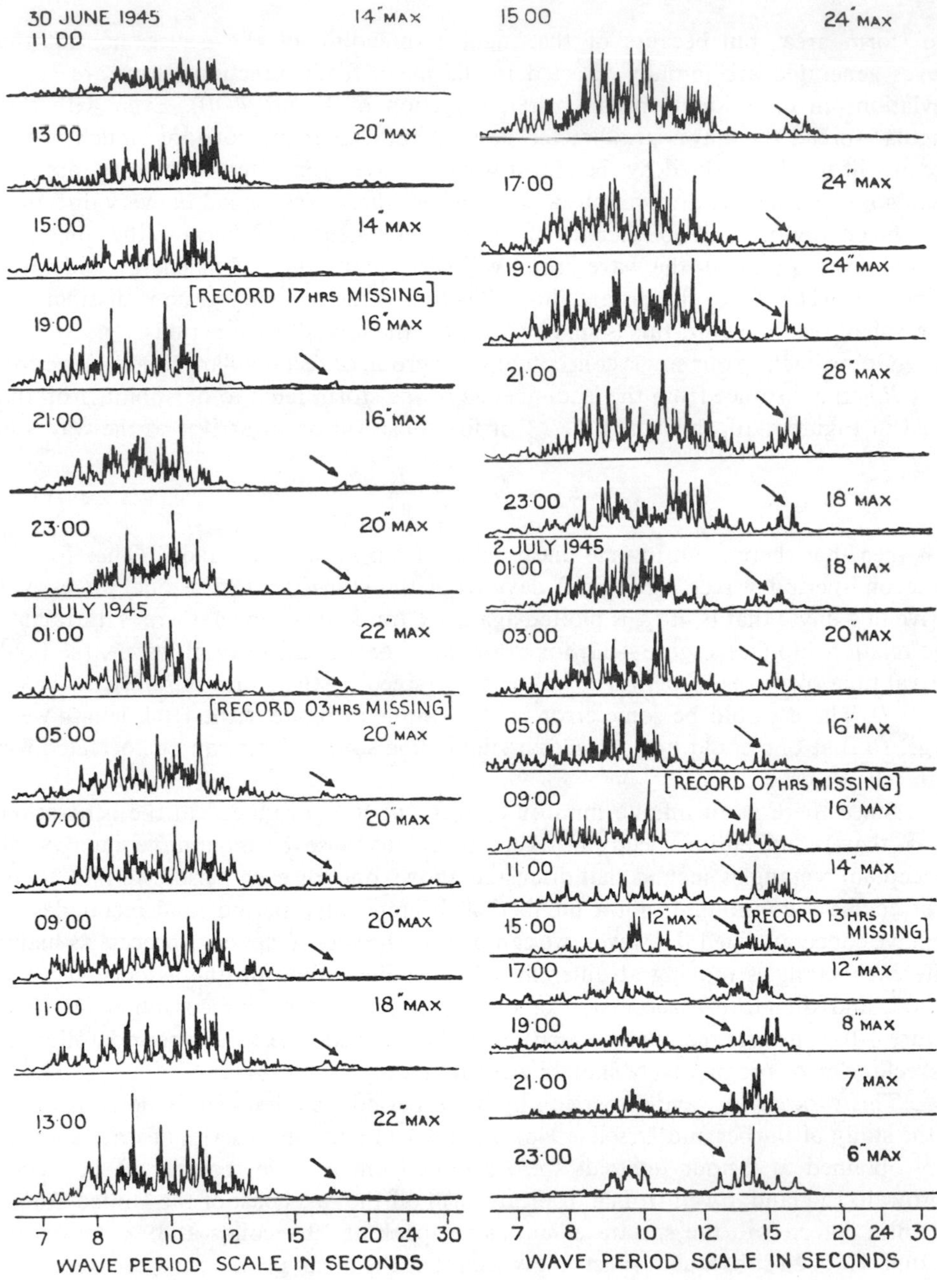

Figure 4-11 Wave spectra obtained at 2-hour intervals by Barber and Ursell (1948) from a wave recorder at Pendeen, Cornwall, England. The arrows point to the spectra peaks resulting from swell generated by a tropical storm off the east coast of the United States. Note the progressive shift in period and changes in the size (energy) of the peak.

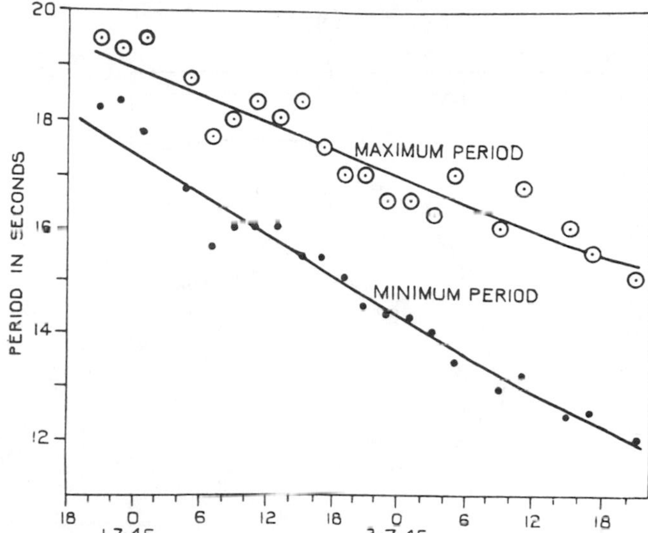

Figure 4-12 The progressive shift of the maximum and minimum periods which limit the width of the spectra peak associated with the storm shown in the spectra of Figure 4-11. [*After* Barber and Ursell (1948)]

Similar results were obtained for other storms, the most distant being off Cape Horn, some 6,000 nautical miles (11,000 km) from Cornwall.

Subsequent observations have confirmed the global nature of swell propagation. Wiegel and Kimberly (1950) demonstrated that southerly swell observed throughout the summer at Oceanside, California, arrives from Southern Hemisphere storms between 40 and 65°S and between 120 and 160°W. Wave heights reached 3 to 4 meters, and periods were from 12 to 18 seconds. Munk and Snodgrass (1957) obtained the energy spectra of the very-low-amplitude, long-period ($T \simeq 20$ sec) *forerunners* of swell, arriving at Guadelupe Island west of Baja California, Mexico, from storms in the Indian Ocean, entering the Pacific along the great circle route between Antarctica and Australia, a travel distance of some 8,000 nautical miles (15,000 km). Snodgrass et al. (1966) traced waves generated by storms near Antarctica, south of New Zealand, by a series of stations in a great circle route across the entire Pacific to Yakutat, Alaska, a distance of 10,000 km. All of these studies demonstrate that deep-ocean swell waves in fact propagate with the classic group velocity appropriate to their period. They also show that the waves follow great circle paths as they travel across the deep ocean, the effects of the earth's rotation on the path being negligible.

As waves travel these immense distances across the open ocean, how much of their energy do they lose? It would appear that little is lost, judging from the size of the swell waves that sometimes reach the west coast of the United States from storms in the far South Pacific. Snodgrass et al. (1966) particularly examined wave attenuation during travel. Figure 4-13 from that study demonstrates typical spectra at distances 0 km (the inferred wave spectra at the end of the storm fetch), 1,000 km, and 10,000 km from the storm area. It is seen that most of the attenuation occurs close to the area of generation and is most pronounced in the short-period (high-frequency) portion of the spectrum. With further travel beyond the first 1,000 km, very little

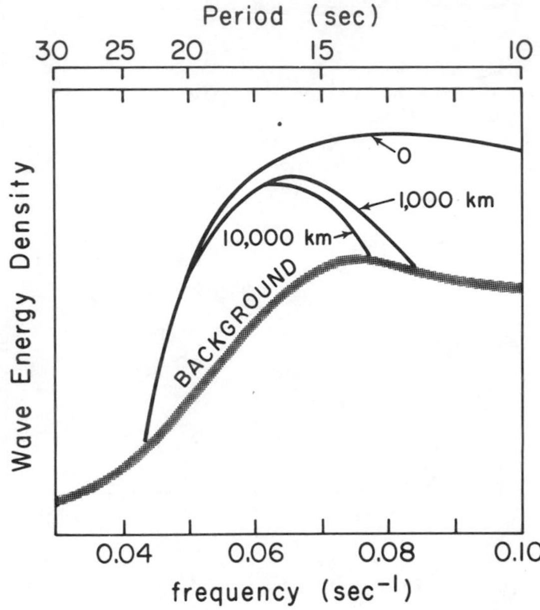

Figure 4-13 Wave spectra obtained by Snodgrass et al. (1966) from a storm in the far South Pacific at distances 0, 1,000, and 10,000 km from the storm front.

additional attenuation occurs. There are several possible ways in which waves may lose energy as they travel across the deep ocean:

1. Internal viscous damping.
2. Angular spreading of the waves as they leave the localized storm area.
3. Contrary winds blowing against the moving waves.
4. Wave-wave interactions as waves of the same storm or of separate storms combine and interact.

These different possible mechanisms will be briefly examined in the next few paragraphs.

The effect of the internal viscous damping of water waves of small amplitude has been studied mathematically by Lamb (1932, article 349) for waves in deep water, and by Hough (1896) for waves of finite depth. Boundary drag on the bottom was ignored in both studies. Keulegan (1950, p. 730) contains a particularly good discussion of internal viscous effects. The expression for the decay in the wave height due to internal friction is

$$H = H_i \exp\left(\frac{-8\pi^2 \nu t}{L^2}\right) \tag{4-5}$$

or

$$H = H_i \exp\left(\frac{-32\pi^4 \nu t}{g^2 T^4}\right) \tag{4-6}$$

where H_i is the initial wave height at time $t = 0$ and H is the wave height at time t; L is the wave length, T the wave period, and ν the kinematic viscosity of water

($\nu = 1.4 \times 10^{-2}$ cm^2/sec). It is seen that the wave height falls off exponentially with time, the rate of decay depending strongly on the wave period T, since it enters the exponent to the fourth power. Viscous damping will preferentially remove the short-period waves, since damping is greatest at that end of the spectrum, and will leave the long-period waves little modified. Viscous damping, then, as well as wave dispersion could be effective in enhancing the regularity of the waves once they leave the generation area. The time required for the wave height to decrease to one-half its original value ($H = H_i/2$) is given by

$$t_{1/2} = 0.0088\frac{L^2}{\nu} \tag{4-7a}$$

$$= \frac{0.0088}{4\pi^2}\frac{g^2T^4}{\nu} \tag{4-7b}$$

Figure 4-14 contains a plot of $t_{1/2}$ versus T based on equation (4-7b). It is seen from this graph that a 5-second wave will travel for some 2,660 hours (a deep-water distance of some 37,400 km) before its height is reduced to half its original value. A wave with a period of 1 second travels for only 4.3 hours (a distance of 12 km) for the same result.

It is apparent, then, that viscous damping is important in removing the very-

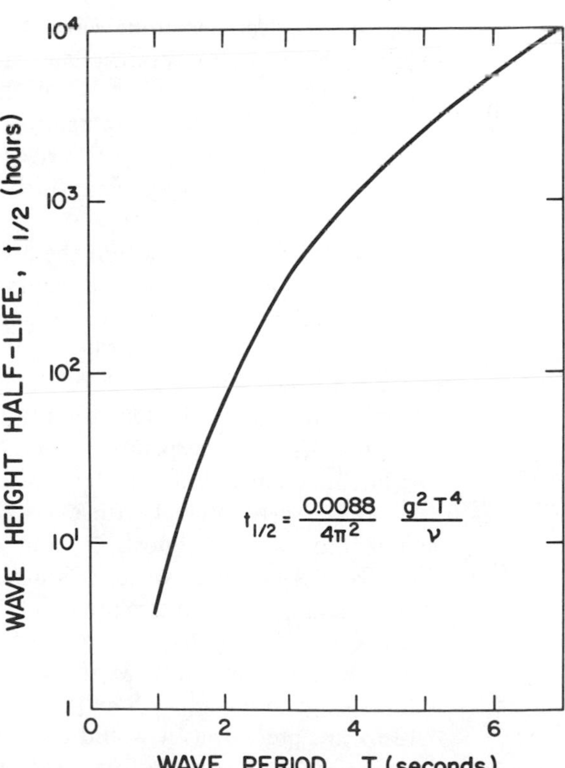

Figure 4-14 The wave half-life (the time required for the wave height to decrease by one-half) due to internal viscous damping.

short-period waves but has little effect on the long-period waves. Snodgrass et al. (1966) conclude that viscous damping could account for only a negligible portion of the attenuation depicted in Figure 4-13. Viscous damping is, however, important in converting complex sea into regular swell.

Not only do waves disperse by wave period in the predominant direction of wave advance, but they also spread sideways over a broadening area because they start out in slightly different directions. Due to the variability in the wind direction about some mean predominant direction, waves are generated with a corresponding variability in direction. Waves in the generation area move not only in the direction of the mean wind, but at various other angles. It has been common practice to account for this variability by assuming that waves up to 30 degrees to the mean wind direction have undiminished wave heights when the isobars are straight, and 45 degrees for curved isobaric patterns. Pierson, Neumann, and James (1955) provide techniques by which angular wave spreading can be accounted for in swell forecasting. One important effect of angular spreading is that positions such as point *B* in Figure 4-10 may receive wave swell from the generation area even though it is to one side of the direction of mean fetch. The other important effect is that like dispersion by wave period, angular spreading will diminish the wave energy arriving at some forecast point. The wave energy arriving at either points *A* or *B* in Figure 4-10 will be only a fraction of what was present in the storm area, and the points will experience diminished wave heights. However, angular spreading cannot account for the attenuation of wave energy observed by Snodgrass et al. (1966) shown in Figure 4-13, most of which occurred near the generation area.

Contrary winds blowing against the moving waves may produce some loss in energy. In deep water, the swell waves are low and gentle and so do not offer much form drag to the opposing wind. Such a loss is probably negligible except in unusual instances.

We have seen that within the area of wave generation the growth of the spectrum is ultimately limited by the nonlinear processes of wave breaking (Phillips, 1958). It affects mainly short-period waves and, if the suggestion of Longuet-Higgins (1969*b*) is correct, the momentum lost by the breaking of the short-period waves may actually enhance the continued growth of the long-period waves. Snodgrass et al. (1966) have attributed the observed attenuation in Figure 4-13 to these same processes continuing for a time after the waves have passed out of the generation area. This would account for the higher attenuation in the shorter-period waves.

Waves may interact with one another to cause the energy of a given wave to scatter into waves of almost the same period but traveling in slightly different directions from the original waves (Phillips, 1960; Hasselmann, 1960, 1963). The overall effect is similar to geometric angular spreading already discussed, but is much more effective in and near the area of wave generation. As wave dispersion separates the waves by period and narrows the spectral peak, scattering by wave-wave interaction will no longer be effective. Snodgrass et al. (1966) attribute the changes seen in Figure 4-13 to this process, as it would also account for the principal attenuation occurring near the generation area before dispersion could be effective.

In summary, as waves leave the generation area they spread out by wave dispersion, the longest period waves traveling the fastest. As a result, at some distant observation point, the longest-period waves will arrive first, followed by progressively shorter-period waves. Viscous damping is important to the short-period waves but has little effect on the long-period waves. Acting together, wave dispersion and viscous damping narrow the wave spectrum and change the complex sea in the generation area to regular swell waves. Nonlinear effects such as wave breaking and wave-wave scattering continue for a short distance outside the generation area and are important in wave energy dissipation. Once waves have reached a regular swell condition they may travel for thousands of kilometers across entire ocean basins with relatively little additional energy loss. Waves generated by storms off Antarctica may be measured on the beaches of Alaska.

WAVES IN SHOALING WATER

In deep water the profile of the ocean swell is very nearly sinusoidal, with long, low crests. As the waves enter shallow water they undergo a transformation (Figure 4-15), starting when they first "feel bottom" at a depth approximately one-half the

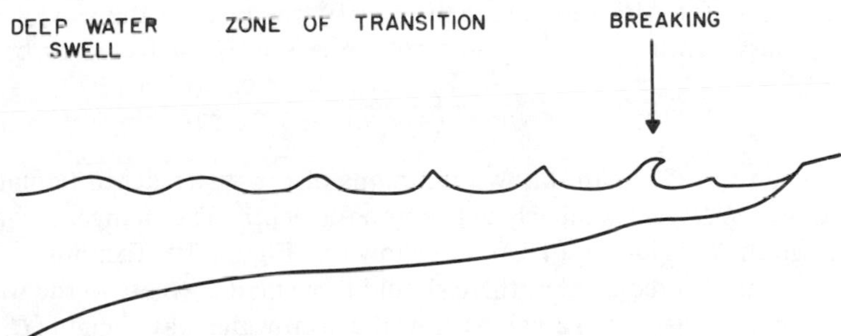

Figure 4-15 Illustrative transformation of surface waves as they pass over progressively shoaling water.

deep-water wave length and becoming significant at one-fourth the deep-water wave length (the deep-water approximation of Airy wave theory in Chapter 3). The wave velocity and length progressively decrease, and the height increases. Only the wave period remains constant. In water depths not far beyond the breaker zone, the wave train consists of a series of peaked crests separated by relatively flat troughs. Finally, the crests become oversteepened and unstable, and they break. The height of the breaking wave may be several times that of the deep-water wave.

The above transformation is most readily apparent in the long-period swell waves, since they "feel bottom" first and generally are the most nearly sinusoidal to begin with. Local storm waves are initially steeper, even in deep water, so that the transformation is not as apparent.

Although solutions are available for waves traveling over a sloping bottom (see Stoker, 1957), they are complicated and seldom used. A simpler approach is to use the wave solutions discussed in Chapter 3, assuming that they apply even though they were developed for a flat horizontal bottom (Rayleigh, 1911). Using Airy wave theory, we have

$$\frac{L}{L_\infty} = \frac{C}{C_\infty} = \tanh\left(\frac{2\pi h}{L}\right) \simeq \left[\tanh\left(\frac{2\pi h}{L_\infty}\right)\right]^{1/2} \tag{4-8}$$

which yields the variations in the wave length L and the phase velocity C with changing water depth h. These variations were plotted in Figure 3-5 (page 44), where it is seen that the wave length and phase velocity decrease with decreasing depth—that is, as the beach is approached. Similarly, we have variations in n (equation 3-23) and hence changes in the group velocity $C_g = Cn$ as the depth decreases.

The variations in the wave height of the shoaling waves can be obtained from a consideration of the energy flux. If the losses of energy due to bottom drag, reflection, and so forth are negligible—generally a safe assumption for waves outside the near-shore region—then the energy flux is constant and we have

$$P = ECn = (ECn)_\infty = \text{constant} \tag{4-9}$$

where $(ECn)_\infty$ is the energy flux evaluated in deep water. Without energy losses the energy flux in the shoaling waves remains equal to its value in deep water. Using the energy relationship for Airy waves (equation 3-20), the ratio of the wave height H in water of arbitrary depth to the deep-water wave height H'_∞ is found to be

$$\frac{H}{H'_\infty} = \left(\frac{1}{2n}\frac{C_\infty}{C}\right)^{1/2} \tag{4-10}$$

Both n and the ratio C/C_∞ are known functions of the water depth (equations 3-23 and 4-8), so that H/H'_∞ will similarly depend on the depth. The changes in wave height with depth, given by equation (4-10), are shown in Figure 3-5. Examining this curve closely, we see that it predicts that there should be a small decrease in the wave height in intermediate water depths to a value below the deep-water wave height ($H/H'_\infty < 1$). This decrease is then followed by a rapid increase in H as still shallower depths are reached. The temporary height reduction in intermediate water depths is brought about by n increasing faster than C decreases, so that $C_g = Cn$ temporarily increases. Since ECn must remain constant, this produces a temporary decrease in the energy E and thus in the wave height H. This decrease in H, followed by a rapid increase, has been observed by Iverson (1951) in laboratory wave studies and can be observed in swell waves approaching a natural beach if they are sufficiently regular. The actual observed increase in the wave height is found to be more rapid and to lead to higher waves than predicted by equation (4-10) and the curve of Figure 3-5. This is partly due to the failure of Airy wave theory at these very shallow depths and large wave heights. A consideration of the finite wave height theories (Stokes, solitary, etc.) yields a better prediction of the final increase in wave height prior to breaking (Munk, 1949*a*).

Commonly waves undergo some refraction, a bending of the crests, as they travel throught shallow water. The above analysis that led to equation (4-10) did not include any consideration of the energy changes brought about by refraction: H'_∞ is the unrefracted deep-water wave height. We shall see in a later section of this chapter, when we examine wave refraction, that the ratio H/H'_∞ obtained from equation (4-10) or the curve of Figure 3-5 can be corrected for refraction effects by multipying the ratio by a refraction coefficient K_r.

The wave steepness H/L also varies in the shoaling waves. The steepness temporarily drops slightly below its deep-water value as the waves pass through intermediate water depths and then rapidly increases. The sudden increase in steepness along with the increase in height in shallow water is probably the most striking feature of the shoaling waves, noticed by even the casual observer. The steepness increases rapidly until a point is reached where the waves become unstable and break.

BREAKING WAVES

When waves reach the beach and enter water approximately as deep as the waves are high, they break, the crest throwing itself forward or disintegrating into bubbles and foam (Figure 4-16). A common notion is that waves break because they drag

Figure 4-16 The spectacular plunging breaker, curling over and throwing its mass forward and downward toward the beach.

on the bottom until they trip and the crest topples forward. This is incorrect; experiments have shown friction to be negligible. Instead, a wave breaks because it becomes overly steep, especially near the wave crest peak, and because the velocity of water particle motion in the crest exceeds the phase velocity of the wave form and so surges ahead.

Three types of breakers are commonly recognized: spilling, plunging, and surging (Figure 4-17). In *spilling breakers* the wave gradually peaks until the crest becomes unstable and cascades down as "white water" (bubbles and foam). In *plung-*

SPILLING BREAKERS

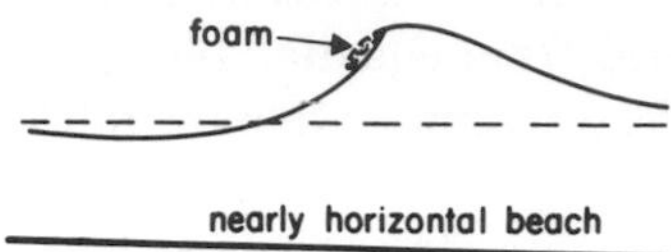

PLUNGING BREAKERS

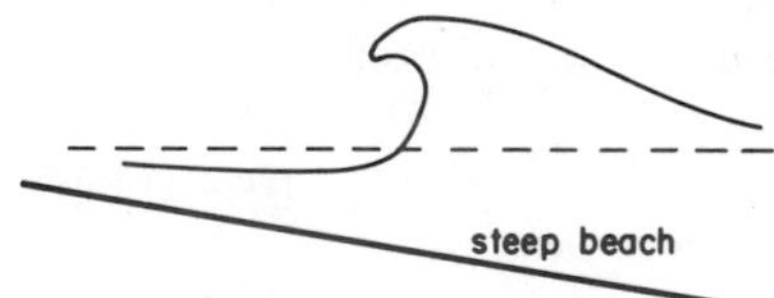

SURGING BREAKERS

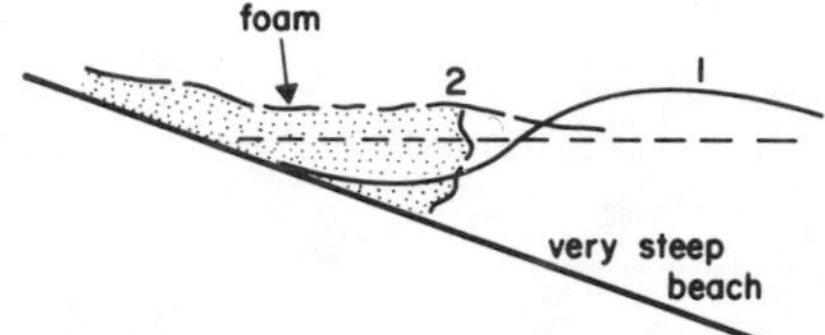

Figure 4-17 The three types of breaking waves that occur at the beach.

ing breakers the shoreward face of the wave becomes vertical, curls over, and plunges downward as an intact mass of water. *Surging breakers* peak up as if to plunge, but then the base of the wave surges up the beach face so that the crest collapses and disappears. There is actually a continuum of breaker types grading from one type to the next, so that application of these classifications is not always easy. In general, spilling breakers tend to occur on beaches of very low slope with waves of high steepness values; plunging waves are associated with steeper beaches and waves of intermediate steepness; surging occurs on high-gradient beaches with waves of low steepness (Patrick and Wiegel, 1955; Wiegel, 1964). Galvin (1968) identifies a fourth type of breaking wave (Figure 4-18), a *collapsing breaker*, which is intermediate between the plunging and surging types. He finds fairly good breaker-type prediction (Figure 4-19) using either $H_\infty/(L_\infty s^2)$ or $H_b/(gsT^2)$, where s is the beach slope. As either of these parameters increases, breaker type changes from surging to plunging to spilling.

In Chapter 3 we saw that there are many theoretical criteria which purport to define the condition at which waves become unstable and break. The most basic of the criteria is that the velocity of the water in the very peak of the crest is equal to the phase velocity of the wave form, the rate at which the crest is moving forward. It is clear that if u, the particle velocity at the crest, exceeds C, the phase velocity, the crest will

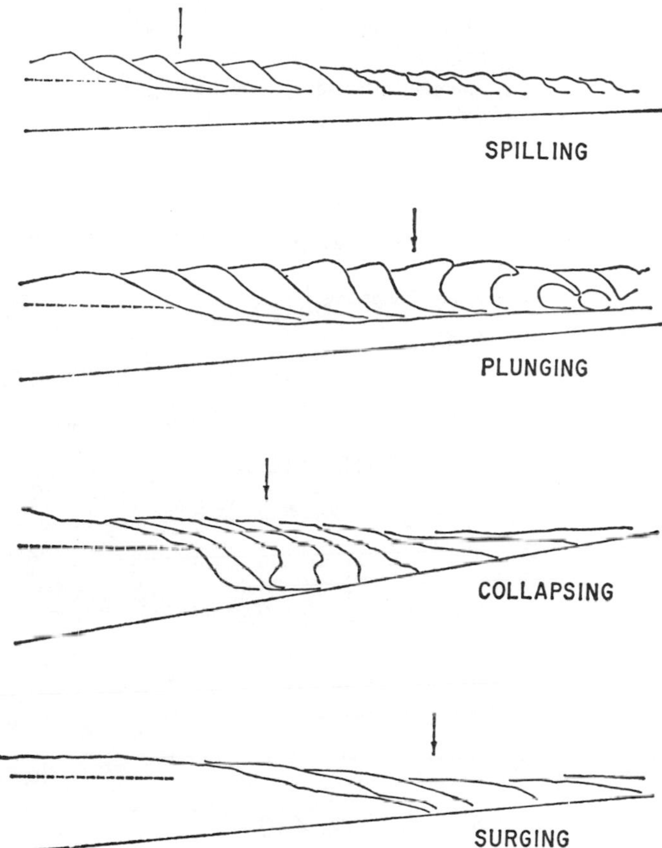

Figure 4-18 The four types of breaking waves on a beach as classified by Galvin (1968). Obtained from high-speed moving pictures, the arrows point to the initial points of breaking. [*After* Galvin (1968)]

topple forward and break. Measurements by Iverson (1951) of breaking waves in the laboratory wave basin confirm that u approaches C near the crest. For solitary waves, instability and breaking occurred at a critical value of the ratio $\gamma_b = H_b/h_b$, although there has been some disagreement as to the exact value. We saw in Chapter 3 that actual measured values of the critical γ_b for solitary waves vary considerably, the value depending principally on the beach slope. Iverson (1951) has shown that the critical value of γ_b for oscillatory waves also varies markedly, depending on the beach slope and the initial wave steepness.

By equating the energy flux at the breaker zone to the energy flux in deep water, applying solitary wave theory to the breakers with $\gamma_b = H_b/h_b = 0.78$ as the breaking criterion, and using Airy wave theory for the deep-water energy flux, Munk (1949*a*) derived the equation

$$\frac{H_b}{H'_\infty} = \frac{1}{3.3(H'_\infty/L_\infty)^{1/3}} \tag{4-11}$$

relating the breaker height to the deep-water wave steepness. This equation is plotted in Figure 4-20 along with the breaking wave data reported in Munk (1949*a*). It is

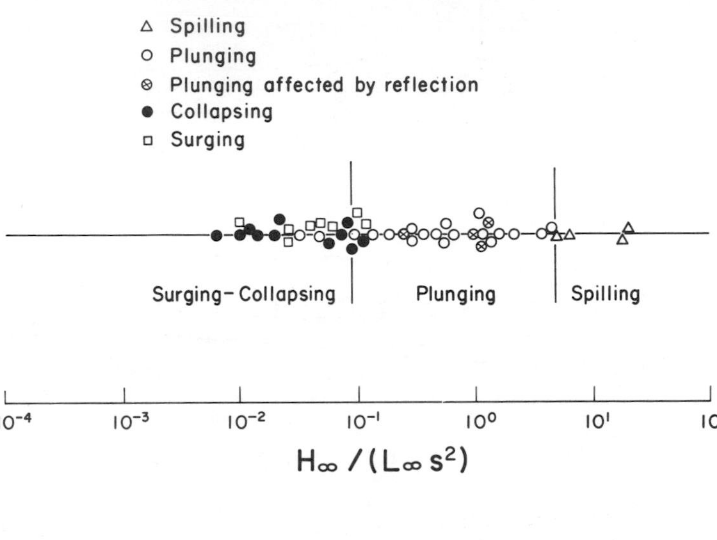

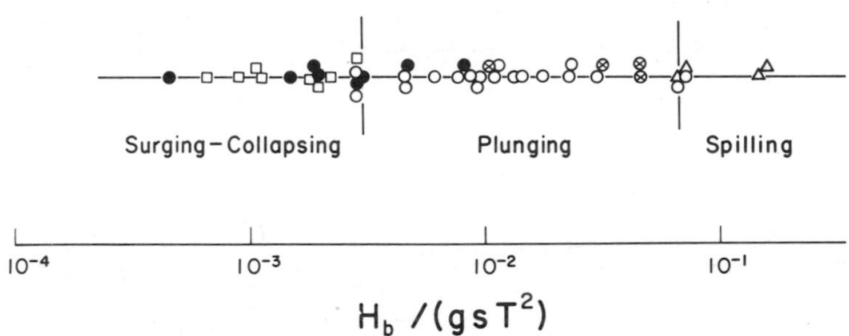

Figure 4-19 The breaker types, as classified in Figure 4-18, as a function of the dimensionless parameters $H_\infty/(L_\infty s^2)$ and $H_b/(gsT^2)$, where s is the beach slope. [*After* Galvin (1968)]

seen that the relationship agrees with the data in the range of low H'_∞/L_∞ values. For steeper waves where equation (4-11) does not apply, Munk (1949*a*) provided an empirical curve which connects with a curve from Airy wave theory at high H'_∞/L_∞ values. Equation (4-11) or its graphical equivalent and the empirical curve of Figure 4-20 have been widely used by coastal engineers to predict breaker heights from deep-water wave conditions. When one reads in the newspaper or hears on the radio that "ten-foot breakers at twelve-second intervals are expected on the coast," the breaker heights were most likely predicted in this way.

Recently Bowen, Longuet-Higgins, and Thornton have applied linear Airy wave theory with considerable success to the generation of longshore currents (see Chapter 7). Such an application encouraged Komar and Gaughan (1973) to examine the applicability to evaluating wave breaker heights in spite of the fact that the theory should not be valid under those conditions. A critical value of $\gamma_b = H_b/h_b$ was accepted as a wave-breaking criterion, but more as an observational fact, not connected with solitary wave theory, which showed poor comparison with laboratory tests (Chapter 3). Using Airy wave theory to evaluate the energy flux in both deep water and at the

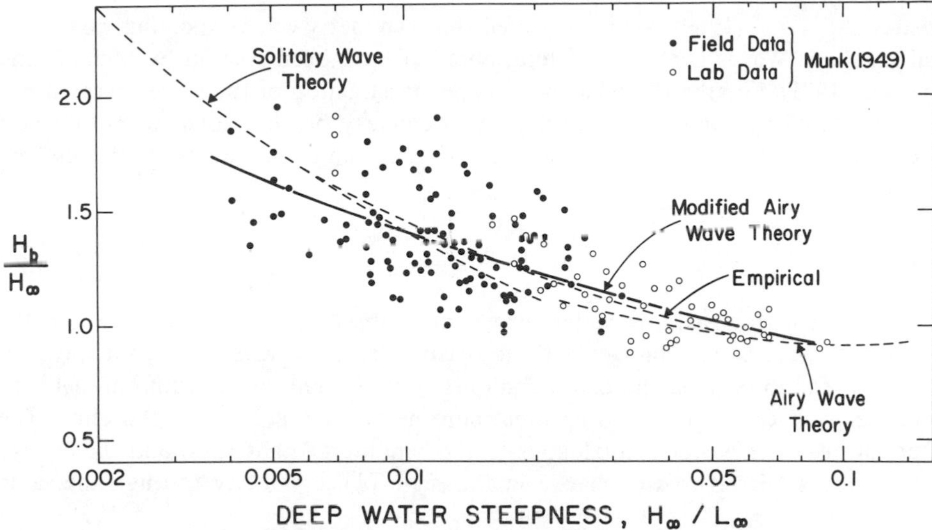

Figure 4-20 The wave breaker height H_b related to the deep-water wave steepness. The dashed curves are from Munk (1949a), the "solitary wave theory" curve being given by equation (4-11). The "modified Airy wave theory" solid curve is from equation (4-12) proposed by Komar and Gaughan (1973).

breaker zone, Komar and Gaughan obtained

$$\frac{H_b}{H'_\infty} = \frac{0.563}{(H'_\infty/L_\infty)^{1/5}} \tag{4-12}$$

the steepness being to the 1/5 power rather than the 1/3 power obtained by Munk (1949*a*) [equation (4-11), above]. The Komar-Gaughan relationship is fitted to the available laboratory and field data so that it is semiempirical, the 0.563 coefficient depending on the data fit. Equation (4-12) is shown in Figure 4-20 as the solid curve. It is seen that this curve fits the data very well over the entire range of H'_∞/L_∞ values, nearly lying atop the empirical curve of Munk (1949*a*).

In his laboratory study Iverson (1951) found that the relationship between H_b/H'_∞ and H'_∞/L_∞ depends on the beach slope. The greater the beach slope, the greater H_b/H'_∞ is for a given value of the wave steepness. Because of this, when a beach slope of 1:20 or greater is involved, its effects on the breaker height must be considered.

Both equations (4-11) and (4-12) apply to waves without refraction. If refraction has occurred, they must be corrected, again by using the refraction coefficient K_r (equation 4-17).

Because of their importance to beach processes in general, the properties of breaking waves have been extensively studied. Only brief mention can be made here of some of these investigations. Besides Iverson (1951), the internal velocity field of breaking waves has been examined by Miller and Zeigler (1965) and Adeyemo (1971). Miller (1973) has considered the effects of surface tension on breaking wave characteristics. Adeyemo (1969) has investigated the asymmetry of the shoaling waves and final breakers. Waves broken by a longshore bar have been studied by McNair and Sorensen (1971). Finally, there has been some investigation of the wave energy decay following breaking, wave reformation, and secondary breaker zones within the surf zone; for example, see Horikawa and Kuo (1967) and Nakamura, Shiraishi, and Sasaki (1967).

WAVE REFRACTION

Upon entering shallow water, waves are subject to *refraction*, in which the direction of wave travel changes with decreasing depth of water in such a way that the crests tend to parallel the depth contours. For straight coasts with parallel contours the wave crests therefore become more nearly parallel to the shoreline. The refraction of water waves is analogous to the bending of light rays, and the change in direction is related to the change in the wave phase velocity through the same Snell's law:

$$\frac{\sin \alpha_1}{C_1} = \frac{\sin \alpha_2}{C_2} = \text{constant} \tag{4-13}$$

where α_1 and α_2 are angles between adjacent wave crests and the respective bottom contours; C_1 and C_2 are the successive phase velocities at the two depths and can be obtained from equation (4-8) or from the curve in Figure 3-5. For the simple case of a straight shoreline with parallel contours, the angle at any depth can be related to the deep-water angle of approach α_∞ with

$$\sin \alpha = \frac{C}{C_\infty} \sin \alpha_\infty \tag{4-14}$$

It can be seen that as the phase velocity C decreases as the shore is approached, the angle α will also decrease from its deep-water value.

Wave refraction can cause either a spreading out or a convergence of the wave energy. This effect can be best examined by concentrating on the *wave rays*, the lines drawn normal to the wave crests and therefore in the direction of wave advance and energy propagation (Figure 4-21). On a straight coast with parallel contours, refraction will cause the rays to spread out, producing a similar spreading of the energy, so that the wave height is reduced at the shoreline below the height that would be achieved without refraction. The reason for this energy reduction is that the energy flux between the rays is constant, and spreading due to refraction requires that the same amount of energy flux be spread out over a greater length of wave crest. The opposite is true for ray convergence. If s_∞ is the spacing of the rays in deep water and s is the spacing later in the shoaling waves (Figure 4-21), then we have

$$P = ECns = (ECns)_\infty = \text{constant} \tag{4-15}$$

Going through the same development that led to equation (4-10), this time we obtain

$$\frac{H}{H_\infty} = \left(\frac{1}{2n}\frac{C_\infty}{C}\right)^{1/2}\left(\frac{s_\infty}{s}\right)^{1/2} \tag{4-16}$$

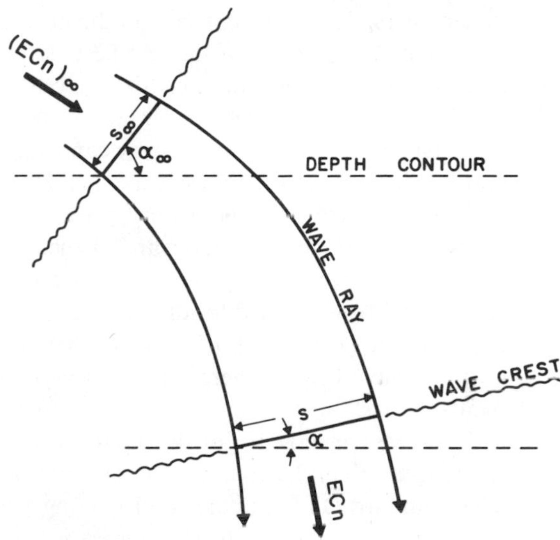

Figure 4-21 The conservation of wave energy flux ECn between two rays orthogonal to the wave crests.

Inclusion of the wave refraction therefore introduces the ratio s_∞/s, which was not present before. In practice, tables (those of Wiegel in Coastal Engineering Research Center, 1973) give the ratio H/H'_∞, which does not include refraction. That value must then be multiplied by the correction coefficient

$$K_r = \left(\frac{s_\infty}{s}\right)^{1/2} \tag{4-17}$$

which may generally be obtained by the construction of a wave refraction pattern diagram for the area of interest (see Figure 4-23). For straight shores with parallel contours, simple geometry considerations give

$$\frac{s_\infty}{s} = \frac{\cos \alpha_\infty}{\cos \alpha} \tag{4-18}$$

so that the refraction coefficient K_r becomes simple to compute.

Where there is an oblique wave approach to the shoreline, there is a longshore component of the radiation stress, the momentum flux associated with the waves. This longshore component is given by

$$S_{xy} = En \sin \alpha \cos \alpha \tag{4-19a}$$

$$= (ECn \cos \alpha)\left(\frac{\sin \alpha}{C}\right) \tag{4-19b}$$

where the x-axis is normal to the shoreline (positive onshore), and the y-axis is parallel to the shoreline; S_{xy} is the onshore flux of momentum that is directed in the longshore direction. Bowen (1969) and Longuet-Higgins (1970) have shown that S_{xy} is important in generating longshore currents in the nearshore under conditions of an oblique wave approach (Chapter 7). Of interest here is the fact that S_{xy} combines equations (4-14) and (4-15) [utilizing equation (4-18)], both of which remain constant during wave shoaling, so that S_{xy} likewise remains constant.

Irregular bottom topography can cause waves to be refracted in a complex way and produce variations in the wave height and energy along the coast. Waves refract and diverge over the deep water of a submarine canyon or other depression so that the waves on the beach shoreward of the canyon are reduced in height while those to either side, where the rays converge, are somewhat higher (Figure 4-22). Waves also refract and bend toward headlands because of the offshore shoal area associated with the headland (Figure 4-22). The wave energy is therefore concentrated on the headland, and the wave heights there may be several times as large as in the adjacent embayments.

Wave convergence or divergence due to wave refraction is important in deciding where to construct a pier or other structure along a particular stretch of coast. The Standard Oil Company's oil-loading wharf at El Segundo, California, was built before wave refraction techniques came into general practice (Dunham, 1951). As a result, it was unwittingly located in an area of strong wave convergence (Figure 4-23).

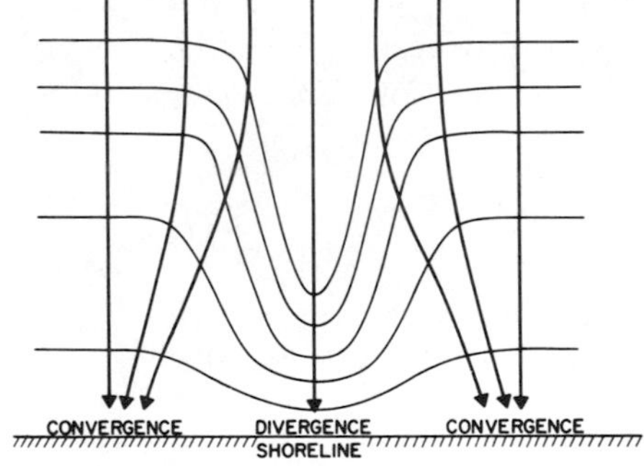

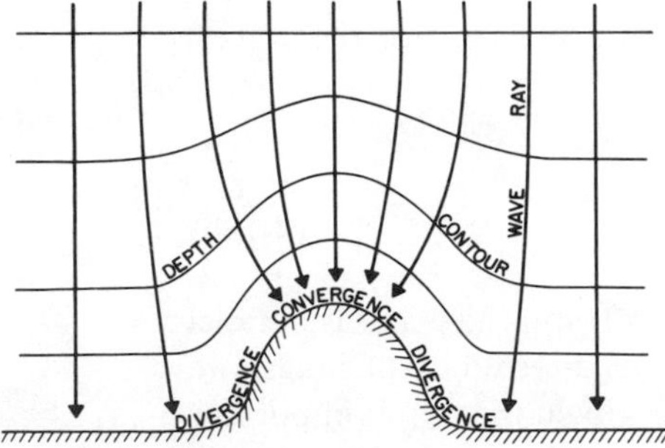

Figure 4-22 The convergence and divergence of wave rays over a submarine canyon and at a headland, resulting from wave refraction.

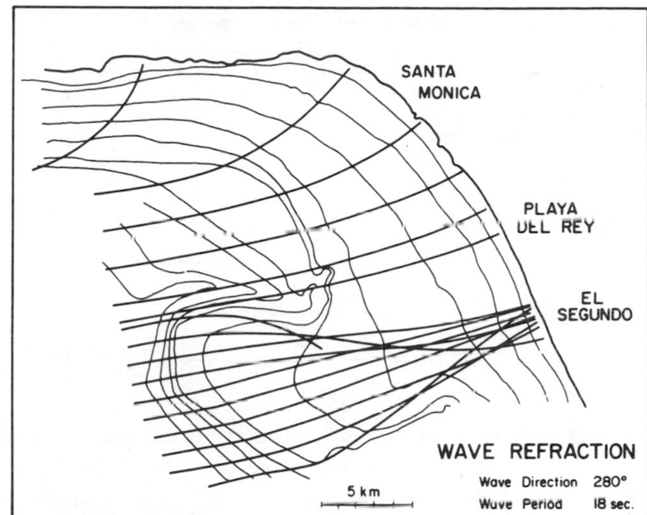

Figure 4-23 Wave refraction near El Segundo, California. Note the convergence of the rays at El Segundo, which produces huge waves at the oil-loading wharf constructed there. [*After* Dunham (1951)]

Under certain wave directions and periods, huge breaking waves were produced at the oil wharf site, closing it for extended periods of time.

It is noted in Figure 4-23 that some of the wave rays cross one another. In this situation the distance s between the adjacent rays decreases to zero, and according to equation (4-16) the wave height H should increase to an infinite value at the crossing point. This obviously does not actually happen. Instead, as the wave height steepens near the crossing point, energy will travel laterally along the wave crest and secondary waves may develop. Such a complex situation cannot be dealt with simply, and the crossing of rays has been one of the principal difficulties in wave refraction considerations. A review of what is presently known concerning the complex wave behavior near wave ray crossings can be found in Pierson (1972).

A review of the procedures to be followed in the construction of a wave refraction diagram can be found in Coastal Engineering Research Center (1973). Harrison and Wilson (1964) have developed a method for utilizing a computer to calculate the wave refraction pattern, and Wilson (1966) has extended the approach to enable the computer to do the actual plotting of the rays.

An adequate treatment of wave refraction requires that the entire wave spectrum be considered. This calls for the construction of a set of refraction diagrams covering the full range of deep-water wave directions and periods within the spectrum. This points up the need for computer routines to perform the extended task.

WAVE DIFFRACTION

Wave diffraction is especially important to engineers in the construction of breakwaters and other structures. Such structures cut off the wave energy, creating a shadow zone, usually desired for harboring boats. The diffraction process causes wave energy to leak into this shadow area (Figure 4-24), perhaps creating undesired

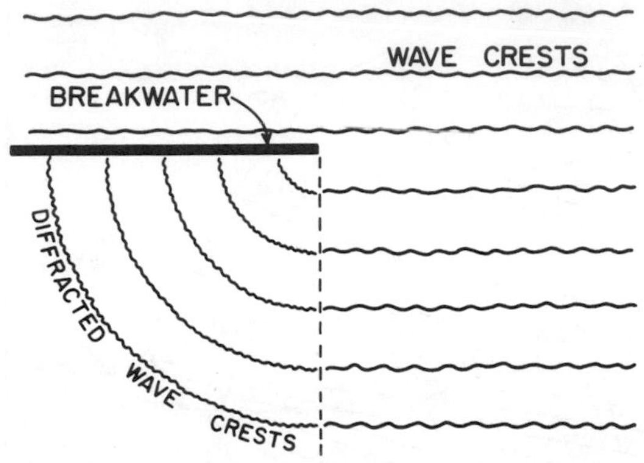

Figure 4-24 Wave diffraction into the shadow zone in the lee of an offshore breakwater.

wave conditions for the boats. In the diffraction process wave energy is transferred laterally along the wave crest from where the height is large to where it is lowest. The analysis of wave diffraction is similar to the analogous case of the diffraction of light. Bretschneider (1966) contains a useful summary of the analysis results. The studies of Putnam and Arthur (1948), Blue and Johnson (1949), and Wiegel (1962) all deal with the problem of wave diffraction in the vicinity of breakwaters.

WAVE INTERFERENCE AND SURF BEAT

When two sets of swell waves are arriving simultaneously from two storm sources, they interact with one another as shown in Figure 4-25, producing a regular variation in the observed wave height. In a systematic fashion the two swell trains add to form large observed waves and then subsequently move out of phase and so subtract to give a low series of waves. In this way a regular variation in wave height is produced, the waves gradually becoming higher until a maximum is reached, the size then decreasing to a minimum. Patterns of this sort are commonly observed in the waves breaking at the beach and have given rise to the popular notion that every seventh wave is the largest. This is approximately true in such situations, although it may actually be every sixth or eighth wave depending on the wave periods of the two swell wave trains.

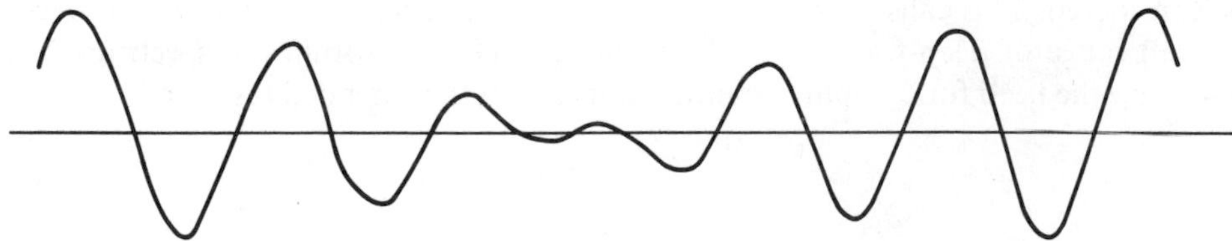

Figure 4-25 The summation of two sinusoidal wave trains of equal height to produce a regular variation in the observed wave height. In this example every sixth wave is the largest. The waves are shown schematically, with the heights greatly exaggerated over the wave length scale.

This fluctuation in height of series of sea waves breaking on the shore has been termed *surf beat* (Munk, 1949*b*; Tucker, 1950). A typical period for the surf beat is 2 to 3 minutes. Munk and Tucker both utilized special wave recorders to detect these variations, those which filter out the shorter-period ordinary swell waves leaving just the long-period envelope of the variation. The "surf beat" they detected in deep water lagged behind the groupings of largest waves, not occurring simultaneously. Munk and Tucker each attributed the waves they measured to surf beat in the nearshore, the variable water level being a line source for these waves radiating seaward away from the beach. However, subsequent studies of such waves show that they actually travel onshore, so that their origin is uncertain but is apparently not connected with water level variations in the surf. Whatever the cause of the waves measured by Munk and Tucker, the term *surf beat* has been adopted for the systematic variations in breaker height resulting from the interference of two swell wave trains as shown in Figure 4-25.

The variable breaker height associated with the surf beat of course has pronounced effects on processes occurring in the surf zone. A succession of high breakers temporarily raises the water level above that which prevails at times of small breakers. The large waves cause the water to run up on the beach, and this is sometimes a hazard to children playing close to the shoreline. This change in water mass in the nearshore in turn causes fluctuations in the speed of the longshore current and strength of rip currents.

PROBLEMS

4-1 Compare the maximum wave heights that can be achieved by 1-sec, 5-sec, and 10-sec waves in deep water. Use Airy wave theory.

4-2 Proportionally how much more power must be derived from the wind to increase a $T = 10$-sec wave from $H = 50$ cm to 100 cm than to increase a $T = 4$-sec wave from $H = 10$ to 20 cm?

4-3 Winds blow across a circular deep-water lake of diameter 2 km. What is the maximum time that the winds could feed energy into waves of $T = 2$ sec, 4 sec, and 10 sec? What does this imply regarding the growth of the 10-sec waves versus the 2-sec waves?

4-4 A storm has a mean wind speed of 34 knots (17.5 m/sec), blowing over a fetch of 800 km (430 nautical miles) for 48 hr. From the Neumann (1953) spectra (Figure 4-6), what is the expected total energy E? Using the relationship $H_{1/3} = 2.83\sqrt{E}$ given by Neumann, what is the corresponding significant wave height? What is the peak period in the spectra, and what is the dominant range of periods? How do these results compare with the significant wave height and period obtained with the Bretschneider curves (Figure 4-5)?

4-5 A wave group of period $T = 10$ sec has an overall length of 1,000 km. What is the approximate life span of a single wave in the group, assuming deep-water conditions? How far will the wave group as a whole have traveled in that time?

4-6 Make some calculations to evaluate the possibility that long-period but low-amplitude forerunners ($T = 20$ sec) might be used to predict the imminent arrival of shorter-period ($T = 10$ to 15 sec) waves of large heights. For example, how much warning time will the forerunners give if the storm distance is 2,000 km? 5,000 km?

4-7 A storm generates waves with periods ranging up to 15 sec. What would be the widths of the spectra at the distances 1,000 km, 2,000 km, and 5,000 km after 125 hr of storm activity, assuming the storm continues for the full time?

4-8 Construct a graph comparable to Figure 4-14 but relating the wave period to the half-life travel distance (the distance over which the waves would travel while decreasing to half their original value because of viscous damping).

4-9 Derive the equation for H/H'_∞ (equation 4-10) from the energy flux relationship (equation 4-9).

4-10 Write a computer program to compute H/H'_∞ as a function of h/L_∞.

4-11 Regular swell waves have a period $T = 12$ sec and a deep-water wave height $H_\infty =$ 100 cm. According to Airy wave theory, what will the wave height be at a water depth of $h = 20$ m? Correct this height for wave refraction if $s_\infty/s = 0.6$.

4-12 Indicate which of the following quantities increase, decrease, or remain essentially constant during wave shoaling: C, $C_g = Cn$, T, E, H, $ECns$.

4-13 Is the radiation stress component S_{xy} (Equation 4-19) always constant in refracting shoaling waves, or only on straight beaches with parallel depth contours?

4-14 Derive equation (4-11) for H_b/H'_∞ using solitary and Airy wave theories as described.

4-15 What is the power or energy flux associated with a moderate ocean swell of 2 m water wave height and period of 10 sec? Ignoring refraction, how many kilowatts of power are delivered to a 100-m length of shoreline (1 erg/sec $= 10^{-10}$ kw)? How does this compare with other energy sources?

REFERENCES

ADEYEMO, M. D. (1969). Effect of beach slope and shoaling on wave asymmetry. *Proc. 11th Conf. Coast. Eng.*, pp. 145–72.

——— (1971). Velocity fields in the wave breaker zone. *Proc. 12th Conf. Coast. Eng.*, pp. 435–60.

BARBER, N. F., and F. URSELL (1948). The generation and propagation of ocean waves and swell, I: wave periods and velocity. *Phil. Trans. Roy. Soc.* (London), series A, 240: 527–60.

BARNETT, T. P., and J. C. WILKERSON (1967). On the generation of wind waves as inferred from airborne radar measurements of fetch-limited spectra. *J. Mar. Res.*, 25, no. 3: 292–328.

BARNETT, T. P. (1970). Wind waves and swell in the North Sea. *EOS, Trans. Am. Geophys. Union*, 51, no. 7: 544–50.

BLUE, F. L., and J. W. JOHNSON (1949). Diffraction of water waves passing through a breakwater gap. *Trans. Am. Geophys. Union*, 30, no. 5: 705–18.

BONCHKOVSKAYA, T. V. (1955). Wind flow over solid wave models. *Akad. Navk SSSR* (Morskot Gidrofizicheskii Institut), 6: 98–106.

BOWEN, A. J. (1969). The generation of longshore currents on a plane beach. *J. Mar. Res.*, 37: 206–15.

BRETSCHNEIDER, C. L. (1952). The generation and decay of wind waves in deep water. *Trans. Am. Geophys. Union*, 33, no. 3: 381–89.

——— (1954). *Generation of wind waves over a shallow bottom.* U.S. Army Corps of Engrs., Beach Erosion Board Tech. Memo. no. 51.

——— (1957*a*). Hurricane design wave practices. *Proc. Amer. Soc. Civ. Engs., J. Waterways and Harbors Div.*, Paper 1238, WW2, 1–33.

——— (1957*b*). Review of Pierson, Neumann, and James, 1955. *Trans. Am. Geophys. Union*, 38, no. 2: 264–66.

——— (1958). Revisions in wave forecasting: deep and shallow water. *Proc. 6th Conf. Coast. Eng.*, pp. 30–67.

——— (1959). *Wave variability and wave spectra for wind generated gravity waves.* U.S. Army Corps of Engrs., Beach Erosion Board Tech. Memo. no. 118, 192 pp.

——— (1963). A one-dimensional gravity wave spectrum. In *Ocean wave spectra*, pp. 41–56. Prentice-Hall, Englewood Cliffs, N.J.

——— (1966). Wave refraction, diffraction, and reflection. In *Estuary and coastline hydrodynamics*, ed. A. T. Ippen, pp. 257–79. McGraw-Hill, New York.

CARTWRIGHT, D. E., and J. DARBYSHIRE (1957). Discussion of chapter 7. In *Proceedings of the symposium on the behavior of ships in a seaway*, ed. G. Neumann and W. J. Pierson, Wageningen, pp. 16–24.

Coastal Engineering Research Center (1973). *Shore protection manual.* U.S. Army, Corps of Engineers, Washington, D.C., 3 volumes.

CORNISH, V. (1934). *Ocean waves and kindred physical phenomena.* Cambridge Univ. Press, London, 164 pp.

DARBYSHIRE, J. (1952). The generation of waves by wind. *Proc. Roy. Soc.* (London), series A, 215: 299–328.

——— (1955). An investigation of storm waves in the North Atlantic Ocean. *Proc. Roy. Soc.* (London), series A, 230: 560–69.

——— (1956). The distribution of wave heights: a statistical method based on observations. *Dock and Harbour Auth.*, 37: 31–34.

——— (1957). A note on the comparison of proposed wave spectrum formulae. *Deutsch. Hydrog. Zeits.*, 10, no. 5: 184–90.

——— (1963). The one dimensional wave spectrum in the Atlantic Ocean and coastal waters. In *Ocean wave spectra*, pp. 27–31. Prentice-Hall, Englewood Cliffs, N. J.

DARBYSHIRE, J., and L. DRAPER (1963). Forecasting wind-generated sea waves. *Engineering*, 195: 482–84.

DUNHAM, J. W. (1951). Refraction and diffraction diagrams. *Proc. 1st Conf. Coast. Eng.*, pp. 33–49.

ECKART, C. (1953). The generation of wind waves over a water surface. *J. Appl. Phys.*, 24: 1485–94.

EWING, J. A. (1971). The generation and propagation of sea waves. In *Dynamic waves in civil engineering*, D. A. Howells, I. P. Haigh, and C. Taylor, eds. pp. 43–56. Wiley, New York.

FLEAGLE, R. G. (1956). Note on the effect of air-sea temperature difference on wave generation. *Trans. Am. Geophys. Union*, 37, no. 3: 275–77.

GALVIN, C. J. (1968). Breaker type classification on three laboratory beaches. *J. Geophys. Res.*, 73, no. 12: 3651–59.

GILCHRIST, A. W. R. (1966). The directional spectrum of ocean waves: an experimental investigation of certain predictions of the Miles-Phillips theory of wave generation. *J. Fluid Mech.*, 25: 795–816.

HARRISON, W., and W. S. WILSON (1964). *Development of a method for numerical calculation of wave refraction.* U.S. Army Coastal Eng. Res. Center Tech. Memo. no. 6, 64 pp.

HASSELMANN, K. (1963). On the non-linear energy transfer in a gravity wave spectrum. *J. Fluid Mech.*, 15; 273–81 (part 2), 385–98 (part 3.). [See references in these papers for earlier works by Hasselmann on nonlinear interactions.]

HELMHOLTZ, H. (1888). *Sitz.-Ber. Akad. Wiss.* (Berlin), p. 647.

HORIKAWA, K., and C.-T. KUO (1967). A study of wave transformation inside the surf zone. *Proc. 10th Conf. Coast. Eng.*, pp. 217–33.

HOUGH, S. S. (1896). On the influence of viscosity on waves and currents. *Proc. London Math. Soc.*, 28: 264–88.

IJIMA, T., and F. L. W. TANG (1967). Numerical calculation of wind waves in shallow water. *Proc. 10th Conf. Coast. Eng.*, pp. 38–45.

IVERSON, H. W. (1951). Studies of wave transformation in shoaling water, including breaking. In *Gravity waves*, National Bureau of Standards Circular no. 521, pp. 9–32.

JEFFREYS, H. (1925). On the formation of water waves by wind. *Proc. Roy. Soc.* (London), series A, 107: 189–206.

——— (1926). On the formation of water waves by winds. *Proc. Roy. Soc.* (London), series A, 110: 241–47.

KELVIN, LORD [W. THOMPSON] (1887). On the waves produced by a single impulse in water of any depth, or in a dispersive medium. *Proc. Roy. Soc.* (London), series A, 42: 80–83.

KEULEGAN, G. H. (1950). Wave motion. In *Engineering hydraulics*, ed. H. ROUSE, pp. 711–68. Wiley, New York.

KOMAR, P. D., and M. K. GAUGHAN (1973). Airy wave theory and breaker height prediction. *Proc. 13th Conf. Coast. Eng.*, pp. 405–18.

LAMB, H. (1932). *Hydrodynamics.* 6th ed. Cambridge Univ. Press, Cambridge, 738 p. [See 1945 reprint edition by Dover, New York.]

LIGHTHILL, M. J. (1962). Physical interpretation of the mathematical theory of wave generation by wind. *J. Fluid Mech.*, 14: 385–98.

LIU, P. C. (1971). Normalized and equilibrium spectra of wind waves in Lake Michigan. *J. Phys. Oceanog.*, 1: 249–57.

LONGUET-HIGGINS, M. S. (1952). On the statistical distribution of the heights of sea waves. *J. Mar. Res.*, 11, no. 3: 245–66.

——— (1969*a*). Action of a variable stress at the surface of water waves. *Phys. Fluids*, 12, no. 4: 737–40.

——— (1969*b*). A nonlinear mechanism for generation of sea waves. *Proc. Roy. Soc.* (London), series A, 311: 371–89.

——— (1970). Longshore currents generated by obliquely incident sea waves. *J. Geophys. Res.*, 75, no. 33: 6778–89 (part 1), 6790–801 (part 2).

LONGUET-HIGGINS, M. S., D. E. CARTWRIGHT, and N. D. SMITH (1961). Observations of the directional spectrum of sea waves using the motions of a floating buoy. In *Ocean wave spectra*, pp. 111–32. Prentice-Hall, Englewood Cliffs, N. J.

McNair, E. C., and R. M. Sorensen (1971). Characteristics of waves broken by a longshore bar. *Proc. 12th Conf. Coast. Eng.*, pp. 415–34.

Miles, J. (1957). On the generation of surface waves by shear flows. *J. Fluid Mech.*, 3: 185–204.

——— (1959). On the generation of surface waves by shear flows, part 2. *J. Fluid Mech.*, 6: 568–82.

——— (1960). On the generation of surface waves by turbulent shear flows. *J. Fluid Mech.*, 7: 469–78.

Miller, R. L. (1973). The role of surface tension in breaking waves. *Proc. 13th Conf. Coast. Eng.*, pp. 433–49.

Miller, R. L., and J. M. Zeigler (1965). The internal velocity field in breaking waves. *Proc. 9th Conf. Coast. Eng.*, pp. 103–22.

Motzfeld, H. (1937). Die turbulente Strömung auf welligen Wänden. *Z. Angewandte Math. Mech.*, 17 (August): 193–212.

Munk, W. H. (1947*a*). A critical wind speed for air-sea boundary processes. *J. Mar. Res.*, 6, no. 3: 203–18.

——— (1947*b*). Tracking storms by forerunners of swell. *J. Meteor.*, 4: 45–47.

——— (1949*a*). The solitary wave theory and its application to surf problems. *N.Y. Acad. Sci. Ann.*, 51: 376–424.

——— (1949*b*) Surf beats. *Trans. Am. Geophys. Union*, 30: 849–54.

——— (1957) Comments on review by C. L. Bretschneider of U.S. Navy Dept. H.O. Pub. no. 603. *Trans. Am. Geophys. Union*, 38, no. 5: 118–19.

Munk, W. H., and F. E. Snodgrass (1957). Measurements of southern swell at Guadalupe Island. *Deep Sea Res.*, 4: 272–86.

Nakamura, M., H. Shiraishi, and Y. Sasaki (1967). Wave decay due to breaking. *Proc. 10th Conf. Coast. Eng.*, pp. 234–53.

Neumann, Gerhard (1953). *On ocean wave spectra and a new method of forecasting wind-generated sea.* U.S. Army Corps of Engrs., Beach Erosion Board Tech. Memo. no. 43, 42 pp.

Neumann, G., and W. J. Pierson (1957). A detailed comparison of theoretical wave spectra and wave forecasting methods. *Deutsch. Hydrog. Zeits.*, 10: 73–92 (part 1), 134–46 (part 2).

Patrick, D. A., and R. L. Wiegel (1955). Amphibian tractors in the surf. *Proc. 1st Conf. Ships and Waves* (Engr. Foundation, Counc. Wave Res., and Am. Soc. of Nav. Archs. and Mar. Engrs.), pp. 397–422.

Phillips, O. M. (1957). On the generation of waves by turbulent wind. *J. Fluid Mech.*, 2, no. 5: 417–45.

——— (1958*a*) Wave generation by turbulent wind over a finite fetch. *Proc. 3rd U.S. Natl. Cong. Appl. Mech.* (A. Soc. Mech. Eng.), pp. 785–89.

——— (1958*b*). The equilibrium range in the spectrum of wind-generated waves. *J. Fluid Mech.*, 4: 426–34.

——— (1958*c*). On some properties of the spectrum of wind-generated ocean waves. *J. Mar. Res.*, 16, no. 3: 231–40.

——— (1960). On the dynamics of unsteady gravity waves of finite amplitude *J. Fluid Mech.*, 9: 193–217.

PIERSON, W. J. (1972). Wave behavior near caustics in models and nature. In *Waves on beaches*, ed. R. E. Meyer, pp. 163–80. Academic Press, New York.

PIERSON, W. J., G. NEUMANN, and R. W. JAMES (1955). *Observing and forecasting ocean waves by means of wave spectra and statistics.* U.S. Dept. of the Navy Hydrog. Offc. Publ. no. 603, 284 pp.

PIERSON, W. J., and L. MOSKOWITZ (1964). A proposed spectral form for fully developed wind seas based on the similarity theory of S. A. Kitaigorodskii. *J. Geophys. Res.*, 69, no. 24: 5181–90.

PUTNAM, J. A., and R. S. ARTHUR (1948). Diffraction of water waves by breakwaters. *Trans. Am. Geophys. Union*, 29, no. 4: 481–90.

PUTZ, R. R. (1952). Statistical distributions for ocean waves. *Trans. Am. Geophys. Union*, 33: 685–92.

RATTRAY, M., and W. V. BURT (1956). A comparison of methods for forecasting wave generation. *Deep Sea Res.*, 3, no. 2: 140–44.

RAYLEIGH, L. (1911). Hydrodynamical notes. *Phil. Mag.*, series 6, 21: 177–87.

ROGERS, L. C. (1966). Blue Water 2 lives up to promises. *Oil and Gas J.*, August 15, pp. 73–75.

ROLL, H. U., and G. FISCHER (1956). Eine kritische Bemerkung zum Neumann-Spectrum des Seeganges. *Deut. Hydrog. Zeits.*, 9: 9–14.

ROSCHKE, W. H. (1954). The propagation of short period air-pressure microoscillations. *Bull. Am. Meteor. Soc.*, 35, no. 1: 20–25.

RUDNICK, P., and R. W. HASSE (1971). Extreme Pacific waves, December 1969; *J. Geophys. Res.*, 76, no. 3: 742–44.

SNODGRASS, D., G. W. GROVES, K. F. HASSELMANN, G. R. MILLER, W. H. MUNK, and W. H. POWERS (1966). Propagation of ocean swell across the Pacific. *Phil. Trans. Roy. Soc.* (London), series A, 259: 431–97.

SNYDER, R. L., and C. S. COX (1966). A field study of the wind generation of ocean waves. *J. Mar. Res.*, 24, no. 2: 141–78.

STANTON, T., D. MARSHALL, and R. HOUGHTON (1932). The growth of waves on water due to action of the wind. *Proc. Roy. Soc.* (London), series A, 137: 283–93.

STEWART, R. W. (1967). Mechanics of the air-sea interface: *Phys. Fluids*, Supplement, 10: 547–55.

STOKER, J. J. (1957). *Water waves.* Interscience, New York, 567 pp.

SVERDRUP, H. U., and W. H. MUNK (1947). *Wind, sea, and swell theory of relationships in forecasting*. Washington, D.C. U.S. Dept. of the Navy Hydrog. Offc. Publ. no. 601.

THIJSSE, J. TH., and J. B. SCHIJF (1949). Penetration of waves and swells into harbours. *Proc. 17th Internatl. Navig. Congr., Section II*, Ocean Navigation Communication IV (Lisbon).

TUCKER, M. J. (1950). Surf beats: sea waves of 1 to 5 min. period. *Proc. Roy. Soc.* (London), series A, 202: 565–73.

WALDEN, H. (1954). Eine Dünungsbeobachtung sowie die Beziehungen zum erzeugenden Windfeld. *Deut. Hydrog. Zeits.*, 7: 59–62.

——— (1957). Methods of swell forecasting demonstrated with an extraordinary high swell off the coast of Angola. *Proceedings of the symposium on the behavior of ships in a seaway*, ed. G. Neumann and W. J. Pierson, Wageingen, Netherlands.

WATTS, J. S., and R. E. FAULKNER (1968). Designing a drilling rig for severe seas. *Ocean Industry*, 3, no. 11: 28–37.

WIEGEL, R. L. (1962). Diffraction of waves by semi-infinite breakwater. *Proc. Am. Soc. Civ. Eng.*, 88, HY 1: 27–44.

——— (1964). *Oceanographical engineering*. Prentice-Hall, Englewood Cliffs, N.J., 532 p.

WIEGEL, R. L., and H. L. KIMBERLY (1950). Southern swell observed at Oceanside, California. *Trans. Am. Geophys. Union*, 31, no. 5: 717–22.

WILSON, B. W. (1955). *Graphical approach to the forecasting of waves in moving fetches*. U.S. Army Corps of Engrs., Beach Erosion Board Tech. Memo no. 73, 64 p.

WILSON, W. S. (1966). *A method for calculating and plotting surface wave rays*. U.S. Army Coastal Eng. Res. Center Tech. Memo no. 17, 31 pp.

Chapter 5

TIDES

Write to Bartolomeo the Turk about the ebb and flow of the Black Sea and ask whether he knows if there is the same ebb and flow in the Hyrcanian or Caspian Sea.

Leonardo da Vinci, 1452–1519
Notebooks

Tides produce a periodic rise and fall of the water level twice daily that has been described romantically by Defant (1958) as "the heartbeat of the ocean, a pulse that can be felt all over the world." The variation in water level so produced is important to processes acting on the beach. On a tideless sea the area of the beach coming under wave attack is small, governed principally by the size of the waves. On the other hand, with a tidal variation the position of wave action migrates continuously, and a wide stretch of beach thereby comes under the action of the waves. In this regard the tidal variation is somewhat of a vexation and makes considerably more difficult the understanding of, for example, sediment distribution on the beach or beach profile configuration.

Tides do have their good points, however. They give rise to currents that are generally strongest in the immediate vicinity of the entrances of bays and lagoons periodically flooded and emptied of tidal waters. These tidal currents are often important in keeping these entrances open. Without them the beach would tend to build across the opening and close off the bay unless river runoff were capable of maintaining the opening. These same tidal currents are also important because they daily flush our coastlines, estuaries, and harbors and help purge them of the pollutants we dump there.

In a very few locations exceptional tidal currents are capable of causing significant coastal erosion. The tidal currents off the citadel of Mont-Saint-Michel, Normandy, France, associated with the highest tides in Europe, attack the rock citadel, which must constantly be repaired and reinforced to resist the erosion. Erosion of coastal rock formations by tidal currents is also observed in the Bay of Fundy, Canada.

In this chapter we shall briefly examine the nature of tides, their generating forces, and the properties of tides in oceans of limited extent and irregular outline. We shall look specifically at the tides found along the coasts of the United States and Britain. More extensive general discussions of tides can be found in Defant (1958) and in Clancy (1969). Defant (1961) and Godin (1972) contain more exhaustive treatments of tidal theory and analysis. The review papers by Hendershott and Munk (1970), Hendershott (1973), and Garrett (1974) provide views on the present state of understanding of ocean tides and discussions of recent investigations.

TIDE-GENERATING FORCES AND THE EQUILIBRIUM THEORY

The early theories regarding the cause of tides were somewhat fanciful. In his geography written in A.D. 902, Ibn al-Fakih retold some of the legends concerning their origin. One legend relates that "verily the Angel, who is set over the seas, places his foot in the sea and thence comes the flow; then he raises it and thence comes the ebb." Another legend has it that tides are due to the breathing cycle of a giant whale. These early hypotheses have proved to be largely without foundation.

It is to Sir Isaac Newton and his publication of *Philosophiae Naturalis Principia Mathematica* in 1686 that we owe our first real understanding of why tides exist and behave as they do. In fact, his ability to explain the tides was strong supporting evidence for his theory of gravitation. Newton demonstrated that the gravitational pull of the moon and sun on the water of the earth's oceans is responsible for the tides.

Newton's law of gravitation states that every element of mass in the universe attracts every other element with a force proportional to their masses and inversely proportional to the square of the distance between them. In mathematical terms the law becomes

$$F = G\frac{m_1 m_2}{r^2} \tag{5-1}$$

where m_1 and m_2 are the masses of the two objects and r is the distance between them; G is the universal gravitational constant and has the value 6.6×10^{11} N·m²/kg².

The force of attraction between the earth and the moon is the vector sum of a great many pairs of forces between the small elements of mass which make up the two bodies. It turns out that for spherical bodies the overall net attraction is the same as if all the mass of the two bodies were concentrated at their respective centers. For the earth-moon system the next force of attraction becomes

$$\hat{F} = G\frac{m_m m_e}{R^2} \tag{5-2}$$

where m_e and m_m are respectively the total masses of the earth and moon and R is the distance between their centers.

Although the net force of attraction between the earth and moon is given by equation (5-2), it is apparent that individual elements of the earth will be attracted by the moon with a slightly different magnitude force due to their varying distances from the moon. A chunk of earth on the surface facing the moon is attracted more strongly than a similar chunk on the opposite side, away from the moon. It is these small departures from the mean net force of attraction that are responsible for the tides.

Because of this force of attraction between the earth and moon, they orbit one another, each moving approximately in a circle about a common center of mass (Figure 5-1). Since the mass of the earth is much greater than that of the moon, this

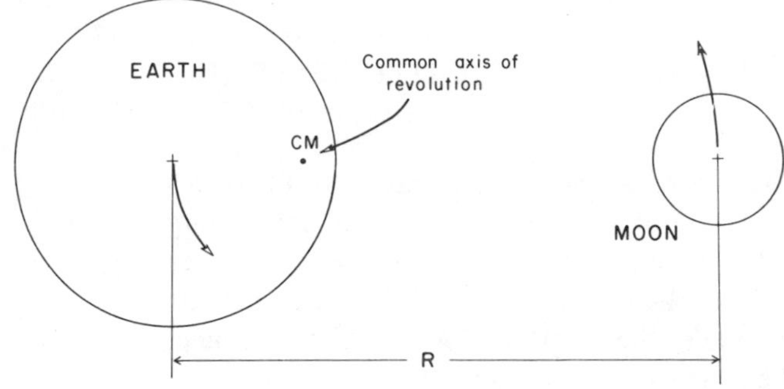

Figure 5-1 Rotation of the earth and moon about a common center of mass. Because of the greater mass of the earth, the center is located within the earth, so that in the mutual rotation the moon swings freely while the earth's motion is more of a wobble.

center of mass is actually located within the body of the earth, so that while the moon sweeps freely around this point, the earth's motion is more of a slight wobble. The force of attraction then provides the centripetal force that produces the circular motion: there is a mutual acceleration toward one another. Some writers introduce a centrifugal force ("fleeing a center") to oppose the force of equation (5-2) "to keep the moon out there"—that is, to maintain it in its circular path. This is a case of faulty reasoning, as a body in equilibrium moves in a straight line through space with a constant speed. The circular motions of the earth and moon show not that their motions are one of a balance of forces, but rather that there is a net force and acceleration toward one another. The centrifugal force is a bogus force for which there is no need.

Forget for the moment that the earth rotates on its own axis. This rotation causes the ocean's water level to be slightly higher at the equator than at higher latitudes, but since the response is uniform along a line of constant latitude, it does not appear as a tidal variation to an observer on earth. It simplifies matters if we initially assume that the earth does not rotate on its own axis but rather maintains its orientation in space. A few moments' consideration of Figure 5-2 will demonstrate that under these conditions, as the earth-moon system revolves around its common center of gravity, each particle of the earth travels simply in a circle, all these circles having exactly the same radius. Therefore, each particle is experiencing a centripetal

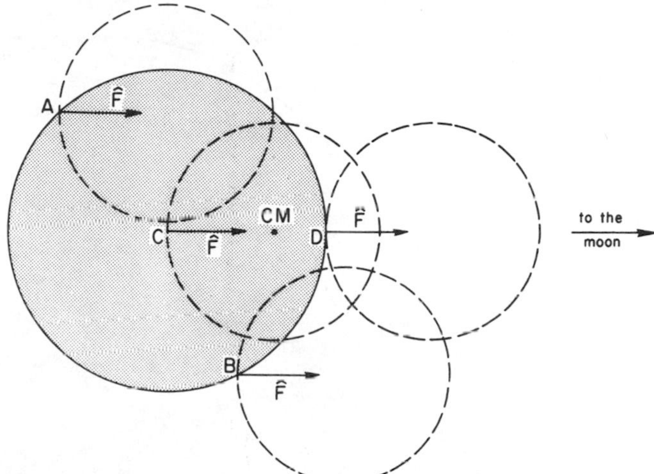

Figure 5-2 Excluding the rotation of the earth about its own axis, in the earth-moon rotation each particle of earth mass moves in a circular path, the centripetal force being equal to $\hat{F}$, the force of mutual attraction between the earth and moon.

acceleration (force) necessary to make it travel in its particular "orbit." It is apparent that this centripetal acceleration is equal in magnitude and direction on every particle and is equivalent to the acceleration associated with the force of attraction of equation (5-2).

We saw that nowhere on the earth's surface will the force of attraction by the moon for a particle be exactly equal in magnitude and direction to the centripetal force (equation 5-2) which governs the motions of the earth and moon about one another. The difference between these forces at any point (Figure 5-3) provides the net force that is responsible for tide generation on the earth.

Consider the earth-moon system of Figure 5-4 (continue to ignore the sun—we shall deal with it later), with the earth initially covered with a uniform layer of water. Acting on an element of water P will be the force of gravitational attraction by the earth, the centripetal force, and the individual attraction of that element by the moon (gravity plus the net force of Figure 5-3). Rather than considering forces, it is better to work with accelerations—that is, forces per unit mass—acting on the water element P of Figure 5-4. The balance of accelerations acting normal to the earth's surface is then given by

$$A_n = -g - \frac{\hat{F}}{m_e}\cos\theta + \frac{Gm_m}{r^2}\cos(\theta+\phi) \tag{5-3}$$

$$= -g - \frac{Gm_m}{R^2}\cos\theta + \frac{Gm_m}{r^2}\cos(\theta+\phi)$$

$$= -g + g\left(\frac{m_m}{m_e}\right)R_e^2\left[\frac{\cos(\theta+\phi)}{r^2} - \frac{\cos\theta}{R^2}\right] \tag{5-4}$$

having used $g = Gm_e/R_e^2$ to remove G in the last step, where R_e is the earth's radius. It is evident that A_n is a maximum when $\theta = 0$ or 180 degrees—when the moon is straight overhead or beneath one's feet on the opposite side of the earth. Under such conditions, according to equation (5-4), a person weighs slightly less because of his

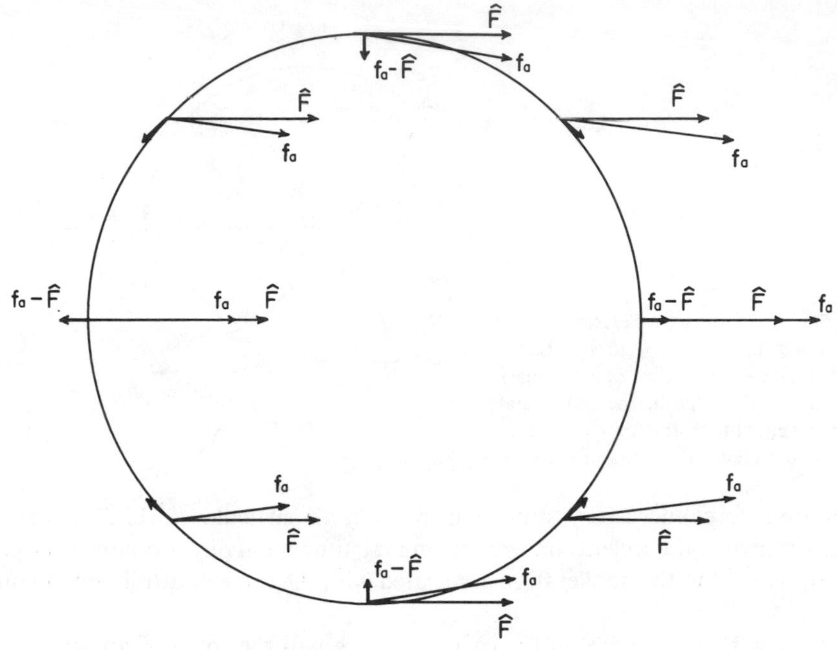

Figure 5-3 The vector subtraction of $\hat{F}$, the total force of attraction between the earth and moon, and f_a, the local attraction of the element of water by the moon.

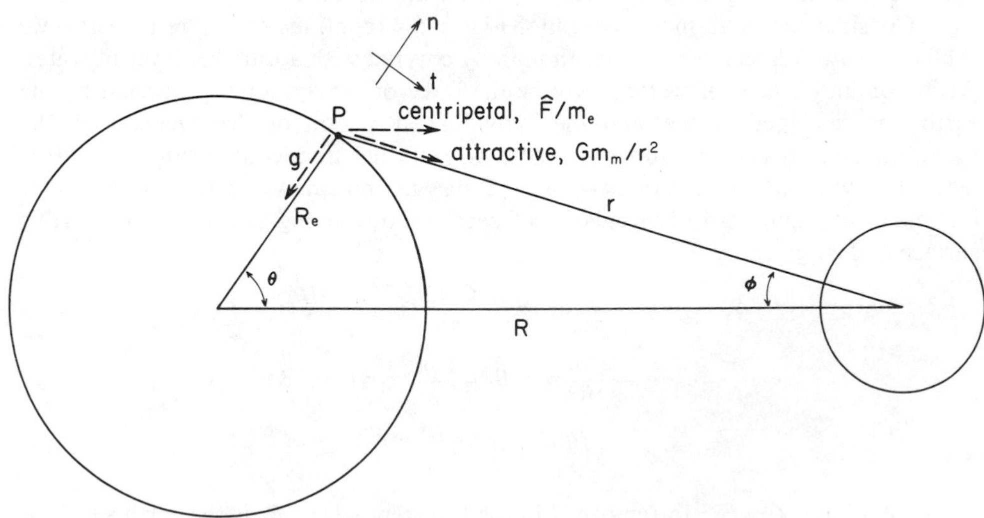

Figure 5-4 The three forces acting on any given water particle P; gravity g, the centripetal acceleration, and the attraction of that particular particle by the moon.

attraction by the moon. For the earth-moon statistics given in Table 5-1, equation (5-4) yields

$$A_n\,[\text{maximum}] = -g + (1.1 \times 10^{-7})g$$

and it is seen that the effect of the moon's attraction (the second term) is negligible in comparison to the attraction of the earth, g, which acts in the same direction. Our person with the moon overhead would have his weight decreased by only a ten-millionth of his original weight—not something one would notice on a bathroom scale.

TABLE 5-1 Earth, Sun, and Moon Dimensions

Diameter of earth	12,753 km
Diameter of moon	3,479 km
Mass of earth	5.98×10^{24} kg
Mass of moon	7.34×10^{22} kg
Mass of sun	1.96×10^{30} kg
Average distance between centers of earth and moon	384,329 km
Average distance between centers of earth and sun	149,360,000 km

Now let us turn our attention to the component of acceleration that is tangent to the surface of the earth. Again referring to Figure 5-4, we have

$$A_t = \frac{Gm_m}{r^2}\sin(\theta + \phi) - \frac{\hat{F}}{m_e}\sin\theta \tag{5-5a}$$

$$= g\left(\frac{m_m}{m_e}\right)R_e^2\left[\frac{\sin(\theta+\phi)}{r^2} - \frac{\sin\theta}{R^2}\right] \tag{5-5b}$$

This relationship can be simplified by expanding the expression within the brackets and neglecting terms which involve $(R_e/R)^2$ or higher-order terms, since they will be very small. Such approximations lead to

$$A_t = \frac{3}{2}g\left(\frac{m_m}{m_e}\right)\frac{R_e^3}{R^3}\sin 2\theta \tag{5-6}$$

We see that A_t has maxima and minima values of $\pm(0.84 \times 10^{-7})g$ when $\theta = 45$ deg and 135 deg and becomes zero at $\theta = 0$ and 180 deg. Even though A_t is small in magnitude, this tangential force or acceleration is not negligible, since it acts at right angles to the earth's gravitational attraction. However small, it has no opposing force, so that it can produce appreciable effects. It is this tangential component that is responsible for tide generation.

The tidal bulges of water are therefore not drawn out by the normal components, but rather by the tangential components as depicted in Figure 5-5. The bulge will continue to grow until the pressure gradient associated with the sloping water surface of the bulge offsets and balances the tangential tide-producing force component. If the earth were completely covered with water, it would be drawn out into an egglike

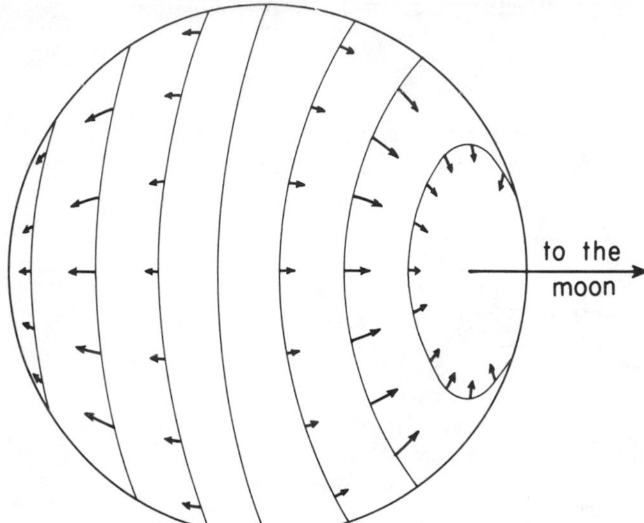

Figure 5-5 The tangential component of $f_a - \hat{F}$ resulting from the moon's attraction for the surface water, the component which gives rise to the tidal bulge.

bulge (more exactly, a prolate spheroid), with the long axis pointing toward the moon. There is indeed a bulge produced on the face opposite the moon as well as on the side directly facing the moon. This is because the vector difference between the attraction of each water element by the moon and the overall attraction between the earth and moon (the centripetal acceleration), shown in Figure 5-3, is directed away from the moon and from the earth's center. This bulge is not due to a centrifugal force resulting from the earth's motions.

How would these tides appear to an observer on earth? Consider a person at point *A* on the equator in Figure 5-6. As the earth rotates daily he would see two high tides of equal height as he passes under the bulges, separated by two low tides. An

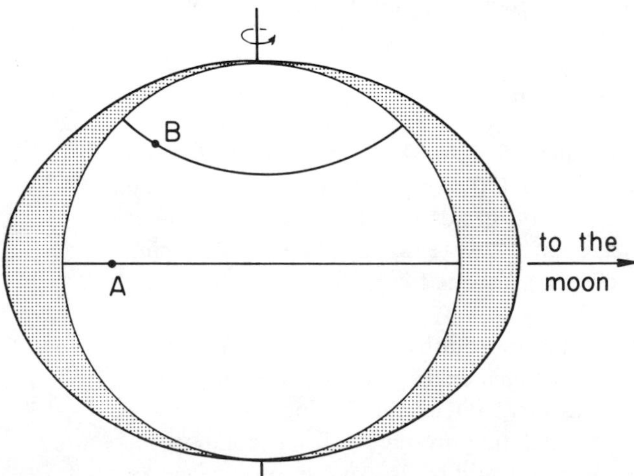

Figure 5-6 A simple tidal bulge as seen by observers *A* and *B* on the rotating earth.

observer at point *B* would similarly see two high tides of equal height, but the height would be somewhat lower than that seen by *A*.

Our observers would find that the periodicity of the occurrence of the high tides is not 12 hours, as one might at first expect. Rather, each high tide occurs about 12 hr 25 min after the preceding one. The observer would pass under the same tidal bulge 24 hr 50.47 min following his previous passage under that particular bulge. This departure from an exact 12 hours is brought about by the orbital motion of the moon around the earth in the same direction as the earth's rotation. The *lunar day*—the time for successive passages of the moon across a given meridian—is therefore longer than the solar day. The lunar day is in fact the 24 hr 50.47 min that our observer noted in the periodicity of the tides. While the earth is turning on its axis, the moon is also revolving around the earth.

If our tide observers *A* and *B* continue on the job for an extended period of time, they will begin to see some variations and would find that the variations are periodic. One variation would be due to the eccentricity of the moon's orbit around the earth. As depicted in Figure 5-7, the moon's path around the earth is elliptical rather

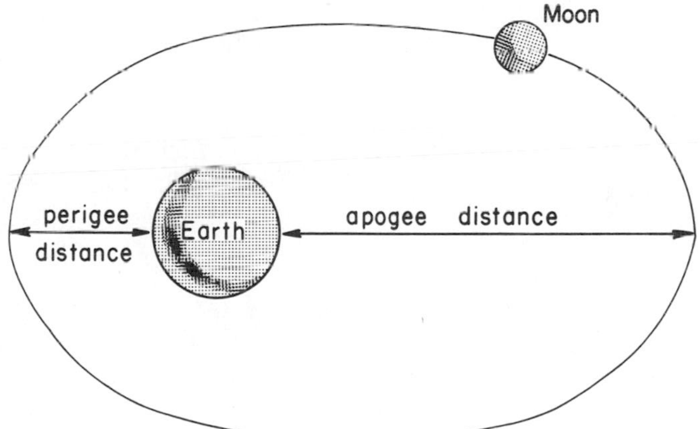

Figure 5-7 The elliptical path of the moon around the earth. The illustration is schematic, as the path is actually nearly circular.

than circular and is closest to the earth at *perigee* (about 357,000 km) and farthest away at *apogee* (407,000 km). Although the departure from a true circle is not great, the varying distance will have an observable effect on the tides, since the tidal force (equation 5-6) is dependent upon the cube of this distance. The periodicity of this variation is 27.55 days. There is a similar variation in the tide associated with the eccentricity of the earth's orbit around the sun. The periodicity of this variation is one *anomalistic year*, which is slightly longer than the "true" or *sidereal year* because the line connecting the earth and sun at perigee progresses around the earth's orbit by an average of 11 seconds of arc per year.

Now that we are considering the sun, let us examine how the presence of the egg-shaped tidal bulge due to the sun's attraction affects our picture of tides on the earth covered with water. The sun's mass is about 27 million times greater than the mass of the moon, so one might expect that it would be far more important in produc-

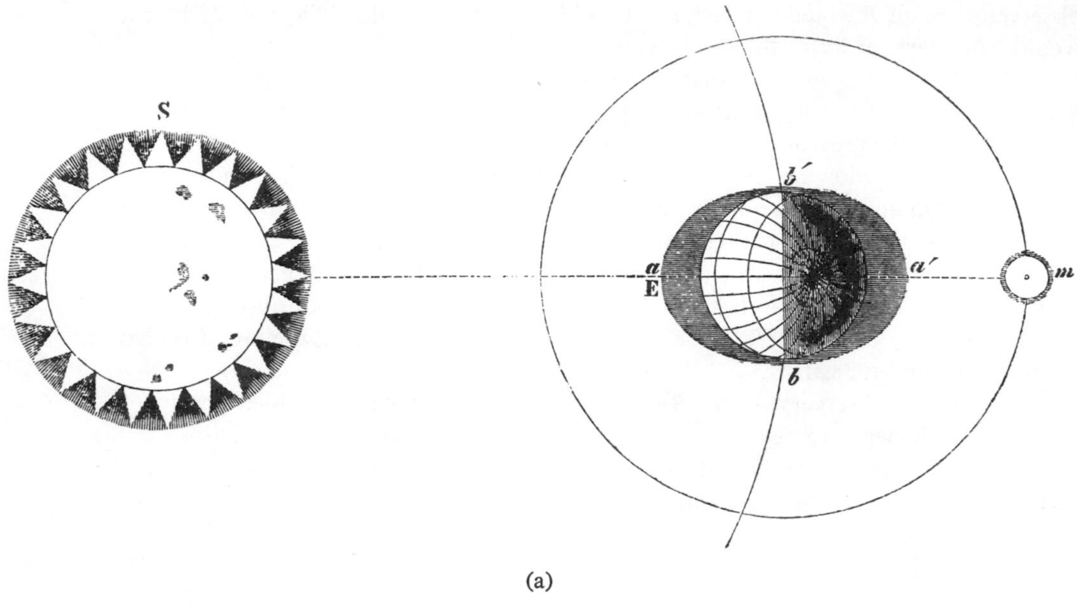

(a)

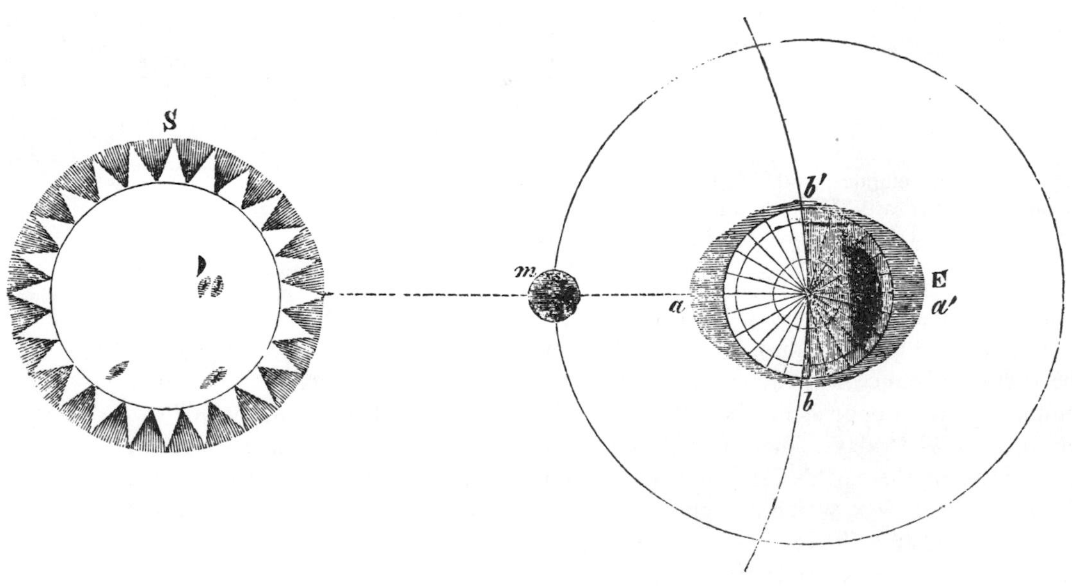

(b)

Figure 5-8 Spring tides produced by the sun, moon, and earth being aligned. (a) Spring tide at the full moon, (b) Spring tide at the new moon. [*From* Rev. Lewis Tomlinson, *Recreations in Astronomy*, (London: Parker, 1858)]

ing tides. But the sun is also much farther away, and we have seen from equation (5-6) that distance is particularly important as it enters as the cube. The sun is 390 times more distant from the earth than is the moon, so that the sun's effects should be lessened by the amount $(\frac{1}{390})^3$, or about 59 million times. Therefore, the sun's tide-generating force is roughly $\frac{27}{59}$ that of the moon—a little less than half. The tidal bulge so produced would not be as large as that caused by the moon.

In Figure 5-8 the earth, moon, and sun are all approximately in line—the conditions of full and new moon. It is apparent that the tidal bulges of the moon and sun are additive in these two cases, so that the high tides so produced will be extreme, the so-called *spring tides*. Their occurrence has nothing to do with the season; the term is derived from the Anglo-Saxon word *springan*, meaning a rising or welling of the water. In Figure 5-9 we see the condition where the sun and moon are in *quadrature*—

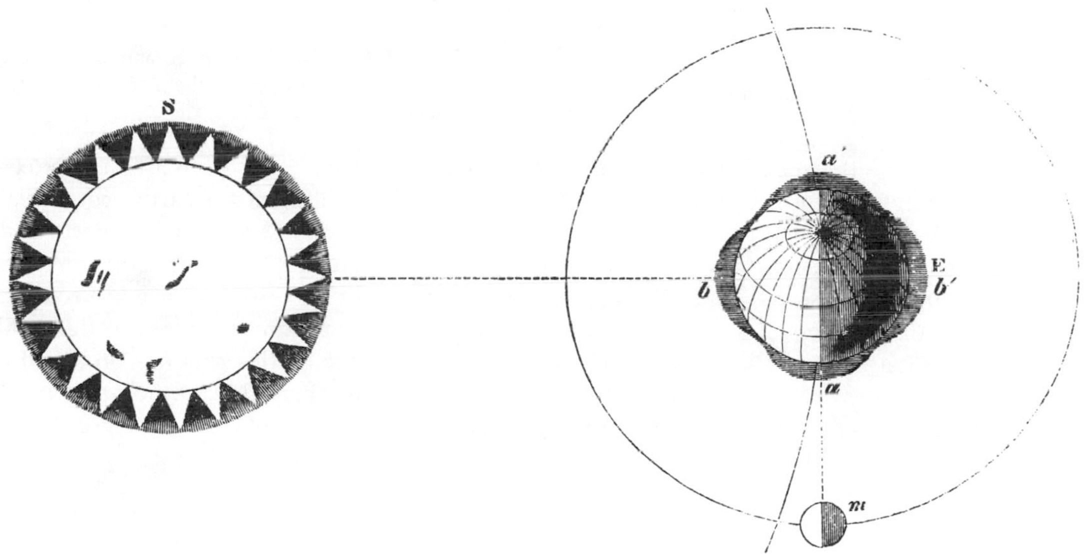

Figure 5-9 Neap tides produced by the sun and moon acting in opposition to one another, producing counterbulges. [*From* Tomlinson, ibid.]

that is, the tide-generating forces are 90 degrees out of phase and operating at right angles to one another. This produces the minimum tidal range—the *neap tides*. Spring tides are about 20% greater than the average tidal range; neap tides are about 20% lower than average. This variation is fortnightly.

So far we have assumed that the moon's orbit about the earth, the earth's orbit about the sun, and the earth's equator all lie in the same plane. In fact they do not, but rather are tilted with respect to one another. The greatest departure is the tilting of the earth's equatorial plane, and hence its axis, with respect to the plane of the earth's orbit around the sun, the two planes making an angle of 23.5 degrees (Figure 5-10). The direction of the earth's axis in space remains nearly constant, as shown. The plane of the moon's orbit is nearly the same as the plane of the earth's orbit about

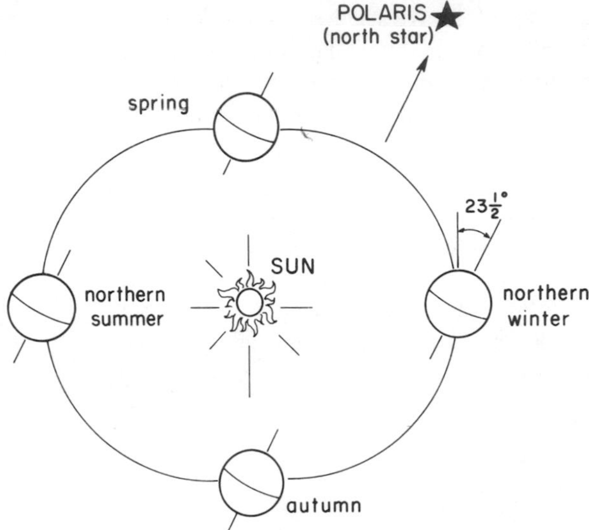

Figure 5-10 The 23.5-deg eccentricity of the earth's axis with respect to the plane of the orbit around the sun.

the sun, the angle between the planes being only 5 degrees, and so to a first approximation it can be considered the same. Because of the tilting of the earth's equatorial plane, the line of the solar tidal force varies by 47 degrees (2 × 23.5 deg) north and south on the earth's surface throughout one year. What effect will this tilting have on our tidal observers on the earth? We can see in Figure 5-11 that as the earth rotates, both observers, *A* and *B*, will still see two high tides, but the heights of the tides will no longer be the same; the declination or tilting of the earth's axis has produced a diurnal variation in the heights of successive tides. This is probably the most noticeable variation our observers would see in the tides.

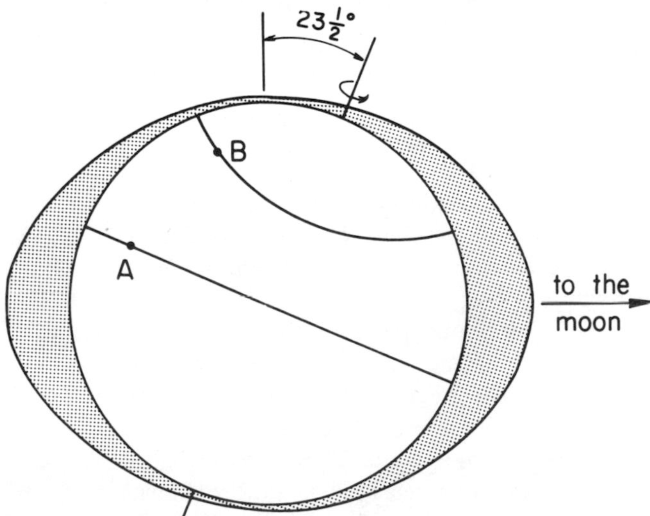

Figure 5-11 The effect of the 23.5-deg eccentricity on the tides as seen by observers *A* and *B*.

The 5-deg tilt of the moon's orbit with respect to the earth's produces a long-term tidal variation. The intersection of the plane of the moon's orbit with the plane of the earth's orbit forms a line. This line of intersection slowly rotates, completing a revolution in 18.6 yr. There are times (most recently being in 1969) that the moon's declination reaches a maximum of $23.5 + 5 = 28.5$ deg. At these times the center line of the moon's tide-generating force varies north and south on the earth's surface by 57 deg over the course of a month. After 9.3 yr the maximum declination of the moon is only $23.5 - 5 = 18.5$ deg, and the north-south variation during the month is only 37 deg.

We have, then, a number of variations in the tide-generating forces and in the resulting tides. These are summarized in Table 5-2 for review. Also given are symbols

TABLE 5-2 The Most Important of the Tide-generating Constituents (*After* Defant, 1958)

	SYMBOL	PERIOD (SOLAR HR)	AMPLITUDE ($M_2 = 100$)	DESCRIPTION
Semidiurnal tides (two tides per day)	M_2	12.42	100.0	Main lunar semidiurnal constituent
	S_2	12.00	46.6	Main solar semidiurnal constituent
	N_2	12.66	19.1	Lunar constituent due to monthly variation in moon's distance
	K_2	11.97	12.7	Solar-lunar constituent due to changes in declination of sun and moon throughout their orbital cycle
Diurnal tides (one tide per day)	K_1	23.93	58.4	Solar-lunar constituent
	O_1	25.82	41.5	Main lunar diurnal constituent
	P_1	24.07	19.3	Main solar diurnal constituent
Long-period tides	M_f	327.86	17.2	Moon's fortnightly constituent

which represent the various forces. We shall see later in this chapter that the complex tides observed at some given location consist of sums of sine waves that exhibit these periodicities but are out of phase with one another and with different amplitudes. The sum of these sine waves gives a predicted tidal variation in water level. The nature of the tide at a given location is governed by which of these constituents are most important (those with the greatest amplitudes).

The greatest tide-generating forces occur in combination to produce the largest tides when, at the same time, the sun is in perigee, the sun and moon are in conjunction or opposition (spring tides), and both sun and moon have zero declination. This combination of circumstances happens about every 1,600 years, the last occurrence having been in about the year A.D. 1400. The tides are now progressively decreasing and will reach a minimum in the year 2300. The previous minimum occurred about A.D. 500, when the Vikings were harassing the peoples of Europe.

DYNAMIC THEORY OF TIDAL BEHAVIOR

A comparison of the observed tides on the earth with the equilibrium theory outlined in the preceding section quickly points up several conflicts. High-tide water often occurs at the wrong time—at times when not in line with the moon's or sun's tide-generating forces. The range of the tide is not usually as predicted by the equilibrium theory, and the observed diurnal inequality often bears little resemblance to the theory. The equilibrium theory does predict the observed periodicities, or nearly so, and should therefore not be abandoned entirely. Rather, it must be modified to account for several complicating factors. The most obvious complications are the irregular shapes and varying depths of the oceans; the world is not uniformly covered with water. The rotation of the earth further produces geostrophic effects that are not considered in the equilibrium theory. The assumption that water has no inertia and therefore responds immediately to gravitational forces is one of the most serious shortcomings of the equilibrium theory. This assumption basically states that the water will not react to accelerations other than those due to gravitational attraction. On the rotating earth, with the tide-producing forces acting with complicated variations, all sorts of accelerations can be expected to exist, so that the inertial effects are significant.

The dynamic theory of tides, first developed by the French mathematician and scientist Laplace, uses the same tide-generating forces but accounts for these complicating factors. In this theory one speaks in terms of *tidal waves* rather than bulges of water. These tidal waves are produced by the tide-generating forces considered above and therefore have the same periods as these forces. Note that we are not referring here to the "tidal waves" of the news media. This is a misnomer, for these catastrophic waves have little to do with tides. Such waves are either *tsunamis*, generated by earthquakes, or *storm surges*, produced by violent onshore winds.

The tidal wave is of such long period, corresponding to the period of the tide-generating force, and of such long wave length that they are all shallow-water waves (Chapter 3). For them, even the deepest ocean is "shallow." We have already seen that for shallow-water waves the velocity of propagation is given by

$$C_s = \sqrt{gh} \tag{5-7}$$

The velocity of the tidal wave is governed by the depth h.

Let us consider the tidal waves generated in a rectangular ocean basin in the Northern Hemisphere. The tidal wave produced by a tide-generating force would advance from east to west until it met the western boundary of the ocean. There it would be reflected and move in the opposite direction, to the east. Now if the reflection is perfect, it will combine with other tidal waves moving to the west in such a way that a type of wave known as a *standing* or *stationary wave* is produced. This type of wave can also be found in an enclosed basin in which the water has been set oscillating by tilting the basin or otherwise disturbing the water. The motion of such a wave is shown in Figure 5-12. The water oscillates about a nodal line where no amplitude

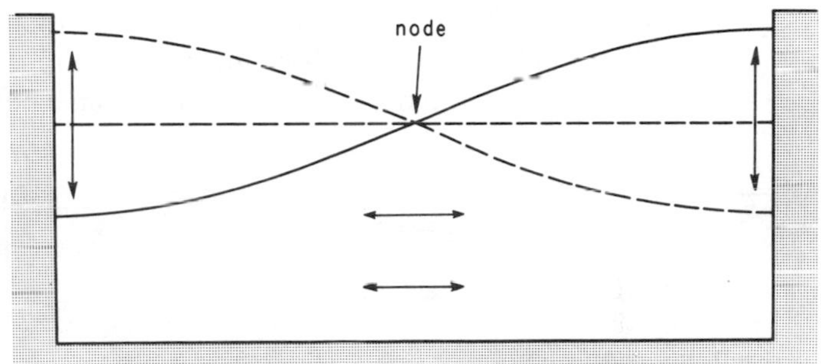

Figure 5-12 A standing wave in an enclosed basin with a single node. The verticle motions are greatest at the end walls, whereas the horizontal current is strongest under the node.

changes are observed but strong periodic oscillating currents are felt. The largest water level changes, the amplitude of the standing wave, are observed at the limiting walls. The period of oscillation of such a basin, having one central nodal line, is given by

$$T_n = \frac{2l}{\sqrt{gh}} \tag{5-8}$$

where l is the east-west length of our ocean or small basin and h is again the water depth. The natural period of oscillation of the water is therefore dependent upon the geometry of the basin—its length and depth.

Our tide-generating force does not simply produce such a standing wave and then cease to be effective. Rather, the force will continue to act on the water in a rather complicated manner, since it is always varying both in direction and in magnitude. Tidal waves are therefore *forced waves*, as opposed to *free waves*. The situation is closely analogous to the swinging of a pendulum. If you give the pendulum a single push, it swings with its natural period, which depends chiefly on the pendulum length. Now if you continually apply forces to the swinging pendulum, it has no choice but to respond: it becomes a forced oscillation. If your applied force happens to correspond to and support the natural period of oscillation, you have a *resonant* condition and the amplitude of swing of the pendulum thereby increases. In a similar way, if the natural period T_n given by equation (5-8) corresponds to the periodicity of the tide-generating force, we have a resonant condition and the amplitude of the standing wave increases. It is in this way that the geometry of the ocean basin governs which of the tide-generating forces and periodicities will be the most effective in producing tides.

Due to the rotation of the earth there are geostrophic forces, the so-called *Coriolis force*, acting on our standing wave oscillation. The Coriolis force in the Northern Hemisphere causes the currents associated with the oscillation to veer off slightly to the right rather than proceed simply in an east-west direction. As the current flows toward the west it will swing somewhat to the north and pile up on the northern boundary of our ocean. Similarly, the eastward flow causes a pile-up at the southern boundary. The overall effect is that the tidal wave, instead of oscillating about a nodal line, now moves around a nodal point called the *amphidromic point* (from the Greek *amphi*, "around," and *dromas*, "running"). The high water of the tidal wave now progresses around the basin in a counterclockwise direction in the Northern Hemisphere, as depicted in Figure 5-13. Such a rotating wave is known as a *Kelvin wave*. The range of the tide will be greatest around the coast, the boundaries of our ocean, and least in the center, near the amphidromic point, where it is zero.

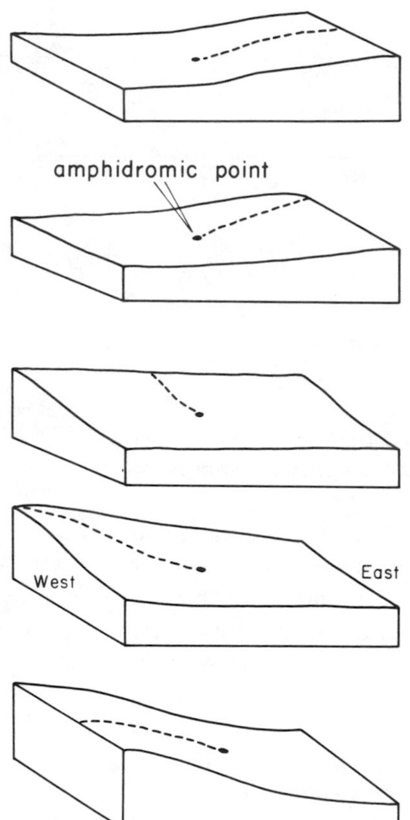

Figure 5-13 The counterclockwise motion of a Kelvin wave (tidal wave) about an amphidromic point.

SHALLOW-WATER EFFECTS

The range of tides in mid-ocean is small, only some 50 cm. However, when the tidal wave invades the comparatively shallow waters of the continental shelf, the height is increased, especially within gulfs and embayments on the coast. The associated tidal currents may also thereby be considerably increased.

Tides can be roughly classified by their spring tidal ranges into the following categories (Davies, 1964):

Microtidal—less than 2 meters
Mesotidal—2–4 meters
Macrotidal—greater than 4 meters

Microtidal and mesotidal ranges are found generally on the open coasts of the world's oceans as well as in the (virtually) landlocked seas such as the Mediterranean, Black, and Red seas (Figure 5-14). The higher, macrotidal ranges are found locally in gulfs and embayments along the coasts. On the Atlantic Ocean margins (Figure 5-14),

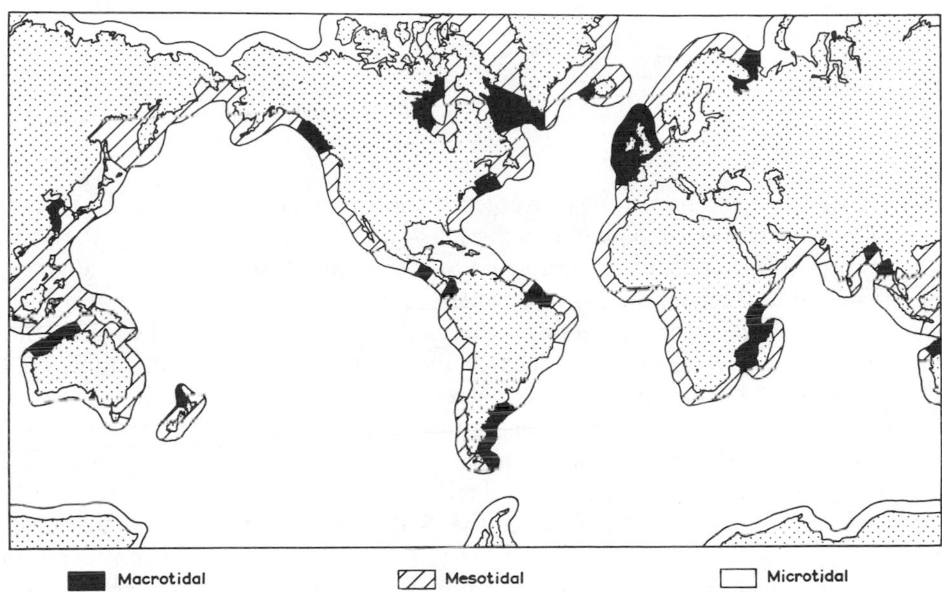

Figure 5-14 Microtides (range <2 m), mesotides (range 2–4 m), and macrotides (range <2 m). [*From* Davies (1964)]

macrotides are found along portions of the British coast, along the northwestern French coast in the vicinity of Mont-Saint-Michel (the Bay of Saint Michel), and in the Bay of Fundy, Canada. The Bay of Fundy in particular demonstrates the augmentation of tides within embayments. The bay is long with a wide mouth, progressively narrowing along its length and developing into two baylets at its landward termination (Figure 5-15). The tidal range at the bay mouth is large but not exceptional (about 3 m at spring tide). A considerable increase in range occurs as one proceeds up the bay: at Saint John the range is increased to 7.6 m, near the end of Chignecto Bay the range is 14.0 m, and at the end of Minas Basin the tidal range has increased to 15.6 m, the largest spring tidal range anywhere in the world.

Around the margins of the Pacific Ocean large tidal ranges are found in the Gulf of California (9 m), in certain bays of Australia (5 to 6 m), and in the Gulf of Siam.

The development of exceptionally high tidal ranges in certain gulfs and embay-

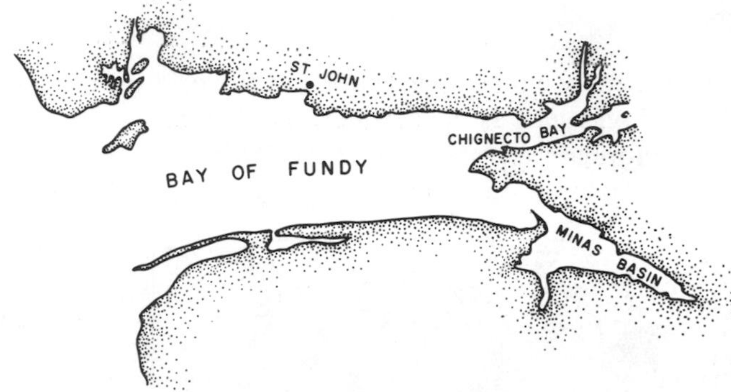

Figure 5-15 The Bay of Fundy, Canada, where the highest tides in the world are found.

ments is due to combinations of convergence and resonant effects. As the tidal front approaches a narrowing indentation in the coastline, like the Bay of Fundy, the enveloping shores constrict its movements and wedge the water together, thereby augmenting the height of the tidal wave. For an enclosed bay or estuary, resonant effects can occur similar to the resonance between the tidal oscillations and the tide-generating forces which we have already examined. If the length of the bay in the direction of tidal wave advance is such that it approximates an integral number of half–wave lengths for the appropriate depth of water, then a resonant standing oscillation can develop within the bay. This also occurs in the Bay of Fundy and, together with constriction, explains the extreme tidal ranges found there.

The tidal wave, as it travels over the shallow continental shelf, undergoes a transformation similar to what we have already seen in swell waves. Considerations of the tidal wave traveling in water of depth h and application of the equations of continuity and Bernoulli (no energy loss) leads to the velocity relationship

$$C = \left(1 + \frac{3}{2}\frac{\eta}{h}\right)\sqrt{gh} \tag{5-9}$$

where η is the local height of the tidal wave surface above the still-water level (Figure 5-16); η/h is small over the deep ocean, so that $C \simeq \sqrt{gh}$, as we have already seen. As the tidal wave approaches the coast, however, η/h increases as the water depth decreases. Since η is greater at the crest than at the trough of the tidal wave, the crest will travel somewhat faster than the troughs. The crest will tend to overtake the preceding trough, causing the tidal wave to become asymmetrical, with a steeper front than backslope to the wave. At a particular site of tidal observation along the coast, this shallow-water effect will be seen as a reduction in the period of flood (tidal rise) and an increase in the time of tidal ebb. Taken to its extreme, if the tidal wave proceeds up an estuary and river, the tidal wave may eventually oversteepen and "break," producing a *tidal bore* traveling up the river. Bores are found throughout the world, but the largest occur on the Amazon in Brazil and on the Chien-tang-kiang in China. When the Amazon River bore moves upstream it looks like a mile-wide waterfall

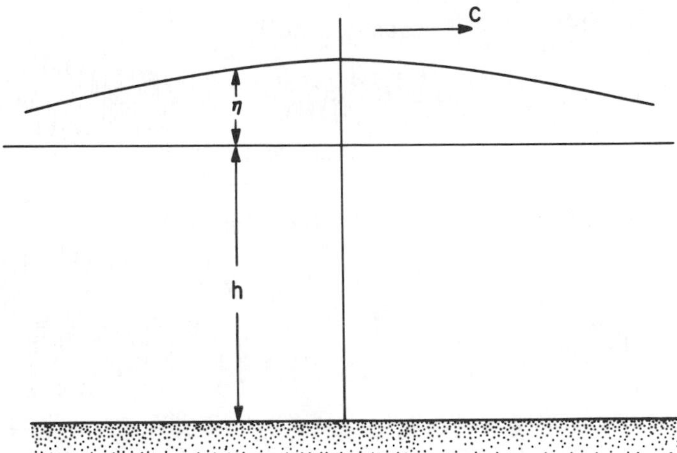

Figure 5-16 Notation of the tidal wave traveling in shallow water.

with a height of some 5 m, traveling at a rate of 10 m/sec. The bore on the Chien-tang-kiang, said to reach a height of up to 7.5 m, has been described by the ancients in graphic terms: "The surge thereof rises like a hill, and the wave like a house; it roars like thunder, and as it comes on it appears to swallow the heavens and bathe the sun" (quoted in Carter, 1966).

TIDES AROUND THE UNITED STATES

Tides along the coasts of the contiguous United States are not particularly interesting with regard to their heights. They do, however, provide good examples of how the natural period of oscillation of the water basin selects out the particular tide-generating force that corresponds most closely to that period and by resonant interactions magnifies the significance of that force over the others.

Because of their lengths and depths, the basins within the Atlantic Ocean have natural periods of oscillation which correspond roughly to the 12-hour tide-generating forces, so that *semidiurnal tides* (two high and two low waters of approximately the same height) prevail there. Figure 5-17 shows a typical variation at New York Harbor, which illustrates a semidiurnal tide with only a slight tendency for mixed tides to occur.

The resonant period for the Gulf of Mexico is closer to 24 hours, so that it responds to and resonates with the tide-producing forces of that period (K_1, O_1 and P_1 of Table 5-2). The result is *diurnal tides*, with generally only a single high tide per day.

The natural period of oscillation of the Pacific basin nearest the United States does not correspond particularly well to either the daily or semidaily tide-generating forces. The result is *mixed tides*; the example in Figure 5-17 from Seattle, Washington, is an illustration of this form. With mixed tides there are two high tides and two low tides per day, with strong inequalities in the heights of successive tides. During one part of the month there is a tendency for the diurnal tides to prevail, with the second tide

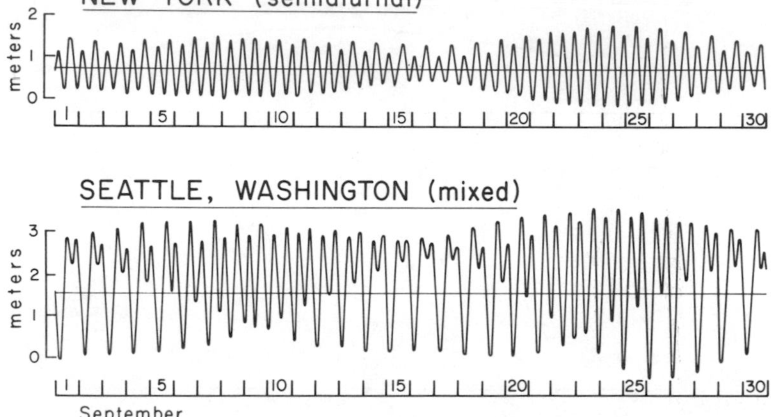

Figure 5-17 Semidiurnal tides at New York and mixed tides at Seattle, Washington. [*After* U.S. Oceanographic Office]

of the day a mere dimple on the water level cycle. At another time in the same month the tides are essentially semidiurnal, with two tides a day of nearly equal height. Such mixed tides prevail over the entire west coast of the United States.

Defant (1958) has shown that the form of the tide can be characterized by the relative magnitudes of the tide-generating constituents M_2, S_2, K_1, and O_1, which sum to form the major part of the tide, as in Figure 5-20. Since these are the major components, Defant proposed a form number

$$N_f = \frac{K_1 + O_1}{M_2 + S_2} \tag{5-10}$$

The tides divide roughly as follows:

$N_f = 0$ to 0.25	semidiurnal form
$N_f = 0.25$ to 1.50	mixed, predominately semidiurnal
$N_f = 1.50$ to 3.00	mixed, predominately diurnal
$N_f = >$ than 3.0	diurnal form

Figure 5-18, from Defant (1958), illustrates tides with form numbers in these different ranges. Immingham, on the south coast of England, has a semidiurnal tide with $N_f = 0.11$. San Francisco has a form number of $N_f = 0.9$ and is mixed with the semidiurnal constituents predominating. Manila ($N_f = 2.15$) is mixed, but with the diurnal predominant. Do-Son, Vietnam, on the Gulf of Tonkin, has a typical diurnal form with $N_f = 18.9$.

Returning to United States tides, the only other aspect that requires comment are the "tidal waves" experienced in the Gulf coast and Caribbean areas, reported in news accounts because they cause widespread destruction along coasts. As already mentioned, a more fitting term for these would be *storm surges*, in that they result from high onshore hurricane winds driving water ashore and greatly augmenting an otherwise mediocre tidal variation. In 1900, for example, hurricane winds of some 60

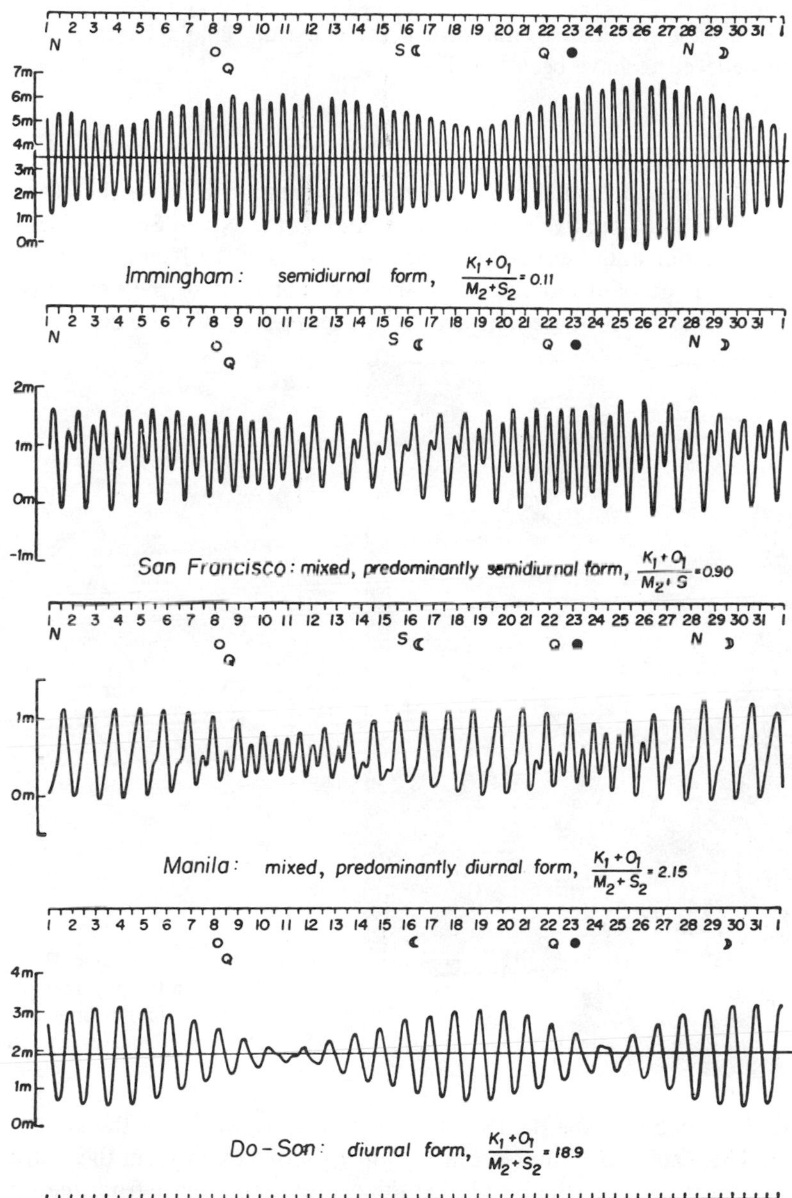

Figure 5-18 The nature of the tidal form as a function of the form number proposed by Defant. [*From* Defant (1958)]

m/sec (120 knots) struck the Gulf area in the vicinity of Galveston, Texas, raising the normal expected tidal range of 60 cm by an additional 4.6 m or more. On top of that were 7-m wind waves produced by the storm. Considerable damage resulted, and more than 5,000 people were killed. In 1970 a similar tropical storm in the Bay of

Bengal produced a storm surge that struck the coast of Bangla Desh. Some 700,000 people are believed to have been killed.

TIDES AROUND BRITAIN

The tides around Britain are particularly interesting because the island is surrounded by a series of relatively small water bodies: the North Sea, two segments of the English Channel, and the Irish Sea. The dimensions of the North Sea are such that it resonates with the tidal movements in the open Atlantic to the north. Three amphidromic systems are produced, as shown in Figure 5-19. This raises an interesting

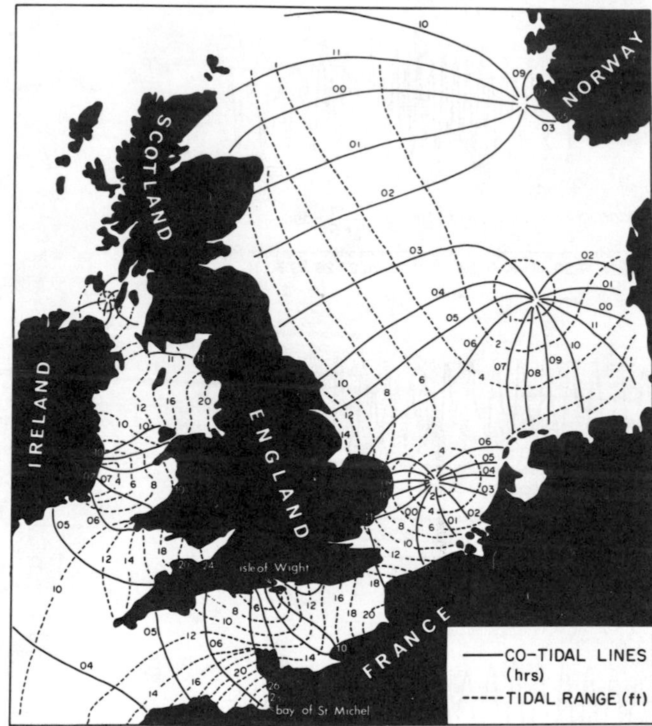

Figure 5-19 Tides around England. The cotidal lines give the positions of the tidal wave crest (high tide) throughout the daily cycle. The dashed curves give the mean tidal range.

point. Friction effects on the tidal waves in the shallow waters of the North Sea are significant. This causes the amphidromic point to move away from the source of the tidal energy. In the case of the North Sea the energy approaches from the north and moves counterclockwise around the basin. Friction therefore shifts the amphidromic point eastward, and we see that the most northerly system has shifted to such a degree that the point is nearly on the coast of Norway. A result of this eastward shift is that the east coast of Britain has sizable tides, since it is distant from the amphidromic point, while the eastern margins of the North Sea (Norway, northern Germany, and Denmark) have comparatively smaller tides.

The English Channel demonstrates an even further stage in the shifting of the amphidromic point, here again probably due to frictional effects. It is seen in Figure 5-19 that the point has shifted entirely onto land and is situated in the south of England. This results in the tidal waves moving in the same direction along both the southern English and northern French coasts, on opposite sides of the Channel. The tidal range in the vicinity of the Isle of Wight, off the south coast of England, is very small since it is near the amphidromic point; in contrast, the tidal range on the northern French coast is the highest in Europe.

In the Irish Sea there is a similar shift of the amphidromic point onto land, it being situated somewhere south of Dublin.

HARMONIC ANALYSIS AND THE PREDICTION OF TIDES

One of the practical purposes of studying tides is to acquire the ability to predict the tide level at any future date. In the year 1213 the abbot of Saint Albans published a rudimentary table from which the time of high tide at London Bridge could be predicted. Regular tide tables, based presumably upon direct observations, were prepared for the main ports of Britain during the seventeenth and eighteenth centuries. The first British Admiralty tide tables were issued in 1833; with these tables one could determine the levels and times of high and low tides for the main ports of the world.

The tidal variations one actually observes at some location are the complex summation of the several constituents caused by the various tide-generating forces. Every tide-generating force fluctuates periodically, but when they are all added together to obtain the actual tide an irregular variation in water level results. This result is closely analogous to the addition of swell waves from two or more sources to obtain irregular waves at the beach (Chapter 3).

The analysis of water waves at a beach was accomplished by wave spectral analysis to sort out the individual swell wave trains. A similar approach is taken with tidal variations; but since the periods of the components are much longer than those of swell waves, a much longer record is required for the analysis.

One might think that since we understand the generation forces of tides we should be able theoretically to predict the tides anywhere in the oceans. Our understanding of tides has not progressed this far, however. Such predictions are precluded by complex influences of the landmasses and submarine topography and by the effects of propagation over the continental slope and shelf. We can predict the periodicities of the various tidal constituents, but we cannot predict their amplitudes or phase relationships. Only long-term observations at the location of interest and harmonic analysis yield this information. Once they are deduced from observations, however, we can predict the tide for that location at any time in the future.

A simple harmonic analysis yields:

1. The periods of the tidal constituents (the time interval between two maxima).

2. The amplitudes of the tidal constituents (half the displacement between maximum and minimum).
3. The phase (the time when the maximum occurs).

These three parameters uniquely define each tidal constituent and can be used to reconstruct the total tidal cycle. Figure 5-20, from Defant (1958), shows seven of the major tidal constituents summed together to yield a predicted tidal fluctuation (solid curve), which compares favorably with the tide actually observed (dashed curve). In general, the first seven to ten largest tidal components are sufficient to provide an accurate predicted tidal variation. Figure 5-21 shows the addition of fifteen components [5-21(b)] to obtain a predicted tide [5-21(c)]. The difference between the observed and the computed water levels [5-21(a) minus 5-21(c)] yields a nonperiodic residue [5-21(d)]. Such nonperiodic residues are the result of winds producing mini-storm surges and of atmospheric pressure fluctuations that raise or push down the local water level. Because of these effects, the actual water level can be considerably different from that given in the tide tables.

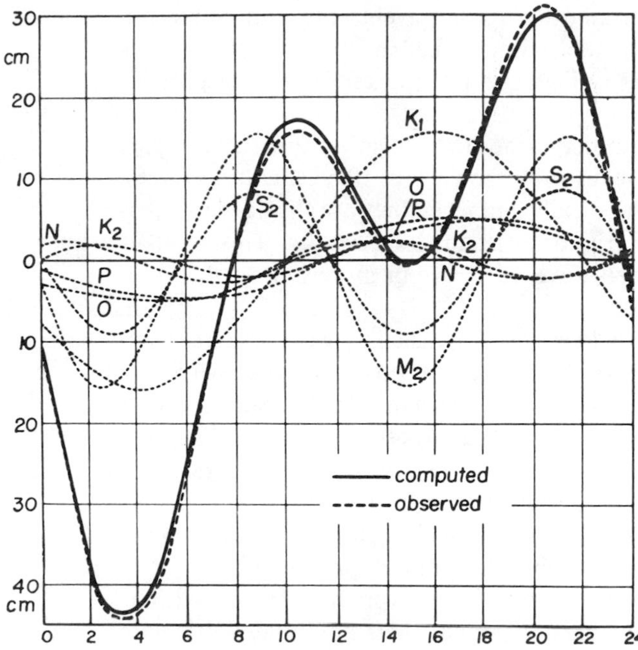

Figure 5-20 Summation of the components of the tidal cycle for one day, compared with the observed tide. [*From* Defant (1958)]

PROBLEMS

5-1 Consider a tidal wave of height 50 cm and period $T = 12$ hr in the deep ocean ($h = 5{,}000$ m). Assuming that its motions are adequately described by Airy wave theory, would it be a deep-water, intermediate, or shallow-water wave? What is its

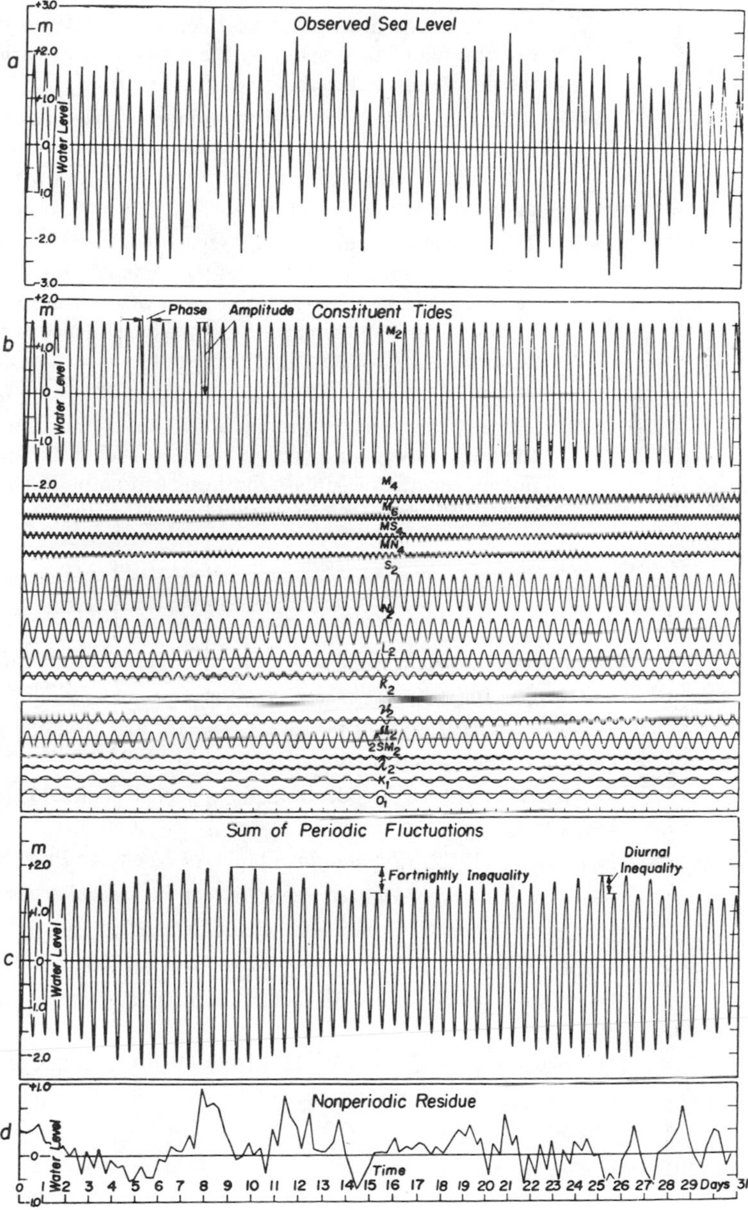

Figure 5-21 Summation of the components to give a predicted tidal variation (c). The observed tide (a) minus the predicted tide yields a nonperiodic residue (d) caused by mini–storm surges, atmospheric pressure fluctuations, and other effects. [*From* Defant (1958)]

energy E and energy flux ECn? How does this compare with a deep-water swell of $H = 50$ cm and $T = 15$ sec?

5-2 If the moon is beneath one's feet on the opposite side of the earth, would this tend to make you weigh more or less? Why?

5-3 Assume that the tide level varies in a simple sinusoidal fashion, the period being 12 hr. With a uniformly sloping beach, would the water level reside longer in the vicinity of the high-tide, mid-tide, or low-tide position on the beach? Evaluate the distribution of time spent by the water level at various increments of distance along the profile of the beach.

5-4 If the tidal wave height is 5 m and the still-water depth is 30 m, according to equation (5-9) how much faster is the tidal wave crest (high tide) traveling than the trough (low tide)? Write a computer program to follow the changing profile and its asymmetrical development under these conditions, assuming it is initially sinusoidal.

5-5 The flow in and out of a bay, separated from the ocean by a sand spit except for one narrow entrance, is largely tide-dominated. What will be the controlling factors in determining the cross-sectional area of the channel connecting the bay and ocean? Devise a graph showing the expected dependence of the section area on the controlling parameters. How would you collect the necessary data to show this dependence and firm up the graph? [See O'Brien (1931) for solution.]

5-6 The ancients believed that an angel moving his foot in and out of the water produces the tides. How would you go about in a scientific fashion to test this hypothesis? (Do not be superficial—this hypothesis can actually explain much of the observed properties of tidal motions, including periodicity.)

REFERENCES

CARTER III, S. (1966). *Kindgom of the tides.* Hawthorn, New York, 160 pp.

CLANCY, E. P. (1969). *The tides.* Anchor, Garden City, N.Y., 228 pp.

DARWIN, G. H. (1962). *The tides.* Freeman, San Francisco, 378 pp. [Lectures delivered by Darwin at the Lowell Institute, Boston, in 1897.]

DAVIES, J. L. (1964). A morphogenic approach to world shorelines. *Zeit. fur Geomorph.*, 8: 127–42.

DEFANT, A. (1958). *Ebb and flow.* Univ. of Michigan Press, Ann Arbor, 121 pp.

——— (1961). *Physical oceanography*, vol. 2: *Tides.* Pergamon Press, New York.

GARRETT, C. (1974). Ocean tides. *Geosci. Canada*, 1, no. 4: 8–14.

GODIN, G. (1972). *The analysis of tides.* Univ. of Toronto Press, Toronto, 264 pp.

HENDERSHOTT, M. C. (1973). Ocean tides. *EOS*; *Trans. Am. Geophys. Union*, 54, no. 2: 76–86.

HENDERSHOTT, M. C., and W. H. MUNK (1970). Tides. *Ann. Rev. Fluid Mech.*, 2: 205–24.

O'BRIEN, M. P. (1931). Estuary tidal prisms related to entrance areas. *J. Civ. Eng.*, 1. no. 8: 738–93.

Chapter 6

CHANGING LEVELS OF SEA AND LAND

Schleiden relates, that an old monk from a convent in the neighbourhood stated that in his youth he had gathered grapes near the monument, in a spot where now the boats of the fisherman are rocked by the waves.

F. A. Pouchet
The Universe (1884)
[The monument mentioned is the ruins of the Temple of Jupiter, Serapis, Italy.]

Now rolls the deep where grew the tree.
Tennyson

Tides produce daily fluctuations in the water level at the coast. Seasonal variations may also be detected, although they usually amount to at most a few tens of centimeters. The latter are caused by the seasonality of coastal rainfall, river discharge, and temperature changes of the water. The other sea level change of importance is associated with the advance and retreat of glaciers over thousands of years. Because of water being alternately locked up within glaciers and then released, the sea level has changed by more than 100 meters within the last 20,000 years; such variations are known to extend for more than 3 million years into the past.

This chapter will examine the details and causes of these annual and long-term sea level changes and will consider what effects they have on coastal erosion and the general coastal geomorphology.

ANNUAL CHANGES IN SEA LEVEL

If you take a record of tidal changes at some coastal station and average it for a one-day span, you will determine a "mean sea level" for that location. Now if you repeat the process on the following day, you might find that the "mean sea level" has changed, perhaps even by as much as a meter. This is because the daily sea level is very susceptible to local weather conditions. A strong onshore wind, for example, will drive water shoreward, raising the mean water level for that day. To eliminate these daily fluctuations, we could extend our averaging period to one month. This would give us a longer-term "mean sea level," but we would still find changes from one month to the next, although the changes would not be so large as those on a daily basis. Within a year the lowest and highest monthly values of sea level might differ by an amount on the order of 50 cm. If we prepare a graph of the monthly sea level average for the entire year, such as those shown in Figure 6-1, in almost all locations we would find a pronounced seasonal variation (Marmer, 1952; Pattullo et al., 1955; Pattullo, 1966). For example, around the United States there is in general a lowest sea level in the spring months and a maximum sea level during the late summer and autumn. This is true throughout most of the world. Pattullo et al. (1955) analyzed tidal records from over 400 stations throughout the world to detect seasonal cycles of sea level compared with the mean for the entire year. The seasonal variations ranged from only a few centimeters in the tropics to amounts on the order of 20 cm or more at higher latitudes. The largest variation, in the Bay of Bengal, exceeded 1 meter. In a review of the then-existing data, Pattullo (1966) found very systematic annual changes in sea level throughout the world. As seen in Figure 6-2, there is a definite switch-over from conditions in February through April to opposite conditions in September. Where deviations from the annual mean are positive in March, they are negative in September, and vice versa. This switch is especially noteworthy around India, where the east and west coasts have opposite deviations at any given time and make a complete switch between March and September.

Patterns of annual sea level change are similar in any given region but vary from one region to another. The causes of the cycles may also differ from one area to another. Figure 6-1, from Marmer (1952), shows a systematic change in the pattern along the length of the Gulf coast of the United States. At Key West, Florida, at the eastern end of the gulf, sea level is lowest around March and highest in October. Proceeding to the west, at Eugene Island, west of the Mississippi River, there is a definite second high sea level occurring in May-June, with a low in July. This two-peak system becomes even more pronounced in the western gulf.

Such annual changes in the sea level, found throughout the world, can be attributed to seasonal variations in climate and ocean water properties. There have been many studies of the causes, of which the papers by LaFond (1939), Pattullo et al. (1955), and Lisitzin (1955) are particularly noteworthy. Reviews are given by Lisitzin and Pattullo (1961), Pattullo (1966), and Rossiter (1962). Even after considerable study, many of the reasons for the annual sea level changes are not well understood.

LaFond (1939) investigated the causes for sea level changes off California. In that area the annual cycle consists of a minimum during March-April and a maximum in August-September (Figure 6-3); the total range is some 25 cm. After examining a number of possible causes, LaFond determined that it is produced by cyclical variations in the surface water density. This density change is brought about mainly by variations in the water temperature; salinity has only a secondary effect. The water

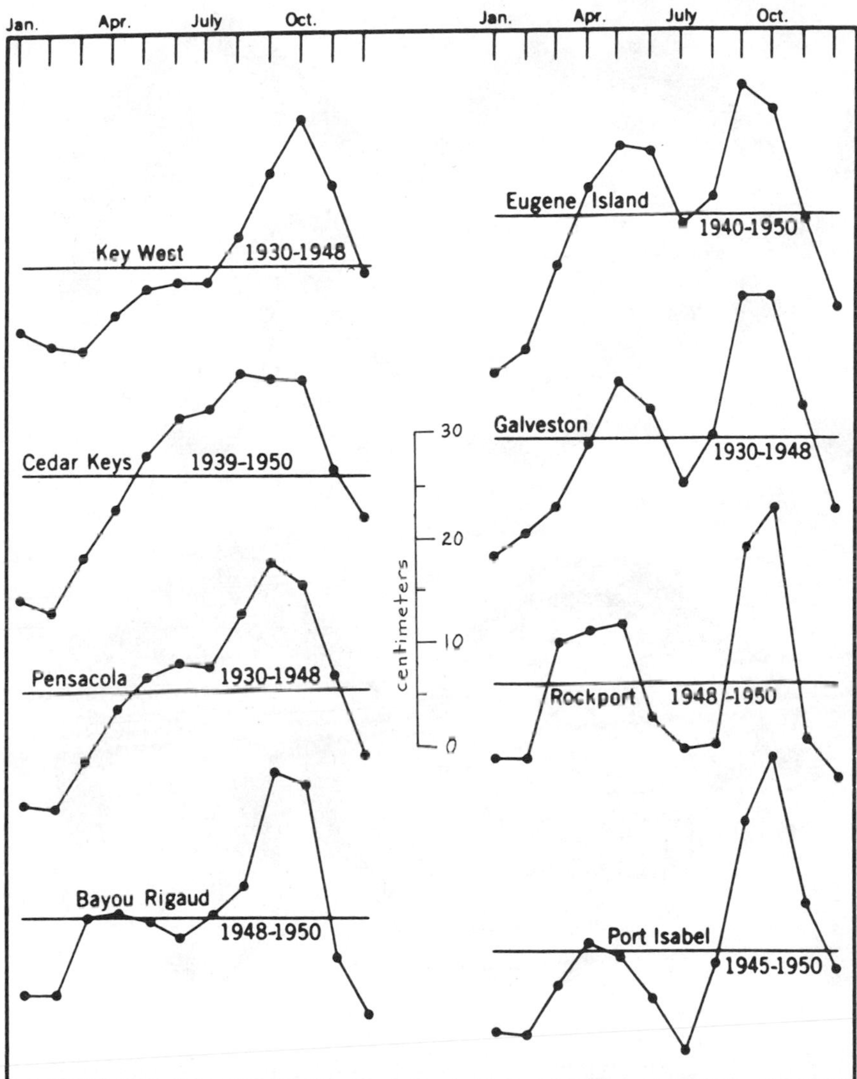

Figure 6-1 Seasonal variations in sea level along the Gulf coast of the United States, determined from tide observations. [*After* Marmer (1952)]

temperature variations are in turn produced mainly by offshore current patterns, and are not simply a warming and cooling of the water from annual changes in solar radiation. Off California the Davidson Current flows from north to south, nearly parallel to the coast. Due to the Coriolis force, this south-flowing current causes a shifting of the surface water offshore and an upwelling of cold deeper water to replace it. This cold water near the coast has a greater density, so that the sea level is reduced. Upwelling is at a maximum in the spring (Figure 6-3), and that accounts for the

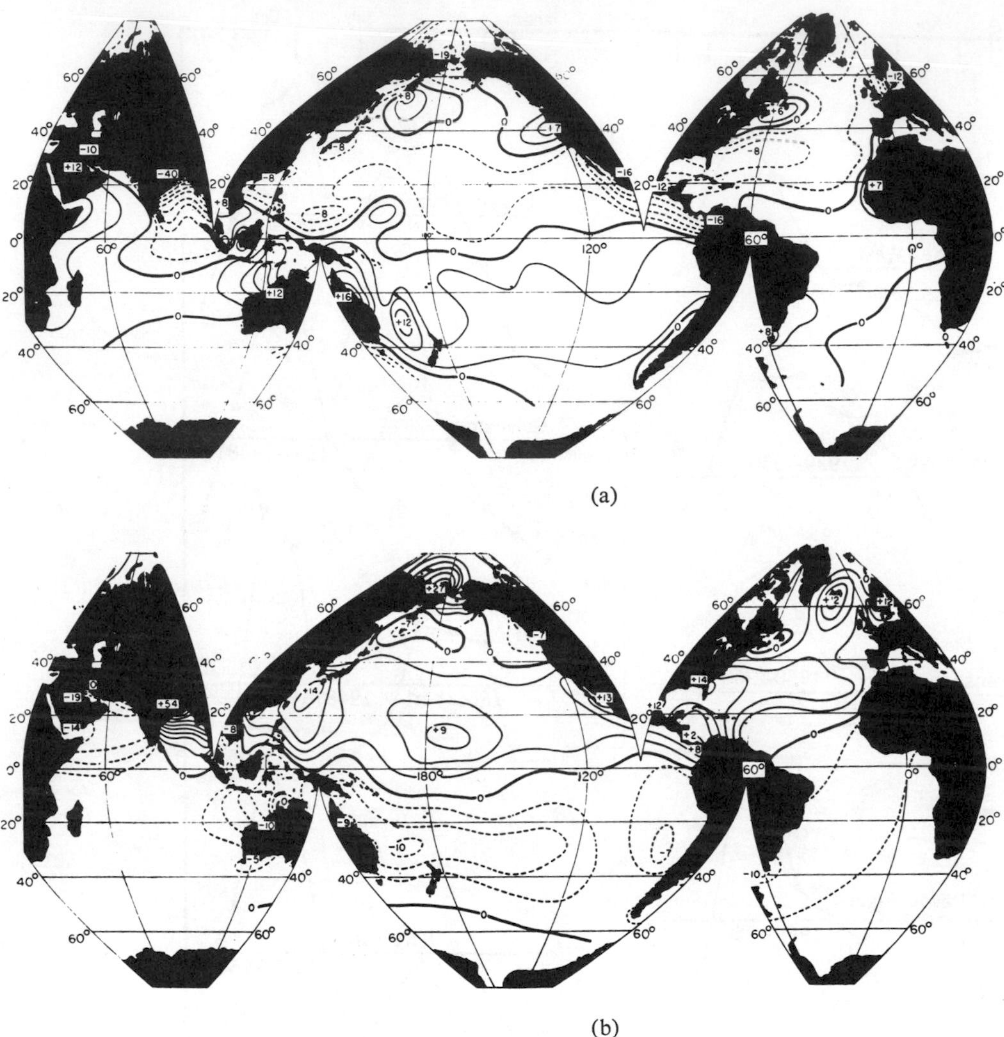

Figure 6-2 Deviations from mean annual sea level. Light solid lines are positive deviations (sea level higher than average); dashed lines are negative deviations. Contour interval is 3 cm. (a) February, March, April, (b) August, September, October. [*From* Pattullo (1966)]

lowered sea level at that time. Throughout the summer upwelling wanes, the water is progressively heated by solar radiation, and the sea level rises, reaching a maximum in about September.

Ocean currents can also cause sea level changes through geostrophic effects, not just by upwelling. This is because an ocean current causes the water surface to slope at right angles to the direction of flow. In the Northern Hemisphere this slope is upward to the right when viewed in the direction of current flow. In the Northern

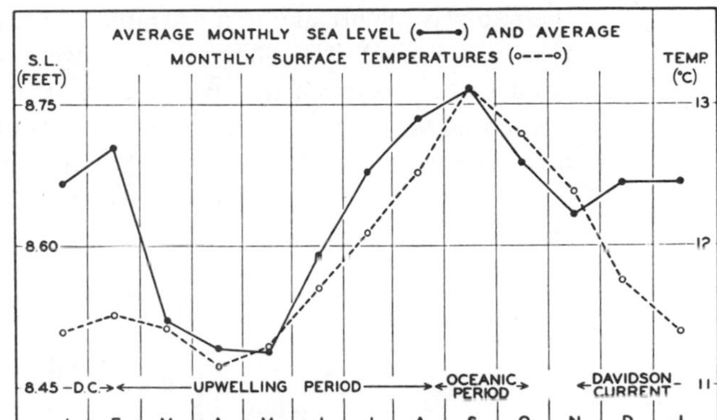

Figure 6-3 Annual variations in sea level off California corresponding to temperature changes in the water surface. The cold water is brought to the surface from the deep ocean by upwelling. [*After* LaFond (1939)]

Hemisphere the Coriolis force tends to turn the flow to the right, and it would do so except that this causes the water level on the right to be higher. A steady-state condition is finally reached when the higher water on the right opposes and balances the Coriolis force. A good example of this phenomenon and its effects on sea level are shown in the study by Wyrtki (1973), who investigated the geostrophic water transport in the Pacific Equatorial Countercurrent. This countercurrent and its associated cross-flow water slope effect sea level differences between pairs of islands situated north and south of the current. Wyrtki found that the sea level difference between island pairs exhibits a marked seasonal variation, with a minimum in March through May and a maximum (18 cm higher) in December. This seasonal variation in slope is caused by seasonal changes in the countercurrent velocity: the greater the current velocity the greater the surface slope. For example, the annual changes at Christmas Island amounted to about 8 cm; Kwajalein Island, on the opposite side of the current, showed a similar 8-cm annual change, but the water level there increased when the level at Christmas Island decreased, and vice versa. Because of the water slope across the width of the flow, when the water rose at Christmas Island, it fell at Kwajalein. Using this fact, Wyrtki was able to use the tidal records to compute the monthly mean discharge in the countercurrent for a 21-year period.

As can be seen in Figure 6-2, sea level changes are generally small in mid-ocean, becoming large in coastal waters. Lisitzin and Pattullo (1961) concluded that over the open ocean the recorded variations in sea level are caused by changes in the atmospheric pressure. Changes in air pressure are followed rather quickly by compensating changes in sea surface level in such a way that the total mass of air plus water resting on the sea floor remains a constant.

The combined effects of several causes of sea level changes are illustrated by the study of Namias and Huang (1972). The changes they investigated spanned a decade rather than a single year and were associated with two large-scale, long-term climatic regimes covering the entire United States. The periods compared were 1948–1957 and 1958–1969, a rather rapid change occurring in 1957 to 1958. In the latter period the sea level off southern California was observed to rise by 5.6 cm over what it maintained in the earlier. Investigating the various possible causes, they found that the rise was mainly due to a warming of the surface waters by 1°C, causing a 3.7-cm rise in sea level. This long-term change resulted from a change in the prevailing wind pattern between the two decades. Changes in the mean atmospheric pressure accounted for a 0.6-cm rise, dynamic effects of ocean currents caused a 1.0-cm change, and the normal long-term sea level rise due to melting of glaciers accounted for another 1.0 cm. The net calculated change amounts to 6.3 cm, which compares favorably with the observed 5.6-cm rise.

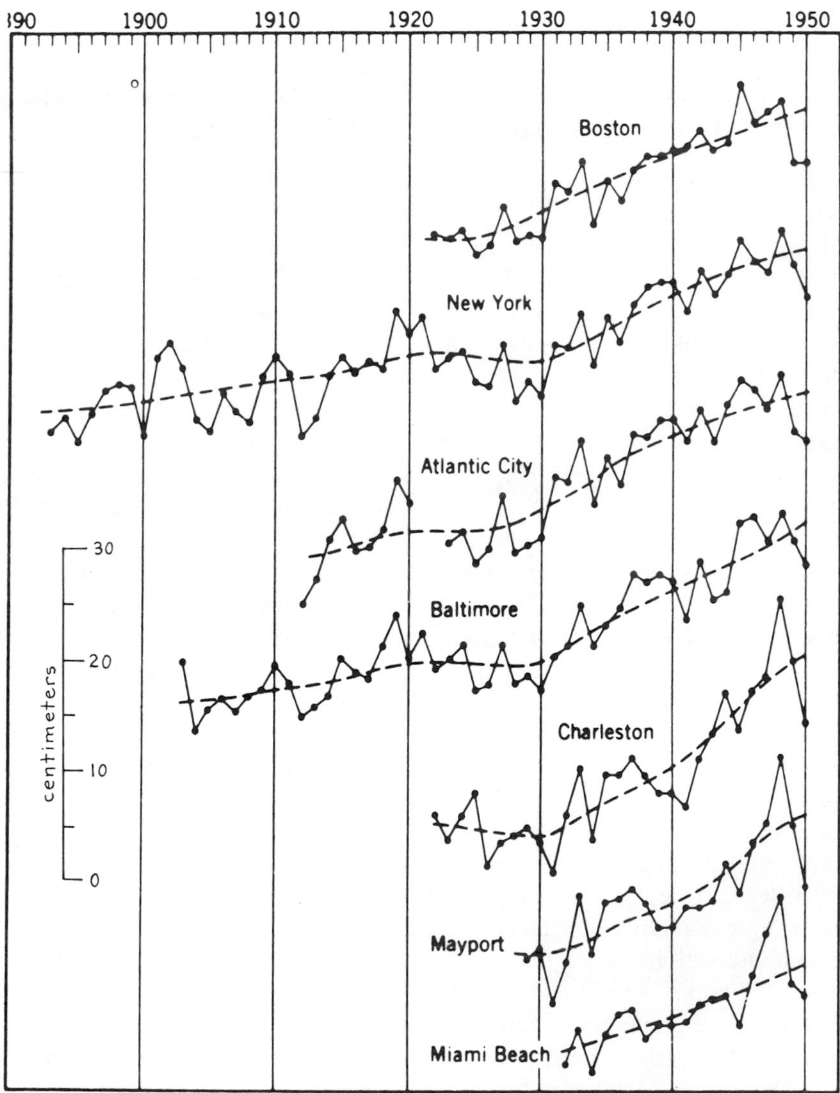

Figure 6-4 Yearly average sea level along the Atlantic coast of the United States demonstrating long-term variations brought about principally by melting of glaciers. [*After* Marmer (1952)]

Meade and Emery (1971) centered their attention on the importance of river runoff on the sea level of the eastern United States and Gulf coastline. They found that the change in sea level due to river flow ranges from 0.01 to 0.05 cm per km^3 change in annual inflow. There is little lag between the runoff and the sea level change; the adjustment of the sea level to a change in river discharge requires much less than a year. The variations in annual river inflow account for 7 to 21% of the total sea level variation from one year to the next along those coasts. This includes the long-term rise in sea level caused by melting of glaciers. If this latter variation is excluded, river runoff accounts for 20 to 40% of the remaining variations.

If sea level is followed year after year at a given tidal station, it will be found to change in a rather steady fashion, generally rising by some 1 to 2 mm/yr (Figure 6-4). This is due to glaciers melting and adding water to the sea, a factor that will be discussed at length in the next section. The observed change at any given site will depend on vertical movements of the land as well as of the water surface. Each location will therefore give slightly different results. As an extreme example, Juneau, Alaska, shows a progressive drop of the sea relative to the land in its tide record. This is because the land is rising faster than the sea, a result of isostatic uplift, the weight of glaciers of the recent geologic past having been removed. Such long-term sea level trends obtained from tidal records around the United States are shown and discussed by Marmer (1952) and by Hicks (1968, 1972).

In summary, superimposed upon long-term sea level changes due to glacier melting are seasonal and decadeal variations caused by climatic and ocean water effects. They include water temperature changes, atmospheric pressure variations, and so on; these are summarized in Table 6-1 with approximations of their relative magnitudes. The importance of such variations to coastal processes will be discussed at the end of this chapter.

TABLE 6-1 Factors Producing Seasonal Deviations in Sea Level (*After* Pattullo, 1966)

	DEVIATION (cm)		
	0	*10*	*20*
Variation in water temperature (mass held constant)			X
Variation in salinity of water (mass held constant)			X
Change in local atmospheric pressure			X
Runoff from rivers			X
Onshore winds			?
Longshore component of winds		?	
Approach of annual or semiannual high of astronomic tides	X		
Increase in total volume of water in oceans by glacial melting	X		

QUATERNARY CHANGES IN SEA LEVEL

Drowned beaches and terraces, mastodon teeth found on the continental shelf (Whitmore et al., 1967), submerged peat beds (Emery et al., 1967) all demonstrate that some 20,000 years ago, at the height of the last large advance of glaciers on the continents, the sea surface stood some 100 meters or more below its present level. During much of the Quaternary ice age (approximately the last 3 million years), a substantial portion of the earth's total supply of water was locked up within ice sheets covering large areas of the continents. The loss of this water from the oceans resulted in a lowered sea level, exposing what is now the continental shelf. A portion of that ice still remains. If this remaining ice were also to melt, sea level would rise by an additional 60 meters, covering New York City and most other coastal areas (Donn et al., 1962).

A timetable of the changes in sea level has been obtained by dating material that has a known narrow relationship to past stands of the sea. The carbon-14 technique has been employed for such dating. Carbon-14 has a half-life of 5,360 years and can provide reliable estimates of ages up to about 20,000 years B.P. (before the present), the reliability decreasing rapidly for greater ages. Radioactive uranium and thorium can be used to date more ancient materials. The application of the carbon-14 technique requires carbon material formed near some former sea level. The most commonly used sea level indicators are shallow-water mollusks (Curray, 1960)—particularly the oyster *Crassostrea virginica*—salt-marsh and freshwater peat, coralline algae, hermatypic corals, beach rock, lagoonal oolites (Milliman and Emery, 1968), and fossil intertidal organisms (van Andel and Laborel, 1964). If these are found on land above the present sea level or underwater on the continental shelf and have not been transported subsequent to their formation, then dating provides an indication of the past sea level. It is apparent that such materials indicate only a relative sea level at that location, not an absolute (eustatic) sea level change, since in the meantime the level of that chunk of earth on which they rest could also have changed. As an extreme example, if the land were rising faster than the sea, the shoreline would retreat seaward even though there is a rise in the eustatic sea level. The possibility of such a tectonic effect is one of the major difficulties in determining the absolute change in sea level. Thus areas identified as stable shelves are ordinarily utilized, while known tectonically active areas have, until recently, been avoided. Even then, much uncertainty enters, as no area can be considered truly stable. As pointed out by Curray (1969), in light of the modern theory of plate tectonics, continental margins lying near plate edges will have a history of large-scale tectonic effects. Continental margins or oceanic islands occupying mid-plate positions may be better suited, but even these will not be entirely stable. High-latitude regions showing isostatic rebound must also be avoided unless the uplift can be accounted for, as must regions near deltas, where there is crustal downwarping from the sediment load. Even the isostatic effects of water loading on the continental shelf must be properly evaluated (Bloom, 1967). Because of these tectonic and isostatic effects, much uncertainty enters, and a great deal of the disagreement as to the fine details of sea level changes results from this.

The chronology for sea level changes over the past 40,000 years is shown in Figure 6-5, compiled from data and proposed curves of several studies in different areas of the world. Although the various studies differ in details, they show that the sea level stood some 130 meters lower than at present about 15,000 to 20,000 years ago, that there was a rapid rise in sea level (8 mm/yr) until about 7,000 years B.P., and then

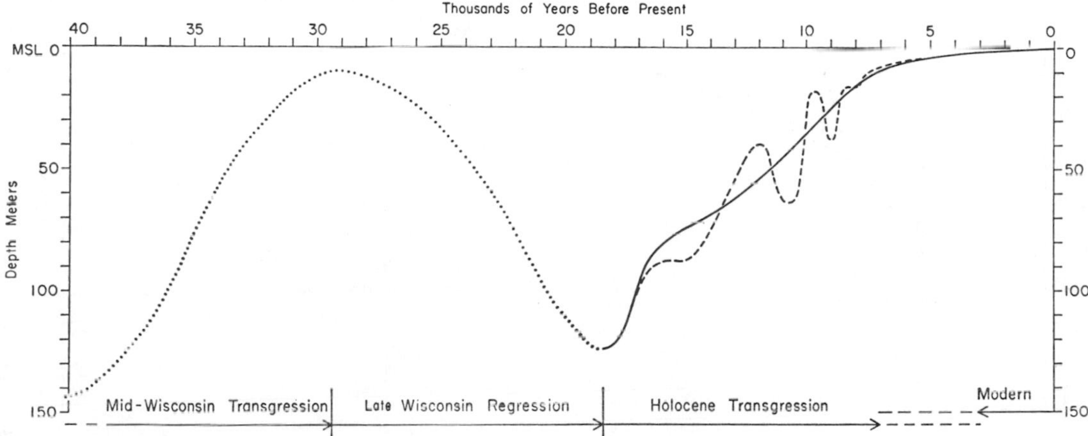

Figure 6-5 Fluctuations in sea level for the past 40,000 years based on published and unpublished carbon-14 dates compiled by Curray (1965). The dotted curve is estimated from only limited data. [*From* Curray (1965)]

a slowing down (1.4 mm/yr) in the rise until the sea reached approximately its present level some 2,000 to 4,000 years B.P. Most controversy concerns the sea level during the past 7,000 years, despite (or because of) the considerable amount of data that exists for that period. About the only agreement is with regard to the present sea level. Studies of Fairbridge (1961) [Australia], van Andel and Laborel (1964) [Brazil], and Fujii and Fuji (1967) [Japan] all indicate that sea level has been higher than at present within the past 7,000 years. Block (1965) provides historic and archeological evidence for these higher stands. However, the studies of Curray (1960, 1961) [Texas], Jelgersma and Pannekoek (1960) [Netherlands], Scholl and Stuiver (1967) [Florida], Redfield (1967) [Florida], Curray, Shepard, and Veeh (1970) [Micronesia], and many others found no evidence for higher stands of sea level during that period. Shepard (1963) and Shepard and Curray (1967) carefully compiled and reviewed the data and information on sea level within the past few thousand years to assess whether there had been previous stands higher than at present. Figure 6-6 shows data from "stable" areas such as the Netherlands and Gulf coast. It is seen that the data gives a smooth trend approaching the present sea level, with no high stands indicated. Even if these areas where the data was obtained are subsiding, correcting Figure 6-6 for subsidence would rotate the curve upward but would still not show the oscillations proposed by Fairbridge (1961). Shepard and Curray find similar results for the U.S. east coast data, but with greater subsidence indicated. Milliman and Emery (1968) interpret the difference between east coast and Gulf coast data differently, calling for a slight uplift of the Gulf

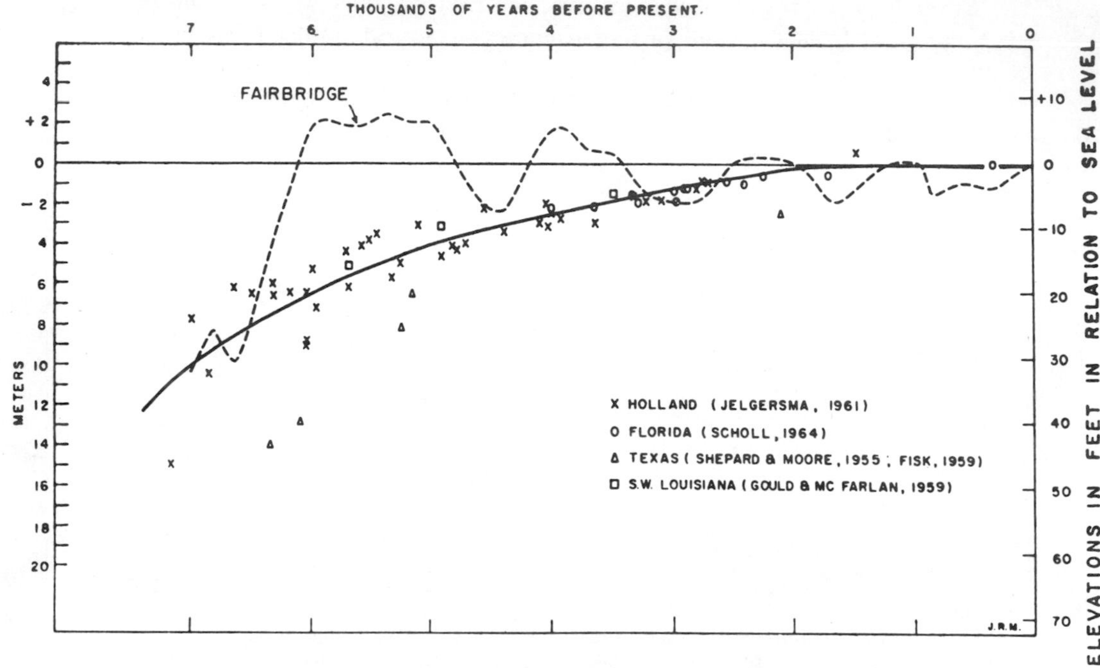

Figure 6-6 Sea level curve for the past 8,000 years based on a compilation of data by Shepard (1963) from tectonically "stable" areas of the world. Also shown is the curve of Fairbridge (1961) based on glacial advances and retreats, suggesting higher stands of sea level than at present. [*From* Shepard and Curray (1967)]

coast rather than a greater subsidence of the east coast. Shepard and Curray (1967) also reviewed the data base which supports higher stands of sea level than at present within the last 6,000 years. This data comes from terraces and intertidal organisms in the tropics. They conclude that many of the terraces are much older than suggested by the dates, originating in earlier interglacial high stands of the sea. Anomalous dates, they claim, are given by utilizing shells from Indian middens, shells transported by birds, and blocks of material thrown up by storm waves. As an example, they repeatedly dated material from terraces on Oahu, Hawaii, at 1.5 and 3.7 meters elevation, and got dates ranging from 7,500 to 50,000 years B.P.

In all cases the proposed higher sea levels amount to only some 2 to 3 meters above the present water level. Whether they did or did not occur is still debatable. Either way, sea level has been roughly at its present position for the past 4,000 years.

Long-term tidal gauge records, over the past 50 years, show a steady rise in sea level, averaging about 1.2 mm/yr, but with higher rates more recently (Marmer, 1952). This gives about a 12-cm rise in sea level per century, which corresponds to the rate at which the glaciers have been melting during the climatic warming spell that has existed since about 1850. Tide records taken in Amsterdam since 1682 give a still longer view of sea level changes (Fairbridge, 1960). The North Sea basin, of which the Netherlands is a part, has been subsiding at a rate of about 10 cm/century. This means that only half the recorded rate of 20 cm/century represents a true rise in sea level. The long record of sea level changes reveals that the fluctuations faithfully reflect climatic history. The level rose between 1725 and 1770 at about the present rate, then fell between 1800 and 1850 during a period of exceptionally cold winters. With the present warming spell the sea level is again rising.

A knowledge of glacial advances and retreats, and of climate fluctuations as revealed by pollen in lakes and by oxygen isotopes in deep-sea cores (Emiliani, 1970), can be used to generate curves for the changing sea level. Detailed fluctuations can thereby be suggested which generally cannot be revealed by the scattered sea level data. Fairbridge (1960, 1961) was one of the first to take this approach, and his resulting curve is that shown in Figure 6-6. We have already seen that there is some data which supports the high-level stands suggested by the Fairbridge curve. Mörner (1969) provides a more recent analysis utilizing this approach. The Fairbridge and Mörner curves are compared in Figure 6-7 along with the Shepard (1963) sea level curve based on "stable area" data. It is seen that the Mörner curve does not predict sea level stands higher than the present, the fluctuations being much smaller than the Fairbridge curve. The agreement between the Mörner curve and the "smoothed" data-based curve of Shepard is striking.

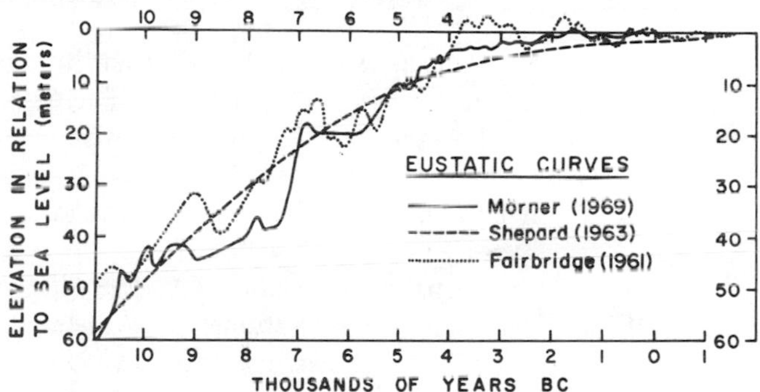

Figure 6-7 The Fairbridge (1961) and Mörner (1969) curves of sea level variations based on glacial advances and retreats and known climate fluctuations, compared with the curve based on data compiled by Shepard (1963), also shown in Figure 6-6. Unlike the Fairbridge curve, the Mörner curve predicts only minor higher stands of the sea than the present level. [*After* Mörner (1969)]

Prior to the low stand of the sea level 15,000 to 20,000 years ago, there is some evidence for a high stand at just about the present level during the period 30,000 to 35,000 years B.P. (Figure 6-5). Some indications for recent higher stands of sea level within the past 7,000 years relate to this much earlier high stand. For example, a beach and associated terrace in western Australia, related to a 3-meter-higher sea level, was first thought to be recent, but Russell (1963) dated it at 35,000 $\pm$ 3,700 years B.P. Similar conclusions were reached by Hails (1965) in eastern Australia. Going still further into the past, there were many cycles of glacial advance and retreat with accompanying changes in sea level. Some of the interglacial high stands are represented by marine terraces and beach ridges high above present sea level (Figure 6-8). The Atlantic coastal plain of the United States bears depositional terraces or ridges at levels 13.5, 6, 7.5, and 4.5 meters above the sea (Oaks and Coch, 1963), and these have been correlated with alternations in the sea level since Miocene times. Cooke (1971) attempted to correlate some of these ridges with terrace levels in Australia.

Figure 6-8 Terraces associated with the water level changes in the Great Salt Lake, Utah. Similar terraces are found along many of the ocean coasts, caused by relative changes in the level of sea and land. [*From* I. C. Russell, *Lakes of North America* (Boston, Ginn and Company, 1895)]

Elevated marine terraces can be found throughout the world. They may be seen along almost the entire length of the U.S. Pacific coast, and have been carefully studied in the Mediterranean and along the British coast by Zeuner (1959) and in Australia by Gill (1961) and Jennings (1961). There is an enormous literature on terraces, a list of which has been compiled by Richards and Fairbridge (1965) and which has been reviewed by Guilcher (1969).

There is some debate as to the relative heights of the successive interglacial sea levels and their relationship to the marine terraces. It is found that there are "stairways" of terraces on most coasts, with the oldest and most poorly preserved-terrace being topmost and the youngest the lowest. Some investigators have interpreted this as an indication that there has been a progressive lowering of successive high stands. Others attribute this to tectonic uplift of the landmass, a gentle uplift preserving successively older terraces at higher levels (Bourcart, 1952). The terrace level would then be attributable to the combined effects of tectonic uplift and actual (eustatic) sea level changes. Most of the best "stairways" of terraces are indeed associated with tectonically active areas such as the U.S. west coast, the Mediterranean, and New Guinea. However, others are found in what is thought to be relatively stable areas such as southern Australia (Gill, 1961), South Africa (Haughton, 1963), and India (Chaterjee, 1961).

Only careful studies that include considerations of the tectonic effects can reveal heights of the previous interglacial sea levels. One such study is that of Broecker et al. (1968), utilizing uplifted coral reefs on Barbados. That island is rising from the sea at a uniform rate, sufficiently fast to separate in elevation coral reef tracts formed at successive high sea level stands, even when the stands were below the present sea level. Three distinct high stands of the sea were found at about 122,000, 103,000, and 82,000 years B.P. Assuming a uniform rate of uplift, the stand at 122,000 years ago was about 6 meters above the present level, in agreement with other investigations, while the next two high levels were each about 13 meters below present sea level. This would explain why terraces of these two later dates are not found in many areas. In a similar study in New Guinea, Veeh and Chappell (1970) found dates of high stands which correspond to those on Barbados. In addition, they found high stands at 180,000–190,000 years B.P. and at 35,000–50,000 years B.P., as well as the present high stand commencing about 6,000 years B.P. Again, only the high stand at about 120,000–140,000 years B.P. was above the present sea level. Chappell (1974) further documents evidence from New Guinea terraces in a careful and thorough study. A well-developed series of terraces were investigated, rising to over 600 meters, progressively uplifted by tectonic activity. Terrace correlation and radiometric dating gave sea level maxima at 30,000, 40,000–50,000, 60,000, 80,000, 105,000, 120,000, 140,000, 185,000, and 220,000 years B.P. It is seen that many of these agree with those from Barbados, but others are added. Using a least-squares search, a "best-estimate" sea level curve and corresponding tectonic uplift pattern were derived from eleven series of terrace elevations. According to the results (Figure 6-9), most of the inter-

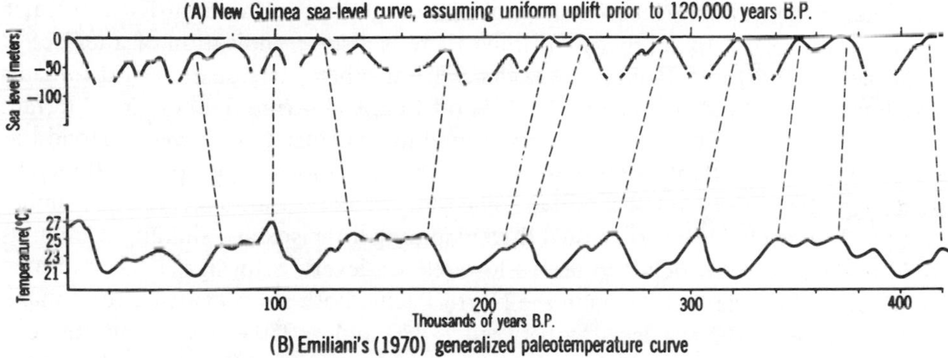

Figure 6-9 Sea level curve based on uplifted terraces in New Guinea, compared to the paleotemperature curve of Emiliani (1970). More recent work on the paleotemperature curve may offer a better correlation than shown and a revision of the dates for the ancient sea level changes (G. R. Heath, pers. communication, 1974). [*From* Chappell (1974)]

glacial high-water levels and their associated terraces stood below the present sea level, even the 30,000–40,000 years B.P. level generally considered to have been close to the present sea level. The present elevations of these terraces above the sea are therefore due to tectonic uplift. Figure 6-9 also correlates the interglacial sea level highs with Emiliani's (1970) paleotemperature curve with convincing agreement.

The Barbados and New Guinea data indicate that most of the interglacial highs in sea level were lower than the present level. The "staircase" of terraces of increasing age upward therefore appears to be due to uplift of the land, bringing the terraces above sea level so that they are preserved. It remains possible that some series of terraces on stable landmasses are due only to the few interglacial highs that exceeded the present sea level, highs dating back more than 120,000 years B.P.

For obvious reasons, not as much information is available on low stands of the sea. The indications of low levels come from nickpoints and terraces observed in echosounders and submersible dives near the outer shelf and upper slope, and from submerged coral reefs. Worldwide occurrences of certain terrace levels indicates a connection with sea level rather than with any local structural control. Terraces are fairly commonly found at −130 meters, which corresponds with the last low at about 15,000 to 20,000 years B.P. [Jongsma (1970), for example]. Other terraces are found at still greater depths. In northern Australia Jongsma (1970) found terraces at 170 to 175 meters' depth and at about 200 meters. These terraces had been earlier identified with lowered sea level by van Andel and Veevers (1967). Veeh and Veevers (1970) date a −175-meter terrace in eastern Australia at 13,600 to 17,000 years B.P., suggesting that perhaps at that time the sea level dropped to below the −130-meter depth usually considered. Jongsma (1970) dates the −200-meter terrace at an age of 170,000 years B.P. or greater. This deepest terrace apparently then corresponds to one of the early glacial lows. Pratt and Dill (1974) also provide convincing evidence for very low glacial sea levels. Their evidence comes from the Bering shelf and from southern Australia. Echosounding records show a continuous terrace at both locations at a depth of 240 meters. In southern Australia there is also an indication of a terrace at −200 meters. Pratt and Dill review evidence from other parts of the world for such considerable low stands of the sea. In dives off Baja California, Dill observed terrace features at a depth of about 220 meters. Sounding evidence from caves in Honduras and on the Bahama Banks indicates that they bottom out at more than 200 meters, again suggesting a very low sea level at some time. Shallow seamounts provide similar evidence. Bloom (1971) has suggested that, assuming little isostatic sinking, seamounts would make ideal "dipsticks" to record lowered sea levels. Schwartz (1972) indicated that Cobb and Bowie seamounts in the North Pacific both show evidence for a maximum lowering of the sea level to between −220 and −250 meters. Thus there is compelling evidence that during some early glacial maximum the sea level dropped a distance on the order of −240 meters, exposing most continental shelves and even part of the continental slope. If the data obtained by Jongsma (1970) at −200 meters is correct, this very low sea level occurred sometime before 170,000 years B.P.

One effect of the sea level changes has been a series of very rapid transgressions and regressions of the shoreline. Each regression took the shore to approximately the edge of the present continental shelf or deeper. It is interesting to contemplate what coastal processes must have been like at such a time. For example, there would have been little opportunity for wave refraction to occur, since the waves would pass over the steep continental slope prior to reaching the shore. In general, the angles of wave approach to the shoreline could therefore be expected to have been greater, and the longshore currents and sediment transport would have been enhanced at such times. Each successive transgression caused the shoreline to migrate across the continental shelf. It is generally considered that the broad, relatively flat coastal plain and continental shelf complex resulted from this series of transgressions (Curray, 1969). Both erosion and sediment deposition are concentrated in the nearshore, and the repeated migration of this region across the continental margin would act to bevel it off to a flat plane, as observed.

During the interval of rapid sea level rise, the shoreline migrated at a sufficiently rapid rate so that the normal rates of sand supply were not sufficient for appreciable beach deposits or ridges to develop. The beach sediments of the transgression are represented by a thin veneer of sand spread over the continental shelves, presently being progressively covered by recent deeper-water sediments (Curray, 1965). It is only in the last 4,000 to 7,000 years that littoral deposits could develop, after sea level had reached nearly its present level and the rate of rise decreased. The sea level about 7,000 years B.P., when the change in rate occurred, corresponds to the bases of many barrier islands (Curray, 1969). Apparently at that time the beaches were able to maintain their positions and build upward at sufficient rates to keep up with the then-slower rise in sea level. The last rapid rise in sea level left the coasts very much out of equilibrium with respect to the sediment supplies, the prevailing wave action, and the shoreline configuration. In the past 4,000 years of near-stable sea level, the coastal processes have been trying to reestablish some sort of equilibrium by cutting back portions of the coast that project too far seaward, blocking bays with spits and other beach forms, and in general working to straighten the coastline. At the same time, estuaries formed by the drowning of river valleys by the risen sea level are now progressively being filled with muds and sand.

In our discussions we have considered that the sea level changes are produced by the advance and retreat of glaciers. Sea level changes could also be produced by temperature variations of the ocean water or by alterations in the total capacity of the ocean basins. It is believed that the total water on and around the earth has remained nearly the same since early geologic time, only a very small amount of "juvenile" water having been added by volcanic action.

A fall of 1°C in the mean temperature of the ocean water would lower sea level by about 2 meters. Estimates of Pleistocene variations in water temperature, based on oxygen isotope measurements on forams in ocean sediments, are within 5°C of the present temperature (Emiliani, 1970). Therefore, temperature changes could account for only about 10 meters' variation in sea level—small (but not negligible) compared to the observed changes (Fairbridge, 1960).

Epeirogenic sinking of an ocean basin would increase its capacity and cause a worldwide lowering of sea level. This is difficult to assess for the Quaternary. It would probably cause some long-term drift or trend in highs or lows of the sea levels caused mainly by glacial advances and retreats. Hays and Pitman (1973) have demonstrated that the worldwide transgressions and regressions of the middle to upper Cretaceous period (approx. 90 million years B.P.) may have been caused by contemporaneous pulses in the rapid spreading at most of the mid-oceanic ridges. Apparently episodes of rapid spreading caused the ridges to expand, reducing the volumetric capacity of the oceans. The Cretaceous was not an ice age, so that changes of sea level arising from that cause did not mask variations because of the ocean's capacity. These variations were on a much longer time scale than were the advances and retreats during the Quaternary.

Deposition of land-derived sediments in the sea would also reduce the capacity of the ocean and would cause a rise in sea level. Present estimates of denundation rates of the continents would account for a sea level rise of 3 mm/century (Fairbridge, 1960), much less than the 10- to 20-cm average rise observed.

Factors other than glacial advance and retreat therefore appear to have only a small influence on the level of the sea, at least over spans of time as short as 3 million years. This is further confirmed by the correspondence between the glacial stages on the continents, ice cores taken in Greenland and Antarctica, and the history of sea level changes. Each time the glaciers advance the sea level drops. The climatic changes that lead to the glacier variations are believed by many to be due to variations in the amount of solar radiation striking the earth. This amount changes in a systematic fashion because of the tilting of the earth's axis, precession of the equinoxes, and variation in eccentricity of the earth's orbit. These variations have periods of 41,000, 21,000, and 90,000 years, respectively. Milankovitch (1938) calculated the fluctuations in radiation received at various latitudes for the past million years and showed that a sequence of warmer and colder periods occurred during that time. These alterations have been correlated with interglacial and glacial phases. For example, the studies of Broecker et al. (1968) on Barbados and of Veeh and Chappell (1970) in New Guinea of interglacial high stands of sea level support the Milankovitch hypothesis of astronomical control over ice age periodicity. Although the Milankovitch hypothesis seems to explain the periodicity, it gives no reason for the initial onset of glaciation.

Block (1965) has hypothesised that volcanic and terrestrial dust fallouts onto the polar ice caps would decrease their albedo (reflectivity) and hence would cause some ice melting. In this way he attempts to explain some of the historic and prehistoric changes in sea level that have been rather erratic and rapid. Other minor fluctuations, such as the sea level drop between 1800 and 1850, appear to be purely of climatic origin whose cause is little understood.

COASTAL READJUSTMENTS TO A SEA LEVEL RISE

Some of the effects of the long-term sea level changes on the coast have already been discussed. The repreated regression-transgression cycles probably produced the relatively flat continental shelves and coastal plains found throughout the world. The last rise in sea level within the past 20,000 years also left the coasts out of equilibrium, the shoreline configuration not being adjusted to the prevailing wave climate. Within the past 7,000 years we have seen an attempt by the coast to make such a readjustment, cutting back headlands and in general straightening the coast. Estuaries have been filling.

In areas where there is tectonic or isostatic uplift of the coast, it is questionable whether any sort of equilibrium can be achieved. This may even be true elsewhere as the water level continues to rise, however slowly. In low-lying coastal areas such as the eastern U.S. and Gulf coasts, this rise can provide a threat to coastal communities in the long run. Apparently, with a rising sea level some of the barrier islands migrate shoreward to "keep their heads above water." Storms remove sand from the beach, wash it over the barrier island, and deposit it on the bay or lagoon side of the barrier.

Under natural conditions the barrier would migrate slowly landward at a rate governed by the sea level rise. This natural process provides a problem to man, since he has unwisely built homes, hotels, and so on immediately behind the beaches, so that the migration becomes "unacceptable" and seawalls, riprap, and groins must be built.

The Great Lakes in North America provide a good example of the effects of a water level rise on the coast. The water level in the lakes fluctuates because of changes in the area's rainfall. Recently the water level reached a particularly high level, submerging many beaches that were unable to build upward. The result has been a considerable acceleration in erosion of the cliffs, composed mainly of loose glacial materials. Again, many homes have been destroyed in the process. It obviously behooves us to know more about variations in water level on the Great Lakes, and on the ocean, and to learn more about the expected response of the coast to such changes.

Bruun (1962) has evolved a conceptual model for beach profile adjustment to a rise in sea level where sand is shifted offshore. Bruun maintains that as the sea level rises, material will be eroded from the upper beach and deposited offshore as shown in Figure 6-10. It is envisioned that there is a shoreward displacement of the beach

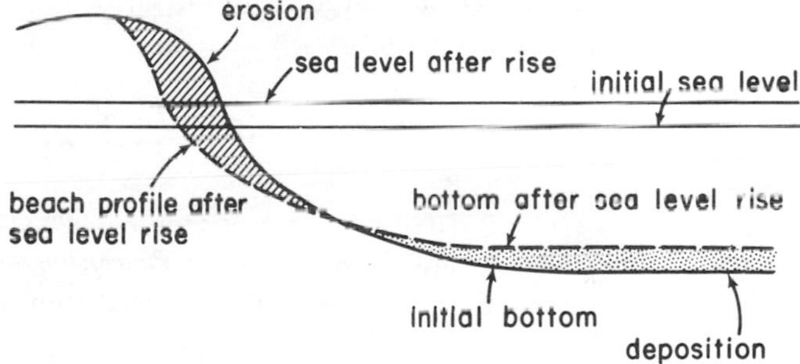

Figure 6-10 Effects of the rise in sea level on coastal erosion and offshore deposition as envisioned by Bruun (1962). He hypothesizes that sediment deposition offshore keeps pace with the rising sea level so that the depth remains constant. [*After* Bruun (1962)]

profile as the upper beach is eroded and that the rise in the nearshore bottom from the deposition keeps pace with the rise in sea level such that the water depth remains constant. Such changes have been partially verified in small-scale laboratory experiments by Schwartz (1965).

There is little information on the effects of the seasonal variations in sea level on the beach profile. LaFond (1938) has demonstrated a remarkable correspondence between the monthly sea level at La Jolla, California, and the level of the beach close to the shore (Figure 6-11). It is seen that a rise in the water level is accompanied by a rise in the sand level on the beach, opposite to the conception of Bruun (1962). The problem with the correspondence shown is that there are seasonal variations in the beach level because of changes in the nature of the waves reaching the beach at different times of the year. Such profile shifts are discussed at length in Chapter 11. The correspondence shown in Figure 6-11 may therefore just be coincidental, the annual cycle of the wave climate causing the changes in sand level, not the variations in the mean sea level. The correspondence is nevertheless remarkable; unfortunately, however, it has not been studied further.

Careful investigation of the effects of sea level changes on the beach are required.

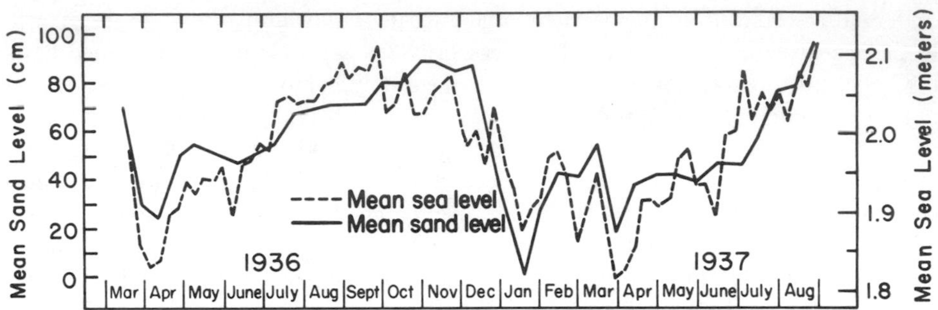

Figure 6-11 A remarkable correspondence between the monthly sea level at La Jolla, California, and the level of the beach near the shoreline. However, as discussed in the text, the correspondence may be only a coincidence. [*From* LaFond (1938)]

This will be difficult on natural beaches, since, in general, variations due to other causes such as varying waves and tides will overshadow the effects from the sea level change unless it is large, as in the Great Lakes. Large-scale wave tank studies would be fruitful.

REFERENCES

BLOCH M. R. (1965). A hypothesis for the change of ocean levels depending on the albedo of the Polar ice caps. *Palaeogeography, Palaeoclimatology, Palaeoecology*, 1: 127–42.

BLOOM, A. L. (1967). Pleistocene shorelines: a new test of isostasy: *Geol. Soc. Am. Bull.*, 78: 1477–94.

——— (1971). Glacial-eustatic and isostatic controls of sea level: in *Late Cenozoic glacial ages*, edited by K. K. Turekian, Yale Univ. Press, New Haven, Conn., pp. 355–78.

BOURCART, J. (1952). *Les frontièrs de l'océan*: Paris.

BROECKER, W. S., D. L. THURBER, J. GODDARD, T.-L. KU, R. K. MATTHEWS, and K. J. MESOLELLA (1968). Milenkovitch hypothesis supported by precise dating of coral reefs and deep-sea sediments. *Science*, 159: 297–300.

BRUUN, P. (1962). Sea level as a cause of shore erosion. *J. Waterways and Harbors Div., ASCE*, 88: 117–30.

CHAPPELL, J. (1974). Geology of coral terraces, Huon Peninsula, New Guinea: a study of Quaternary tectonic movements and sea-level changes. *Geol. Soc. Am. Bull.*, 85: 553–70.

CHATERJEE, S. P. (1961). Fluctuations of sea level around the coast of India during the Quaternary period. *Zeit. Geomorph.*, Supp., 3: 48–56.

COOKE, C. W. (1971). American emerged shorelines compared with levels of Australian marine terraces. *Geol. Soc. Am. Bull.*, 82: 3231–34.

CURRAY, J. R. (1960). Sediments and history of Holocene transgression, continental shelf, northwest Gulf of Mexico. *Recent sediments Northwest Gulf of Mexico*, edited by F. P. Shepard, F. B. Phleger, and Tj. H. van Andel, *Am. Assoc. Petrol. Geol.*, pp. 221–66.

——— (1961). Late Quaternary sea level: A discussion. *Geol. Soc. Am. Bull.*, 72: 1707–12.

——— (1965). Late Quaternary history, continental shelfs of the United States: in *The quaternary of the United States*, edited by H. E. Wright and D. G. Frey, Princeton Univ. Press, pp. 723–35.

——— (1969). History of continental shelves. In *The new concepts of continental margin sedimentation*, edited by D. J. Stanley, AGI Short Course Lecture Notes, 8 pp.

CURRAY, J. R., F. P. SHEPARD, and H. H. VEEH (1970). Late Quaternary sea-level studies in Micronesia: CARMARSEL expedition. *Geol. Soc. Am. Bull.*, 81: 1865–80.

DONN, W. L., W. R. FARRAND, and M. EWING (1962). Pleistocene ice volumes and sea-level lowering. *Jour. of Geol.*, 70: 206–14.

EMERY, K. O., and L. E. GARRISON (1967). Sea levels 7,000 to 20,000 years ago. *Science*, 157, no. 3789: p. 684–87.

EMERY, K. O., R. L. WIGLEY, A. S. BARTLETT, M. RUBIN, and E. S. BARGHOORN (1967). Freshwater peat on the continental shelf. *Science*, 158: pp. 1301–7.

EMILIANI, C. (1970). Pleistocene paleotemperatures. *Science*, 168; 822–25.

FAIRBRIDGE, R. W. (1960). The changing level of the sea. *Scientific American*, 202: 70–79.

——— (1961). Eustatic changes in sea level. *Physics and chemistry of the earth*, 4: 99–185.

FUJII, S., and N. FUJI (1967). Postglacial sea level in the Japanese Islands, *Jour. Geoscience*, Osaka City Univ., 10: 43–51.

GILL, E. D. (1961). Changes in the level of the sea relative to the level of the land in Australia. *Zeit. Geomorph.*, Supp., 3: 73–79.

GUILCHER, A. (1969). Pleistocene and Holocene sea level changes. *Earth-Science Review*, 5: 69–97.

HAILS, J. R. (1965). A critical review of sea-level changes in eastern Australia since the last glacial. *Austral. Geog. Studies*, 3: 63–78.

HAUGHTON, S. H. (1963). *The stratigraphic history of Africa south of the Sahara*: Hafner, New York, 365 pp.

HAYS, J. D., and W. C. PITMAN III (1973). Lithospheric plate motion, sea level changes and climatic and ecological consequences. *Nature*, 246, no. 5427: 18–22.

HICKS, S. D. (1968). Long-period variations in secular sea level trends. *Shore and Beach*, 36, no. 1, 32–36.

——— (1972). On the classification and trends of long period sea level series. *Shore and Beach*, 40, no. 1: 20–23.

JELGERSMA, S., and A. J. PANNEKOET (1960). Post-glacial rise of sea-level in the Netherlands (a preliminary report.) *Geol. en Mijnbouw*, 39. 201–7.

JENNINGS, J. N. (1961). Sea level changes in King Island, Bass Strait. *Zeit. Geomorph.*, Supp., 3: 80–84.

JONGSMA, D. (1970). Eustatic sea level changes in the Arafura Sea. *Nature*, 228, no. 5267: 150–151.

LAFOND, E. C. (1938). Relationship between mean sea level and sand movements. *Science*, 88, no. 2274: 112–13.

——— (1939) Variations of sea level on the Pacific Coast of the United States. *Jour. of Marine Res.*, 2: 17–29.

LISITZIN, E. (1955). Les variations annuelles du niveau des océans. *Bull. d'Inform.*, Comité d'Océanographie et d'Etudes des Côtes, no. 6.

——— (1974) *Sea level changes.* Elsevier, Amsterdam, 288 pp.

LISITZIN, E., and J. PATTULLO (1961). The principal factors influencing the seasonal oscillation of sea level. *J. Geophys. Res.*, 66: 845–52.

MARMER, H. A. (1952). Changes in sea level determined from tide observations. *Proc. 2nd Conf. Coastal Engr.*, pp. 62–67.

MEADE, R. H., and K. O. EMERY (1971). Sea level as affected by river runoff, Eastern United States. *Science*, 173: 425–427.

MILANKOVITCH, M. (1938). *Handbuch der Geophysik*, edited by Koppen and Geiger, 9: 593.

MILLIMAN, J. D., and K. O. EMERY (1968). Sea levels during the past 35,000 years. *Science*, 162: 1121–23.

MÖRNER, N-A (1969). Eustatic and climatic changes during the last 15,000 years. *Geologie en Mijnbouw*, 48: 389–99.

NAMIAS, J. and J. C. K. HUANG (1972). Sea level at southern California: a decadal fluctuation. *Science*, 177: 351–53.

OAKS, R. A., and N. K. COCH (1963). Pleistocene sea levels, southeastern Virginia. *Science*, 140: 979–84.

PATTULLO, J., W. MUNK, R. REVELLE, and E. STRONG (1955). The seasonal oscillation in sea level. *J. Marine Res.* 14: 88–155.

PATTULLO, J. G. (1966). Seasonal changes in sea level. In *The Sea*, ed. by M. N. Hill, Interscience Publ., New York, pp. 485–96.

PRATT, R. M., and R. F. DILL (1974). Deep eustatic terrace levels: further speculations: *Geology*, 2, no. 3: 155–59.

REDFIELD, A. C. (1967). Postglacial change in sea level in the western North Atlantic Ocean. *Science*, 157: 687–92.

RICHARDS, H. G., and R. W. FAIRBRIDGE (1965). Annotated bibliography of Quaternary shorelines, 1945–64. Academy of Natural Science, Philadelphia, Publ. no. 6.

ROSSITER, J. R. (1962). Long-term variations in sea level. In *The Sea*, edited by M. N. HILL, Interscience Publ., New York, 1: 590–610.

RUSSELL, J. R. (1963). Recent regression of tropical cliffy coasts. *Science*, 139: 9–15.

SCHOLL, D. W., and M. STUIVER (1967). Recent submergence of Southern Florida: a comparison with adjacent coasts and other eustatic data. *Geol. Soc. Am. Bull.*, 78: 437–54.

SCHWARTZ, M. (1965). Laboratory study of sea-level rise as a cause of shore erosion: *J. of Geol.*, 73: 528–34.

——— (1972). Seamounts as sea-level indicators. *Geol. Soc. Am. Bull.*, 83: 2975–80.

SHEPARD, F. P. (1963). Thirty-five thousand years of sea level. In *Essays in marine geology, in honor of K. O. Emery*, Univ. Southern Calif. Press, Los Angeles, pp. 1–10.

SHEPARD, F. P., and J. R. CURRAY (1967). Carbon-14 determination of sea level changes in stable areas. In *Progress in Oceanography*, "The Quaternary History of the Ocean Basins," Pergamon Press, Oxford, 4: 283–91.

VAN ANDEL, Tj. H., and J. LABOREL (1964). Recent high relative sea level strand near Recife, Brazil. *Science*, 145, no. 3632: 580–81.

VAN ANDEL, Tj. H., and J. J. VEEVERS (1967). Morphology and sediments of the Timor Sea. *Bull. Bureau Miner. Resour. Geol. Geophys. Australia*, 83: 173 pp.

Veeh, H. H., and J. Chappell (1970). Astronomical theory of climatic change: support from New Guinea. *Science*, 169: 862–65.

Veeh, H. H., and J. J. Veevers (1970). Sea level at −175 m off the Great Barrier Reef 13,600 to 17,000 years ago. *Nature*, 226, no. 536: 536–37.

Whitmore, F. C., K. O. Emery, H. B. S. Cooke, and D. J. Swift (1967). Elephant teeth from the Atlantic continental shelf. *Science*, 156, no. 3781: 1477–81.

Wyrtki, K. (1973). Teleconnections in the equatorial Pacific Ocean. *Science*, 180: 66–68.

Zeuner, F. E. (1959). *The pleistocene period*, Hutchinson, London, 447 pp.

Chapter 7

NEARSHORE CURRENTS

It [a rip current] is a sort of river, a stream showing on the surface deep and powerful, easily perceptible, running with the velocity of a mill race. So swift and powerful is it that a motorboat could not stem its sweeping current. It will carry brick, large rocks and even chunks of lead far out to sea. The most powerful swimmer will find himself helpless as a babe in its rushing grasp.

M. P. Hite
Science, 62 (1925)

Currents in the nearshore zone are important in several respects. When of considerable strength they can be a distinct hazard to swimmers, being responsible for many drownings each year. However, the nearshore currents have their beneficial aspects as well, causing a flushing of the nearshore waters, replacing them with generally cleaner offshore water. It will be seen in the following chapter that nearshore currents are also responsible for the longshore transport of sand on beaches.

There are two wave-induced current systems in the nearshore zone which dominate the water movements in addition to the to-and-fro motions produced by the waves directly. These are: (1) a cell circulation system of rip currents and associated longshore currents, and (2) longshore currents produced by an oblique wave approach to the shoreline.

The cell circulation is shown schematically in Figure 7-1. The most apparent feature of this circulation is the *rip currents*: strong, narrow currents that flow seaward from the surf zone. Rip currents receive considerable attention in the news; they are the so-called riptides often responsible for swimming fatalities, as they may sweep a person out to deep water. Rip currents are relatively narrow, so that anyone caught in one should swim parallel to shore for a time to escape the current before attempting to swim back to shore. Surfers may be observed using the rip currents for a free ride

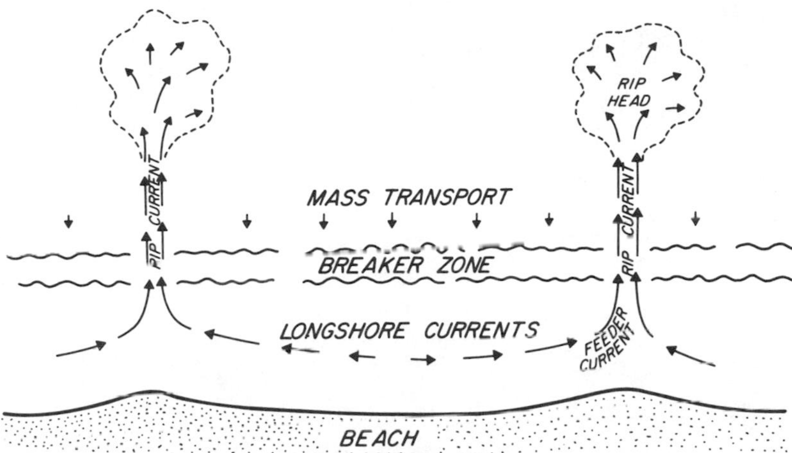

Figure 7-1 The nearshore cell circulation consisting of (1) feeder longshore currents, (2) rip currents, and (3) a slow mass transport returning water to the surf zone. [*After* Shepard and Inman (1950)]

seaward of the breakers. Rip currents or rips are fed by a system of longshore currents (Figure 7-1) which increase in velocity from zero about midway between two adjacent rips, reaching a maximum just before turning seaward into the rip. To make up for the water moving seaward out the rip currents, there must also be a slow mass transport of water moving shoreward through the breaker zone in the area between the rips. The slow mass transport, the feeding longshore currents, and the rip currents taken together form a cell circulation system in the nearshore zone (Shepard and Inman, 1951). In addition to being a hazard to swimmers, the cell circulation may be important in renewing the water in the nearshore zone and removing pollutants such as mud or sewage that may be dumped there. Inman, Tait, and Nordstrom (1971) present models for the flushing of the nearshore by rip currents. The first half of this chapter will examine the mode of generation of this important cell circulation system.

It is well known that when waves approach a straight coastline at an oblique angle, a longshore current is established flowing parallel to the coastline in the nearshore zone. The velocity of the current decreases quickly to zero outside the breaker zone, so it is clearly wave-induced and cannot be attributed to ocean currents or tides. This current is particularly significant in that it is responsible for the net transport of sand or other beach material along the shore. There have been a dozen of so theories proposed to account for the generation of these currents; the theories will be reviewed later in this chapter.

Commonly both the cell circulation and the longshore currents due to an oblique wave approach are present simultaneously. The current pattern actually observed then is, to a first approximation, the vector sum of the two current systems. This summation to obtain the observed current is diagramed in Figure 7-2. It is seen in Figure 7-2(c) that, under an oblique wave approach with both systems present, the current pattern

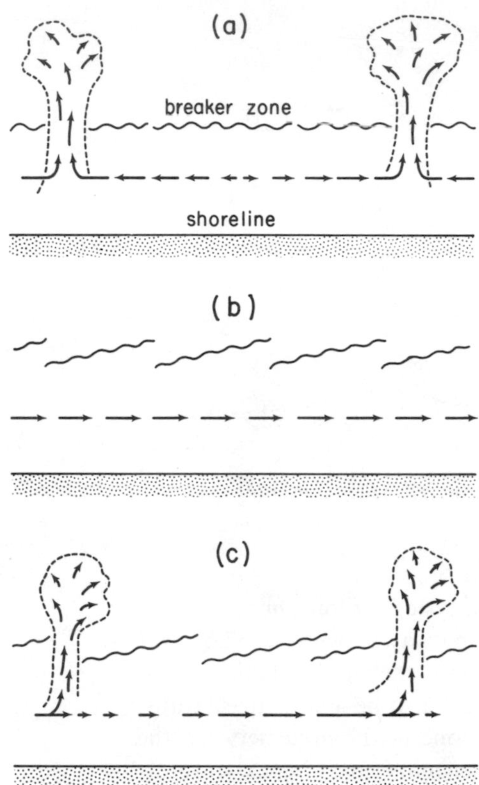

Figure 7-2 The summation of a cell circulation (a) with a longshore current from an oblique wave approach (b) resulting in the observed current pattern in the nearshore (c). [*After* Komar and Inman (1970)]

becomes asymmetrical, with the zero velocity node just updrift from the rip current and a velocity increase extending from there to the next rip current. It has been observed that under such conditions with an oblique wave approach, the cell circulation system sometimes moves slowly alongshore, the motion being made apparent by the migration of the rip currents (Bowen and Inman, 1969, p. 5,488).

NEARSHORE CELL CIRCULATION

The first scientific observations of rip currents were made by Shepard, Emery, and LaFond (1941). They found that the velocity of rip currents and the distance they flow seaward are related to the height of the incoming waves. They further recognized that the positioning of rip currents could be governed by offshore topography, the rip currents generally occurring away from areas of wave convergence—that is, away from the areas of highest waves.

Shepard and Inman (1950, 1951) showed that the water motion could be described in terms of the circulation cell with longshore currents feeding rip currents, as discussed above. This study also obtained the first comprehensive series of field measurements. It was recognized that although offshore topography and its effects on wave refraction may govern the positioning of rip currents along a beach, such circulation cells with rip currents could exist on long, straight beaches with regular bottom relief.

McKenzie (1958) demonstrated in his field studies that each incident wave system arriving at the beach forms a characteristic pattern of longshore currents and rip currents. He noted that with large waves only a few strong rips are produced, whereas when the waves are smaller the rips are weaker and more numerous.

The first attempts at explaining the generation of rip currents were based on the existence of an onshore mass transport of water associated with the waves. This water would pile up on the beach and provide a head for the outflowing currents. Due to the difficulty of making quantitative estimates of the onshore mass transport near the breakers as a function of the wave parameters, this approach was never particularly successful.

Bowen (1969*a*) and Bowen and Inman (1969) have demonstrated both theoretically and experimentally that variations in the *wave set-up*—the rise in the mean water level above the still-water level due to the presence of waves—may provide the necessary longshore head of water to drive the feeder longshore currents and produce the rip currents. The variation in the wave set-up is due to longshore variations in the incoming wave height—the higher the waves the greater the wave set-up. Therefore, the currents would flow from positions of highest breaker height and turn seaward as rip currents at positions of lowest wave height. The original variations in wave height could be produced by wave refraction; lacking this, Bowen and Inman (1969) demonstrated that interactions between the incoming waves with edge waves trapped within the nearshore region could also produce a systematic variation in wave height along the shore.

The following review summarizes Bowen's (1969*a*) development of the generation mechanism; his paper may be examined for the details. It will be recalled from Chapter 3 that associated with the wave motion is a *radiation stress*, an excess flux of momentum due to the presence of the waves (Longuet-Higgins and Stewart, 1964). If one considers waves arriving at the beach with their crests parallel to the shoreline, there will be a shoreward flux of momentum S_{xx} given by equation (3-25) as

$$S_{xx} = E\left[\frac{2kh}{\sinh(2kh)} + \frac{1}{2}\right] = E\left(2n - \frac{1}{2}\right) \tag{7-1}$$

where $k = 2/\pi L$, E is the wave energy density, L is the wave length, and h is the water depth. The x-coordinate is positive in the onshore direction, normal to the shoreline. In shallow water this becomes

$$S_{xx} = \frac{3}{2}E = \frac{3}{16}\rho g H^2 \tag{7-2}$$

since $n = 1$, and having used $E = \rho g H^2/8$ for the wave energy. This momentum flux entering the nearshore zone with the incoming waves cannot simply disappear but rather must be balanced by opposing forces which will dissipate the momentum. Part of the momentum may be reflected from the beach and travel seaward associated with the reflected wave. Another fraction may be dissipated as the force of frictional drag on the bottom. Important to our present considerations, Longuet-Higgins and Stewart (1963) have shown theoretically and Bowen, Inman, and Simmons (1968) have demonstrated experimentally that the radiation stress produces negative and positive changes in the mean water level in the nearshore region known respectively as wave *set-down* and *set-up*. The pressure gradient of the sloping water surface balances the change (gradient) of the incoming momentum according to

$$\frac{dS_{xx}}{dx} + \rho g(\bar{\eta} + h)\frac{d\bar{\eta}}{dx} = 0 \tag{7-3}$$

where $\bar{\eta}$ is the difference between the still-water level and the mean water level in the presence of the waves (the set-down or set-up). Seaward of the breaker zone the energy flux is constant (ECn = constant); Longuet-Higgins and Stewart (1963), using this along with equation (7-1), integrated equation (7-3) to obtain

$$\bar{\eta} = -\frac{1}{8}\frac{H^2 k}{\sinh(2kh)} \tag{7-4}$$

for the departure of the mean water level from the still-water level outside the breaker zone. It is seen that seaward of the breakers $\bar{\eta}$ is negative, so that there will be a *set-down*, a lowering of the mean water level to below the still-water level. Inside the breaker point it was assumed that the height of the broken wave, or bore, remains an approximately constant proportion of the mean water depth—that is,

$$H = \gamma(\bar{\eta} + h) \tag{7-5}$$

where $\gamma \simeq 0.8$. Using this along with equation (7-2) for S_{xx}, equation (7-3) yields the relationship

$$\frac{d\bar{\eta}}{dx} = \left(\frac{1}{1 + 8/3\gamma^2}\right) \tan\beta \tag{7-6}$$

The gradient of the water depth below the still-water level, $-dh/dx$, is identical to the beach slope, which is denoted by $\tan\beta$, β being the angle of beach slope. Since γ is nearly constant (Bowen, Inman, and Simmons, 1968), equation (7-6) indicates that in the surf zone the mean sea level slope $d\bar{\eta}/dx$ will be constant and proportional to the beach slope.

The laboratory measurements made by Saville (1961) were found to be in reasonable agreement with the above theoretical development of Longuet-Higgins and Stewart (1963). The more detailed laboratory measurements of Bowen, Inman, and Simmons (1968) show that the theory predicts remarkably well the set-down outside the breaker zone and the set-up inside the surf zone (Figure 7-3).

As the waves travel into shallow water they break at the critical instability ratio $\gamma_b = H_b/h_b \simeq 0.8$ to 1.25, depending on the beach slope (Chapter 3). Because of this the larger waves will break in deeper water than the smaller waves and the set-up begins further seaward where the larger waves occur. Although the gradients of the set-up (equation 7-6) are approximately the same for both the large and small breakers, since the set-up begins further seaward where the larger breakers are found the actual water level rise associated with the high waves is considerably greater than

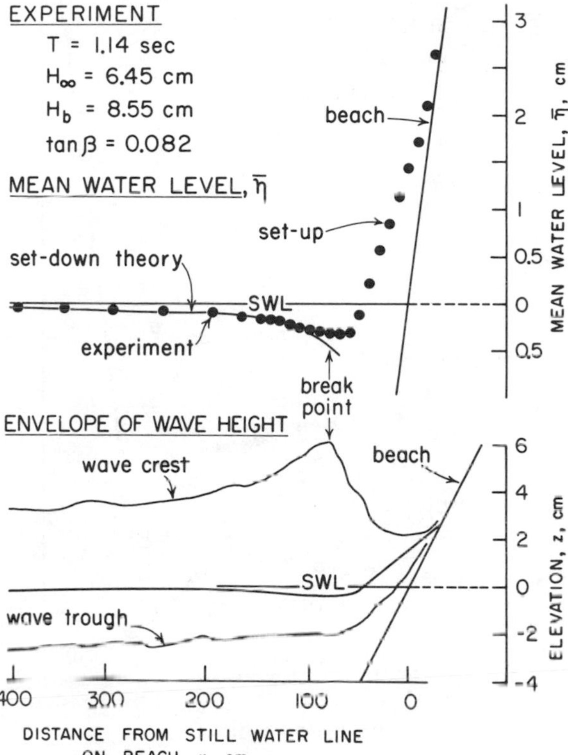

Figure 7-3 Wave set-down and set-up in the nearshore, departures of the mean water level from the still-water level (SWL), produced by the radiation stress (momentum flux) of the incoming waves. The theoretical curve shown for the set-down refers to equation (7-4). [*After* Bowen, Inman, and Simmons (1968)]

the rise due to the lower waves. This dependence on the wave height is shown in Figure 7-4, taken from Bowen (1969*a*). Inside the surf zone, then, the mean water level is higher shoreward of the larger breakers than it is shoreward of the small waves. Therefore a longshore pressure gradient exists which will drive a longshore current from positions of high waves to adjacent positions of low waves.

It would appear from Figure 7-4 that since the set-down outside the larger waves is greater, there would be a similar tendency for water to flow from positions of low breakers to positions of high breakers just outside the break point, opposite in direction to the current generated within the surf zone. It turns out, however, that the y-component of the radiation stress, which has not been included until now, prevents this latter current and at the same time enhances the current within the surf zone. It will be recalled from Chapter 3 that in addition to the S_{xx}-component of the radiation stress, there is a S_{yy}-component—that is, a momentum flux acting parallel to the wave crests, in this case acting parallel to the shoreline. This component was seen to be given by

$$S_{yy} = \frac{1}{8}\rho g H^2\left[\frac{kh}{\sinh(2kh)}\right] \tag{7-7}$$

Since the wave height H varies alongshore, S_{yy} will similarly vary and there will exist

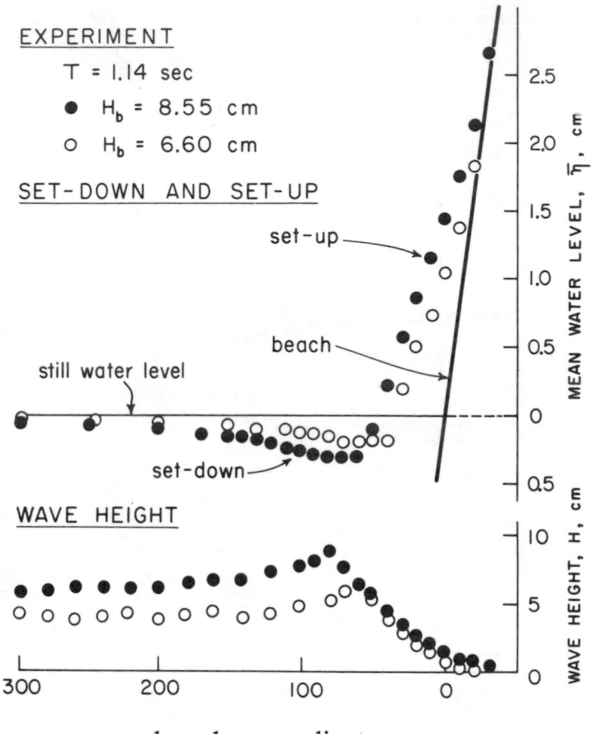

Figure 7-4 Wave set-down and set-up for two different initial wave heights. Note that the larger wave height causes a greater set-down, and that although the water slope of the set-up is approximately the same for both wave heights, the water level is higher shoreward of the larger breakers, since the set-up begins in somewhat deeper water. [*After* Bowen (1969*a*)]

a longshore gradient

$$\frac{\partial S_{yy}}{\partial y} = \frac{1}{4}\rho g H \frac{\partial H}{\partial y}\left[\frac{kh}{\sinh(2kh)}\right] \tag{7-8}$$

The longshore pressure gradient outside the breaker zone, due to the differences in the mean water level brought about by contrasts in wave set-down, is given by

$$\rho g(\bar{\eta} + h)\frac{\partial \bar{\eta}}{\partial y} = \rho g(\bar{\eta} + h)\frac{1}{4}\frac{kH}{\sinh(2kh)}\frac{\partial H}{\partial y}$$

$$\simeq -\frac{1}{4}\rho g H \frac{\partial H}{\partial y}\left[\frac{kh}{\sinh(2kh)}\right] \tag{7-9}$$

having used equation (7-4) to determine $\partial\bar{\eta}/\partial y$ and using $\bar{\eta} \ll h$ outside the surf zone. Comparing equations (7-8) and (7-9), it is seen that they differ only in sign. Therefore, as determined by Bowen (1969*a*), the longshore variation in the wave set-down outside the breaker zone is balanced by the gradient in the *y*-directed radiation stress ($\partial S_{yy}/\partial y$), so that no net force exists outside the breaker zone to produce circulation. It is a different matter within the surf zone where both the longshore variation in the set-up and the longshore gradient in the radiation stress act in the same direction, away from positions of high waves and toward areas of low waves. Within the surf zone the two forces combine to produce the observed flow of water away from the regions of high waves and toward the positions of low waves. The flow turns seaward as a rip current where the waves and the set-up is lowest and the longshore currents converge.

Applying the equations of motion and continuity, Bowen (1969*a*) demonstrated that the above driving forces within the surf zone would produce a circulation pattern similar to the observed cell circulation. In his solutions he obtained strong longshore currents confined to the surf zone, rip currents in regions of low wave height, and a weak return flow beyond the breaker zone to complete the cell. The introduction of nonlinear terms into the equations of motion, as first suggested by Arthur (1962), causes the outward-flowing rip current to become narrower with increasing Reynolds number.

The cell circulation therefore depends primarily on the existence of variations in the wave height along the shore. There are two principal ways in which such variations can be produced. The most obvious is by wave refraction, which may concentrate the wave rays in one area of the beach, causing high waves, and at the same time spread the rays in an adjacent area of the beach and so produce low waves. The positions of the rips and the overall cell configuration will then be governed by wave refraction and hence by the offshore topography. A good example of such a control occurs at Scripps Beach near Scripps Institution of Oceanography, La Jolla, California. As shown in Figure 7-5, a submarine canyon is present a short distance offshore.

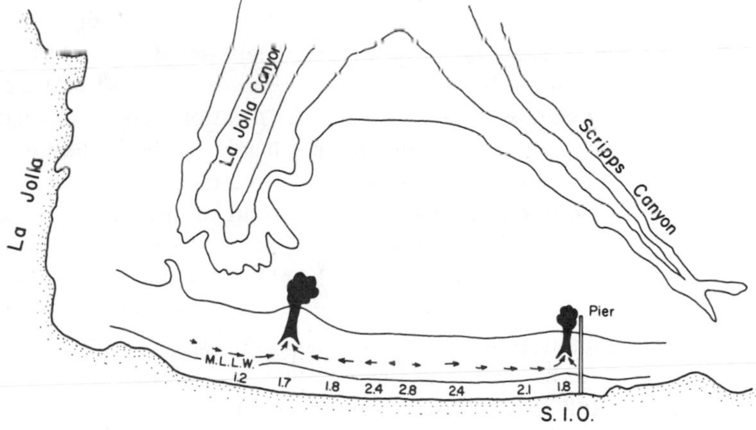

Figure 7-5 Rip currents at Scripps Beach, La Jolla, California, produced by a longshore variation in wave breaker heights caused by wave refraction over the offshore submarine canyons. The numbers along the shoreline refer to measured values of the breaker heights in meters. [*After* Shepard and Inman (1951)]

This canyon is important in producing wave refraction, causing a spreading of the wave rays on the beach shoreward of the canyon position. This results in low waves in the lee of the canyon and larger waves to either side. A strong rip current is always present in the canyon lee, shifting position only slightly as waves arrive from different deep-water directions.

Shepard and Inman (1950, 1951) noted that circulation cells with rip currents exist on long, straight beaches with regular bottom topography, clearly not the product of wave refraction. Bowen and Inman (1969) have shown both theoretically and experimentally in a wave basin that the ordinary incident swell waves may generate standing edge waves on the beach that have the same period as the incoming waves. The interaction or summation of the incoming and edge waves produces alternately high and low breakers along the shoreline and therefore gives rise to a regular pattern of circulation cells with evenly spaced rip currents.

Let us examine the motions of the standing edge waves, first by themselves, and then in interaction with the incoming ocean swell according to Bowen and Inman (1969). Edge waves are generally standing waves with crests normal to the shoreline and wave lengths from crest to crest parallel to the shoreline; that is, they are strung out along the length of the beach, opposite in orientation to the incoming swell waves. As shown in Figure 7-6, there will be alternate positions of nodes where there is no observable up-and-down motion of the water surface due to the edge waves, and antinodes where the full edge wave height is observed as the up-and-down motion. Edge wave oscillations may be best observed as a "run-up" on the sloping beach face. Several offshore variations in edge wave height are possible, depending on which edge wave is present; but in general the height is a maximum at the shore and decreases rapidly offshore, becoming negligible a short distance outside the breakers.

The important factor in the interactions between the edge waves and the incoming ocean swell waves in the generation of cell circulation is that they have the same period. Because of this, the positions where the two wave sets are in phase—or conversely, completely out of phase—remains stable. Consider the superposition of a simple uniform wave train, with a constant height alongshore, with the edge waves of the same period in the nearshore region. This superposition is diagramed in Figure 7-7 for the breaking point of the incoming waves. At the instant of breaking, the standing edge wave at one antinode position may be in phase with the incoming wave, so that they add to enhance the height of the breaking wave. At the same time, at the next antinode along the beach the incoming wave and the edge wave would be 180 degrees out of phase and so would subtract to produce a lower breaker height. Only at the positions of the nodes, where the edge wave system makes no contribution to the height, would the true height of the incoming wave be observable. Since the input waves and edge waves have the same period, the large or small breakers persistently remain at fixed antinode positions, every other antinode position being the site of large breakers and the alternate antinodes being the locations of small breakers. The result is a consistent longshore variation in the breaker height, with a regular alternation of high and low breakers. This in turn will produce a regular pattern of rip currents and cell circulation (Figure 7-8), the rips being found in the positions of low breakers—that is, at every other antinode of the causative edge waves (Bowen and Inman, 1969).

Eckart (1951) and Ursell (1952) investigated edge waves theoretically but obtained somewhat different solutions. Both considered edge waves generated on a uniformly sloping beach. For low beach slopes the two solutions give equivalent results, but as the slope increases the assumptions made by Eckart become less applicable and the formulation of Ursell can be expected to give more accurate results. For the length of the edge waves on a beach of slope angle β, with an edge wave period

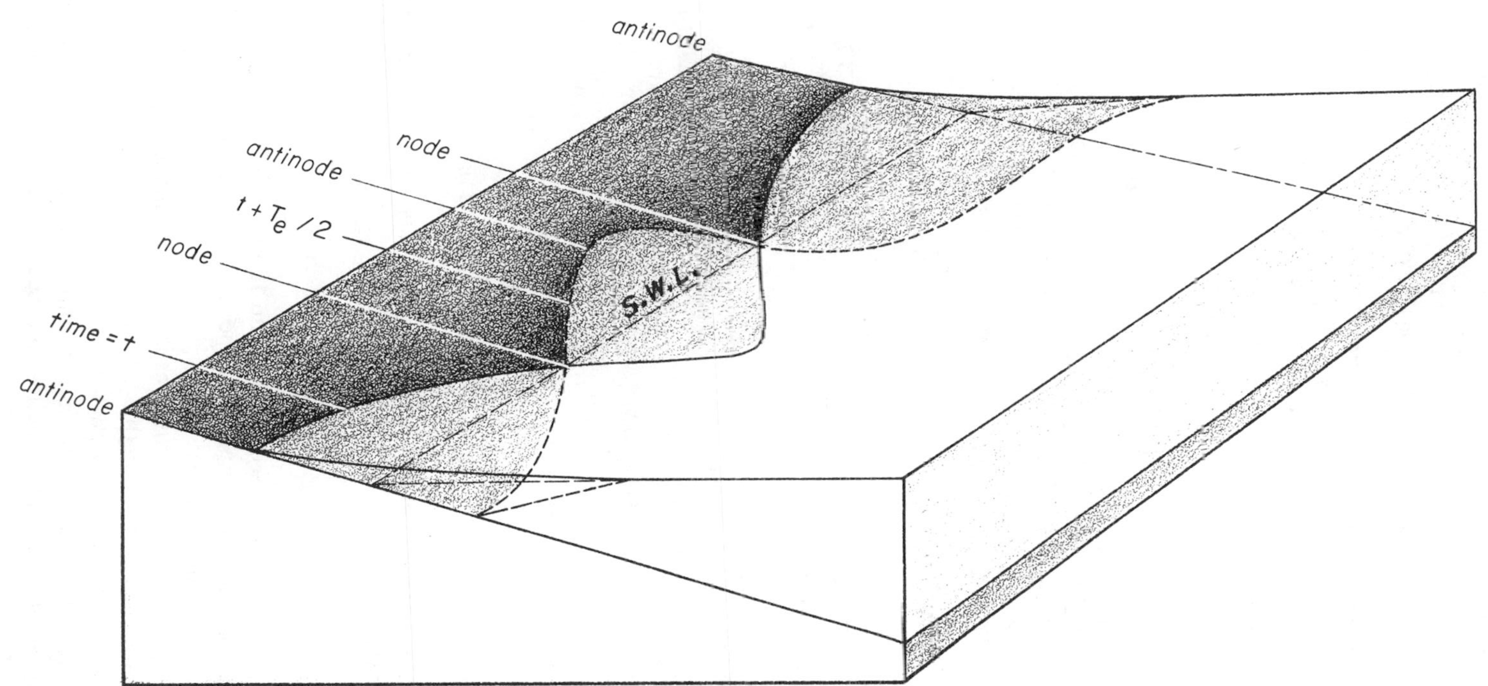

Figure 7-6 Schematic diagram of one type of standing edge wave ($n = 0$) consisting of nodes where no motion is observed and antinodes where the full motion occurs. The dashed line shows the runup on the beach due to the edge wave, one-half period ($T_e/2$) later than the runup given by the solid line.

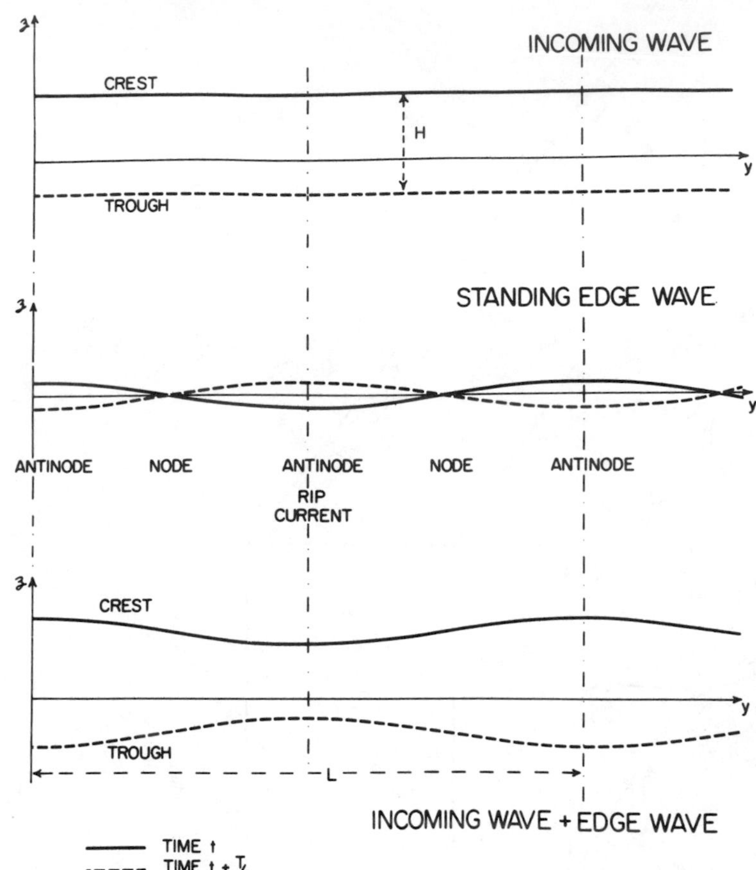

Figure 7-7 Addition of the incoming swell wave and the standing edge wave at the breaker position to give a longshore variation in the observed breaker height. The net height is greatest where the edge wave and incoming wave are in phase and lowest where they are 180 deg out of phase. [*From* Bowen and Inman (1969)]

T_e, Ursell obtained

$$L_e = \frac{g}{2\pi} T_e^2 \sin [(2n + 1)\beta] \tag{7-10}$$

where n is the offshore modal number, which may have integral values 0, 1, 2, . . . , limited by the cutoff

$$(2n + 1)\beta < \frac{\pi}{2} \tag{7-11}$$

As has already been discussed, Bowen and Inman (1969) found that for rip current generation the edge waves must have the same period as the incoming swell waves, so that $T_e = T$ in this instance.

Since the rip currents occur at every other antinode, the rip current spacing will be equal to the edge wave length L_e. It is seen that this spacing is mainly dependent upon the wave period and, to a smaller degree, the beach slope. Because of the presence of the modal number n in the relationship, there are several possible values of L_e for each given combination of period and beach slope. The theoretical development

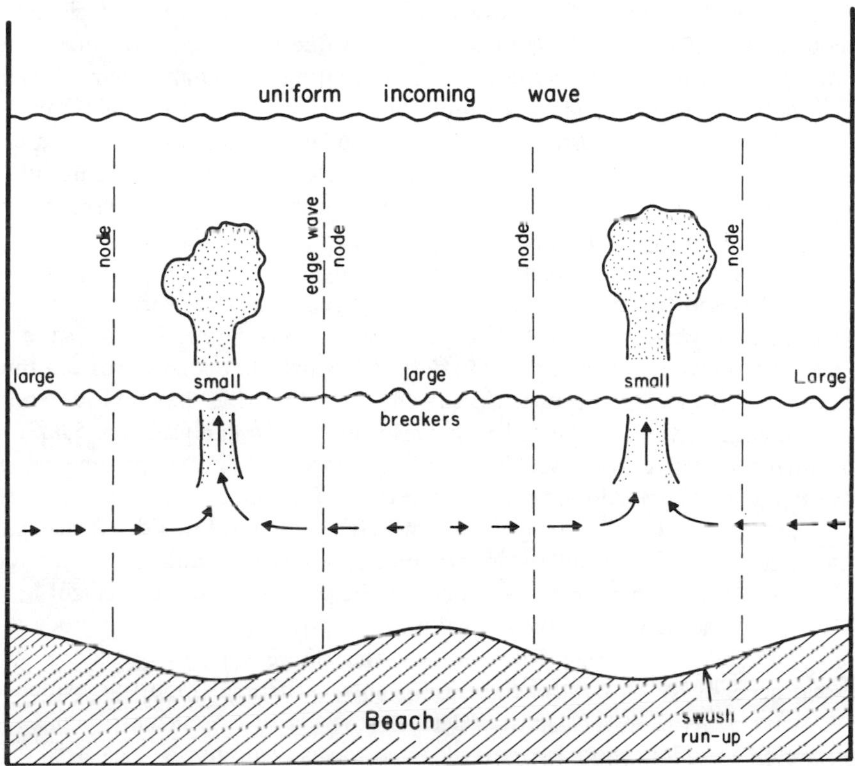

Figure 7-8 Positioning of the rip currents where the breaker height is smallest—that is, where the edge wave and incoming wave are 180 deg out of phase.

and observations of Bowen and Inman (1969) suggest that it may be possible to determine which mode will predominate for a given set of wave conditions from the value of the dimensionless ratio

$$\frac{X_b(2\pi/T)^2}{g \tan \beta} \tag{7-12}$$

where X_b is the width of the surf zone.

Extra boundary conditions are imposed on the edge wave system in the laboratory wave basin experiments, such as those conducted by Bowen and Inman (1969), that are not normally present on natural beaches. Walls at both ends of the limited beach require that

$$L_e = \frac{2\Lambda}{m} \tag{7-13}$$

where Λ is the longshore length of the beach (the distance between the walls), and m is the longshore modal number which may have integral values 1, 2, 3, ... In the laboratory the edge waves therefore have both offshore and longshore modal numbers.

If the period of the generating waves in the wave basin is such that it yields an L_e from equation (7-10) that corresponds to an L_e value that is possible according to equation (7-13), then the edge waves are at a resonating period and their interactions with the incoming waves produce a simple but pronounced pattern of rip currents. If the L_e values do not so correspond for the input wave period, then they are non-resonating and that L_e which corresponds to the closest resonating T_e usually predominates. Under such circumstances an additional mode may produce secondary rip currents simultaneously and therefore a rather complex circulation.

Some doubts have been expressed concerning the existence of edge waves on natural beaches and therefore on their ability to account for cell circulation patterns not produced by wave refraction. The theoretical considerations and wave basin experiments of Bowen and Inman (1969) definitely show that edge waves can interact with normal swell waves to produce rip currents. The question then boils down to the existence of edge waves on natural beaches. Bowen and Inman (1969, p. 5,488) made measurements of rip current spacings and accompanying wave parameters at El Moreno Beach on the northwest coast of the Gulf of California. The beach there has a steep face ($\tan \beta = 0.148$), rising above a nearly horizontal low-tide terrace, so that possible effects of wave refraction on rip current formation could be eliminated. In the four sets of field measurements obtained, there was very close agreement between the rip current spacing and the theoretical wave length of edge waves of mode $n = 1$. This correspondence, together with the fact that there was no beach topography that could account for or control the rip currents, is strong supporting evidence for an edge wave generation mechanism. Because the entire rip current system moved alongshore on a net longshore current, the edge waves could not be measured directly with the wave-recording staffs, so that the evidence for their presence remained circumstantial. Huntley and Bowen (1973) have measured edge waves on the south coast of England, and in their one set of measurements they found that the edge waves were a subharmonic of the incoming waves (the edge wave period T_e was 10 seconds, while the incoming waves were $T = 5$ seconds). Similar subharmonics were produced by Bowen and Inman (1969) in their wave basin experiments. A cell circulation could not develop under these circumstances due to the difference in the wave period; indeed, no rip currents were present. However, the measurements do demonstrate the existence of edge waves on natural beaches, and it remains only for them to have the same period as the swell waves to generate rip currents.

An alternative explanation to edge waves as the origin of rip currents on plane smooth beaches has been developed by Hino (1975). His approach is a hydrodynamic instability theory, where there is a small initial disturbance to the wave set-up which is otherwise constant in the longshore direction. Rip currents form with a preferred longshore spacing that is about 4 times the distance from the shoreline to the breaker zone, a ratio of rip spacing to surf zone width that is commonly observed on natural beaches. This interesting theory by Hino (1975) therefore shows promise, but requires experimental and field testing.

Cell circulation can redistribute beach sediments and therefore have a profound effect on beach configuration (Chapter 10). This becomes a complicating factor with regard to the generation of the cell circulation in that the troughs scoured by the rip currents may act to stabilize the rip current positions. The cell circulation is then strongly affected by the beach topography and is not completely free to respond to changing conditions of swell waves and edge waves. A number of studies have found shoreward-moving currents over bars or shoals, a longshore current confined to a trough extending along the length of the beach, and narrow seaward-flowing rip currents passing through troughs that cut across the bar [see, for example, McKenzie (1958)]. The bottom configuration and accompanying cell circulation change markedly only during high-wave conditions of a storm.

Sonu (1972) has conducted an especially thorough study of the nearshore circulation in an area of irregular bottom topography at Seagrove, Florida. The nearshore consisted of alternate shoals and troughs along the shoreline. His detailed measurements revealed (Figure 7-9) that shoreward currents in the circulation cell typically occurred over the shoals and that the rip currents occurred over the troughs. The outflowing rips attained velocities as great as 2 m/sec. The interesting discovery was that the breaker heights were uniform along the beach, so that one would not at first expect a cell circulation. However, over the shoals the waves broke by spilling and tended to maintain the broken crest through the surf zone, while those entering the rip broke by plunging in a narrow strip near the bar and traveled the remainder of the surf zone with relatively unbroken crests. Because of this difference in breaking, shoreward of the shoals and the spilling breakers there was a higher set-up than shoreward of the plunging breakers at the rip current position, since the spilling waves continuously dissipated their energy, a requirement for set-up development. Actual measurements by Sonu demonstrated higher set-up over the bars than over the rip troughs, and the currents flowed in the direction of the water surface gradient. Taking a similar approach to Bowen (1969*a*) but assuming a constant wave breaker height alongshore and an undulatory surf zone bed, Sonu demonstrated theoretically that cell circulation can be produced under these conditions. The conclusion is that rip currents can be maintained even when the breaking waves are uniform in height alongshore while the surf zone bed has undulations consisting of alternate bars and seaward-trending troughs. Further study of this has been undertaken by Noda (1974).

Based on observations such as those of Sonu (1972), a school of thought has formed that argues that irregular topography always exerts complete control over the water circulation; that is, the topography came first and generated the circulation. But as pointed out by Noda (1974), there remains the basic question of how the original bottom deformation was formed if not by a cell circulation. In addition, rip currents can develop on smooth beaches without any influence of bottom irregularities. Whatever the cause of cell circulation—whether wave refraction effects on the distribution of breaker heights, edge waves, or the mechanism proposed by Hino (1975)—the rip currents most probably come first, causing sediment transport and producing the bottom topography. At some later stage the irregularities of the bottom may be sufficient to provide the principal control over the nearshore circulation.

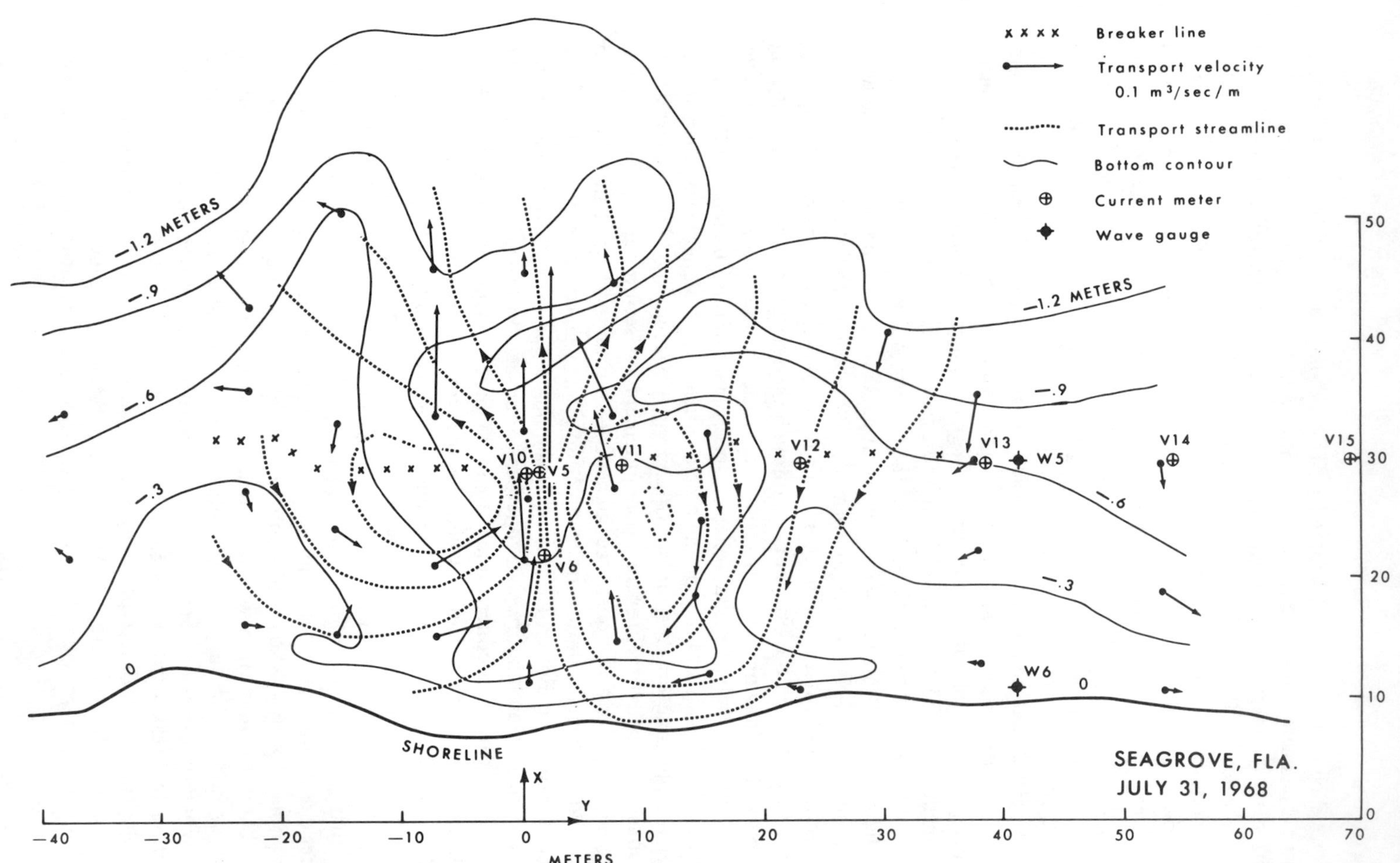

Figure 7-9 Longshore current flow from bar positions to channel where the current flows seaward as a rip current. The flow streamlines shown have a separation of 0.4 m³/sec; $H_b = 39.5$ cm, $T = 5$ sec, and $\alpha_b = 0$. [*From* Sonu (1972)]

LONGSHORE CURRENTS DUE TO AN OBLIQUE WAVE APPROACH

There has been considerable interest in the generation of longshore currents by waves breaking with an angle to the shoreline. This interest is reflected in the number of theories that have been devised to account for the generation of such currents and to forecast their magnitudes. Galvin (1967), giving a complete review of the early theories, arrived at the justifiable conclusion that a satisfactorily proven prediction was not available. Since that time there has been considerable progress through the studies of Bowen (1969*b*), Longuet-Higgins (1970*a*, 1970*b*), and Thornton (1971), all of whom attributed the generation of these currents to the longshore component of the radiation stress (momentum flux) of the water waves. We shall quickly review the early theories and then concentrate our attention on the radiation stress approach.

Considering the interest in these currents, it is somewhat surprising that there are relatively few adequate measurements of the currents and the simultaneous wave parameters. Part of the reason for this is the difficulty in obtaining satisfactory field measurements of the wave parameters, especially the angle of wave approach, which is important in the generation of these currents. The recent use of digital wave sensors and spectral analysis techniques may rectify this. Another problem, not generally recognized in obtaining many of the earlier measurements of longshore currents, is the presence of the cell circulation in addition to the currents resulting from an oblique wave approach. As shown in Figure 7-2, the observed current is the sum of the two current systems. If care is not taken, the measurements obtained would be the current resulting from the combined systems, not just the current due to the oblique wave approach alone. To rid oneself of the portion due to the cell circulation—that is, the current due to a longshore variation in wave height—one can take many measurements along an extended length of beach and then average them. Averaging, paying due attention to the current directions, would tend to cancel out the portion due to the cell circulation. This was carefully done in the study of Komar and Inman (1970), and since that study obtained all the wave parameters with a digital wave sensor array and applied spectral analysis techniques, the data is probably the best presently available. One difficulty is that the wave measurements were obtained outside the breaker zone; since the theories involve the wave breaker parameters, these must first be calculated from the offshore wave data. Galvin and Nelson (1967) reviewed the other available field and laboratory measurements. One shortcoming of the laboratory studies is that an extended length of beach is required over which the current must accelerate to reach the required steady-state velocity. Commonly the laboratory beaches are not of sufficient length for this to be achieved. The most comprehensive of the laboratory studies are those of Galvin and Eagleson (1965) and Brebner and Kamphuis (1963).

Following Galvin (1967), we can roughly classify the theories of longshore current generation according to the general approach taken:

1. Considerations of the continuity of water mass.

2. Energy flux considerations.
3. Momentum flux considerations.

The various approaches will be examined briefly in this order.

Continuity of Water Mass

Inman and Bagnold (1963) developed a longshore current model based on the shoreward mass transfer of water in solitary waves advancing into the nearshore zone. It was assumed that this water entering the nearshore region moves alongshore, progressively increasing in velocity as more and more water is added until some critical velocity is reached, at which point the whole current turns seaward as a rip current. The input of water by the solitary waves would be balanced by the seaward-flowing rip current. The entire system would start anew updrift of the rip current. On the basis of this model they derived the relationship

$$\bar{v}_l = 4\sqrt{\frac{\gamma_b}{3}}\,\frac{l}{T}\tan\beta\sin\alpha_b\cos\alpha_b \qquad (7\text{-}14)$$

for the average longshore current (midway between the rips), where l is the distance between the rip currents.

Bruun (1963) has similarly formulated two longshore current models based mainly on the conservation of water mass in the nearshore zone. The mass input is presumed to be the volume of the incoming waves which dump water into the surf zone, thereby giving rise to longshore currents and rip currents. Only the highest one-sixth of all waves are assumed to contribute to the currents. One of Bruun's models is examined in Figure 7-10 with the available field data; the result is a considerable scatter. The second model of Bruun and the relationship from Inman and Bagnold (1963) cannot be tested with the data because of the lack of measurements of l, the distance between the rip currents.

The principal potential shortcoming of these models is that they attribute the generation of the rip currents to continuity of water flow, returning water seaward which was discharged into the nearshore region by the mass of the waves. We now know that the rips are in general a part of the cell circulation which is produced by longshore variations in wave height. These models therefore incorrectly combine the two systems of longshore current generation. However, it is possible that in certain circumstances a continuity model applies. Offshore bars are sometimes observed to build up nearly to the water's surface (Chapter 11), at which point the waves plunge over the bar into a trough; the trapped water then flows alongshore until a gap in the bar is reached where the stream returns seaward as a "rip current." Such a development more closely fits the assumptions inherent in the continuity models.

Galvin and Eagleson (1965) have also attempted a continuity approach but calculated the mass flux of water into the nearshore zone as

$$\begin{aligned} Q &= [\text{Cross-sectional area of breaker}]\; C_b \sin\alpha_b \\ &\simeq (0.5H_bL_b\cos\alpha_b)C_b\sin\alpha_b \qquad (7\text{-}15) \end{aligned}$$

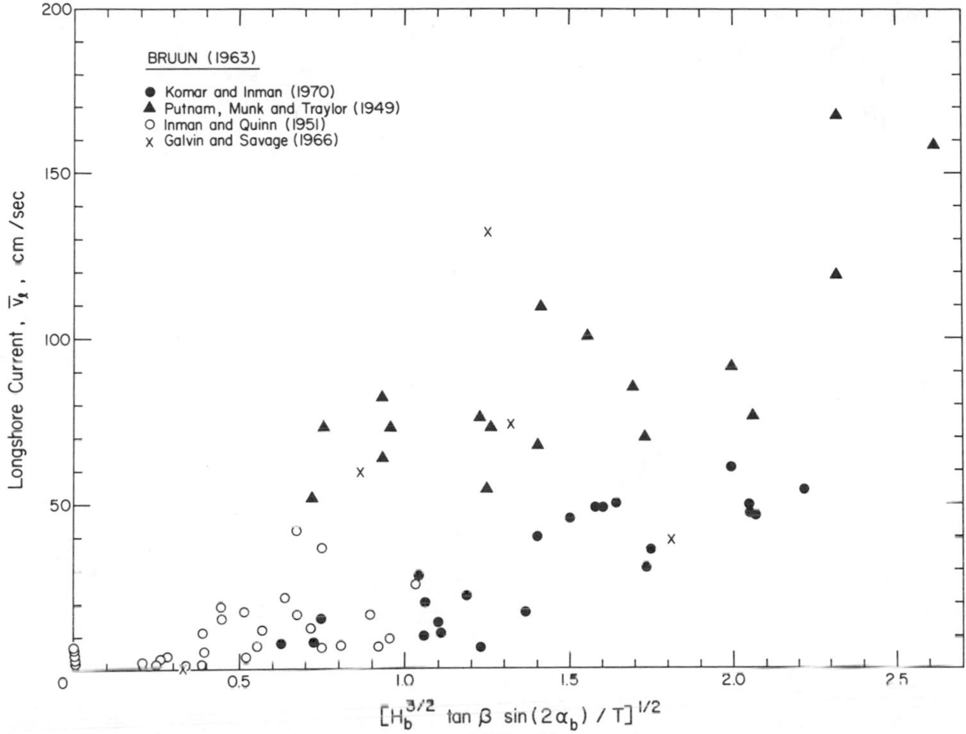

Figure 7-10 Brunn's (1963) longshore current model, based on continuity considerations, tested against field data.

where C_b and L_b are respectively the wave phase velocity and wave length at the breaker zone. Equating this with the longshore discharge associated with the longshore current, they obtained

$$\bar{v}_l = 2k_1 gT \tan\beta \sin\alpha_b \cos\alpha_b \tag{7-16}$$

in which $k_1 \simeq 1$. This relationship is tested in Figure 7-11. Their model is much the same as that devised by Inman and Bagnold (1963), except that it differs in the way the wave discharge ashore is evaluated. It suffers from some of the same shortcomings. No direct mention is made as to how the water brought ashore is returned seaward on real ocean beaches. It may be as rip currents, in which case the two longshore current systems are again mistakenly combined; or the water could return seaward as the offshore backwash between the onshore moving crests. This latter picture of a return flow would produce a longshore current which consists of pulses, a pulse occurring each time a wave crest comes ashore. The longshore current would then be the longshore component of each pulse. This would be very like the drift of water alongshore in the swash zone when the incoming swash strikes the shoreline at an angle. However, the model would not produce steady longshore currents in the surf zone, as are observed. It is interesting to note from equations (7-14) and (7-16) that the formulations of the continuity models tend to yield relationships which suggest that

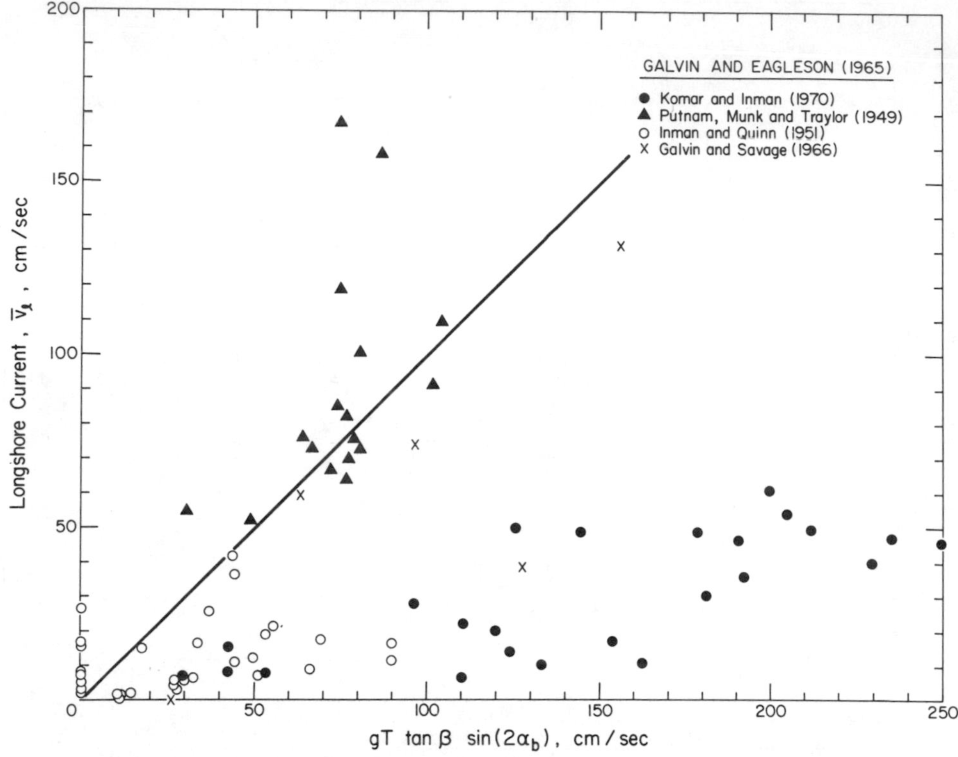

Figure 7-11 The longshore current model of Galvin and Eagleson (1965), based on continuity considerations, tested against field data.

there is no direct dependence of the longshore current on the height of the incoming waves, only on their period and angle of approach to the beach. This of course strikes one as intuitively implausible. Galvin (1967, p. 298) indicates, however, that the value of k_1 in equation (7-16) will depend on the value of H_b.

Energy Flux Approach

Putnam, Munk, and Traylor (1949) developed a model in which the "longshore component of the wave energy flux," $ECn \sin \alpha_b \cos \alpha_b$, into the nearshore zone is assumed to drive a longshore current which has just the right velocity to dissipate this energy flux by frictional bottom drag. Using solitary wave theory to evaluate the energy flux, they arrived at the formulation

$$\bar{v}_l = \left(0.871 \frac{H_b^2}{T} g \frac{s}{c_f} \tan \beta \sin 2\alpha_b\right)^{1/3} \qquad (7\text{-}17)$$

where c_f = coefficient of friction ($\tau = c_f \rho \bar{v}^2$),
s = the fraction of the total energy flux that is responsible for producing the longshore current.

This model, utilizing the energy flux, has never met with much success in predicting longshore currents and is not utilized today.

Momentum Flux Approach (Putnam, Munk, and Traylor, 1949)

Any of the momentum approaches are more promising than the above energy approach. This is because momentum is conserved when the waves break, whereas an appreciable portion of the energy is dissipated in breaking. There are two distinct ways in which the momentum of the waves has been evaluated. Bowen (1969*b*), Longuet-Higgins (1970*a*, 1970*b*), and Thornton (1971) have used the radiation stress concept to evaluate the momentum flux associated with the waves in examining longshore current generation. The resulting model will be reviewed in a moment. Putnam, Munk, and Traylor (1949) considered the momentum associated with the moving mass of water of the broken wave as it travels ashore. Let us examine their model first.

With the breaking of the wave a certain mass of water is thrown into motion in the direction of wave propagation. Associated with this moving water mass is a momentum. Putnam, Munk, and Traylor (1949) attempted to formulate a model for the generation of longshore currents on the basis that the longshore component of this momentum provides the driving force. Again applying the solitary wave theory, they obtained a steady, uniform velocity

$$\bar{v}_l = \frac{a}{2}\left(\sqrt{1 + \frac{4C_b \sin \alpha_b}{a}} - 1\right) \tag{7-18}$$

where

$$a = 2.61 \frac{H_b}{T} \frac{\tan \beta \cos \alpha_b}{c_f}$$

For all the available data (field and laboratory), Inman and Quinn (1952) found that the drag coefficient c_f in equation (7-18) would have to vary over a range of 3.5 orders of magnitude. This is intolerably large; one would expect at most one order range for c_f. Inman and Quinn (1952) found that the drag coefficient c_f is in turn related to the velocity of the longshore current by

$$c_f = \frac{0.24}{\bar{v}^{3/2}} \tag{7-19}$$

Using this in place of c_f in equation (7-18), they obtained the semiempirical relationship

$$\bar{v}_l = \left[\left(\frac{1}{4\epsilon^2} + C_b \sin \alpha_b\right)^{1/2} - \frac{1}{2\epsilon}\right]^2 \tag{7-20}$$

where

$$\epsilon = 108.3 \frac{H_b}{T} \tan \beta \cos \alpha_b \tag{7-21}$$

Because of its semiempirical development, the units of equation (7-20) will not work out properly. The relationship is set up in such a way that English units must be used throughout: if H_b is given in feet and C_b in feet per second, then the calculated value of $\bar{v}_l$ is in feet per second. Equation (7-20) is tested in Figure 7-12 with the available field measurements. It is seen to be much more successful than the models previously examined, and for that reason it has been used widely in past applications to estimate longshore current magnitudes.

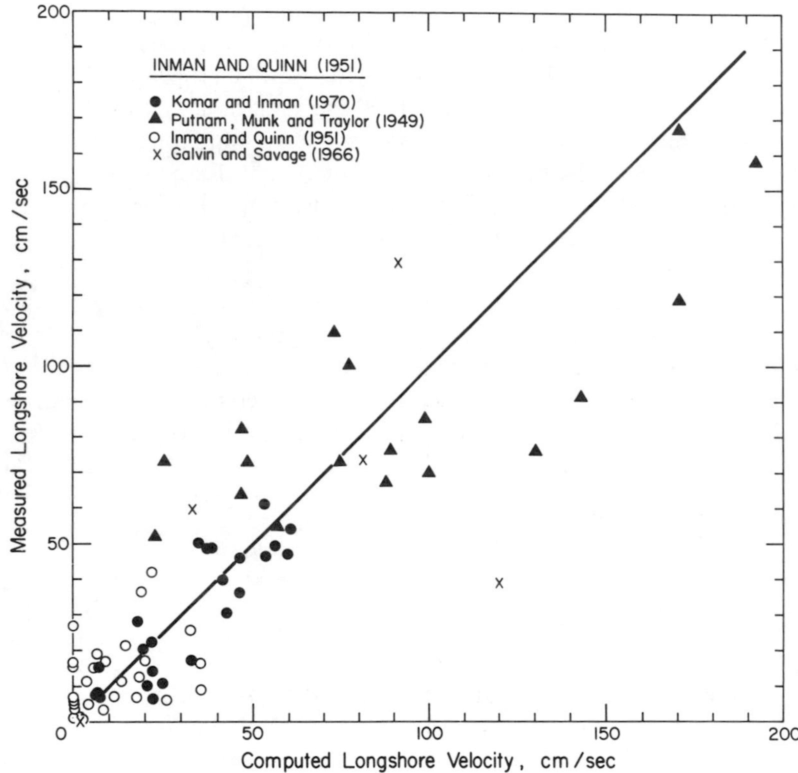

Figure 7-12 The semiempirical longshore current model of Inman and Quinn (1951), given by equation (7-20), tested against field data.

Radiation Stress Approach

Bowen (1969*b*) has suggested that the longshore current due to an oblique wave approach may be generated by the longshore component of the radiation stress

$$S_{xy} = En \sin \alpha \cos \alpha \tag{7-22}$$

This is the flux toward the shoreline (the x-direction) of momentum directed parallel to the shoreline (the y-direction). It is recalled from Chapter 3 that S_{xy} remains constant (or essentially so) during shoaling of waves over a uniform shelf and is dissipated only when the waves reach the nearshore region. Following Bowen's suggestion and development, Longuet-Higgins (1970*a*) derived the relationship

$$\bar{v}_l = \frac{5\pi}{8}\frac{\tan\beta}{c_f}u_m \sin\alpha_b \tag{7-23}$$

where u_m is the maximum value of the horizontal orbital velocity evaluated at the breaker zone, with

$$u_m = \left(\frac{2E_b}{\rho h_b}\right)^{1/2} \tag{7-24}$$

The derivation of equation (7-23) is based on the assumption that S_{xy} is dissipated in the nearshore zone by the production of a longshore current and its associated bottom drag. The onshore gradient $\partial S_{xy}/\partial x$ is balanced against the bottom drag on the current.

Using a classical control volume approach for the nearshore, Eagleson (1965) examined the growth of a longshore current away from a barrier across the surf such as a groin or jetty. At large distances from the barrier Eagleson's solution becomes nearly equivalent to equation (7-23) determined by Longuet-Higgins (1970*a*). However, since Eagleson's solutions were included only in an M.I.T. internal report, they were not generally known.

The sand transport studies of Komar and Inman (1970) had earlier suggested that

$$\bar{v}_l = 2.7u_m \sin\alpha_b \tag{7-25}$$

This was prompted by the agreement between two seemingly independent estimates of the littoral sand transport, the agreement only being possible if the longshore current due to an oblique wave approach is given by the relationship of equation (7-25). This relationship was therefore first indicated by data on littoral sand transport, although it was substantiated by the field data on longshore currents (see further discussion in Chapter 8). Komar and Inman (1970) found no dependence on the beach slope $\tan\beta$ as would be suggested by the theoretical relationship derived by Longuet-Higgins (equation 7-23). Figure 7-13 and 7-14 contain available field measurements of longshore currents; Figure 7-13 examines equation (7-23), and Figure 7-14 tests equation (7-25). A comparison between the two suggests that

$$\frac{\tan\beta}{c_f} \simeq \text{constant} \tag{7-26}$$

that is, the drag coefficient c_f increases proportionally with an increase in beach slope. The constancy of this ratio has also been suggested on theoretical grounds by Komar (1971*a*). However, Longuet-Higgins (1972) has indicated that equation (7-26) may be more apparent than real, since with increasing beach slope the increasing horizontal mixing by eddies within the surf zone would be more important than an increase in c_f to keep the ratio constant. This does not prove to be the case. In a comparision between equation (7-25) and the complete solution for the velocity distribution across the surf zone, I found (Komar, 1975) that an increase in beach slope resulted in a corresponding increase in drag coefficient c_f, just as suggested by equation (7-26):

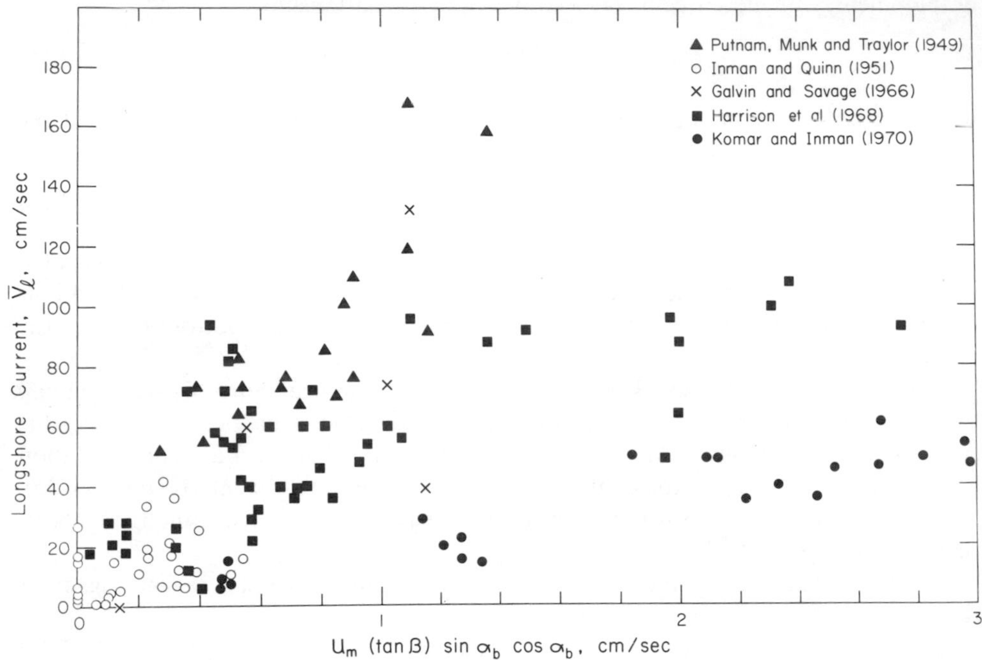

Figure 7-13 Examination with all available field data of the longshore current relationship (equation 7-23) deduced theoretically by Longuet-Higgins (1970*a*) applying radiation stress concepts. Compare with Figure 7-14. [*After* Komar and Inman (1970)]

The horizontal eddy mixing is not as important to the mid-surf velocity; instead, it is most significant to the currents existing beyond the surf zone. Whatever the reason, the field data does indicate that equation (7-26) is correct such that the longshore current midway through the surf zone is given by equation (7-25).

Komar (1975) has undertaken a complete review of the longshore current relationship proposed in Komar and Inman (1970) and derived theoretically by Longuet-Higgins (1970*a*). This examined the agreement with the field data as well as with all of the available laboratory data. The agreement between the field data and equation (7-25) is actually somewhat better than indicated in Figure 7-14. This is because the figure combines data from five separate field studies and, in addition to being at different locations, there might also be systematic differences as to how the data was collected and analyzed. For example, the data of Harrison et al. (1968) measured the maximum longshore current just shoreward from the breakers rather than the lesser current at mid-surf. The result, as can be seen in Figure 7-14, is that their data plots somewhat higher than the measurements obtained in the other studies. Taking the Harrison et al. (1968) data by itself, it is found (Komar, 1975, figure 5) to agree with the form of equation (7-25), but with a coefficient of 3.5 rather than 2.7. In the laboratory tests much greater angles of wave breaking are achieved, and the

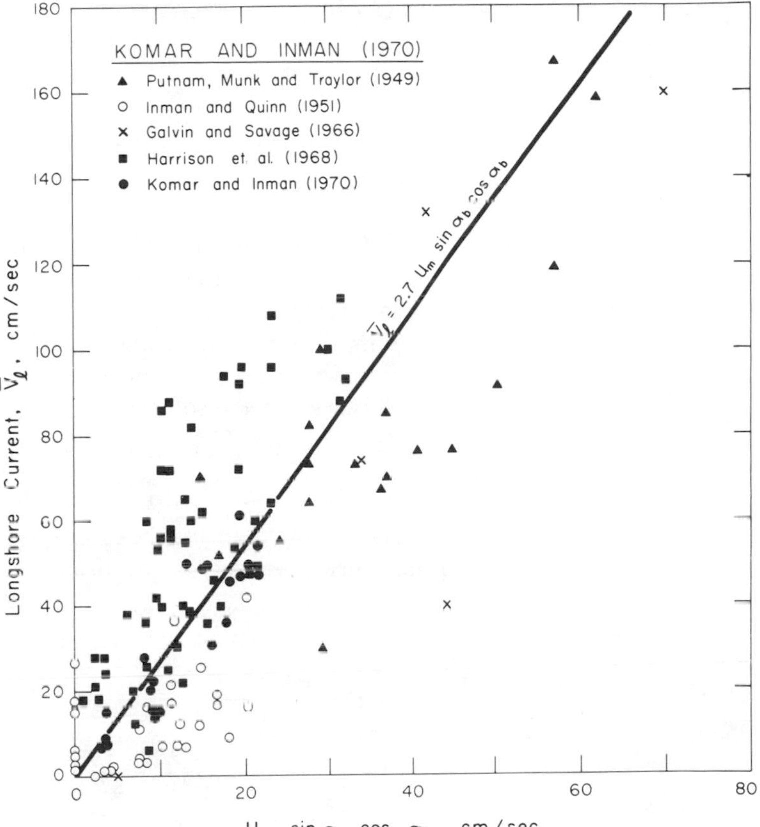

Figure 7-14 The longshore current relationship deduced by Komar and Inman (1970) from the results of their sand transport studies (see Chapter 8). A comparison with Figure 7-13 indicates that $\tan\beta/c_f$ is approximately constant. [*After* Komar and Inman (1970)]

results indicate that equation (7-25) should be modified to

$$\bar{v}_l = 2.7u_m \sin\alpha_b \cos\alpha_b \tag{7-27}$$

In the field the breaker angles are small, so that $\cos\alpha_b \simeq 1$ and equation (7-25) was satisfactory. Inclusion of the cosine factor is also supported by theory (equation 7-30). With the exception of the laboratory data of Galvin and Eagleson (1965), it was found that the remaining laboratory measurements as well as all the field data are in agreement with equation (7-27). Since the agreement is much better than with the other existing theories, and since it has the best theoretical basis, being derivable from radiation stress concepts (Longuet-Higgins, 1970*a*), equation (7-27) offers the best available prediction of the longshore current velocity at mid-surf generated by waves breaking at an angle to the shoreline.

If we compare Figure 7-14 with Figure 7-11 for the Inman and Quinn (1952) model, it is seen that they bear a marked similarity with respect to the positions of the data points. It turns out that for essentially all the data $(C_b \sin\alpha_b)^{1/2}$ is several times larger than $1/2\epsilon$ in equation (7-20), so that the relationship effectively reduces to

$$\bar{v}_l \simeq C_b \sin\alpha_b \tag{7-28}$$

Utilizing equation (7-24), $C_b = \sqrt{g(H_b + h_b)}$, and $\gamma_b = H_b/h_b$, this result can be modified to

$$\bar{v}_l \simeq \frac{2\sqrt{\gamma_b + 1}}{\gamma_b} u_m \sin \alpha_b \tag{7-29}$$

For $\gamma_b = 1.0$ the coefficient is 2.8, almost exactly that obtained by Komar and Inman (1970) in equation (7-25). This apparently is the reason for the success of the Inman and Quinn (1952) empirical model.

Bowen (1969*b*), Longuet-Higgins (1970*b*), Thornton (1971), and Komar (1975) have examined the variations of the longshore current across the width of the nearshore region, again using the radiation stress approach. To do this they consider the local $\partial S_{xy}/\partial x$ as the driving thrust, and include a horizontal eddy viscosity, a horizontal mixing effect which produces an onshore-offshore transfer or "smearing" of the momentum. This leads to the solution

$$V = AX + B_1 X^{P_1} \quad \text{for} \quad X < 1 \text{ (inside surf zone)} \tag{7-30a}$$

$$V = B_2 X^{P_2} \quad \text{for} \quad X > 1 \text{ (outside breaker zone)} \tag{7-30b}$$

where $X = x/X_b$ and $V = v/v_0$, X_b being the distance from the shoreline to the breaker zone, v the longshore current velocity at the distance x from the shoreline, and where

$$v_o = \frac{5\pi}{16} \frac{\gamma\zeta^2}{(1+\gamma)^{1/2}} \sqrt{gh_b} \frac{\tan\beta}{c_f} \sin\alpha_b \cos\alpha_b \tag{7-30c}$$

$$P = \frac{\pi N \tan\beta}{\gamma c_f} \tag{7-30d}$$

$$P_1 = -\frac{3}{4} + \sqrt{\frac{9}{16} + \frac{1}{\zeta P}}$$

$$P_2 = -\frac{3}{4} - \sqrt{\frac{9}{16} + \frac{1}{\zeta P}}$$

$$A = \frac{\zeta^2}{(1 - \frac{5}{2}P\zeta)}$$

$$B_1 = \frac{P_2 - 1}{P_1 - P_2} A$$

$$B_2 = \frac{P_2 - 1}{P_1 - P_2} A$$

$$\zeta = \frac{1}{(1 + \frac{3}{8}\gamma^2)}$$

and where γ is the ratio of the wave height to water depth with values in the range 0.8 to 1.2. This solution is basically the same as that presented by Longuet-Higgins (1970*b*), except that the effects of wave set-up on the water depth are included.

The dimensionless parameter P (equation 7-30d) reflects the significance of the horizontal eddy transfer: the larger the value of P the more important this effect. The term N is a numerical constant with limits $0 < N < 0.016$ in the expression for the horizontal eddy viscosity, $\rho N x (gh)^{1/2}$, used by Longuet-Higgins (1970*b*). A family of longshore current profiles obtained with equation (7-30) is shown in Figure 7-15. The inclusion of horizontal mixing effectively couples together the adjacent elemental water columns, resulting in a diffusion in a direction perpendicular to shore. As seen in Figure 7-15, this lateral diffusion of momentum also enables the momentum flux inside the surf zone to drive longshore currents outside the surf zone. The straight-line solution for $P = 0$ is where there is no horizontal eddy mixing, giving the discontinuity at the breaker line with no current outside the surf zone. Comparison of the solutions with the laboratory measurements of Galvin and Eagleson (1965) of the variations in velocity across the nearshore suggests a range of P from 0.1 up to 0.4. It is seen in Figure 7-15 that for these values of P the current should be largely confined to the surf zone, decreasing rapidly outside the breaker zone. This conforms with what is observed in nature.

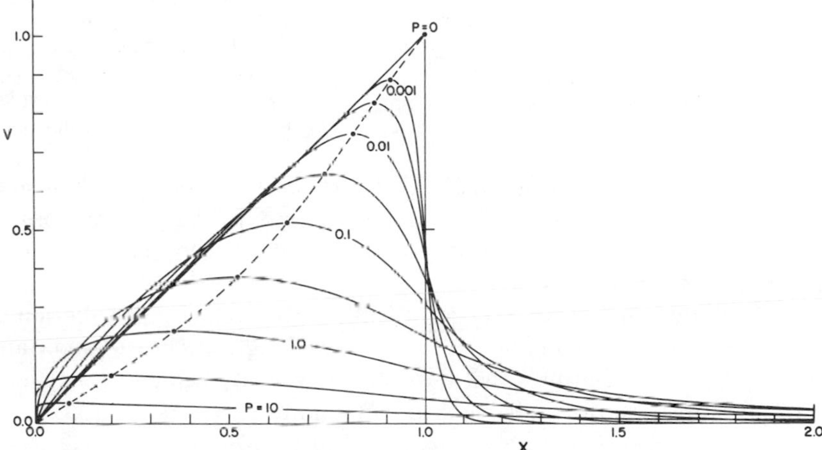

Figure 7-15 A family of longshore current profiles across the surf zone obtained with the solution of equation (7-30), given in dimensionless form, where $X = x/X_b$ and $V = v/v_0$, the local value of the current v dividend by v_0 of equation (7-30c). The larger the value of P (equation 7-30d), the greater the effect of the horizontal eddy transfer. [*After* Longuet-Higgins (1970*b*)]

There are no satisfactory measurements in the field of the velocity distribution across the nearshore. Equation (7-27) gives the velocity midway through the surf zone. Since this relationship has been tested with the available field and laboratory data, it can be used in turn to test or "calibrate" the complete distribution given by equation (7-30). This is accomplished by equating equation (7-27) to the value obtained from equation (7-30) when $X = 0.5$ (the mid-surf solution). This leads to a relationship between $(\tan \beta)/c_f$ and N or P, shown in Figure 7-16 for $\gamma = 0.8$ and 1.0. It is seen that over the expected range of P values, $(\tan \beta)/c_f$ shows relatively little change. For example, for a beach slope of $\tan \beta = 0.100$, the drag coefficient c_f could possibly vary only from 0.0175 to 0.0133. This confirms the findings and conclusions of Komar and Inman (1970) on the near-constancy of the ratio $(\tan \beta)/c_f$, a conclusion that led to equation (7-26) for the longshore current velocity. Although the results indicate

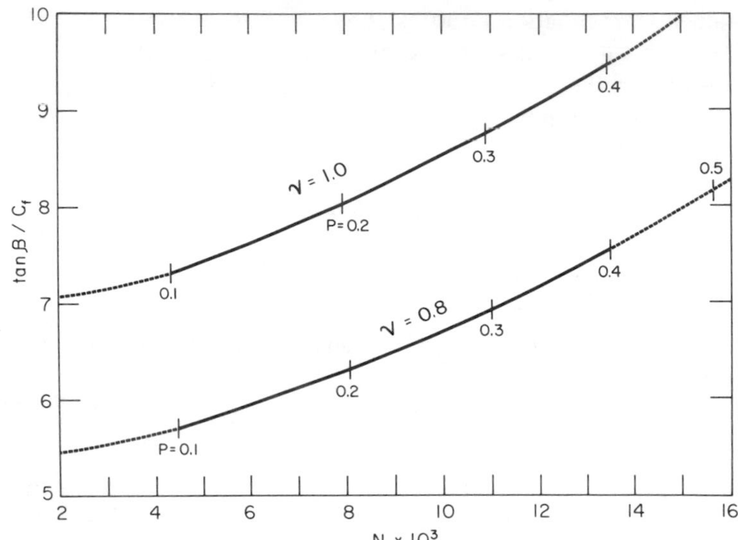

Figure 7-16 Relationship between $(\tan\beta)/c_f$ and N, with values of P also shown, obtained by equating the longshore current velocity of equation (7-27) to the mid-surf solution of equation (7-30). This effectively calibrates the solution for the complete velocity distribution against the value of equation (7-27) that has been tested with the available field and laboratory data.

that the magnitude of the velocity at the mid-surf position does not show a strong dependence on N if it is to agree with the available data, the value of N does exert a strong control on the velocities elsewhere in the nearshore, especially close to shore and outside the breaker zone.

In utilizing the solution of equation (7-30) for the velocity distribution across the surf zone, Figure 7-16 should be employed in selecting values for P and $(\tan\beta)/c_f$ (and therefore c_f), as this insures that the mid-surf velocity will agree with equation (7-27) and therefore with the available field and laboratory data.

CURRENTS DUE TO BOTH A LONGSHORE VARIATION IN WAVE HEIGHT AND AN OBLIQUE WAVE APPROACH

At the beginning of the chapter we categorized the nearshore currents into (1) a cell circulation, and (2) currents produced by an oblique wave approach. The two systems were then summed in Figure 7-2 to obtain a pattern resembling the currents actually observed in the nearshore. This can be improved upon by examining the longshore current produced by both driving mechanisms: a longshore variation in wave height plus waves arriving at an angle to the beach. Komar (1975) has considered such situations and arrived at

$$\bar{v}_l = 2.7u_m \sin\alpha_b \cos\alpha_b - \frac{\pi\sqrt{2}}{c_f\gamma_b^3}\left(1 + \frac{3\gamma_b^2}{8} - \frac{\gamma_b^2}{4}\cos^2\alpha_b\right)u_m\frac{\partial H_b}{\partial y} \qquad (7\text{-}31)$$

for the velocity at the mid-surf position. Note that if $\partial H_b/\partial y = 0$, then the relationship reduces to equation (7-27) for the current from an oblique wave approach alone.

If $\partial H_b/\partial y$ is negative (decreasing H_b in the positive y-direction), then the net current $\bar{v}_l$ is larger than could be generated either by the longshore variation in wave height or by the oblique wave approach acting alone, since the driving forces act together and so enhance one another. If $\partial H_b/\partial y$ is positive, then the driving forces act in opposite directions to produce a net current of reduced magnitude. The formulation of equation (7-31) is inadequate in that it does not account for the presence of rip currents.

It is apparent that under certain circumstances the thrust due to the oblique wave approach may be opposed and balanced by the longshore variation in wave height such that $\bar{v}_l = 0$ in equation (7-31). This condition was of special interest to Komar (1971*b*, 1975) in that it accounted for the development of an equilibrium cuspate shoreline. Cusps were observed to form in a laboratory wave basin, each cusp developing in the lee of a rip current. Once the cusps projected seaward a sufficient degree, it was found that rather suddenly the rip currents and associated longshore currents ceased to exist. The oblique wave approach to the cusp flanks opposed and balanced the local $\partial H_b/\partial y$ caused by the edge waves. This provides one example of the joining of the two longshore current–generating mechanisms. It also demonstrates that rip currents can generate a cuspate shoreline but may no longer be present at the time of cusp observation. This aspect will be discussed further in Chapter 10.

Although there may be a balance at mid-surf in the forces of longshore current generation such that $\bar{v}_l = 0$ in equation (7-31), there could still be nonequilibrium in the other portions of the surf zone width. This was examined in Komar (1975), which develops the complete solution of the velocity distribution across the surf. This solution is an extension of that obtained by Longuet-Higgins (1970*b*), including a forcing term due to the longshore variation in wave set-up, $\partial\bar{\eta}/\partial y$. Examples are shown in Figure 7-17. When $\partial\bar{\eta}/\partial y = 0$, the solution is the same as that of Longuet-Higgins (1970*b*), given in equation (7-30), caused by an oblique wave approach alone. With $\partial\bar{\eta}/\partial y = -0.0005$, the water slopes downward in the positive y-direction, the direction of the longshore component of the oblique wave approach, so that the two forces combine to produce stronger longshore currents. For positive values of $\partial\bar{\eta}/\partial y$ the water slope opposes the oblique wave approach and the current is diminished. For $\partial\bar{\eta}/\partial y = 0.0025$ (Figure 7-17), the forces are close to balancing and the resulting longshore current is small. The strongest residual current is close to the shoreline and is flowing in the opposite direction to the current beyond the mid-surf zone. Although not complete, the results do indicate that a near-balance can be achieved, yielding only a greatly diminished current. This is confirmed by the laboratory cuspate shoreline observations described above.

On real ocean beaches of complex topography and with longshore variations in wave height as well as an oblique wave approach, the only means to analysis is by numerical computer solutions. The study of Sonu (1972) at Seagrove, Florida, discussed earlier, aptly demonstrates this approach. Noda (1974) provides additional analytical models under such conditions. Our understanding of nearshore circulation under the complex conditions of nature is best advanced by studies such as these.

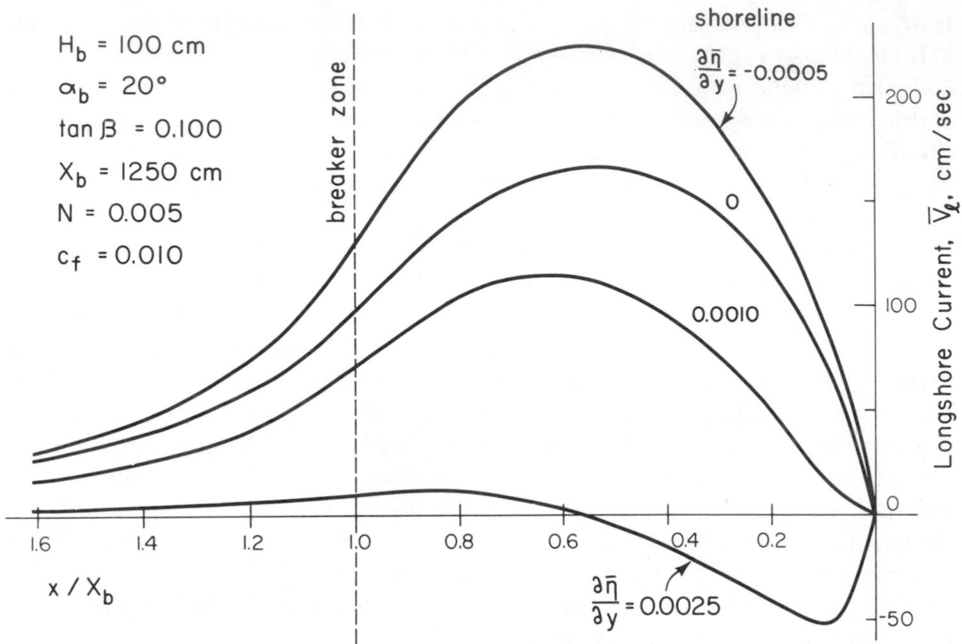

Figure 7-17 Examples of complete solutions of the distribution of longshore current velocities through the surf zone for a series of values for the longshore variation in the wave set-up ($\partial\bar{\eta}/\partial y$). With $\partial\bar{\eta}/\partial y = 0.0025$, the set-up slope in the longshore direction nearly opposes and balances the thrust due to the oblique wave approach, and the velocities are greatly weakened. [*From* Komar (1975)]

EMPIRICAL PREDICTION OF LONGSHORE CURRENTS

Harrison and Krumbein (1964), Harrison et al. (1965), Sonu, McCloy, and McArthur (1967), and Harrison (1968) have attempted to apply multiregression analysis to the prediction of longshore currents. Harrison and Krumbein (1964) and Harrison et al. (1965) screened offshore wave data in predicting longshore currents at Virginia Beach, Virginia, and found that the six variables T, H_∞, tan β, onshore wind speed, offshore wind speed, and wave direction were the strongest factors accounting for current variability, their relative importance being in that order. Only 1 to 2% of the variability was due to the offshore wave direction, the parameter that should theoretically have been most important. The reason for this is that correlating the longshore current to the deep-water wave direction does not take into consideration wave refraction, which controls the angle at which the waves finally break at the shoreline. It is doubtful whether empirical equations involving offshore wave data can have predictive value (Galvin, 1967, p. 299).

Sonu, McCloy, and McArthur (1967) at the Outer Banks, North Carolina, and

Harrison (1968) at Virginia Beach, Virginia, correlated longshore current velocities with the breaking wave parameters and obtained similar results. The multiregression analysis findings are of principal interest in that they indicate which wave parameters appear to have the most significance in longshore current generation. Both studies indicate that nearly all the variability of the current velocity can be accounted for by the magnitude of the breaker angle: the breaker height, period, and beach slope only accounted for secondary variations according to their results. They differed in the effect of the wave period: with an increase in wave period Sonu, McCloy, and McArthur (1967) found a reduction in the current, whereas Harrison (1968) found a positive correlation. This points to one of the problems of utilizing regression analysis techniques: the results are always dependent on the data base and really only apply to that data. No physical reasoning goes into their formulation. The results also depend on the techniques employed. For example, Sonu, McCloy, and McArthur (1967) used both multilinear regression and nonlinear regression with the same data and got very different results. With nonlinear regression the angle of wave breaking was the least important of all the independent variables, whereas it was the most significant in the linear regression analysis.

The resulting equations from regression analysis must be applied with reservation and only to conditions that are very similar to those under which the data base was collected. One might want to employ them in an extensive study of processes at one particular beach, especially a beach with complicated topography where application of the physical models of longshore current generation would be difficult. Even here, the results of regression analysis would be of doubtful validity at other beaches and at other times. Longshore current models with a physical basis are generally greatly to be preferred over such statistical models.

Brebner and Kamphuis (1964) present two empirical equations for forecasting the longshore current velocity from the deep-water wave parameters H_∞, T, and α_∞. The correlations are based on their laboratory data. Relating the currents to the deep-water wave parameters was felt to be advantageous, since wave-forecasting methods yield estimates of the deep-water wave height, period, and direction of wave advance. It is doubtful whether such a correlation can be made with deep-water parameters, except perhaps to obtain only the grossest estimate of the longshore current velocity. Such a procedure does not account for the degree of wave refraction that occurs between deep water and the beach where the waves finally break. For example, waves with a deep-water angle $\alpha_\infty = 35$ deg. could strike a steep beach with a breaker angle $\alpha_b = 25$ deg. and a low-sloping beach where wave refraction is important with $\alpha_b = 1$ deg. The resulting longshore currents at the two beaches would be expected to be very much different, even though the deep-water wave parameters were the same and therefore the current velocities predicted by the equations of Brebner and Kamphuis (1964) would be the same. The empirical formulations of Brebner and Kamphuis have one advantage over the other empirical attempts in that they have energy and momentum considerations in mind and the results are therefore at least dimensionally correct.

WIND AND TIDE EFFECTS ON THE NEARSHORE CIRCULATION

In addition to longshore currents being generated by an oblique wave approach or by longshore variations in the wave height (or set-up), winds blowing in the longshore direction and tides draining from the beach may also contribute to the observed flow. The importance of winds and tides remains largely unevaluated.

Shepard and Inman (1950) pointed out the significance of winds in generating longshore currents, although they state that it is difficult to separate the currents generated directly by the wind stress from the currents associated with the short-period wind-generated waves. With a wind stress acting in the longshore direction, the resulting current should be greater than the current predicted with equation (7-27) utilizing the breaker angle and energy of the wind-generated waves. In this manner the two effects might be separated. Such a study, to this author's knowledge, has not been undertaken. Fox and Davis (1973), working on Lake Michigan, determined a successful empirical model which relates the generated longshore current directly to the barometric pressure time variations which generate the winds. No attempt was made to separate the wind stress effects from the wind-generated waves. Such an empirical model was particularly successful on Lake Michigan, since at that location swell from distant storms is not generally important and longshore currents are instead generated by local weather fronts that pass over the area. Wind velocity in the longshore direction was one of the variables used in the studies by Harrison and Krumbein (1964) and by Sonu, McCloy, and McArthur (1967), but it exhibited only weak correlations with the resulting longshore current. On the Oregon coast powerful winds commonly blow in the longshore direction, and it is this author's impression that the direct wind stress on the water within the surf zone can have a profound effect on the longshore currents under these conditions. More study should be done on this phenomenon.

As the tide ebbs, water drains from the beach and may enhance the rip currents and feeder longshore currents. This increase is particularly strong as the low tide is approached and water is trapped within troughs running parallel to the shoreline, prevented from direct offshore flow by the bar that is now partially exposed. The only escape is to flow alongshore within the trough and then seaward through the rip channel cutting the bar. The effect may be further enhanced in that under these conditions waves breaking over the bar may spill their water over into the trough and also contribute to the flow. Sonu (1972) found at Seagrove, Florida, that the tide level controlled the intensity of wave breaking and therefore the strength of the circulation. He noted that rip currents were generally stronger during low tide than at high tide. In addition, Sonu found that the rip currents at high tide fluctuated with the period of the incoming swell, whereas the stronger rips at low tide tended to fluctuate with a beat of period 25 to 50 seconds. Shepard and Inman (1950) similarly found pulsations in the longshore current velocity resulting from surf beat due to alternating sets of high and low breaking waves.

This increase in rip current and longshore current velocities at low tide is of importance to swimmers and other users of the beach. On Oregon beaches there is a greater frequency of drownings at low tide than at high tide, partly due to more people being on the beach at low tide for beachcombing and clamming, but perhaps also due to the increased danger from the stronger rips.

SUMMARY

The nearshore cell circulation consisting of rip currents and feeding longshore currents (Figure 7-1) is normally generated by variations in the wave breaker height along the length of the beach (Bowen, 1969*a*; Bowen and Inman, 1969). The water level shoreward of the higher breakers is raised above that shoreward of the small waves, so that the water flows toward the position of small waves, where it converges and turns seaward as a rip current. This is enhanced by the longshore gradient of the radiation stress ($\partial S_{yy}/\partial y$). The longshore variations in wave height may be produced either by wave refraction causing divergence and convergence of the wave rays or by edge waves trapped in the nearshore interacting with the normal swell waves. Hino (1975) has developed an alternative explanation to edge waves as the cause of some rip currents. Sonu (1972) has also found that cell circulation can be maintained on an irregular beach with longshore bars cut by troughs, without variations in breaker height. Waves breaking on the shoals are of the spilling variety and produce a higher set-up than the plunging waves within the rip trough. The flow is thus in the longshore direction from the shoal to the trough, where it returns seaward as a rip current.

Longshore currents may also be produced by waves breaking at an angle to the shoreline. Of all the equations formulated for the generation of these currents, those based on radiation stress (momentum flux) concepts have the firmest theoretical basis (Bowen, 1969*b*; Longuet-Higgins, 1970*a*, 1970*b*; Thornton, 1971). The equation formulated by Longuet-Higgins (1970*a*) is basically the same as that deduced by Komar and Inman (1970) based on the equivalence of two models for littoral sand transport. Considering comparisons with all the available field and laboratory longshore current data, the best formulation for the current at the mid-surf position is

$$\bar{v}_\ell = 2.7 u_m \sin \alpha_b \cos \alpha_b$$

where $u_m = (2E_b/\rho h_b)^{1/2}$ is the maximum orbital velocity evaluated at the breaker zone and α_b is the breaker angle. Equation (7-30) gives the complete velocity distribution across the entire surf zone and beyond the breaker zone, modified after Longuet-Higgins (1970*b*).

The theory of longshore current generation has been considerably advanced over the past few years, thanks chiefly to the development and application of radiation stress concepts for water waves. At this stage careful field studies of the circulation systems are needed. Nearly all the available data predated this theoretical advance, so that studies on long, regular beaches would be valuable, relating the longshore current to the variations in breaker height along the beach and to the oblique wave approach angle. Information is desired on the distribution of longshore current velocities across the surf zone, to be compared with the theoretical formulations.

A more complex and difficult problem is that of the effects of the nearshore currents on the beach topography leading to longshore and rip troughs and a generally irregular beach, which in turn affects the pattern of the circulation. The study of Sonu (1972) is a good beginning to investigations of nearshore currents on irregular beaches. Finally, the contributions of tides and wind-induced currents to the nearshore circulation require further study.

PROBLEMS

7-1 What is the cause of wave set-down and set-up in the nearshore? How are they related to the wave height?

7-2 The edge wave period is the same as that of the swell waves: $T_e = T = 10$ sec. If the beach slopes with an angle $\beta = 5$ deg, what might be possible rip current spacings generated by these edge waves?

7-3 Edge waves of period $T_e = 12$ sec and $n = 1$ are present on a beach of slope angle $\beta = 5$ deg. Groins are built across the beach (Figure 13-4) which may interact with these edge waves. What groin spacings, Λ in equation (7-13), might cause resonant conditions and therefore adverse currents between the groins?

7-4 Waves break on the beach with a height of 3 m in water of depth 3.5 m, and with an angle of 7 deg to the shoreline. What will be the resulting longshore current velocity?

7-5 For the conditions of Problem 7-4, what longshore variation in breaker height, $\partial H_b/\partial y$, would prevent this current from developing such that $\bar{v}_l = 0$ if the beach slope is 0.100? What ΔH_b would give this slope over a longshore distance of 100 m?

REFERENCES

ARTHUR, R. S. (1962). A note on the dynamics of rip currents. *J. Geophys. Res.*, 67, no. 7: 2777–79.

BOWEN, A. J. (1969*a*). Rip currents, 1: theoretical investigations. *J. Geophys. Res.*, 74: 5467–78.

——— (1969*b*). The generation of longshore currents on a plane beach. *J. Mar. Res.*, 37: 206–15.

BOWEN, A. J., D. L. INMAN, and V. P. SIMMONS (1968). Wave "set-down" and "set-up:" *J. Geophys. Res.*, 73, no. 8: 2569–77.

BOWEN, A. J., and D. L. INMAN (1969). Rip currents, 2: laboratory and field observations: *J. Geophys. Res.*, 74: 5479–90.

BREBNER, A., and J. W. KAMPHUIS (1963). *Model tests on the relationship between deep-water wave characteristics and longshore currents.* Queen's Univ., Kingston, Ontario, Research Report no. 31, 25 pp.

——— (1965). Model tests on the relationship between deep-water wave characteristics and longshore currents. *Proc. 9th Conf. Coast. Eng.*, pp. 191–96.

BRUUN, P. (1963). Longshore currents and longshore troughs. *J. Geophys. Res.*, 68: 1065–78.

EAGLESON, P. S. (1965). *Theoretical study of longshore currents on a plane beach.* M.I.T. Hydrodynamic Lab Tech. Report no. 82, Cambridge, Mass., 31 pp.

ECKART, C. (1951). *Surface waves in water of variable depth.* Univ. of California, Scripps Inst. of Oceanography, Wave Report no. 100, SIO Ref. 51–12, La Jolla, 99 pp.

FOX, W. T., and R. A. DAVIS (1973). Simulation model for storm cycles and beach erosion on Lake Michigan. *Geol. Soc. Am. Bull.*, 84: 1769–90.

GALVIN, C. J. (1967). Longshore current velocity: a review of theory and data. *Revs. in Geophys.* 5, no. 3: 287–304.

GALVIN, C. J., and P. S. EAGLESON (1965). *Experimental study of longshore currents on a plane beach.* U.S. Army, Coastal Engr. Research Center Tech. Memo. no. 10, 80 pp.

GALVIN, C. J., and R. A. NELSON (1967). *Compilation of longshore current data.* U.S. Army, Coastal Engr. Research Center, Misc. Paper no. 2–67, 19 pp.

HARRISON, W. (1968). Empirical equation for longshore current velocity. *J. Geophys. Res.*, 73, no. 22: 6929–36.

HARRISON, W., and W. C. KRUMBEIN (1964). *Interactions of the beach-ocean-atmosphere system at Virginia Beach, Virginia.* U.S. Army Coastal Engr. Research Center, Tech. Memo. no. 7, 102 pp.

HARRISON, W., N. A. PORE, and D. R. TUCK (1965). Predictor equations for beach processes and responses. *J. Geophys. Res.*, 70: 6103–09.

HARRISON, W., E. W. RAYFIELD, J. D. BOON III, G. REYNOLDS, J. B. GRANT, and D. TYLER (1968). *A time series from the beach environment.* Land and Sea Interactions Lab. Norfolk, Virginia, ESSA Research Lab. Tech. Memo. no. 1, 28 pp.

HINO, M. (1975). Theory on formation of rip current and cuspidal coast. *Proc. 14th Conf. Coast. Eng.*, pp. 901–19.

HUNTLEY, D. A., and A. J. BOWEN (1973). Field observations of edge waves. *Nature*, 243: 160–61.

INMAN, D. L., and R. A. BAGNOLD (1963). Littoral processes. In *The sea*, ed. by M. N. Hill, 3: 529–53. Interscience, New York.

INMAN, D. L., and W. H. QUINN (1952). Currents in the surf zone. *Proc., 2nd Conf. Coast. Eng.*, pp. 24–36.

INMAN, D. L., R. J. TAIT, and C. E. NORDSTROM (1971). Mixing in the surf zone. *J. Geophys. Res.*, 76: 3493–514.

KOMAR, P. D. (1971*a*). The mechanics of sand transport on beaches. *J. Geophys. Res.*, 76, no. 3: 713–21.

——— (1971*b*). Nearshore cell circulation and the formation of giant cusps. *Geol. Soc. Am. Bull.*, 82: 2643–50.

——— (1975). Nearshore currents: generation by obliquely incident waves and longshore variations in breaker height. In *Proceedings of the symposium on nearshore sediment dynamics*, ed. J. R. Hails and A. Carr, Wiley, London, pp. 17–45.

KOMAR, P. D., and D. L. INMAN (1970). Longshore sand transport on beaches. *J. Geophys. Res.*, 75, no. 30: 5914–27.

LONGUET-HIGGINS, M. S. (1970*a*). Longshore currents generated by obliquely incident sea waves, 1. *J. Geophys. Res.*, 75, no. 33: 6778–89.

——— (1970*b*). Longshore currents generated by obliquely incident sea waves, 2. *J. Geophys. Res.*, 75, no. 33: 6790–801.

——— (1972). Recent progress in the study of longshore currents. In *Waves on beaches*, ed. by R. E. MEYER, pp. 203–48. Academic Press, New York.

LONGUET-HIGGINS, M. S., and R. W. STEWART (1963). A note on wave set-up: *J. Mar. Res.*, 21: 4–10.

——— (1964). Radiation stress in water waves; a physical discussion with applications. *Deep-Sea Res.*, 11, no. 4: 529–63.

MCKENZIE, R. (1958). Rip current systems. *J. Geol.*, 66, no. 2: 103–13.

NODA, E. K. (1974). Wave-induced nearshore circulation. *J. Geophys. Res.*, 79, no. 27: 4097–106.

PUTNAM, J. A., W. H. MUNK, and M. A. TRAYLOR (1949). The prediction of longshore currents. *Trans. Am. Geophys. Union*, 30: 337–45.

SAVILLE, T. (1961). Experimental determination of wave set-up. *Proc. 2nd Tech. Conf. Hurricanes*, pp. 242–52.

SHEPARD, F. P., K. O. EMERY, and E. C. LAFOND (1941). Rip currents: a process of geological importance. *J. Geol.*, 49; no. 4: 337–69.

SHEPARD, F. P., and D. L. INMAN (1950). Nearshore circulation related to bottom topography and wave refraction. *Trans. Am. Geophys. Union*, 31, no. 4: 555–65.

——— (1951). Nearshore circulation. *Proc. 1st Conf. Coast. Eng.*, pp. 50–59.

SONU, C. J. (1972). Field observation of nearshore circulation and meandering currents. *J. Geophys. Res.*, 77, no. 18: 3232–47.

SONU, C. J., J. M. MCCLOY, and D. S. MCARTHUR (1967). Longshore currents and nearshore topographies. *Proc. 10th Conf. Coast. Eng.*, pp. 524–49.

THORNTON, E. B. (1971). Variations of longshore current across the surf zone. *Proc. 12th Conf. Coast. Eng.*, pp. 291–308.

URSELL, F. (1952). Edge waves on a sloping beach. *Proc. Roy. Soc.* (London), series A, 214: 79–97.

Chapter 8

LONGSHORE TRANSPORT OF SEDIMENTS ON BEACHES

The empiricist . . . thinks he believes only what he sees, but he is much better at believing than at seeing.

G. Santayana
Skepticism and Animal Faith (1955)

The longshore movement of sand on beaches manifests itself whenever this natural movement is prevented through the construction of jetties, breakwaters, and groins (Chapter 12). Such structures act as dams to the littoral drift, causing a build-up of the beach on its updrift side and simultaneous erosion in the downdrift direction (Figures 8-1 and 12-5). This often has severe consequences in the erosion of coastal property, and for this reason coastal engineers have shown considerable interest in the quantities of littoral drift and the processes that produce this movement. Littoral transport is also of interest to the geologist as a sand-transporting agent and to the geomorphologist because of its role in the formation of sand spits and other coastal sediment features.

In the nineteenth century it was commonly believed that tidal currents and ocean currents that approach close to the shore are primarily responsible for littoral sand transport. We now know that the wave-induced longshore currents (Chapter 7) are the chief cause of the sediment movement; the other currents are effective only under exceptional circumstances. For example, near the mouth of a bay where tidal currents become strong, they may become significant in sediment transport on the beach. Strong winds may also generate longshore currents that, combined with wave action, produce a sediment transport. In this chapter our attention will focus mainly

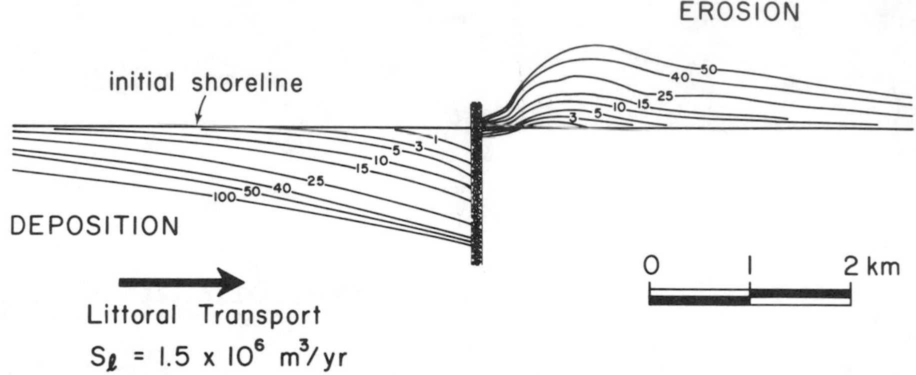

Figure 8-1 The blockage of the littoral sand transport by the emplacement of a breakwater or groin transverse to the drift. This example is from a physical laboratory model of a proposed breakwater at Cotonou, Dahomey; included are the shoreline changes, given in years, as well as the drift rate pertaining to the prototype. Twenty minutes of model operation is equivalent to a year of prototype conditions according to the model. [*After* Sireyjol (1965)]

on sand transport by wave-induced currents. However, the more general sand transport relationships can easily be applied to evaluate sand transport where wave action combines with the other types of nearshore currents.

WAVE POWER APPROACH

A long-standing desire of the coastal engineers has been to be able to estimate the *littoral transport rate*, the rate at which sediment moves alongshore, from a knowledge of the wave and current parameters that cause the transport. To attempt this they have relied most heavily on empirical correlations between the rate of sand drift and the expression

$$P_l = (ECn)_b \sin \alpha_b \cos \alpha_b \tag{8-1}$$

where $(ECn)_b$ is the wave energy flux evaluated at the breaker zone, and α_b is the angle the breaking wave crests make with the shoreline; P_l has been variously called the "longshore component of wave power" and the "longshore component of the wave energy flux." Objections have been raised to this terminology (Eaton, 1951; Longuet-Higgins, personal communication), since the $\cos \alpha_b$ factor should not be included in any "longshore component." The reasoning behind the inclusion of the $\cos \alpha_b$ factor is that $(ECn)_b$ is the energy flux per unit wave crest length, and the $\cos \alpha_b$ converts this to a unit shoreline length basis. However, to avoid any semantic problems we shall call the expression of equation (8-1) the "P_l-parameter."

The empirical approach of correlating the sand transport rate to the P_l-parameter basically dates back to the investigations early in this century of the Danish engineer Munch-Petersen (see discussion in Beach Erosion Board, 1950), who demonstrated that the drift rate could be qualitatively predicted as to direction and amount by determining the balance of the longshore component of the wave energy per unit crest length. Scripps Institution of Oceanography (1947) first suggested that the work (power or energy flux) performed by the waves in the nearshore zone might be a useful parameter for predicting the littoral transport rate from the wave action. Following this suggestion, the Los Angeles District Corps of Engineers (see Eaton, 1951) demonstrated an approximate correlation between the littoral drift rate and the P_l-parameter of equation (8-1). The wave parameters were determined by hindcast and refraction techniques. The results did not warrant adoption of any empirical relationships.

Watts (1953*b*) obtained the first field measurements by which the sand transport rate could be related to the local wave characteristics. Based on the rate at which a bypassing plant had to pump sand past the jetties at South Lake Worth Inlet, Florida, he proposed the use of the empirical relationship

$$S_l = 0.0011P_l^{0.9} \tag{8-2}$$

where S_l is the longshore volume transport rate of sand in cubic yards per day, P_l has units of foot-pounds per day per foot of beach length, and pound is a unit of force (the acceleration g is included).

Caldwell (1956) correlated P_l to the rate of dispersion of an artificial accumulation of sand placed on the beach at Anaheim Bay, California. Combining his data with that collected by Watts (1953*b*), Caldwell obtained the approximate empirical relationship

$$S_l = 210P_l^{0.8} \tag{8-3}$$

where P_l is now in millions of foot-pounds per day per foot of shoreline.

Savage (1959, figure 4) summarized the available data from field and laboratory studies in a graph of S_l versus P_l. Inman and Bagnold (1963) reinterpreted the data and obtained

$$S_l = 125P_l \tag{8-4}$$

as a good fit of the data plotted by Savage (1959), but with an exponent of unity and where the units are the same as for equation (8-3).

The empirical approach may provide an approximate engineering "answer" with regard to the quantity of sand moving along a beach under a given set of wave conditions. However, besides providing little consideration of the real processes that produce the transport, the empirical approach also gives rise in this instance to difficulties concerning the physical units. Consider, for example, equation (8-2). For the units to properly balance in the relationship, the empirical coefficient 0.0011 must necessarily have the units $(\text{ft}^3/\text{day})/(\text{ft-lb}/\text{ft}\cdot\text{day})^{0.9}$. The value of the coefficient is therefore entirely dependent upon the units chosen in the correlation; the value would be different for any other possible units.

Inman and Bagnold (1963) have pointed out that the littoral transport rate would be better expressed as an immersed-weight transport rate I_l rather than as a volume transport rate S_l. The two are related by

$$I_l = (\rho_s - \rho)ga'S_l \tag{8-5}$$

where ρ_s and ρ are respectively the sand and water densities, and a' is the correction factor for the pore space of the beach sand (approximately 0.6 for most beach deposits). The use of the immersed-weight transport rate is based on a consideration of the problem of sediment transport in general from an energetics point of view by Bagnold (1963). There are two distinct advantages in relating I_l and P_l, rather than S_l and P_l. The first is that I_l has units of work (energy flux, power), being dyn/sec or ergs/cm · sec in cgs units, the same as P_l, so that they can be related by

$$I_l = K(ECn)_b \sin \alpha_b \cos \alpha_b = KP_l \tag{8-6}$$

where K is a dimensionless proportionality coefficient. The second advantage in using the immersed-weight transport rate is that, as can be seen in equation (8-5), it takes into consideration the density of the sediment grains. Therefore, a correlation between I_l and P_l could be used for beaches composed of coral sand and so forth, as well as the more usual quartz sand.

Komar and Inman (1970) obtained simultaneous measurements of the wave and current parameters in the surf zone and the resulting longshore transport of sand. Unlike the previous studies, the measurements represent short periods of time, on the order of 2 to 4 hours. Quantitative measurements of the littoral transport rate were obtained from the time history of the center of mass of sand tracers, natural sand tagged with a thin coating of fluorescent dye. The use of such sand tracers is discussed in detail by Ingle (1966). The direction and flux of the wave energy was measured from an array of digital wave sensors placed in and near the surf zone. Two beaches were involved in the study: El Moreno Beach, a coarse-sand beach with a steep face located on the Gulf of California, Mexico, and Silver Strand Beach, California, a fine-grained beach of low profile with an extensive surf zone. The two locations provide a strong contrast in the wave regime as well as in the beach configuration.

To compare the results of Komar and Inman (1970) with the measurements of Watts (1953*b*) and Caldwell (1956), it is necessary to place the computations of energy on the same basis. The computations of Watts and Caldwell were obtained from significant wave height analyses of analog records, whereas those of Komar and Inman were based on spectral analyses. It may be recalled from Chapter 3 that Longuet-Higgins (1952) predicted theoretically, and measurements have since substantiated, that the significant wave height is a factor 1.42 greater than the root-mean-square wave height. This means that the wave energy calculated from the significant wave height would be about a factor 2 ($= 1.42^2$) too high, twice that obtained by spectral analysis. Therefore, to place the data of Watts (1953b) and Caldwell (1956) on the same basis as that of Komar and Inman (1970), and to give it the correct levels of energy, the values of the energy flux reported by Watts and Caldwell must be halved. This has been done in compiling the measurements in Figure 8-2, correlating I_l and P_l. No such correction is required for the laboratory measurements included in the figure, as these are based on simple sinusoidal waves.

The straight line of unit slope fitted to the field data in Figure 8-2 yields a coefficient $K = 0.77$, so that equation (8-6) becomes

$$I_l = 0.77(ECn)_b \sin \alpha_b \cos \alpha_b = 0.77P_l \tag{8-6a}$$

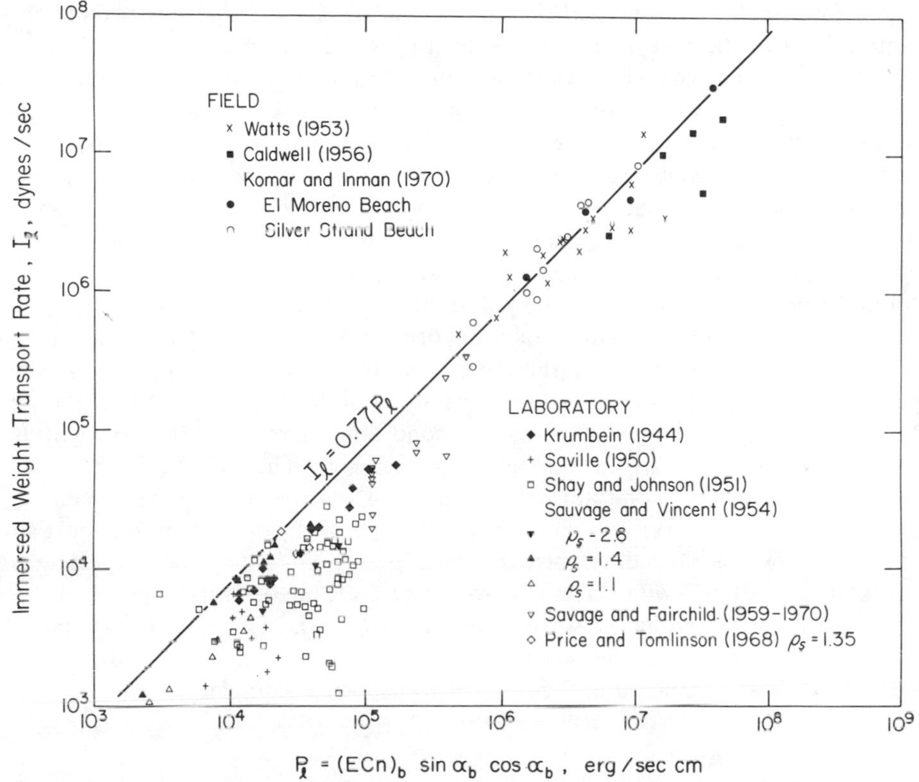

Figure 8-2 The available data on the immersed-weight sand transport rate as a function of the P_l-parameter. [*After* Komar and Inman (1970)]

In making estimates of the transport rate from the wave parameters, either Figure 8-2 can be used directly or equation (8-6a) can be utilized.

Figure 8-2 raises other points of interest. The agreement between the two beaches involved in the study of Komar and Inman (1970) suggests a lack of significant dependence of the transport rate on the beach slope. The slope at El Moreno Beach is some four times greater than that at Silver Strand Beach, so any dependence, unless very small, should make itself apparent. This lack of variation with beach slope can be attributed to a similar lack of dependence of the longshore current velocity on the beach slope (Chapter 7). We shall look at this again later in the chapter.

El Moreno and Silver Strand beaches also represent a contrast in the sediment grain size: 600 microns mean diameter at El Moreno and 175 microns at Silver Strand. Suspended-load transport depends largely on the settling velocity and hence on the sediment grain size. The agreement between the data from the two localities therefore suggests that the suspended-load transport of sand in the surf zone is less important than the bed-load transport. The relative significance of bed-load and suspended-load transport on beaches will be discussed further in a moment.

Notice in Figure 8-2 how nearly all the laboratory measurements fall below the straight line established from the field data: the straight line actually appears to form an upper limit to the plot. It has been shown for sand transport in rivers (Bagnold, 1966) that the immersed-weight transport rate is proportional to the available power of the flowing water, with a constant proportionality factor, only when the transport conditions are fully developed (sheet sand movement under river flood flow). Under lower flow regimes, those which are not fully developed and where the sediment bottom is rippled, the available power to transport sand is less and the proportionality factor is lower and no longer constant. A similar effect can also be expected to occur in the transport of sand on beaches and could account for the observed distribution of plotting positions of the laboratory data. This would suggest that the transport conditions are not fully developed for a majority of the laboratory experiments and that the transport mechanism is therefore less efficient than for conditions which prevail on sand beaches in the field. The few laboratory measurements that were obtained under fully developed transport conditions agree with the straight line extrapolated in Figure 8-2 from the field measurements. The straight line represents the maximum transport conditions that can occur with a given P_l-parameter, and therefore exists as an upper boundary to the plotting of the data. Any transport conditions that are not fully developed, whether in the laboratory or natural gravel beaches, would fall below the line and K would be less than 0.77. If the material is too coarse even to be moved by the wave action, then $K = 0$. In a sense, K incorporates an efficiency factor between the available power P_l for moving beach sediment and the actual work accomplished in the form of transported sediment, I_l.

ENERGETICS MODELS OF SAND TRANSPORT ON BEACHES

The above correlation between the sand transport rate and the P_l-parameter, although apparently successful, is basically empirical, and its formulation involved no real considerations of the mechanics of sand transport. A more fundamental examination of the processes involved is therefore needed.

An alternative, seemingly independent relationship connecting the immersed-weight transport rate to the wave and current parameters, one which better accounts for the processes involved, is given by Bagnold (1963, p. 518). Under the orbital motion of waves, sediment is moved with a back-and-forth motion with no net transport even though wave energy is expended (Figure 8-3). This expenditure of energy supports and suspends the sand above the bottom, so that no additional stress is required to transport it. The presence of any unidirectional current superimposed on the to-and-fro motion therefore could produce some net drift of sediment. The magnitude of this current does not matter in that it does not have to provide the stress to support the sediment above the bottom, this having already been accomplished by the waves. This model is shown conceptually in Figure 8-3, and with it Bagnold (1963) derived the relationship

$$i_\theta = K'\omega\frac{u_\theta}{u_o} \tag{8-7}$$

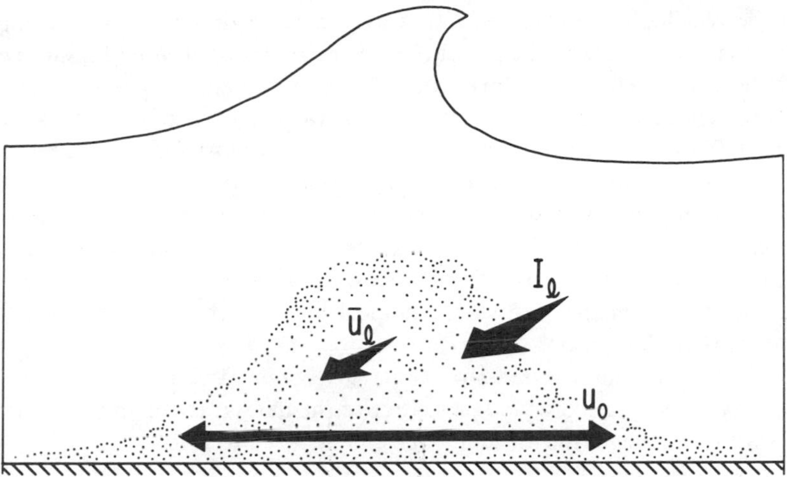

Figure 8-3 Schematic of model for sand transport wherein the orbital velocity u_o due to the waves places the sand in motion, and the current $\bar{u}_l$ provides a net transport of sand, I_l.

for the sediment transport per unit width, i_θ, in the direction θ determined by the unidirectional current u_θ. The available power from the wave motion, ω, supports the sediment above the bottom, and u_o is the orbital velocity of the wave motion such that ω/u_o becomes the stress exerted by the waves, K' is a dimensionless coefficient. Inman and Bagnold (1963, p. 545) have applied this model to the littoral zone, where they assumed that a portion of the wave energy flux $(ECn)_b \cos \alpha_b$ is dissipated in placing the sand in motion, so that the mean stress applied to the beach bed is proportional to $[(ECn)_b \cos \alpha_b]/u_o$, where u_o is now the mean speed of the wave motion relative to the bed within the surf zone and is assumed to be proportional to the orbital velocity near the bottom just before the wave breaks. Once the sediment is in motion, it becomes available for transport by the longshore current $\bar{v}_l$, so that the total immersed weight of sediment transported in unit time past a section of beach becomes

$$I_l = K'(ECn)_b \cos \alpha_b \frac{\bar{v}_l}{u_o} \tag{8-8}$$

The model of Bagnold (1963), equation (8-7) above, was partially verified by Inman and Bowen (1963), who superimposed currents on wave motion in a wave channel and measured the resulting sand transport along the channel. The results of that study will be examined in greater detail in Chapter 11. The model applied to the nearshore region, equation (8-8), was tested by Komar and Inman (1970) with the field measurements they obtained. The average longshore current $\bar{v}_l$ needed in the model was measured while the sand transport was in progress. To obtain a representative $\bar{v}_l$ value, measurements were made along an extended length of beach, along the complete length of beach over which the sand tracer was transported, and across the entire nearshore zone. All individual measurements were then averaged to obtain

$\bar{v}_l$. It will be recalled from Chapter 7 that averaging in this way removes any longshore currents that are due to the presence of cell circulation, those longshore currents that feed the rip currents. If the average is based on adequate measurements, the $\bar{v}_l$ that remains will be that longshore current generated by waves arriving at an angle to the shore. This is the current that is of interest in the unidirectional transport of sand along a beach. The cell circulation can produce only local sediment transport, and if the cells migrate alongshore, as they are commonly observed to do, the sediment transport due to the cell circulation currents would alternate in direction as the current direction changes. In Chapter 10 we shall examine the transport of sand due to circulation cells that are stable in position and the resulting modifications of the shoreline configuration.

Figure 8-4 contains the measurements of Komar and Inman's (1970) testing equation (8-8). Although there is some scatter to the data, it is apparent that it follows a trend to which a straight line of unit slope can be fitted to give

$$I_l = 0.28(ECn)_b \cos \alpha_b \frac{\bar{v}_l}{u_m} \tag{8-9}$$

where u_m is the maximum orbital velocity under the breaking wave and is calculated with

$$u_m = \left(\frac{2E_b}{\rho h_b}\right)^{1/2}$$

Therefore, equation (8-9) is successful in relating the littoral transport rate to the wave and current parameters. There must be some connection between this model and the

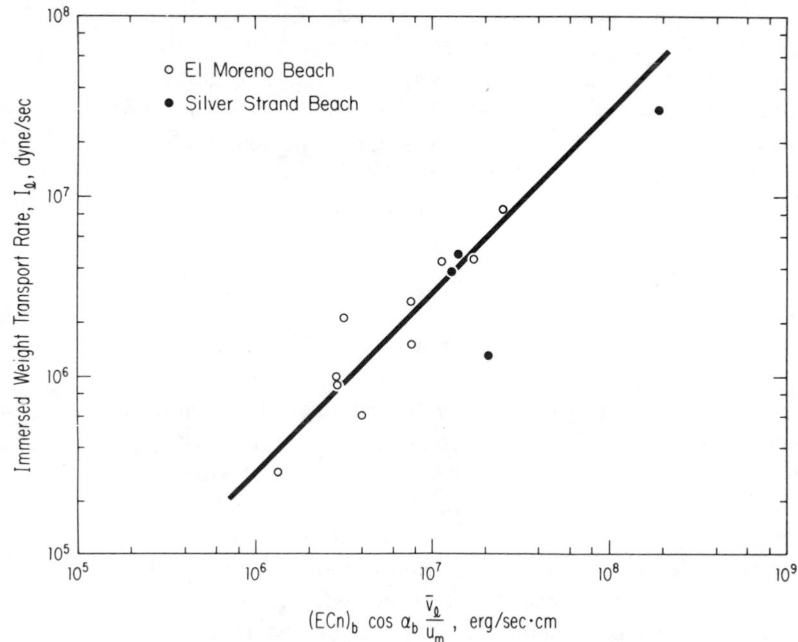

Figure 8-4 Relationship between the immersed-weight longshore sand transport rate and the model of equation (8-8) deduced by Bagnold (1963) and Inman and Bagnold (1963). [*After* Komar and Inman (1970)]

empirical correlation between the sand transport and the P_l-parameter, equation (8-6a), otherwise we would have two separate equations relating the same physical parameters, something nature abhors. Komar and Inman (1970) concluded that the connection between the two is through the mode of generation of the current $\bar{v}_l$ in equation (8-9). They simply solved equations (8-6a) and (8-9) simultaneously to obtain

$$\bar{v}_l = 2.7 u_m \sin \alpha_b \tag{8-10}$$

for the longshore current. We saw in Chapter 7 that this equation successfully relates the longshore current to the wave parameters, even though it originally was based on the sand transport data. Furthermore, the relationship is basically the same as that derived theoretically by Longuet-Higgins (1970). In fact, it was the results of Komar and Inman that suggested such a correspondence to Longuet-Higgins. If we take the theoretical equation for $\bar{v}_l$ obtained by Longuet-Higgins [Chapter 7, equation (7-23)] and substitute it into equation (8-9), we obtain

$$I_l = 0.55\left(\frac{\tan \beta}{c_f}\right)(ECn)_b \sin \alpha_b \cos \alpha_b \tag{8-11}$$

which is the same as the empirical equation (8-6a) except for the presence of the ratio $\tan \beta / c_f$. It will be recalled that in Chapter 7 we saw that this ratio is approximately constant. One can conclude from this that the empirical correlation between the sand transport and the P_l-parameter, equation (8-6a), is successful primarily because the more basic model, equation (8-9), which considered the mechanics of transport involved, is correct and that the longshore current $\bar{v}_l$ is of the form deduced by Komar and Inman (1970) and derived by Longuet-Higgins (1970).

This would suggest, then, that the correlation between I_l and P_l is a matter of luck and that there is nothing fundamental in the relationship. However, Komar (1971) derives the $I_l = KP_l$ relationship directly for the special mode of littoral transport known as swash transport. *Swash transport* occurs on steep-faced beaches where the waves break and swash at an angle to the shoreline, the littoral drift resulting from the sawtooth motions of the sediment along the shore (Figure 8-5). No true longshore currents exist, so that the model of equation (8-9) does not apply. Instead, the drift is the longshore component of the sediment that is swept up the beach face under the advancing swash. Longuet-Higgins (1973) provides an alternative derivation by considering the thrust in the longshore direction from the waves arriving to the shore at an angle. This is the same thrust that produces the longshore current, but Longuet-Higgins equates it directly to the sand transport rather than having the thrust produce the longshore current which in turn causes the sand transport. Longuet-Higgins would therefore dispense with equation (8-9). I prefer to view the radiation stress as producing the current that in turn causes the sand transport by interacting with the wave orbital motions according to the model of Bagnold (1963). The other objection to Longuet-Higgins's approach is that equation (8-9) is more versatile; it can be applied to cases where $\bar{v}_l$ is generated by tidal currents, wind-driven currents, and currents due to the cell circulation, not just to the conditions where the current is due to the oblique wave approach to the shoreline.

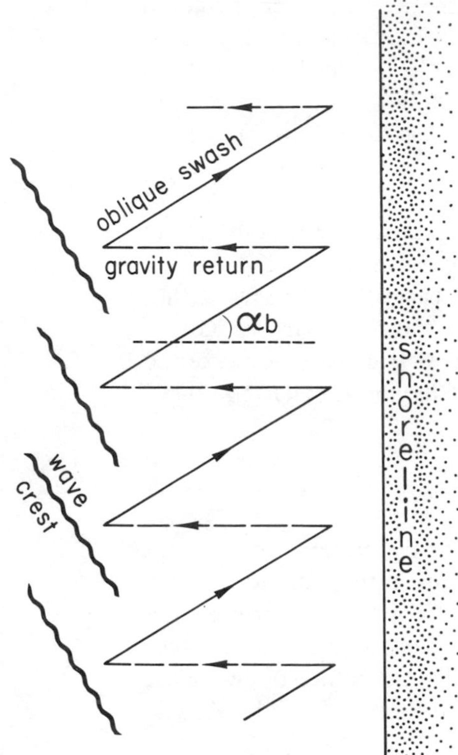

Figure 8-5 The zigzag motion of the sediment along a steep beach face under the wave swash. The incoming wave swash drives the sand up the beach at an oblique angle and the return gravity flow washes it back to its original level.

Komar (1971) has used the model of equation (8-7) to examine the variation in sand transport across the width of the nearshore zone. The distribution is basically proportional to the product of the local stress exerted by the waves, that which puts the sand into motion, and the longshore current velocity. It was shown that with certain reasonable assumptions, integration of the relationships across the surf zone would lead to equation (8-8), so that the averaging approach used by Inman and Bagnold (1963) to obtain that equation would appear to be satisfactory. Figure 8-6 shows an example of the resulting distribution of the longshore sand transport, based on an extension of the approach of Komar (1971). The longshore current distribution of Figure 8-6 is given by equation (7-30) of Chapter 7. Summation of the sand transport distribution across the surf zone given in Figure 8-6 is equal to the total sand transport as given by equation (8-6a). This equivalence of the summation to the total transport serves to "calibrate" the solution for the sand transport distribution. It is seen that the sand transport rate reaches a maximum shoreward of the breaker zone but seaward of the maximum of the longshore current distribution. This is because the sand transport rate is proportional to the product of the longshore current and the local stress exerted by the wave motion, this stress increasing from the shoreline to a maximum at the breaker line.

Thornton (1973) has also examined variations in sand transport, both across

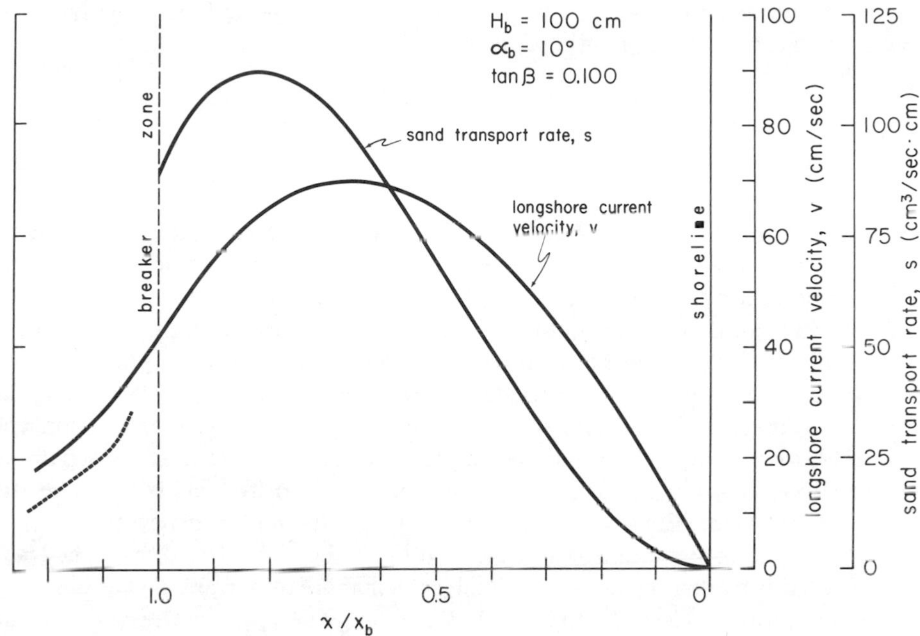

Figure 8-6 The distribution of the longshore sand transport across the width of the nearshore obtained by having the local transport proportional to the product of the bottom stress exerted by the waves and the local value of the longshore current velocity. The proportionality coefficient is determined by equating the summation of the sand transport distribution to the total transport given by equation (8-6a).

the surf zone and beyond the breaker zone. His theoretical treatment is similar to that developed by Komar (1971), but Thornton obtained measurements of the transport to test the theory. These measurements were collected with bed-load traps, which are basically small catchments buried in the sand with a slot oriented such that they catch sand as it moves alongshore close to the bed. Measurements with such traps are highly uncertain, especially within the surf zone where they function poorly and have a tendency to become buried. Thornton estimated that the traps are probably only about 40 to 100% efficient in trapping the transported sand, doing most poorly under high-transport conditions. Although the absolute amounts are therefore in question, the measurements gave reasonable trends across the surf, with a maximum transport in the breaker zone, decreasing rapidly both offshore and shoreward, where at the test site there was a longshore trough. Agreement between theory and measurement was resonably good.

In summary, equation (8-9) remains the most general of the sand transport relationships and can be used where $\bar{v}_l$ is a current other than that produced by an oblique wave approach to the shoreline. For example, $\bar{v}_l$ could be generated as a local wind-driven current, a tidal current, or as a longshore current associated with the nearshore cell circulation. In the special case where $\bar{v}_l$ is due entirely to an oblique wave approach and thus is given by equation (7-27), the longshore sand transport is given by equation (8-6a), relating I_l to P_l.

SUSPENDED- VERSUS BED-LOAD TRANSPORT

The total littoral transport consists of the sum of the suspended load and the bed load. The *suspended load* is transported within the water column, maintained above the bottom by the turbulence of the water. The *bed load* comprises the concentrated sediment that moves on or in close proximity to the bottom, maintained in a dispersed state by grain-to-grain contacts. Under these definitions, grains that are saltating across the bottom are included in the *bed-load transport*. Although this distinction between bed load and suspended load is precise, following the concepts of Bagnold, from a practical standpoint of measurement they are nearly inseparable. The above considerations and measurements of littoral drift include *both* the suspended and bed loads. Equations (8-6) and (8-9) give the *total* transport rate.

Which is more significant, the suspended or the bed load transport? The traditional view has been that the suspended load is much more important, a view especially held in engineering circles. The first impression one gets by wading through the surf zone is that there are considerable quantities of sand in suspension. However, most of this sand usually moves as a smooth sheet close to the bottom, typical of sheet bed-load movement, rather than following the irregular paths of turbulent eddies which support any suspended load. This is shown by the study of Brenninkmeyer (1975). His observations were made with almometers (a vertical series of photoelectric cells opposite a light source) which detect the light transmission that is a function of the sand concentration in the water. With proper calibration, the almometers yield a continuous measure of sand concentration in the waters of the surf. Brenninkmeyer's observations extended across the entire surf and breaker zones on the beach at Point Magu, California. With wave periods between about 10 to 15 seconds, he found that within the breaker zone sand thrown above the bottom lasted on an average of only 4.5 seconds. Even more important, significant uplifts occurred only once every 35 minutes. This means that the primary breakers seldom penetrate entirely to the bottom; instead, their plunge is absorbed by the underlying water column. Sand movement was found to be restricted primarily to within 10 cm of the bottom. In the outer surf zone Brenninkmeyer found that sand movement is small, suspension being almost nonexistent. Within the inner surf zone sand transport increases in frequency and elevation above the bottom, reaching a maximum in the "transition zone" (Chapter 11), where the return swash from the beach face collides with incoming wave bores to generate considerable turbulence and, in some cases, even standing hydraulic jumps. Brenninkmeyer found that in this transition zone the air-saturated bore penetrates to the bottom, lifting concentrations up to 500 grams of sand per liter of water to heights of 50 cm above the bottom. The conclusion from the study of Brenninkmeyer is that suspended sediment is important only within the narrow transition zone but not in the remainder of the surf zone and breaker zone. Because of this, and since the longshore currents are strong in the outer surf zone and weak within the transition zone, it can be expected that bed-load transport will dominate over suspended-load transport.

We can make a very rough estimate of the suspended-load transport rate from measured concentrations of suspensions that have been obtained within the surf and breaker zones, chiefly by Watts (1953*a*) and Fairchild (1973). In contrast to the study of Brenninkmeyer (1975), Watts and Fairchild both obtained their measurements of sediment concentration by pumping samples from the surf. If c is the average volume concentration of sand in suspension (the volume of sand per unit volume of sand-water mixture), assuming that this suspended sand is carried alongshore by the longshore current $\bar{v}_l$, the longshore flux of sediment load is given by

$$\text{Flux} = c\bar{v}_l A \tag{8-12}$$

where A is the total cross-sectional area of the nearshore region. Placing the flux into terms of a static volume of transported sediment due to suspended load, we have

$$S_l\,[\text{suspension}] = \frac{\text{Flux}}{a'} = \frac{c\bar{v}_l A}{a'} \tag{8-13}$$

where $a' = 0.6$ is the pore space factor, as before. Putting this in turn on an immersed-weight basis, we have

$$I_l\,[\text{suspension}] = (\rho_s - \rho)ga'S_l\,[\text{suspension}] = (\rho_s - \rho)gc\bar{v}_l A \tag{8-14}$$

The cross-sectional area of the nearshore region can be approximated by the triangle shown in Figure 8-7, whose sides are X_b, the width of the nearshore zone, and h_b, the depth at breaking. The area then becomes

$$A = \frac{1}{2}X_b h_b = \frac{1}{2\gamma_b^2}\frac{H_b^2}{\tan\beta} \tag{8-15}$$

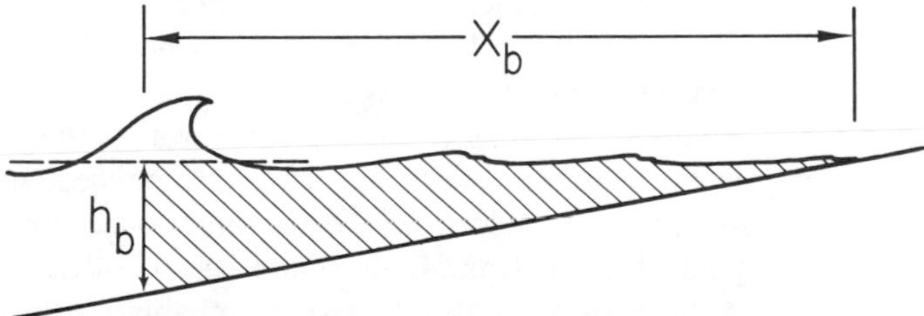

Figure 8-7 The area of the nearshore zone approximated as a triangle with sides X_b and h_b.

Using equation (7-27) for the longshore current $\bar{v}_l$, the form based on the results of Komar and Inman (1970) and Longuet-Higgins (1970) and, as we have seen, the relationship that gives the best prediction, equation (8-14) becomes

$$I_l\,[\text{suspension}] = \frac{2.7}{2\gamma_b^2\tan\beta}(\rho_s - \rho)gcu_m H_b^2 \sin\alpha_b \cos\alpha_b \tag{8-16}$$

This relationship is similar to those derived by Dean (1973) and Galvin (1973). It differs from Galvin's in that he utilized the longshore current relationship of Galvin and Eagleson (1965), equation (7-16), which shows little agreement with the available field data. Dean's development is more thorough than that presented here.

Let us now compare equation (8-16) for the suspended-load transport with the *total* tranport rate given by

$$I_l = 0.77(ECn)_b \sin \alpha_b \cos \alpha_b = 0.77(\tfrac{1}{8}\rho g H_b^2 C_b) \sin \alpha_b \cos \alpha_b \tag{8-17}$$

our interest being the transport from an oblique wave approach. The ratio of suspension transport to the total transport becomes

$$\frac{I_l\,[\text{suspension}]}{I_l} = \frac{\text{Suspended}}{\text{Total}} = \frac{7.0c}{\gamma_b \tan \beta}\left(\frac{\rho_s - \rho}{\rho}\right) \tag{8-18}$$

where $u_m/C_b = \gamma_b/2$ has been used to elimiate that ratio. We can use equation (8-18) to obtain an approximate estimate of the suspended-load transport rate in comparison to the total transport rate. The evaluation depends almost entirely on the value of c, the measured concentration of the suspended sediment in the surf and breaker zones.

From samples pumped from the surf zone, Watts (1953*a*) obtained concentrations of sand consistently about $c = 4.6 \times 10^{-4}$ (volume of sediment per unit total volume). Similar but somewhat lower values were obtained by Fairchild (1973). Using this concentration in equation (8-18), and setting $\tan \beta = 0.035$ as an average slope of a fine-sand beach, we obtain

$$\frac{I_l\,[\text{suspension}]}{I_l} = \frac{7 \times 4.6 \times 10^{-4}}{0.80 \times 0.035}\left(\frac{2.65 - 1.03}{1.03}\right) \simeq 0.2$$

This indicates that the suspended-load transport rate I_l [suspension] is approximately one-fifth of the total transport; the bed load must make up the other four-fifths. This rough estimate therefore supports the conclusion of Komar and Inman (1970) and Brenninkmeyer (1975) that the bed-load transport is much more significant than the suspended load. The above calculation may actually overestimate the importance of the suspended load in that in the development of the model it was assumed that the suspended sediment moves alongshore with the velocity $\bar{v}_l$ of the longshore current; no lag of the sediment behind the flowing current is permitted. In addition, the measurements of c by Watts (1953) and by Fairchild (1973) were obtained in close proximity to piers extending out through the nearshore zone, the pilings of which produce appreciable scouring and enhancement of the amount of sand placed in suspension. In addition, the maximum concentrations measured were collected within 8 cm of the bottom. This would be within the sheet-movement carpet of sand close to the bottom. Since this carpet probably moves by grain-grain interactions, it should be considered as part of the bed load rather than a suspended load. More realistic values of c may be expected to be below the 4.6×10^{-4} measurement used in the above estimate of the suspended load. Incidently, in his model of littoral transport, Galvin (1973) used $c = 3.2 \times 10^{-4}$, lower than the average we have used but probably still too high in light of the above evaluations of the measurements of c. Using Galvin's average c in equation (8-18) gives a suspended load less than 14%, the bed load being the remaining 86%.

The main argument in favor of suspended-load transport over a bed-load transport is the measurements by Thornton (1973) using bed-load traps. These traps capture only a small portion of what must have been the total sand transport, and this has been taken to indicate that the suspended load must be large, making up the bulk of the untrapped sand transport. However, we have seen that the efficiency of bed-load traps is in question, in that they tend to clog and bury and thus miss much of the transport. In addition, since the bed load may extend for as much as 10 cm or so above the bottom, only a very small portion of the bed load would be sampled, even if the sampler were 100% efficient.

The indications are, then, that the bed-load transport dominates over the suspended-load transport on beaches. The prevalence of the bed load can in part be explained by the loss of fine sediments offshore. Any material that is able to stay in suspension tends to be carried out of the nearshore zone by rip currents or by simple offshore diffusion. This is the reason for the deficiency on beaches of grains in the silt and finer sizes and the reason for the absence of slowly settling particles such as the micas (Chapter 13). Only material coarse enough to remain in the bed load can maintain its position on the beach.

NET LITTORAL TRANSPORT

The net longshore movement of sediment at any beach is the sum of the transport under all the individual wave trains arriving at the shore from countless wave-generation areas. For example, on a north-south–trending beach, the sand may move northward for a time due to waves arriving from the south and then later move to the south under waves coming from the north. The net transport of sediment under these two wave trains will be the difference between the north and south movements. This net transport is generally small, much smaller than the total transport up and down the beach, and on some beaches may be essentially zero.

This change in direction may be seasonal. For example, on the beaches of southern California the transport during the winter is predominately to the south, storm waves arriving from the Gulf of Alaska and other areas to the north. During the summer, swell waves arrive from the Southern Hemisphere, which is then in its storm-producing winter months. Thus there is a north transport of sand on the California beaches during the summer. However, the south transport during the winter is larger, so that the net transport is to the south, although small compared to the total transport the sand has undergone.

To estimate the net transport, one can evaluate I_l for each wave train arriving at the beach and then obtain a net I_l by balancing the individual values according to direction. Alternatively, one can balance the individual P_l values of the storm waves to obtain a net P_l at the beach and then use this value in equation (8-6a) to calculate the net I_l.

The long-term net sediment transport along a beach is reflected in the accumulation of sediment at jetties and breakwaters. Johnson (1956, 1957) has compiled in this way estimates of the littoral drift at a number of locations along the coasts of the world (Table 8-1). It is seen that the littoral drift can produce a movement of up

TABLE 8-1 Littoral Drift Rates Along Coasts [*From* Johnson (1956, 1957)]

LOCATION	DRIFT RATE (m^3/yr)	PREDOMINANT DIRECTION	YEARS OF RECORD
Atlantic Coast			
Suffolk Co., N.Y.	255,000	W	1946–55
Sandy Hook, N.J.	377,000	N	1885–1933
Sandy Hook, N.J.	334,000	N	1933–51
Asbury Park, N.J.	153,000	N	1922–25
Shark River, N.J.	255,000	N	1947–53
Manasquan, N.J.	275,000	N	1930–31
Barnegat Inlet, N.J.	191,000	S	1939–41
Absecon Inlet, N.J.	306,000	S	1935–46
Ocean City, N.J.	306,000	S	1935–46
Cold Springs Inlet, N.J.	153,000	S	—
Ocean City, Md.	115,000	S	1934–36
Atlantic Beach, N.C.	22,600	E	1850–1908
Hillsboro Inlet, Fla.	57,000	S	—
Palm Beach, Fla.	115,000 to 172,000	S	1925–30
Gulf of Mexico			
Pinellas Co., Fla.	38,000	S	1922–50
Perdido Pass, Ala.	153,000	W	1934–53
Galveston, Texas	334,700	E	1919–34
Pacific Coast			
Santa Barbara, Calif.	214,000	E	1932–51
Oxnard Plain Shore, Calif.	756,000	S	1938–48
Port Hueneme, Calif.	382,000	S	1938–48
Santa Monica, Calif.	207,000	S	1936–40
El Segundo, Calif.	124,000	S	1936–40
Redondo Beach, Calif.	23,000	S	—
Anaheim Bay, Calif.	115,000	E	1937–48
Camp Pendleton, Calif.	76,000	S	1950–52
Great Lakes			
Milwaukee Co., Wis.	6,000	S	1894–1912
Racine Co., Wis.	31,000	S	1912–49
Kenosha, Wis.	11,000	S	1872–1909
Ill. state line to Waukegan	69,000	S	—
Waukegan to Evanston, Ill.	44,000	S	—
South of Evanston, Ill.	31,000	S	—
Outside the United States			
Monrovia, Liberia	383,000	N	1946–54
Port Said, Egypt	696,000	E	—
Port Elizabeth, South Africa	459,000	N	—
Durban, South Africa	293,000	N	1897–1904
Madras, India	566,000	N	1886–1949
Mucuripe, Brazil	327,000	N	1946–50

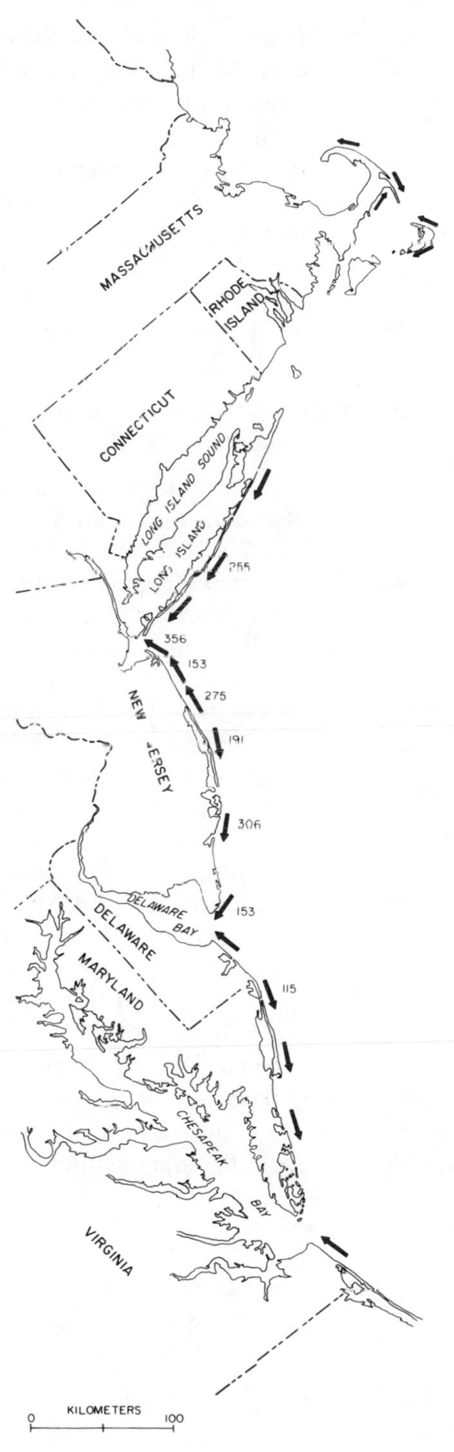

Figure 8-8 Littoral drift directions and magnitudes (given in thousands of cubic meters per year) for the northeastern United States. [*Data from* Johnson (1956, 1957)—also given in Table 8-1]

to nearly a million cubic meters of sand along the beach in a single year. Johnson's data for the northeastern United States is diagramed in Figure 8-8 along with additional transport-direction arrows suggested by sand spit extension and entrapment behind groins. This area demonstrates the important influence of the coastal orientation on the net transport direction and rate. Beach sand on Long Island, trending roughly east-west, is moving to the west, while in northern New Jersey the sand is moving northward. There is a node on the New Jersey coast such that in southern New Jersey the sand transport is to the south. In general, along the northeast coast, beach sediments are converging toward estuaries and bays such as Chesapeake Bay due to the changes in coast orientation.

THRESHOLD OF SEDIMENT MOTION

A problem related to the littoral drift in the nearshore is that of the threshold of sediment motion under wave action. This will govern, for example, whether shingle is being transported on a beach under certain wave conditions. It also determines to what water depths sand is in motion and could therefore be carried onshore to add to the beach volume, or carried alongshore to contribute basically to a sediment drift parallel to the shoreline. In Chapter 11 we deal more completely with sediment transport beyond the breaker zone, especially the onshore-offshore transport which affects the beach profiles, but let us now examine the sediment threshold under waves.

As the orbital velocity of water flow over a bed of sediment is increased, a stage is reached when the water exerts a force or stress on the particles sufficient to cause them to move from the bed and be transported. This stage is generally known as the *threshold of sediment movement*, or as the *critical stage for erosion or entrainment*.

Many equations have been proposed for the threshold of sediment motion under waves; Silvester and Mogridge (1971) present thirteen different equations gathered from the literature. Komar and Miller (1973) review much of the data on threshold under waves, which comes from harmonically oscillating a cradle holding the sediment within still water (Bagnold, 1946; Manohar, 1955), from an oscillating-flow water tunnel (Rance and Warren, 1969), and from laboratory wave channels (Horikawa and Watanabe, 1967). Due to the difficulties of observation and control, to date no measurements have been obtained from the field.

The review by Komar and Miller (1973) found that for grain diameters less than about 0.5 mm (medium sands and finer) the threshold is best related by the equation

$$\frac{\rho u_t^2}{(\rho_s - \rho)gD} = 0.21\left(\frac{d_o}{D}\right)^{1/2} \tag{8-19}$$

where u_t and d_o are respectively the near-bottom threshold velocity and orbital diameter of the wave motion, related to the wave height H, water depth h, wave length L, and period T by

$$u_t = \frac{\pi d_o}{T} = \frac{\pi H}{T \sinh(2\pi h/L)} \tag{8-20}$$

obtained from Airy wave theory by setting $z = -h$ (Chapter 3). The density of water is ρ, and ρ_s and D are respectively the density and diameter of the sediment grains. This relationship of equation (8-19) is modified after an empirical equation obtained by Bagnold (1946).

For grain diameters greater than 0.5 mm (coarse sands and coarser), the threshold is best predicted with an empirical curve relating d_o/D to $\rho u_t/(\rho_s - \rho)gT$, where T is the wave period, first presented by Rance and Warren (1969). This latter dimensionless number represents the ratio of the acceleration forces to the effective gravity force acting on the grains. The empirical curve can be expressed as the equation

$$\frac{\rho u_t^2}{(\rho_s - \rho)gD} = 0.46\pi \left(\frac{d_o}{D}\right)^{1/4} \tag{8-21}$$

which is similar to equation (8-19) above for finer grain sizes.

Equation (8-19) can be used to evaluate the threshold for grain sizes at least as fine as the lower silt range (where cohesive effects can be expected to cause departures from the established relationships), and equation (8-21) gives good results for grain sizes larger than 0.5 mm and as coarse as 5 cm (the coarsest material studied). For a given grain density and diameter, the threshold under waves can therefore be established by a certain wave period T and orbital velocity u_t or diameter d_o. Since $u_t = \pi d_o/T$, only two of these three parameters need be established in defining the threshold. Utilizing equations (8-19) and (8-21), Komar and Miller (1975*a*) produced Figure 8-9 for the threshold values of T and u_t. This graph assumes $\rho_s = 2.65$ g/cm^3 and so applies only to ordinary quartz sands. Given a grain diameter D, the required combinations of T and u_t are easily selected. It is seen that there are multiple combinations of T and u_t for any given grain size, the higher the wave period T the greater the required orbital velocity u_t. At low wave periods and for grain sizes around 0.5 mm there is disagreement between equations (8-19) and (8-21), so that the curves in Figure 8-9 do not join smoothly. In this transition region Komar and Miller drew smooth dashed curves as a compromise.

Once the threshold wave period and orbital velocity u_t are determined, there are of course many combinations of water depth h and wave height H that could yield the required u_t. The linearized relationship for u_t is given in equation (8-20) and discussed in Chapter 3. To expedite application further, Komar and Miller (1975*b*) prepared a computer program that for a given D and ρ_s combination prints incremental values of T, H, and h which define the threshold of sediment motion. Within the program itself, combinations are eliminated which specify wave conditions that are impossible (the criteria are discussed in Chapter 3).

By evaluating the drag coefficients with the results of Jonsson (1967, figure 6), Madsen and Grant (1975) and Komar and Miller (1975*a*) were able to compare the threshold under waves with the threshold under unidirectional steady currents. The oscillatory threshold data was found to fit the curve given by Shields (1936) for unidirectional threshold, with about the same degree of scatter as the unidirectional threshold data. Therefore, the Shields threshold curve could be used for oscillatory water motions as well as for unidirectional flow. However, this requires evaluation

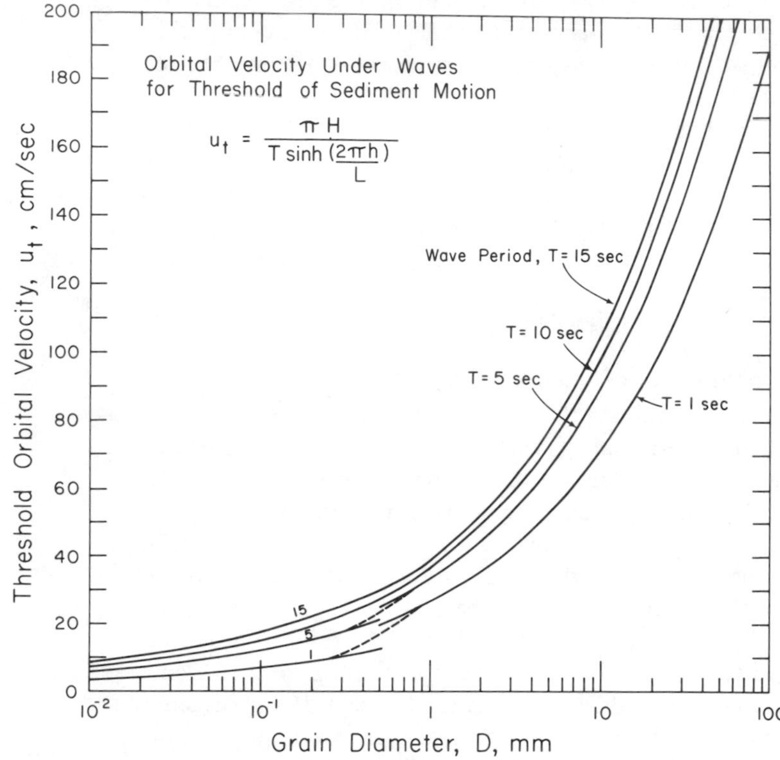

Figure 8-9 The wave period T and near-bottom orbital velocity u_t required for threshold of motion of sediment grain size D and density $\rho_s = 2.65$ g/cm³ (quartz). The orbital velocity is in turn related to the wave height H and water depth h by the equation shown. [*From* Komar and Miller (1975a)]

of the drag coefficient, so that it is much simpler to utilize equations (8-19) and (8-21) to determine the threshold conditions or Figure 8-9 based on these equations.

The threshold evaluated with equations (8-19) and (8-21) have been used to calculate water depths to which a range of sediment grain sizes could be set in motion by waves with a period $T = 15$ seconds. The results are shown in Figure 8-10 for a range of reasonable wave heights. It is seen that waves of this period would be capable of moving sediments to depths of 100 meters and more. Longer period waves would cause sediment motions to still greater water depths. This conforms with the observations of oscillatory ripple marks on continental shelves to these depths (Chapter 13).

Any results using equations (8-19) and (8-21) can be expected to be conservative. As discussed by Silvester and Mogridge (1971), the velocities of water particles near the bottom may be higher than derived by first-order Airy wave theory, the interactions of wave trains of slightly differing period can generate higher instantaneous velocities, and small protuberances on the bed could cause sediment motion at lower velocities and therefore greater water depths than indicated by the analysis. One result of this would be that oscillatory ripple marks could be generated at still greater water depths than indicated by the above threshold equations. Due to the higher instantaneous velocities caused by the above effects, sediment may undergo some motion for a brief span of time, long enough to cause rippling. But since the sediment motion

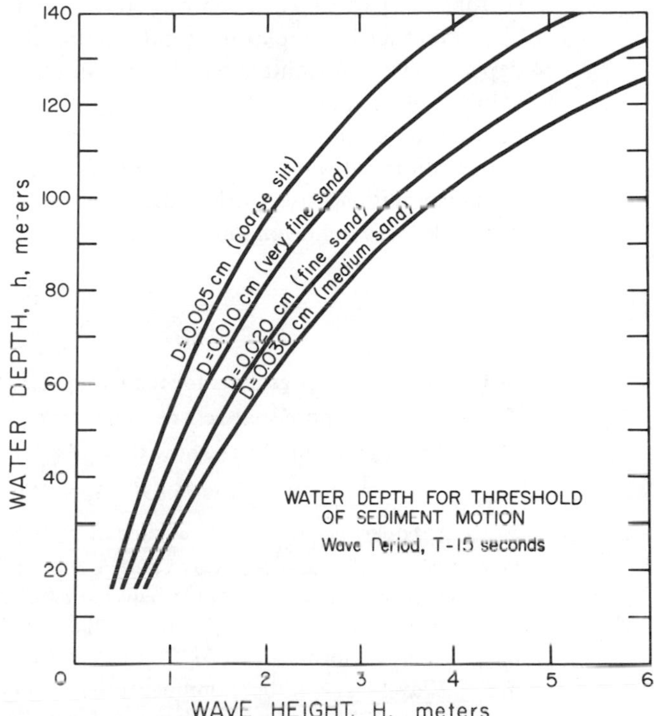

Figure 8-10 The water depth to which sediments can be set in motion by surface waves of period $T = 15$ seconds. [*From* Komar and Miller (1975a)]

may take place for only, say, five minutes out of each hour—that is, only when the waves sum to give the high orbital velocities—this may not be sufficient to be considered as a general exceeding of the threshold conditions.

The severest limitation to the above equations for threshold under waves is that they refer to a sinusoidal wave motion. Very often in deep water under ocean swell wave conditions this is an adequate description of the orbital motions. As discussed above, under real ocean wave conditions, with a wide wave spectrum, the orbital motions are irregular, being the summation of many sinusoidal motions. In addition, as waves enter intermediate and shallow water the wave crests peak and the troughs become wide and flat. Under such waves the orbital motions give strong onshore velocities under the crests and weaker offshore velocities under the troughs (Chapter 3). With such orbital motions the above threshold equations are inadequate except for perhaps a rough estimate as to whether threshold is reached sometime during the wave cycle. Under such circumstances, the approach of Ippen and Eagleson (1955), Eagleson, Dean, and Peralta (1958), and Johnson and Eagleson (1966) may be attempted. They relate the stresses exerted on the sediment particle to the instantaneous near-bottom water motions. These stresses include hydrodynamic drag and lift, the virtual or "apparent" mass force, a force due to the instantaneous pressure gradients under the waves, the inertia force, and the force of gravity. Threshold occurs when the moment of the forces due to the water flow exceeds the moment of gravity, so that the particle is tipped out from among the other particles that form the bed. This approach is complex in that it requires the evaluation of various coefficients which depend on the instantaneous Reynolds number and so continuously change as the velocity of the flow changes. Although the approach is much more difficult, it could be used to examine the threshold under irregular wave motions and also the threshold under combined wave and unidirectional flows. With such an analysis, no difficulties of interpretation arise as they did above, where ripple marks were generated before threshold was achieved on the average according to equations (8-19) and (8-21).

PROBLEMS

8-1 Work out the cgs units for the parameters S_l, I_l, and P_l.

8-2 Why is the pore-space correction factor, a', included in equation (8-5)?

8-3 From equation (8-6a), show that

$$S_l = (6.85 \times 10^{-5})(ECn)_b \sin \alpha_b \cos \alpha_b$$

for quartz sand beaches, where $(ECn)_b$ is given in erg/cm·sec and S_l is in m³/day.

8-4 Waves of deep-water wave height 2 m and period 12 sec have a deep-water angle of approach of 15 deg (angle between crests and shoreline) and break at the shoreline with an angle of 7 deg. Estimate the immersed-weight sand transport rate I_l and the volume transport rate S_l if the beach is composed of quartz sand.

8-5 Waves break at the shoreline with a height of 4 m in a water depth of 5 m and with an angle of 7 deg. Evaluate I_l and S_l for this quartz sand beach.

8-6 Waves break at the shore with a height of 4 m at a water depth of 5 m. An average longshore current of 30 cm/sec is measured in the surf. Evaluate I_l and S_l for this quartz sand beach.

8-7 If you need to know the long-term net sand transport at some coastal site, how would you go about making such an estimate?

8-8 Starting with an initially straight shoreline and a cell circulation with rip currents as depicted in Figure 7-1, what will be the sand transport pattern and how will the shoreline be modified? If you knew the pattern of longshore current velocities and wave conditions, how would you evaluate the pattern of sediment transport along and across the beach?

REFERENCES

BAGNOLD, R. A. (1946). Motion of waves in shallow water: interaction between waves and sand bottom. *Proc. Roy. Soc.* (London), series A, 187: 1–15.

——— (1963). Mechanics of marine sedimentation. In *The sea*, ed. M. N. Hill, 3: 507–28. Interscience, New York.

——— (1966). *An approach to the sediment transport problem from general physics.* U.S. Geol. Survey, Washington, D.C., Prof. Paper 422-I, 37 pp.

BEACH EROSION BOARD (1950). Munch-Peterson's littoral drift formula. *Bull. U.S. Army Corps of Engrs., Beach Erosion Board*, 4: 1–31. [Redraft of speech given in 1938.]

BRENNINKMEYER, B. M. (1975). Frequency of sand movement in the surf zone. *Proc. 14th Conf. on Coast. Eng.* pp 812–27.

CALDWELL, J. M. (1956). *Wave action and sand movement near Anaheim Bay, California.* U.S. Army Corps of Engrs., Beach Erosion Board Tech. Memo. no. 68, 21 pp.

DEAN, R. G. (1973). *Heuristic models of sand transport in the surf zone.* Conf. on Eng. Dyn. in the Surf Zone, Sidney, Australia, 7 pp.

EAGLESON, P. S., and R. G. DEAN (1961). Wave-induced motion of bottom sediment particles. *Trans. Am. Soc. Civ. Engrs.*, 126: 1162–89.

EAGLESON, P. S., R. G. DEAN, and L. A. PERALTA (1958). *The mechanics of the motion of discrete spherical bottom sediment particles due to shoaling of waves.* U.S. Army Corps of Engrs., Beach Erosion Board Tech. Memo. no. 104, 41 pp.

EATON, R. O. (1951). Littoral processes on sandy coasts. *Proc. 1st Conf. on Coast. Eng.*, pp. 140–54.

FAIRCHILD, J. C. (1973). Longshore transport of suspended sediment. *Proc. 13th Conf. on Coast. Eng.*, pp. 1069–88.

GALVIN, C. J. (1973). A gross longshore transport rate formula. *Proc. 13th Conf. on Coast. Eng.*, pp. 953–70.

GALVIN, C. J., and P. S. EAGLESON (1965). *Experimental study of longshore currents on a plane beach.* U.S. Army Coastal Eng. Research Center Tech. Memo. no. 10, 80 pp.

HORIKAWA, K., and A. WATANABE (1967). A study on sand movement due to wave action. *Coast. Eng. in Japan*, 10: 39–57.

INGLE, J. C. (1966). *The movement of beach sand.* Elsevier, New York, 221 pp.

INMAN, D. L., and R. A. BAGNOLD (1963). Littoral processes. In *The sea*, ed. M. N. Hill, 3: 529–33. Interscience, New York.

INMAN, D. L., and A. J. BOWEN (1963). Flume experiments on sand transport by waves and currents. *Proc. 8th Conf. on Coast. Eng.*, pp. 137–50.

IPPEN, A. T., and P. S. EAGLESON (1955). *A study of sediment sorting by waves shoaling on a plane beach.* U.S. Army Corps of Engrs., Beach Erosion Board Tech. Memo. no. 63, 83 pp.

JOHNSON, J. W. (1956). Dynamics of nearshore sediment movement. *Bull. Am. Soc. Petrol.* Geologists, 40, no. 9: 2211–32.

——— (1957). The littoral drift problem at shoreline harbors. *J. Waterways and Harbors Div., ASCE*, 83, WW1, paper 1211: 1–37.

JOHNSON, J. W., and P. S. EAGLESON (1966). Coastal processes. In *Estuary and coastline hydrodynamics*, ed. A. T. Ippen, pp. 404–92. McGraw-Hill, New York.

JONSSON, I. G. (1967). Wave boundary layers and friction factors. *Proc. 10th Conf. on Coast. Eng.*, pp. 127–48.

KOMAR, P. D. (1971). The mechanics of sand transport on beaches. *J. Geophys. Res.*, 76, no. 3: 713–21.

KOMAR, P. D., and D. L. INMAN (1970). Longshore sand transport on beaches. *J. Geophys. Res.*, 75, no. 30: 5914–27.

KOMAR, P. D., and M. C. MILLER (1973). The threshold of sediment movement under oscillatory water waves. *J. Sediment. Petrol.*, 43: 1101–10.

——— (1975a). Sediment threshold under oscillatory waves. *Proc. 14th Conf. on Coast. Eng.*, pp. 756–75.

——— (1975b). On the comparison between the threshold of sediment motion under waves and unidirectional currents with a discussion of the practlcal evaluation of the threshold. *J. Sediment. Petrol.*, 45: 362–67.

LONGUET-HIGGINS, M. S. (1952). On the statistical distribution of the height of sea waves. *J. Mar. Res.*, 11: 245–66.

——— (1970). Longshore currents generated by obliquely incident sea waves, part 1. *J. Geophy. Res.*, 75, no. 33: 6778–89.

——— (1973). *The mechanics of the surf zone.* 13th Int. Cong. Appl. Mech., Moscow.

MADSEN, O. S., and W. D. GRANT (1975). The threshold of sediment movement under oscillatory waves: a discussion. *J. Sediment. Petrol.*, 45: 360–1.

MANOHAR, M. (1955). *Mechanics of bottom sediment movement due to wave action.* U.S. Army Corps of Engrs., Beach Erosion Board Tech. Memo. no. 75, 121 pp.

RANCE, P. J., and N. F. WARREN (1969). The threshold movement of coarse material in oscillatory flow. *Proc. 11th Conf. on Coast. Eng.*, pp. 487–91.

SAVAGE, R. P. (1959). *Laboratory study of the effect of groins in the rate of littoral transport.* U.S. Army Corps of Engrs., Beach Erosion Board Tech. Memo. no. 114, 55 pp.

SCRIPPS INSTITUTION OF OCEANOGRAPHY (1947). *A statistical study of wave conditions at five sea localities along the California Coast.* Univ. of California Wave Report no. 68, LaJolla, Calif., 34 pp.

SHIELDS, A. (1936). *Anwendung der Ahnlichkeits Mechanik und der Turbulenzforschung auf die Geschiebe Bewegung.* Preuss. Versuchanstalt für Wasserbau und Schiffbau, Berlin, 20 pp.

SILVESTER, R., and G. R. MOGRIDGE (1971). Reach of waves to the bed of the continental shelf. *Proc. 12th Conf. on Coast. Eng.*, pp. 651–67.

SIREYJOL, P. (1965). Communication sur la construction du port de Cotonou (Dahomey). *Proc. 9th Conf. on Coast. Eng.*, pp. 580–95.

THORNTON, E. B. (1973). Distribution of sediment transport across the surf zone. *Proc. 13th Conf. on Coast. Eng.*, pp. 1049–68.

WATTS, G. M. (1953*a*). *Development and field tests of a sampler for suspended sediment in wave action.* U.S. Army Corps of Engrs., Beach Erosion Board Tech. Memo. no. 34, 41 pp.

——— (1953*b*). *A study of sand movement at South Lake Worth Inlet, Florida.* U.S. Army Corps of Engrs., Beach Erosion Board Tech. Memo. no. 42, 24 pp.

Chapter 9

THE BUDGET OF LITTORAL SEDIMENTS

The Walrus and the Carpenter
Were walking close at hand:
They wept like anything to see
Such quantities of sand:
"If this were only cleared away,"
They said, "it would be grand!"

"If seven maids with seven mops
Swept it for half a year,
Do you suppose," the Walrus said,
"That they could get it clear?"
"I doubt it," said the Carpenter,
And shed a bitter tear.

Lewis Carroll
Through the Looking-Glass (1871)

The application of the budget of littoral sediments has proved to be an extremely useful approach in evaluating the relative importance of the various sediment sources and losses to the nearshore zone and in accounting for regions of beach erosion or deposition. The *budget of sediments* is nothing more than an application of the principle of continuity or conservation of mass to the littoral sediments. The time rate of change of sand within the system is dependent upon the rate at which sand is brought into the system versus the rate at which sand leaves the system. The budget involves assessing the sedimentary contributions (credits) and losses (debits) and equating these to the net gain or loss (balance of sediments) in a given sedimentary compartment (Bowen and Inman, 1966). The balance of sediments between the losses and gains is reflected in local beach erosion or deposition. Table 9-1 summarizes the possible *sources* and *sinks* (losses) of sand for a littoral sedimentary budget. In general, longshore movements of sand into a littoral compartment, river transport, and sea cliff erosion provide the major credits; longshore movements out of the compartment, offshore transport (especially through submarine canyons), and wind transport shoreward to form sand dunes are the major losses or debits. Much of this chapter will involve an examination of the relative importance of these several sources and sinks of beach sand and how they can be evaluated for specific cases.

TABLE 9-1 The Budget of Littoral Sediments [*After* Bowen and Inman (1966)]

CREDIT	DEBIT	BALANCE
Longshore transport into area	Longshore transport out of area	Beach deposition or erosion
River transport	Wind transport out	
Sea cliff erosion	Offshore transport	
Onshore transport	Deposition in submarine canyons	
Biogenous deposition	Solution and abrasion	
Hydrogenous deposition	Mining	
Wind transport onto beach		
Beach nourishment		

In some locations the coastal area may be naturally divided up into a series of compartments or cells which are essentially entities. For example, Inman and Chamberlain (1960) have shown that the southern California coast can be divided into four discrete sedimentation cells (also see Inman and Frautschy, 1966). Each cell contains a complete cycle of littoral transport and sedimentation (Figure 9-1), with rivers being the principal sources of sediments for the cells and the chief sinks being the submarine canyons which bisect the California continental shelf and intercept the sand as it moves along the beach. A typical cell, Figure 9-1, begins with a rocky promontory or rocky stretch of coast where the supply of sand is restricted. The littoral drift of sand is downcoast, determined by the prevailing wave approach from

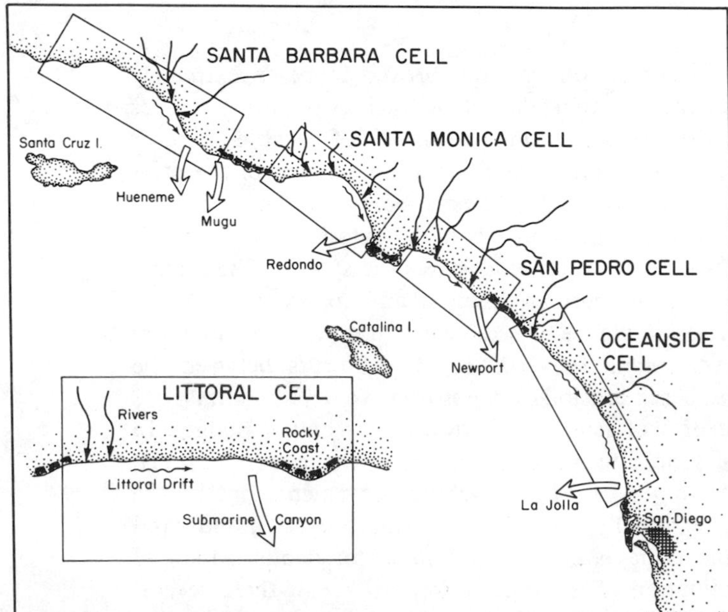

Figure 9-1 The southern California coast divided into a series of cells, each cell being a system where rivers add sand to the beaches and the sand moves south as littoral drift to be trapped within submarine canyons and lost offshore to deep water. [*After* Inman and Frautschy (1966)]

the northwest. Downdrift from the rocky coast the beach gradually widens as sand is supplied by the rivers. Each littoral cell is terminated by a submarine canyon which captures the beach sand and funnels it out to very deep water, where it is permanently lost from the beach. This causes the next littoral cell to begin anew with a rocky coast devoid of beach sand, and the system is repeated. A budget of sediments can be developed for each of these compartments. Along other coasts, however, there may be no natural limits as found in California, and in formulating the sand budget it is necessary to choose somewhat arbitrary boundaries.

One application of the budget of sediments involves estimating what effects man might have on the system by attempting to alter it. If we dam a certain river, how much sand will be lost to the beach near its mouth? Will placing riprap on an eroding sea cliff cause a major depletion of the adjoining beach? How great is the littoral drift of sand along a particular beach, and what would be the effects of building a jetty there?

STREAM SOURCES

The principal sources of nearshore sediments for most coastal areas are the streams which transport large quantities of sand directly to the ocean. The supply of material reaching the beach from the inland depends upon the nature of the hinterland: its elevation, the types of rocks found there, the density of vegetation, and the climate are all important factors in determining the amount and type of load carried by the rivers.

Another important factor is whether estuaries are present, separating the rivers from the nearshore. It appears that not only do estuaries sometimes trap the river sand so that it does not reach the beach, but tidal currents within the estuary may carry sand inward from the beach so that the estuary becomes a deficit or sink of littoral sediments (Kulm and Byrne, 1966; Meade, 1969).

The evaluation of sand transport by rivers is a major subject in itself, and we can only briefly summarize the methods. There are two rather distinct approaches to estimating the sediment supplied to the beach by a particular river. The first involves empirical correlations between the sediment supply, the drainage area of the river basin, and the effective annual precipitation. The second approach is that of estimating the sand transport from measurements of the river discharge or velocity, utilizing published equations which relate the two.

Langbein and Schumm (1958) have correlated the effective annual precipitation to the sediment yield from river drainage basins throughout all the climatic regions of the United States (Figure 9-2). The *effective precipitation* eliminates that portion of the total precipitation that is lost through evaporation and transpiration; it is the amount of precipitation that is reflected in the *runoff*, the water that drains from the land into streams and rivers. The loss to *evapotranspiration* in moist areas where water is constantly available is governed principally by the temperature, and is given by the curve of Figure 9-3. The runoff and effective precipitation can be used inter-

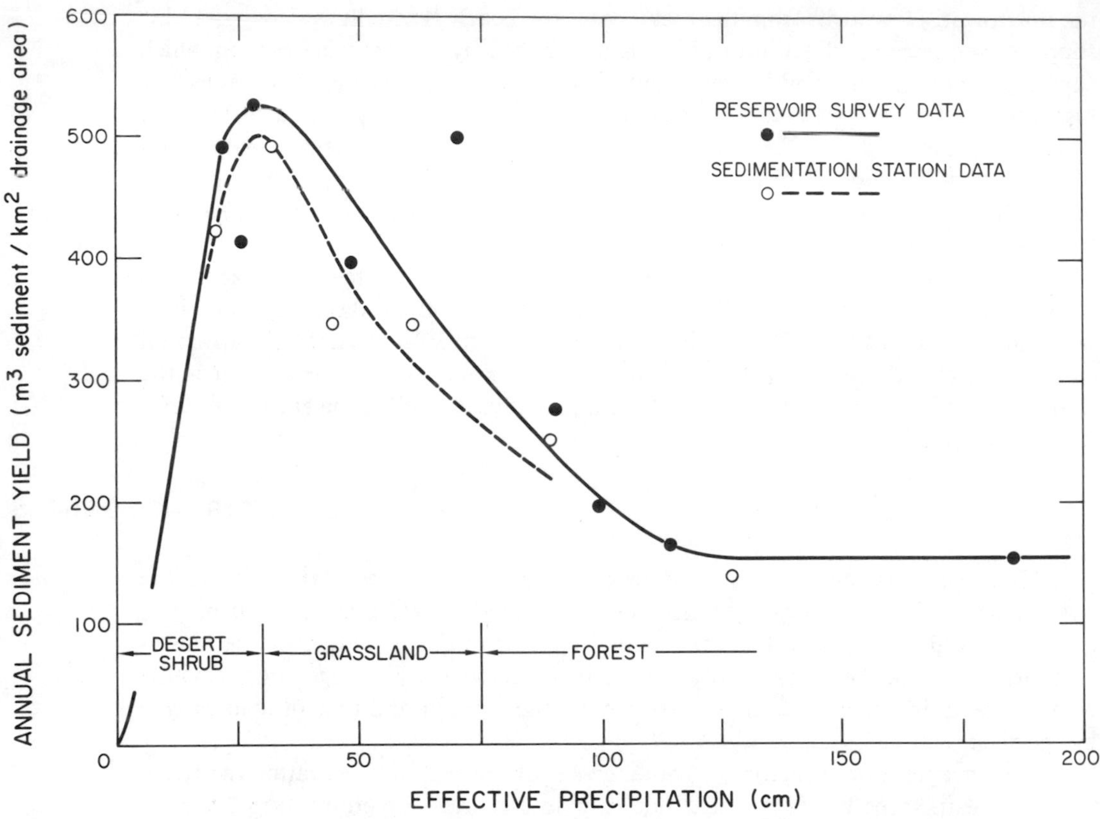

Figure 9-2 The river sediment yield per unit area of drainage basin as a function of the effective precipitation. The curve is normalized to a drainage area of 100 km^2. For other drainage areas, utilize the rule that the sediment yield is inversely proportional to the 0.15 power of the drainage area. [*After* Langbein and Schumm (1958)]

changeably, as is shown in the curves of Figure 9-4. If the stream discharge is known instead of the precipitation, then dividing the discharge by the drainage area yields the mean runoff. As defined by Langbein and Schumm (1958), the corresponding effective precipitation is the amount of precipitation required to produce that runoff at a temperature of 50°F; the 50°F curve of Figure 9-4 must be used to determine the effective precipitation that is empirically correct for Figure 9-2 no matter what the actual temperature of the area is.

Two sources of sediment-yield data were used to construct Figure 9-2: records collected by the U.S. Geological Survey gauging stations, which measure sediment transport, and surveys of sediment trapped by reservoirs. The two approaches yield somewhat different results, as is seen in Figure 9-2, with the curve based on the reservoir data being somewhat above that of the gauging stations. The reason for this is that the sediment station records do not include bed load, the coarser fraction of the

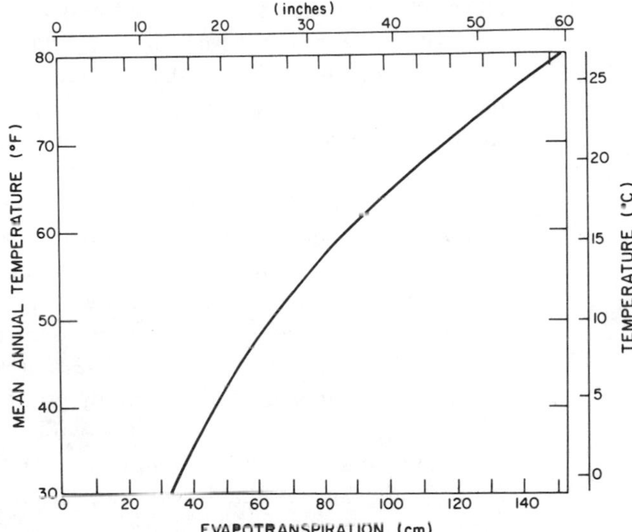

Figure 9-3 The loss of water to evapotranspiration, dependent on the temperature. Subtracting the evapotranspiration from the total precipitation gives the effective precipitation. [*After* Langbein (1949)]

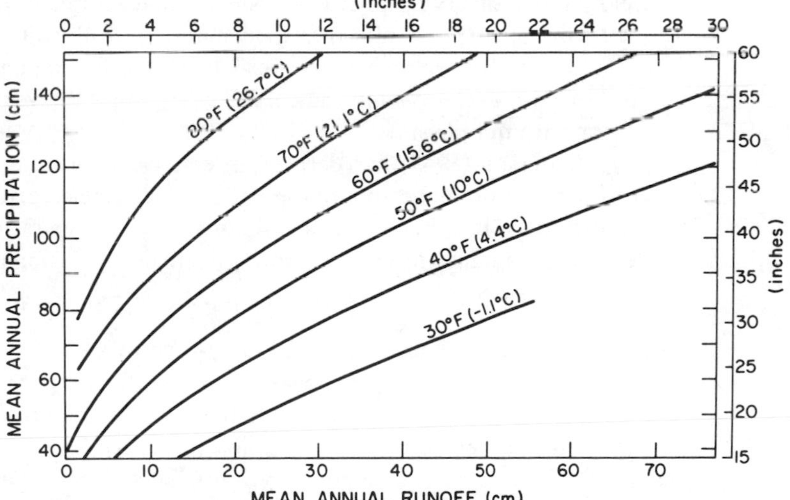

Figure 9-4 The river runoff related to the mean annual precipitation. Through these curves and Figure 9-3 the actual river runoff can be equated to the effective precipitation. [*After* Langbein (1949)]

river sediment found near the bed; the reservoirs trap both the bed and suspended loads.

Both curves of Figure 9-2 show the sediment yield reaching a maximum at an annual precipitation of about 25 to 35 cm (10 to 14 inches), decreasing sharply on both sides of this maximum. This maximum occurs at a rainfall where desert scrub gives way to grasslands. With increasing precipitation above 35 cm there is progressively more vegetation protection, which helps prevent erosion and therefore produces less sediment yield. With decreasing precipitation below 25 to 35 cm, one has the effects of decreasing stream runoff and hence diminishing sediment yield; the curve must of course predict zero sediment yield for zero precipitation.

Several studies have shown that sediment yield per unit area of drainage decreases with increasing drainage area, a result of the flattening gradients and lowered probibility that an intense storm will cover the entire drainage basin. The curves of Figure 9-2 have been normalized to a drainage area of 100 km^2 utilizing the graphs of Brune (1948), which show that the sediment yield is inversely proportional to the 0.15 power of the drainage area. In application, this effect of drainage basin area must be taken into account. For example, if the drainage area of the river in question is 750 km^2, then the sediment yield rates given by Figure 9-2 must be divided by the factor $(750/100)^{0.15} = 1.35$ to obtain the proper yield rate for that area.

Each point shown in Figure 9-2 represents an average of anywhere from five to thirty-eight individual correlations grouped according to effective precipitation. This averaging has the effect of removing some of the variability due to differences in geology, degree of cultivation, and vegetation type and distribution. For example, cultivation in an area that is receiving 150 cm annual precipitation would considerably increase the sediment yield above that predicted by the curves of Figure 9-2. Forest cutting and forest and brush fires can similarly be expected to increase the sediment yield. Anderson (1955, p. 121) has shown that sediment yield in the Gibraltar watershed, southern California, increased markedly following brush and forest fires and decreased as the watershed recoverd from the fire damage. Such effects should be borne in mind in applying the curves of Figure 9-2, even though in most cases corrections are not possible.

Anderson (1949) has derived an empirical formula which relates the sediment yield per unit drainage area to the stream discharge, vegetation cover density, and the channel area. It is based on records of many reservoirs in the San Gabriel and San Bernardino mountains in southern California. Regression analysis yielded

$$\log S_r = 1.041 + 0.866 \log q + 0.370 \log A_{ch} - 1.236 \log c$$

where S_r = annual sediment yield (acre-ft/sq mile),
q = maximum yearly peak discharge (ft^3/sec per sq mile),
A_{ch} = area of main channel of the watershed (acres/sq mile),
c = cover density of vegetation (percent).

This formula is of course good only for the southern California area to which it is empirically related and perhaps to other areas that are physically very similar. It does point the way to refinements of the estimate of sediment yield based on the geography of the drainage basin. Similar formulations have been developed by Lustig (1965) and Scott, Ritter, and Knott (1968).

Evaluations of the sediment yield from the curve of Figure 9-2 must be corrected for that proportion of the total load that actually finds its way onto the beach. If the local beaches retain only cobble-size material and the nearby river transports no cobbles, the river obviously makes no contribution to the beach no matter how much sand it is carrying. In any river much of the load is fine clay and silt which inevitably is lost to deep water and does not supply the beach. It is apparent that estimating the actual gain to the beach from a river source can be difficult. It is generally considered that roughly 75 to 80 percent of the total river sediment load is fine enough to be transported in suspension. Unless the beach environment is of exceptionally low energy, this river-suspended material will be lost offshore. Only the remaining 25 percent or less could supply the beach.

Building a major dam across a river will greatly reduce the effective area of a drainage basin and therefore diminish the sediment contribution of that river (Norris, 1964). No material eroded in the region of the headwaters can pass through the reservoir behind the dam and reach the coast. This effect on beaches should be borne in mind when dam construction is being considered. A similar effect is produced when sand is mined from the riverbed. Sand removed in this way of course does not reach the beach. The presence of dams on the river or any sand mining must be taken into account in estimating the river sediment yield.

The second approach to estimating sediment contributions by rivers is to utilize sediment transport equations and measurements of river discharge or velocity. Complete reviews of sediment transport in rivers can be found in the books by Graf (1971) and Yalin (1972). Sediment transport equations attempt to relate the transport rate to measurable river parameters: flow depth, width, channel slope, and flow velocity or discharge. Only a brief review will be given here, with stress placed on the approaches that are simple and most directly suited to our application.

The methods of Einstein (1950, 1964) are those most commonly employed by engineers in estimating sediment transport by rivers. The calculation of the load must be performed for each separate grain-size fraction into which the bed material is divided; the total sediment transport is then the sum of the discharges of the individual grain sizes. For this reason the method is relatively lengthy and difficult to apply and requires reference to several graphs for evaluation of coefficients. A good summary and the necessary graphs can be found in Einstein (1964).

Colby and Hembree (1955) provide a modification to the approach of Einstein (1950) which yields the total sediment transport directly. Several graphs of coefficients which are required may be found in their publication or in Einstein (1964).

Bagnold (1966) presents an approach to sediment transport in rivers that is based on general physics principals. The comparable approach to sand transport on beaches according to Bagnold was presented in Chapter 8. In river flow Bagnold considers mainly fully developed transport conditions which occur under high flow rates; less consideration is given to moderate and low flow where the bottom is rippled and the transport becomes more complex as a result.

Shields (1936) considered the sediment transport rate to be proportional to the river flow shear stress that is in excess of the critical shear stress at which sediment begins to move. The semiempirical equation he obtained is

$$\frac{\rho_s q_s}{\rho S q} = 10 \frac{\tau_o - (\tau_o)_{cr}}{(\rho_s - \rho) g D} \tag{9-1}$$

where q_s = sediment discharge per unit stream width,
q = water discharge per unit stream width,
ρ_s = density of sediment grains,
ρ = density of water,
S = axial gradient of stream channel,
D = diameter of sediment grains,
and where the river flow shear stress is given by

$$\tau_o = \rho g h S \tag{9-2}$$

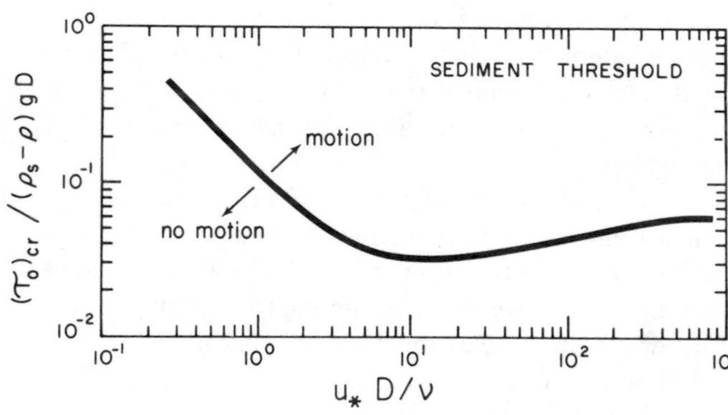

Figure 9-5 The critical stress $(\tau_o)_{cr}$ for the threshold of sediment motion for grains of diameter D. The stress velocity $u_* = \sqrt{(\tau_o)_{cr}/\rho}$. [*After* Shields (1936)]

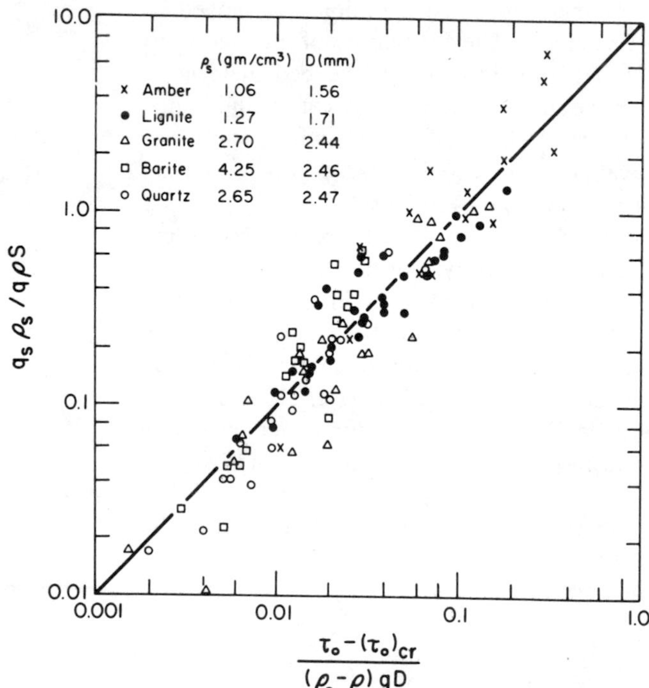

Figure 9-6 The bed-load sediment discharge rate q_s dependent upon the stress τ_o exerted by the river flow. [*From* Shields (1936) and Graf (1971)]

h being the water depth (or *hydraulic radius*—channel section area divided by the wetted perimeter). The critical shear stress $(\tau_o)_{cr}$ at which sediments begin to be transported is given in Figure 9-5. In Figure 9-6 it is seen that the relationship of equation (9-1) holds for a wide range of grain sizes and densities. The range of scatter is equivalent to a factor of 10, but this is reasonable for sediment transport studies. Equation (9-1) is dimensionless and thus can be used with any system of units.

Equation (9-1), presented as an example, is just one of many relationships that have been proposed for evaluation of the sediment transport in rivers. Several are presented in Graf (1971) and Yalin (1972) with explanations for their application. These references should be consulted for the details of the approaches.

Equation (9-1) includes only bed-load transport in the river; the finer-grained suspended load is not evaluated with this equation (Graf, 1971). With respect to the actual contribution to the beach, only a portion of the bed load determined from the equations might be sufficiently coarse to remain on the beach. On the other hand, the proportion could approach 100 percent because, unlike estimates from the Langbein and Schumm (1958) method, the suspended load is already excluded. Only by comparing the beach sediments with direct samples from the river bed can one properly evaluate what proportion might end up on the beach.

CLIFF EROSION

In general, second in importance to rivers for supplying sediments to beaches is sea cliff erosion. Of course, in some restricted locations, such as pocket beaches or where the rivers enter estuaries that trap the river sands, cliff erosion may be the primary source. In general, however, cliff erosion probably does not account for more than about 5 to 10 percent of the material on most beaches.

The long-term importance of the supply of sediments to the beach through cliff erosion is sometimes made apparent by the emplacement of riprap or other structures which eliminate this supply. At Bournemouth Bay on the south coast of England the construction of a promenade on the seafront since 1900 has halted cliff erosion and cut off this supply of sand (Bird, 1968). The result has been a gradual diminishing of the beach as the sand is carried away by longshore transport and not replenished.

The erosion of sea cliffs was considered in Chapter 2. The highest rates of cliff erosion occur in unconsolidated sediments such as glacial deposits. Such cliffs in the Holderness, England, are retreating at rates of up to 1.75 m/yr (Valentine, 1954). On the other hand, Shepard and Grant (1947) found that the wave erosion of rocky coasts in southern California had been negligible during the preceding fifty years. They did find retreat rates of as much as 0.3 m/yr (1 ft/yr) in unconsolidated formations. In general, cliff erosion rates appear to be less than 1 m/yr. The exception to this is when jetties or breakwaters are constructed which strip away the protective beach so that the waves strike directly against the cliffs (Chapter 12). Under these circumstances sea cliffs are known to retreat by tens of meters in a single year.

Cliff retreat rates are commonly determined by comparing series of aerial photographs or repeated ground surveys. In countries such as England, old maps can be used to determine long-term erosion rates (very accurate maps there date back to the nineteenth century, less accurate ones much earlier). For the more recent but shorter-term erosion rates, monitoring the cliff retreat with a series of stakes is most effective.

The cliff retreat rate must be converted into an estimate of the actual amount of material supplied to the beach. The total volume of sediments eroded from the cliff is obtained by multiplying the retreat rate by the height and length of the cliff. This total volume must in turn be corrected for the percentage of the sediment that actually remains on the beach. If the cliffs are of clay and silt, no matter how much cliff erosion occurs there will be little gain to the beach. For example, Valentine (1954) has shown that only about 3 percent of the sediment eroded from the boulder-clay cliffs of the Holderness reaches the sandy spit of Spurn Head to the south.

WIND TRANSPORT

Wind-blown sand may be a source of beach sands, but more commonly winds blow onshore and therefore are more effective in removing sand from the beach and transporting it inland. The resulting inland sand dunes are a familiar feature along coasts, and their volumes attest to their importance in accounting for substantial losses of sand from the beach.

The total volume of sand in the dune field divided by the total time of formation would give an estimate of the rate of loss from the beach. The more usual approach is to measure the rate at which dunes are advancing away from the beach. This can be accomplished by direct field measurements or with aerial photographs. In this way Cooper (1958) obtained yearly migration rates of 0.44 cm/day (0.17 cm/day in winter, 0.71 cm/day in summer) for dunes on the Oregon-Washington coast. Bowen and Inman (1966) obtained an average rate of 0.15 cm/day (2 ft/yr) for the dune fields between Pismo Beach and Point Sal, California.

ONSHORE-OFFSHORE TRANSPORT

Sediments that are too fine to remain on the beach will be placed in suspension and carried offshore into deep water and lost from the beach. Beach sands carried alongshore under an oblique wave approach into an area of progressively increasing wave height will diminish in volume as coarser and coarser material drifts offshore. If the wave energy is decreasing, it is possible that finer sand may find its way from the offshore and onto the beach. Such losses and gains are sometimes made apparent by longshore variations in the grain size and can be evaluated volumetrically on this basis. However, not all longshore variations in grain size are caused by this selective removal or addition of certain grain sizes on the beach (Chapter 13), so that such considerations must not be indiscriminate.

Occasionally sediments may be eroded from unconsolidated offshore sources on the continental shelf and drift shoreward and onto the beach. For example, beach sediments on the coast of the Netherlands are derived in part from the shallow waters of the North Sea.

This potential source or loss of beach sediments is the most difficult to evaluate quantitatively. The presence of the source can sometimes be judged through an examination of the petrology of the shelf and beach sands. Offshore relict sands are often coated with a red iron oxide (Emery, 1968) whose presence in the beach sands may indicate an offshore contribution. Similarly, heavy minerals or shell fragments unique to the offshore sands may be used as natural tracers of onshore movements of sediments (the approach will be discussed later in this chapter).

Hydrographic surveys of the adjacent shelf together with beach surveys have been used (Bowen and Inman, 1966) to detect losses of sediments from the offshore. The difficulty of the approach is that a small change in sediment level can supply a large quantity of sand if it occurs over a wide shelf area. This approach therefore requires accurate very-long-term surveys to detect the slight changes in depth normally involved.

In the overall budget of sediments scheme the offshore source (or sink) generally remains unknown. Sometimes it can be roughly estimated indirectly by drawing the balance between the known losses and gains and comparing these with the beach erosion or deposition. In the few published budgets that have been developed for various coastal areas, it has been concluded that the offshore source is of minor importance in comparison to the other sources.

DEPOSITION DOWN SUBMARINE CANYONS

The importance of submarine canyons in southern California in trapping littoral sands and funneling them offshore into deep water has already been mentioned. In this area the heads of the submarine canyons are in shallow water very near to the beach and therefore can intercept the beach sands. Similar situations occur off the coast of Senegal, West Africa, where the Cayar Canyon originates near the shoreline (Dietz, Knebel, and Somers, 1968), and off the southern tip of Baja California (Shepard, 1964). One unusual canyon, the Congo Canyon, extends up into the estuary of the Congo River and traps the sand directly from the river so that it does not reach the beaches. On most coasts, however, the canyon heads terminate in deeper water near the continental shelf edge and so do not play a direct role in the beach sand budget.

The La Jolla–Scripps submarine canyon system, offshore from the Scripps Institution of Oceanography in California (Figure 7-5) has been more closely studied than any other canyon. A review can be found in Shepard and Dill (1966, pp. 31–59). These canyons are found at the downdrift end of the Oceanside littoral cell (Figure 9-1). Shepard (1951) and Chamberlain (1964) repeatedly surveyed the head of Scripps Canyon at regular intervals to determine sand accumulation and removal. Chamberlain determined that the mean yearly sand accumulation and subsequent loss to deep water amounted to about 2×10^5 cubic meters. This is approximately equivalent to the total littoral drift reaching the adjacent beach and accounts for the fact that sand does not accumulate on the beach.

BIOGENOUS DEPOSITION

Shell and coral fragments are important sediment sources in some beaches, especially those in the tropics where biological productivity is high (Chapter 13). This contribution is generally insignificant in extratropical areas, except where terrigenous sources are nonexistent.

The biogenous source can be estimated from the known productivity rates of the organisms of interest. A better estimate of the contribution can be gained by dividing the total volume of the material in the beach by the time elapsed in beach deposition, providing both of these quantities are known. Inman, Gayman, and Cox (1963) evaluated the biogenous contribution from a fringing reef on the Island of Kauai, Hawaiian Islands, by measuring the rate at which a known quantity of terrigenous sand supplied by a river was diluted by the reef material.

HYDROGENOUS DEPOSITION

Some beaches, generally found in the tropics, are composed of oolites, nearly spherical sand-sized calcium carbonate grains (Chapter 13). They are generally formed by direct precipitation from seawater but may be reworked from older deposits. Their occurrence is rare.

BEACH NOURISHMENT

An economical method for engineers to stabilize and protect long reaches of coastline is to place sand fill on the beach. The new beach serves as a buffer, protecting sea cliffs and coastal property from the intense wave action of the ocean and of course providing a wide beach for recreational uses. This approach has been taken at a number of beach sites along the U.S. coast and elsewhere.

The sand supply for beach nourishment is most often taken from the adjacent offshore, pumped through a hydraulic pipeline directly onto the beach. In this manner, for example, 530,000 cubic meters of sand fill was placed on a beach at Bridgeport, Connecticut (Vesper, 1965). At other locations the material is pumped or dredged from lagoonal areas landward of the beach proper, or hauled from nearby sand dunes or other inland source. The grain size of the added sand must obviously be appropriate to the beach that is to be nourished; if too fine, it will soon be lost offshore.

MINING OF BEACH SEDIMENTS

Removal of beach sediments has the opposite effect of beach nourishment, tending to undermine the buffering role of the beach and therefore promoting erosion of the sea cliffs and coastal property. For this reason, the removal of beach sands is generally prohibited unless the shoreline is rapidly accreting. Mining of shingle in the nineteenth century from the beach at Hallsands on the southern Devon coast, England, produced rapid erosion of the sea cliffs and destruction of the village.

Any mining of beach sands must be included in the sand budget. The mining of river sands and its effects on the beach has already been discussed.

SOLUTION AND ABRASION

Solution and abrasion of sand reduces its grain size so that one would expect that it might become too fine to remain on the beach, instead being carried off into deep water. Studies of quartz-sand abrasion in the surf have concluded, however, that abrasion is insignificant and that loss of sand from the beach in this way is very small (Chapter 13). Only with coarser material (gravels and cobbles) might abrasion become a significant factor. Similarly, the solubility of quartz in seawater is negligible, and thus there will be no loss of beach sand for this reason.

Beach sediments composed of calcium carbonate (shell and coral debris) are more susceptible to abrasion and solution. This is apparent in that shell fragments are inevitably smooth and well rounded (Chapter 13). This area has not been adequately studied to make any quantitative estimates of sediment loss from these effects.

LONGSHORE TRANSPORT

It is not generally feasible to make direct field measurements of the littoral drift, the movement of sand along the beach under the wave action. Such measurements are difficult to make; moreover, in the budget of sediments the long-term net transport is required, necessitating a prohibitively large number of measurements. Instead, some indirect approach must be taken in evaluating the littoral drift. Three methods have been used: (1) measuring rates of accretion or bypassing of sand at a littoral barrier such as a jetty or breakwater; (2) computing the littoral drift from statistical wave data utilizing an equation which relates the two; and (3) measuring the rate of dilution of heavy minerals within the beach sands.

The direction of littoral transport can sometimes be determined from observations of the shoreline configuration in the vicinity of coastal structures, natural or man-made (Figure 9-7). The entrapment of sand behind groins can indicate the direction of the littoral drift during the immediately preceding period. This can vary, however, so that to determine the long-term predominant direction of littoral transport requires observations of the groins at regular intervals for at least a year. Jetties and breakwaters trap larger quantities of sand, so that the shoreline configuration in their vicinity is a better indication of the long-term direction of littoral drift. Even here caution is required, as the shoreline shape may be misleading with regard to the drift direction (Chapter 12). If the jetty or breakwater traps all of the littoral drift, then rates of accretion or erosion (or bypassing) in its vicinity can provide a quantitative estimate of the littoral drift rate. For example, the rate of entrainment of sand

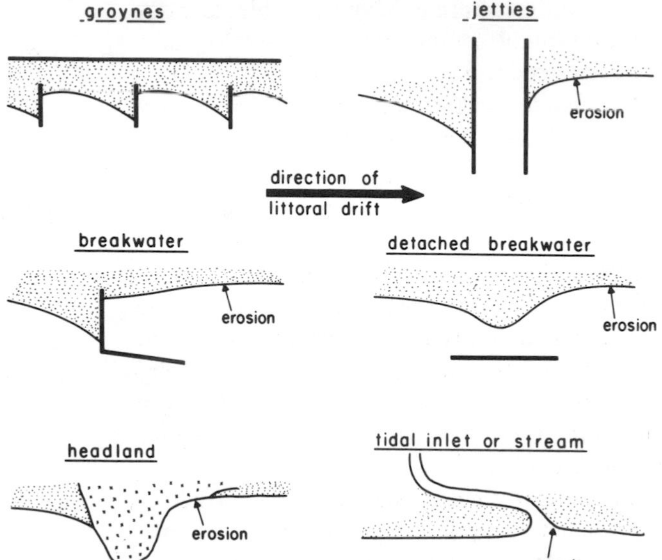

Figure 9-7 A variety of indicators of the direction of the prevailing littoral sand drift along the beach.

at the Santa Barbara breakwater, California, enabled Johnson (1953) to determine that the drift along that portion of coast is 215,000 cubic meters per year (Chapter 12).

Rocky headlands and other natural littoral barriers can similarly be utilized to indicate the transport direction. However, they generally cannot be used for quantitative estimates of the transport rate, as the beaches in their vicinity tend to reach a condition of equilibrium where the sand supply balances the sand loss.

The migration of tidal inlets and stream mouths alongshore have been used to determine the littoral drift direction. This evidence should be used with caution.

In Chapter 8 we saw that the sand transport rate can be related to the wave energy flux ECn and the wave breaker angle α_b through the equation

$$I_l = 0.77(ECn)_b \sin \alpha_b \cos \alpha_b \tag{9-3}$$

Rather than measuring the littoral drift directly, it is apparent that we can instead measure the wave parameters and evaluate the drift using this equation. One can either obtain the wave measurements directly with wave-recording instruments on the coast or use *hindcasting* techniques whereby past weather charts over the ocean are used to evaluate heights, periods, and directions of waves generated by storms [see discussion in Chapter 4 and the details of the approach in Coastal Engineering Research Center (1973)]. In this way the National Marine Consultants (1960) prepared deep-water wave statistics on a monthly basis for seven offshore locations on the California coast. Using wave-refraction diagrams (Chapter 4), one can use this data to yield breaking wave statistics from which to compute the littoral drift. Bowen and Inman (1966) and Anderson (1971) provide examples of evaluating the littoral drift by this approach.

For the budget of sediments the long-term net drift is required, obtained by adding the daily estimates of the littoral drift, taking into account the direction along the coast (Chapter 8). This net drift may be small compared to the total longshore movement of sand on the beach, as the drift may be upcoast for a time and then later downcoast as the direction of the wave approach changes. If this is the case, the evaluation of the net drift may not be very reliable, as it would be the difference of two large values—the total drifts up and down the coast—whose individual uncertainties are larger than the net drift value.

Bowen and Inman (1966), using the 215,000-m^3/yr value for the drift at Santa Barbara, California, and applying the data of Trask (1952, 1955) on the content of augite, a naturally occurring heavy mineral, were able to compute the littoral drift rates at Surf and Gato, California. The basis of the computation is the progressive dilution and decreasing amount of augite in the beach sands as more sand (without augite) is added to the beach, increasing the total littoral drift. The use of heavy minerals as tracers of littoral drift will be considered at the end of this chapter.

THE BALANCE: BEACH EROSION OR DEPOSITION

For a given littoral compartment the total volume of sand added to the beach (credits) from the various sources can be balanced against the total losses (debits). If the losses are greater than the gains, then there will be a net deficit, which will be reflected as a decrease in the total volume of beach sediment: beach erosion will occur. Similarly, if the credits outweigh the debits, there will be beach deposition. The lack of either beach erosion or deposition indicates that a state of equilibrium exists between the sources and losses.

If there is beach erosion or deposition, it can generally be evaluated by comparing series of beach profiles. Therefore, the balance in the budget of littoral sediments is known beforehand. The art of the approach is in evaluating the sources and losses such that their balance agrees (reasonably) with this measured erosion or deposition.

A HYPOTHETICAL EXAMPLE

Bowen and Inman (1966) and Anderson (1971) provide real examples of the application of the budget of littoral sediments. Rather than discussing the details of their results, we shall instead consider a simple hypothetical example.

Figure 9-8 shows a portion of a beach terminated on the south by a rocky headland. Arriving from the north is a littoral drift S_l due to the prevailing wave approach from that direction. Added to that quantity is sand eroded from the sea cliffs at a rate S_{cf} and sand transported from a river at the rate S_r. The principal losses of sand appear to be as wind-blown dunes and a submarine canyon. Recent erosion on the beach indicates that within the past five years there has been a net deficit in the balance

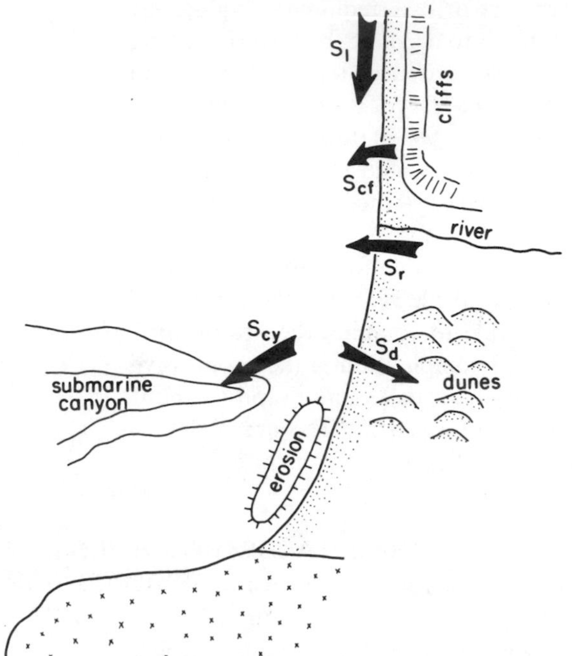

Figure 9-8 Schematic of hypothetical example of budget of sediments, including a littoral drift S_l; sediment gains from a river source, S_r, and from cliff erosion, S_{cf}; and losses down a canyon, S_{cy}, and shoreward into sand dunes, S_d.

between sources and losses. Prior to that time the sources and losses approximately balanced one another.

Littoral drift

With offshore wave statistics and wave-refraction diagrams we determine from equations (8-6) and (9-3) that

$$\begin{aligned} S_l\,[\text{south}] &= 105{,}000\ \text{m}^3/\text{yr} \\ S_l\,[\text{north}] &= 35{,}000\ \text{m}^3/\text{yr} \\ \hline \text{Net } S_l &= 70{,}000\ \text{m}^3/\text{yr to the south} \end{aligned}$$

This net S_l value is large compared to the uncertainties in S_l [south] and S_l [north] and so should be a reasonable estimate.

Cliff erosion

A combination of field measurements and surveys indicates that the sea cliff has been retreating at a rate of 0.8 m/yr. The cliff has an average height of 6 meters, and a total length of 1,270 meters is exposed to wave erosion. The cliff consists of interbedded sand and clays, and it is found by sampling that about 40% of the material would remain on the beach once it has been eroded. The yield from the cliff erosion is then

$$S_{cf} = 0.8\ \text{m/yr} \times 6\ \text{m} \times 1{,}270\ \text{m} \times 0.40 = 3{,}050\ \text{m}^3/\text{yr}$$

River source

Fifteen years previously a dam had been built on the river, decreasing its original drainage area of 1,400 km² to 600 km² (below the dam). No discharge measurements are available for the river. The effective precipitation in the area is 75 cm/yr; the graph of Figure 9-2 indicates that for this rainfall the sediment yield per square kilometer of drainage area will be 310 m³/yr. This rate must be corrected for the drainage area size by dividing by the factors $(1400/100)^{0.15} = 1.49$ and $(600/100)^{0.15} = 1.31$. This would give the total sediment yield of the river, suspended and bed loads; sediment samples from the river suggest that only about 15% of the total load is coarse enough to remain on the beach. The river source is then

$$S_r \text{ [before dam]} = 310 \text{ m}^3/\text{yr}\cdot\text{km}^2 \times (1/1.49) \times 1{,}400 \text{ km}^2 \times 0.15$$
$$= 43{,}690 \text{ m}^3/\text{yr}$$
$$S_r \text{ [after dam]} = 310 \text{ m}^3/\text{yr}\cdot\text{km}^2 \times (1/1.31) \times 600 \text{ km}^2 \times 0.15$$
$$= 21{,}300 \text{ m}^3/\text{yr}$$

The estimate of the yield after construction of the dam may be too high, as the 600 km² of drainage area remaining below the dam is principally flat outwash plains whose yield rate is certainly lower than indicated. In the empirical correlation of Langbein and Schumm (1958) this calculated yield rate would be for an unaltered small drainage basin, with typically high land and river channel slopes.

Wind-blown losses

A small dune field exists shoreward of the beach. Five years of measurements indicate that the dunes advanced shoreward at an average rate of 0.10 cm/day (0.365 m/yr). If the dunes have a mean height of 5 meters, the volume transport becomes 1.82 m³/yr per meter of coast length. Dunes extend for 1,100 meters along the coast, so that the total rate of loss is approximately $S_d = 2{,}000$ m³/yr.

Losses down the submarine canyon

No long-term measurements of sand accumulation and loss in the canyon head are available. Repeated surveys over a two-year period indicate a 200,000-m³ loss of beach sand to the canyon, giving a rate of $S_{cy} = 100{,}000$ m³/yr. This rate must be viewed as very uncertain.

Other losses and gains

Onshore-offshore movements of sand could not be determined, but it is thought (hoped) that the quantities involved are small. There are no significant biogenous or hydrogenous sources. There has been no mining of sand from the beach or river and no beach nourishment.

The balance

Before dam construction on the river the balance of sources and losses was:

Littoral drift:	+ 70,000 m^3/yr
Cliff erosion:	+ 3,000
River transport:	+ 44,000
Sand dunes:	− 2,000
Submarine canyon:	−100,000
Balance	+ 15,000 m^3/yr

This positive balance indicates that there should have been some deposition on the beach, but we have seen that at that time the beach was stable, neither eroding or advancing. This means that in our budget we have either overestimated our sources, underestimated our losses, or both. Much of the error could be in our evaluation of the littoral drift, but more likely we are mistaken in our estimate of losses down the submarine canyon.

Following construction of the dam, the river contribution is decreased to 21,000 m^3/yr, or probably less. Assuming the other sources and losses remain about the same, the new balance would be −8,000 m^3/yr, indicating slight beach erosion. Beach erosion did actually occur, but repeated surveys of the beach indicate that the actual amount of beach sand lost to erosion is about 40,000 m^3/yr, much greater than indicated by our negative balance. The reason for this discrepency would be the same as for the lack of balance before dam construction plus the fact that the new 21,000-m^3/yr yield rate for the river is certainly an overestimate. Although the accuracies of the evaluations of sources and losses do not permit a more refined budget of sediments, the shift to a negative balance following dam construction plus the onset of beach erosion soon after does indicate that the dam is responsible for the beach erosion.

HEAVY MINERALS AS TRACERS OF SEDIMENT SOURCE AND TRANSPORT

Before we conclude our consideration of the budget of littoral sediments, a short discussion is needed on the utilization of naturally occurring heavy minerals in the sediments as indicators of the source and movements of the sands to the beach. Within the normal continental beach sands composed principally of light-colored grains of quartz and feldspar there can be found small quantities of heavy minerals: hornblende, epidote, tourmaline, augite, zircon, and many others. In general these accessory minerals are much denser than the quartz and feldspar grains and are darker in color. Often these heavy minerals are concentrated because of their differing hydraulic behavior from the quartz and feldspar and can be seen as dark laminae (Figure 13-14). Certain rock types will yield different suites of these heavy minerals than other rock types. A granite may provide different heavy minerals than will a schist, gneiss, or basic volcanic rock, and one granite could provide heavy minerals not supplied by still another granite. The presence in a beach sand of a particular heavy mineral or suite of heavy minerals can point to the erosion of a certain rock type or recycled sediment deposit as the partial source of the sand. Sometimes the proportion of those heavy minerals versus the other varieties can yield information as to the relative importance of that rock type as a sediment source to the beach. Also, the distribution of those heavy minerals in beach sands along the coast can indicate the direction of littoral transport; the heavy minerals can sometimes be followed back to their sources.

There are many examples of the use of heavy minerals in this way to answer questions regarding the source of beach sediments. One of the first and best studies of this sort is that of Trask (1952), who was interested in the problems that arose at Santa Barbara, California, following the construction of the breakwater (Chapter 12). Trask demonstrated that a significant proportion of sand filling the Santa Barbara harbor comes from a distance of more than 160 km up the coast. In this example Trask used the mineral augite as a tracer, a black ferromagnesian silicate found commonly in basic igneous rocks. In that portion of the California coast this mineral has its source rocks near Morro Bay, to the north of Santa Barbara. This mineral, along with the lighter quartz and feldspar and other heavy minerals, moves to the south as littoral drift for the 160 km before being trapped within the Santa Barbara breakwater. The streams between Morro Bay and Santa Barbara supply no additional augite, so that its proportion among the heavy minerals progressively decreases. As has already been discussed, Bowen and Inman (1966) in turn used this decreasing augite proportion to calculate the increasing littoral transport as one moves south along the coast. Of additional interest is the fact that the sand is able to move around Point Conception before reaching Santa Barbara. This again is proven by tracing augite contained in the sands around this point. Trask (1955) goes on to study other examples of sand moving around California headlands.

Although it involves considerable effort, the application of heavy minerals to tracing sediment movements along the coast and determining their sources can lead to a much better understanding of the budget of littoral sediments.

PROBLEMS

9-1 The effective precipitation in a drainage basin of area 350 km² is 75 cm/yr. What is the annual sediment yield for that river? What would it be if the drainage area were 75 km²?

9-2 Other conditions being the same, why does a smaller river drainage area provide a greater sediment yield per unit area than does a large drainage area?

9-3 In Figure 9-2, why does the sediment yield data from sedimentation stations plot lower than the reservoir data? Which set of data is better for the evaluation of the budget of sediments for beaches, and why?

9-4 Why will sand mining in a river diminish the quantity of sand being supplied to the beach? After all, the river discharge q in equation (9-1) remains the same, so why won't q_s, the supply of sand, remain the same? What changes in the river parameters from mining will cause q_s to decrease according to equation (9-1)?

9-5 The major sediment source to the southern California beaches (Figure 9-1) has been the rivers. Nearly all of these rivers have been dammed. What will be the impact on the beaches? Within any one compartment or cell, where will the impact be felt first?

9-6 Under what circumstance will river sediment mining or river damming not have a detrimental effect on the local beaches?

9-7 The loss of sand from the beach by winds blowing it onshore into dunes is evaluated by the rate at which the dunes migrate shoreward. What possible errors could be involved in this estimate?

9-8 Under what circumstances might the mining of beach sediments be acceptable?

9-9 What is wave hindcasting, how is it done, and how is it used to evaluate the net sediment transport at a given beach?

9-10 Sand on California beaches is lost principally down submarine canyons. It is suggested that we dam up these canyons to prevent this loss. What do you think of such a suggestion? What would such damming do to the local budget of sediments in the immediate vicinity of the canyon? In light of your answer to Problem 9-5, above, what should be done with the sand that is prevented from going down the canyon? Work out a feasible scheme for such an attempt.

REFERENCES

ANDERSON, H. W. (1949). Flood frequencies and sedimentation from forest watersheds. *Trans. Am. Geophys. Union*, 30: 576–84.

——— (1955). Detecting hydrologic effects of changes in watershed conditions by double-mass analysis. *Trans. Am. Geophys. Union*, 36: 119–25.

ANDERSON, R. G. (1971). *Sand budget for Capitola Beach, California.* Master's thesis, Naval Postgraduate School, Monterey, Calif., 57 pp.

BAGNOLD, R. A. (1966). *An approach to the sediment transport problem from general physics.* U.S. Geological Survey Professional Paper no. 422–I, U.S. Government Printing Office, Washington, D.C., 37 pp.

BIRD, E. C. F. (1969). *Coasts.* M.I.T. Press, Cambridge, 246 pp.

BOWEN, A. J., and D. L. Inman (1966). *Budget of littoral sands in the vicinity of Point Arguello, California.* U.S. Army Coastal Engineering Research Center Tech. Memo. no. 19, 56 pp.

BRUNE, G. M. (1948). *Rates of sediment production in midwestern United States.* U.S. Dept. Agr. Soil Consv. Service SCB-TP-65, 40 pp.

CHAMBERLAIN, T. K. (1964). Mass transport of sediment in the heads of Scripps Submarine Canyon, California. In *Papers in marine geology* (Shepard Commemorative Volume), ed. R. L. Miller, pp. 44–58. Macmillan, New York.

COASTAL ENGINEERING RESEARCH CENTER (1973). *Shore protection manual.* U.S. Army Corps of Engineers, Washington, D.C., 3 vols.

COLBY, B. R., and C. H. Hembree (1955). *Computations of total sediment discharge, Niobrara River near Cody, Nebraska.* U.S. Geological Survey Water Supply Paper no. 1,357, U.S. Government Printing Office, Washington, D.C., 187 pp.

COOPER, W. S. (1958). *Coastal sand dunes of Oregon and Washington.* Geol. Soc. Am. Memoir no. 72, 169 pp.

DIETZ, R. S., H. J. KNEBEL, and L. E. SOMERS (1968). Cayer submarine canyon. *Geol. Soc. Am. Bull.*, 79: 1821–28.

EINSTEIN, H. A. (1950). *The bed-load function for sediment transportation in open channel flows.* U.S. Dept. Agr. Tech. Bull. no. 1,026, 70 pp.

——— (1964). River sedimentation. In *Handbook of hydrology*, ed. V. T. Chow, McGraw-Hill, New York, Section 17-II, 67 pp.

EMERY, K. O. (1968). Relict sediments on continental shelves of world. *Bull. Am. Assoc. Petrol. Geol.*, 52: 445–64.

GRAF, W. H. (1971). *Hydraulics of sediment transport.* McGraw-Hill, New York, 513 pp.

INMAN, D. L., and T. K. CHAMBERLAIN (1960). Littoral sand budget along the southern California coast [Abstract]. *Rpt. 21st Int. Geol. Cong.* (Copenhagen), *Vol. of Abstracts*, pp. 245–46.

INMAN, D. L., W. R. GAYMAN, and D. C. COX (1963). Littoral sedimentary processes on Kauai, a subtropical high island. *Pac. Sci.*, 17, no. 1: 106–30.

INMAN, D. L., and J. D. FRAUTSCHY (1966). Littoral processes and the development of shoreline. *Proc. Coast. Eng. Speciality Conf., ASCE* (Santa Barbara, Calif.), pp. 511–36.

JOHNSON, J. W. (1953). Sand transport by littoral currents. *Proc. 5th Conf. State Univ. of Iowa Studies in Eng.*, Bulletin no. 34, pp. 89–109.

KULM, L. D., and J. V. Byrne (1966). Sedimentary response to hydrography in an Oregon estuary. *Mar. Geol.*, 4: 85–118.

LANGBEIN, W. B. (1949). *Annual runoff in the United States.* U.S. Geological Survey Circular no. 52, U.S. Government Printing Office, Washington, D.C., 14 pp.

LANGBEIN, W. B., and S. A. SCHUMM (1958). Yield of sediment in relation to mean annual precipitation. *Trans. Am. Geophys. Union*, 39: 1076–84.

LUSTIG, L. K. (1965). *Sediment yield of the Castaic watershed, western Los Angeles County, California; a quantitative geomorphic approach.* U.S. Geological Survey Professional Paper no. 422-F, U.S. Government Printing Office, Washington, D.C., 23 pp.

MEADE, R. H. (1969). Landward transport of bottom sediments in estuaries of the Atlantic Coastal Plain. *J. Sediment. Petrol.*, 39: 222–34.

National Marine Consultants (1960). *Wave statistics for seven deep water stations along the California coast.* Paper prepared for the Los Angeles and San Francisco Districts, U.S. Army Corps of Engineers.

NORRIS, R. M. (1964). Dams and beach-sand supply in southern California. In *Papers in marine geology* (Shepard Commemorative Volume), ed. R. L. Miller, pp. 154–71. Macmillan, New York.

SCOTT, K. M., J. R. RITTER, and J. M. KNOTT (1968). *Sedimentation in the Piru Creek watershed, southern California.* U.S. Geological Survey Water Supply Paper no. 1,798-E, U.S. Government Printing Office, Washington, D.C., 48 pp.

SHEPARD, F. P. (1951). Mass movements in submarine canyon heads. *Trans. Am. Geophys. Union*, Tulsa, Okla., 32: 405–18.

——— (1964). Sea-floor valleys of Gulf of California. In *Marine geology of the Gulf of California*, ed. Tj. H. van Andel and G. G. Shor, pp. 157–92. Amer. Assoc. Petrol. Geol., Memoir no. 3.

SHEPARD, F. P., and R. F. DILL (1966). *Submarine canyons and other sea valleys.* Rand McNally, Chicago, 381 pp.

SHEPARD, F. P., and U. S. GRANT IV (1947). Wave erosion along the southern California coast. *Geol. Soc. Am. Bull.*, 58: 919–26.

SHIELDS, A. (1936). *Anwendung der Ahnlichkeitsmechanik und Turbulenzforschung auf die Geschiebebewegung*. Mitteil. Preuss. Versuchsanst. Wasser, Erd, Schiffsbau, Berlin, n. 26.

TRASK, P. D. (1952). *Source of beach sand at Santa Barbara, California, as indicated by mineral grain studies*. U.S. Army Corps of Engrs., Beach Erosion Board Tech. Memo. no. 28, 24 pp.

——— (1955). *Movement of sand around southern California promontories*. U.S. Army Corps of Engrs., Beach Erosion Board Tech. Memo. no. 76, 60 pp.

VALENTIN, H. (1954). Der Landverlust in Holderness, Ostengland, von 1852 bis 1952. *Die Erde,* 3, no. 4: 296–315.

VESPER, W. H. (1965). *Behavior of beach fill and borrow area at Seaside Park, Bridgeport, Connecticut*. U.S. Army Coastal Engineering Research Center Tech. Memo. no. 11, 24 pp.

YALIN, M. S. (1972). *Mechanics of sediment transport*. Pergamon Press, Oxford, 290 pp.

Chapter 10

SHORELINE CONFIGURATION

. . . It is astonishing and incredible to us, but not to Nature, for she performs with utmost ease and simplicity things which are even infinitely puzzling to our minds, and what is very difficult for us to comprehend is quite easy for her to perform.

Galileo
Dialogue Concerning the Two World Systems (1630)

Drawing an analogy with the graded stream, Tanner (1958) introduced the concept of an *equilibrium beach* as one whose curvature in plan view and profile are adjusted in such a way that the waves impinging on the shore provide precisely the energy required to transport the load of sediments supplied to the beach. The time element for such an equilibrium condition is long-term—years and decades—rather than instantaneous. This chapter examines the processes that govern the curvature of an equilibrium beach, while Chapter 11 will consider the equilibrium beach profile. Analytical and computer simulation models, which are used to understand the overall curvature of the coastline, will be considered first. Later in the chapter a variety of rhythmic shoreline features will be examined: giant cusps, crescentic bars, and beach cusps.

WAVE REFRACTION AND THE SHAPE OF THE POCKET BEACH

The most elementary equilibrium shoreline is that in which there is no net long-term littoral drift. This condition is most closely approached by a pocket beach where there is little or no additional sand being supplied to the beach (Figure 10-1).

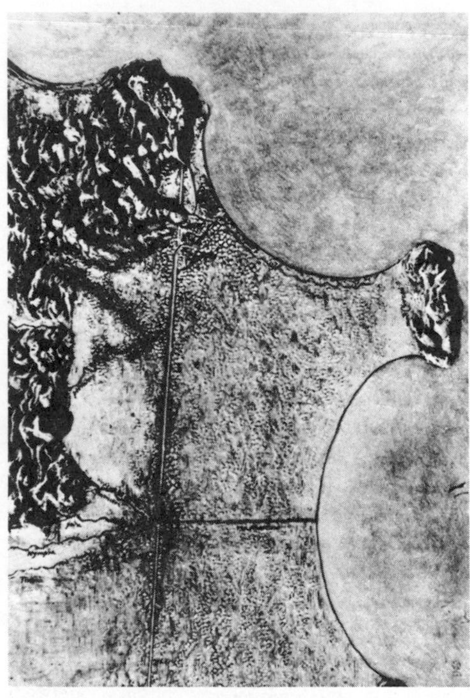

Figure 10-1 Leonardo da Vinci's plan for draining the Pontine Marshes, Italy. Notice the arcuate curvature of the shoreline between the rocky promontories.

Consider first a simple wave train arriving at the pocket beach; these waves could be the prevailing or dominant waves in the area. Wherever the wave crests strike the shoreline at an angle α_b they produce a longshore sediment transport I_l given by (Chapter 8)

$$I_l = 0.77(ECn)_b \sin \alpha_b \cos \alpha_b \tag{10-1}$$

where $(ECn)_b$ is the energy flux of the breaking waves. Sediment will be shifted alongshore until we have the condition in which $\alpha_b = 0$ everywhere along the shoreline, at which time sediment transport ceases. This equilibrium condition, then, is one in which the shoreline is everywhere parallel to the crests of the incoming waves. It will be recalled (Chapter 4) that as waves shoal they refract or bend, the actual shape depending on the nature of the offshore topography. Refraction, then, controls the shape of the wave crests and therefore the exact shape and orientation of the beach that is experiencing little or no net littoral drift. This was realized in part by Lewis (1938) who, using a number of examples of beaches in Britain, demonstrated that they orient themselves according to the prevailing waves. He clearly recognized that the long-term average conditions are significant. Lewis did not extensively develop the role played by refraction, although he appears to have realized its importance. Davies (1958) has more precisely demonstrated the significance of wave refraction. Figure 10-2, from his paper, is an excellent example of the way in which local beaches orient themselves parallel to the refracted wave crests and develop the same curvature.

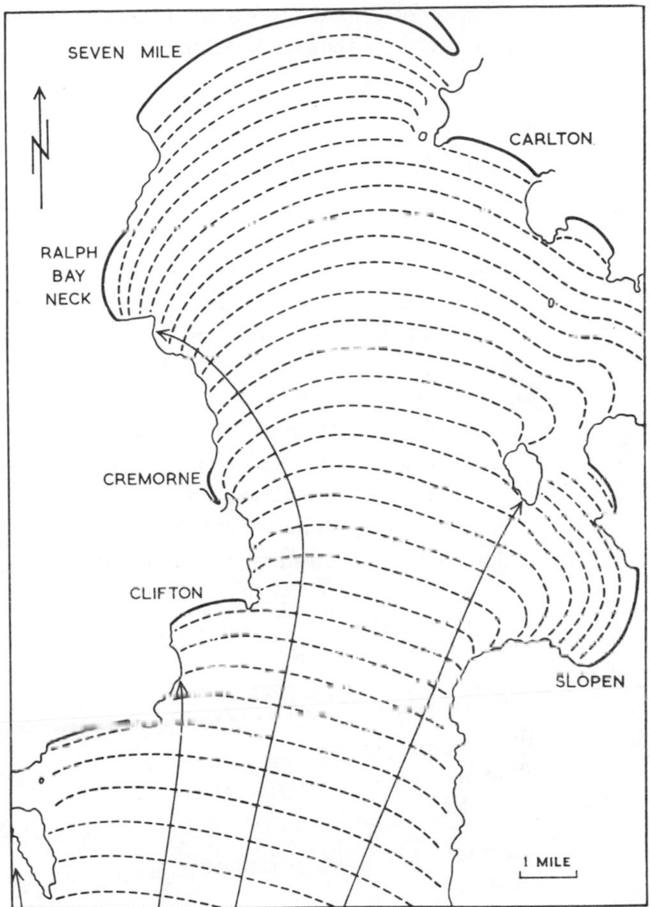

Figure 10-2 Wave refraction diagram of Frederick Henry Bay, Tasmania, for a 14-second southwesterly swell. Every fifth wave crest is shown as a broken line, and the wave orthogonals are equidistant in deep water. Sand beaches are represented by thickened lines. [*From* Davies (1958)]

If there were a significant source of sand to the pocket beach, then the equilibrium shoreline would have to make an angle with the refracted wave crests in order to redistribute the sand alongshore away from the source. This correspondence between the refracted wave crests and the shoreline configuration will be made more apparent later in the chapter when we examine simulation models of shoreline configuration. At that time the effects of many different wave trains arriving at a beach will be examined, rather than those of a single "prevailing" wave train. As might be expected, the shoreline then wobbles about under the different wave trains arriving from slightly different directions.

Because the exact curvature of the beach is governed by wave refraction, which in turn is dependent upon conditions well out to sea from the beach itself, it is clear that we cannot expect the beach shape to be that of some prescribed geometric curve. We must then view as futile or at least as simple chance the attempts of Hoyle and King (1958) to fit such beaches to arcs of great circles that subtend central angles of 0.25 radian, or the comparisons of beaches to a cycloid (Bruun, 1953), or the attempt of Yasso (1965) to fit the beach curvature to logarithmic spirals, or any combinations thereof as attempted by Silvester (1970).

ANALYTICAL SOLUTIONS OF SHORELINE CONFIGURATION

The analytical approach attempts to obtain mathematical solutions for the shape of shorelines in the longshore direction. Two principal problems have been approached in this way: (1) the shoreline configuration where a groin or jetty is blocking the littoral drift, and (2) the shape of the cuspate delta where a river is supplying sand to the beach. In brief, the approach is to solve a relationship such as equation (10-1), which relates the sand transport to the wave parameters, together with a continuity equation for sediment movement in the longshore direction. The exact problem itself is defined by the imposed boundary conditions.

The continuity equation relates the rate at which a section of beach retreats or advances to the change in the quantity of littoral drift in the longshore direction. Consider the section of shoreline shown in Figure 10-3, which has a width Δx. If

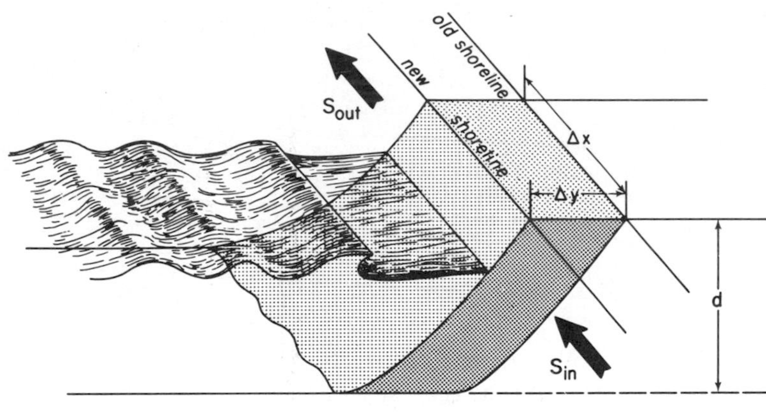

Figure 10-3 Volume change of sediment within a cell of width Δx, expressed as a change in the shoreline position Δy. The volume change is related to the longshore sediment drift into the cell versus the drift out.

S_{in} is the rate of sediment drift into this compartment (m³/day) and S_{out} is the rate of drift out, then the net accumulation or erosion ΔV (m³) in the compartment in the lapsed time Δt (days) is given by

$$\Delta V = (S_{\text{in}} - S_{\text{out}})\,\Delta t \tag{10-2}$$

If S_{in} is greater than S_{out}, then ΔV will be a positive quantity, which denotes deposition within the compartment; if S_{in} is less than S_{out}, then $\Delta V = -$ and the beach erodes.

The volume change ΔV must be reflected in an advance or retreat of the shoreline. If Δy is the change in the shoreline position in the onshore-offshore direction in that increment of time, Δt, then from the geometry of the wedge depicted in Figure 10-3 we have

$$\Delta V = d\,\Delta y\,\Delta x \tag{10-3}$$

where d is a linear dimension such that $d\,\Delta y$ gives the section area of beach eroded or deposited. Combining equations (10-2) and (10-3) yields

$$\Delta y = (S_{\text{in}} - S_{\text{out}})\frac{\Delta t}{d\,\Delta x} \tag{10-4}$$

If we place $\Delta S = S_{\text{out}} - S_{\text{in}}$ and allow all the increments to decrease in magnitude, we obtain

$$\frac{dy}{dt} = -\frac{1}{d}\left(\frac{dS}{dx}\right) \tag{10-5}$$

which is the continuity equation for sand movement alongshore. It states that the rate of change of shoreline position (dy/dt) is governed by the longshore change in the quantity of sediment being transported (dS/dx). If dS/dx is positive, the quantity of sand being transported is increasing in the longshore direction (the x-direction), so that dy/dt is negative, which means that the shoreline must retreat or erode to supply this increasing quantity of littoral drift. Similarly, if the quantity of sand being transported in the longshore direction is decreasing ($dS/dx = -$), then dy/dt is positive and the shoreline is advancing. If $S =$ constant, then $dS/dx = 0$ and $dy/dt = 0$; the shoreline position is then stable.

Using equation (8-5), which relates the immersed-weight sediment transport rate I_l to the volume transport rate S_l, equation (10-1) becomes

$$S_l = \frac{0.77}{(\rho_s - \rho)ga'}(ECn)_b \sin\alpha_b \cos\alpha_b \tag{10-6}$$

or

$$S_l = (6.85 \times 10^{-5})(ECn)_b \sin\alpha_b \cos\alpha_b \tag{10-7}$$

having set $\rho_s = 2.65$ g/cm^3 (quartz sand), $a' = 0.6$, and including a factor change such that if $(ECn)_b$ is given in units of ergs/cm·sec, the value of S_l obtained is in units of m^3/day, the form most suitable in the present context (Komar, 1973*a*). The analytical methods attempt to solve equation (10-7) together with the continuity equation (10-5) to obtain mathematical expressions for the shoreline configuration. However, even with the simplest boundary conditions, one obtains a very complicated differential equation whose solution is impossible except by numerical methods on a computer (Grijm, 1961, 1965).

In order to make the problem tractable to mathematical treatment, the equations must be simplified. Pelnard-Considere (1954) considered variations in S_l resulting only from variations in the breaker angle α_b [$(ECn)_b$ was assumed to be everywhere constant] so that S_l could be expanded in a Taylor series as a function of α_b alone. This yields the equation

$$\frac{\partial y}{\partial t} = \frac{1}{d}\left(\frac{\partial S_l}{\partial \alpha_b}\right)\frac{\partial^2 y}{\partial x^2} \tag{10-8}$$

which states that the rate of deposition ($\partial y/\partial t$) is linearly dependent upon the curvature of the coast ($\partial^2 y/\partial x^2$). Equation (10-8) involves certain approximations such that the difference between the breaker angle and deep-water incident angle must be small at all times. Although the original continuity relationship, equation (10-5), is exact, if the Pelnard-Considere modification is used, only approximate results can be expected. More critical, only problems in which S_l varies with α_b alone can be considered; $(ECn)_b$ must remain constant.

Pelnard-Considere (1954) considered the shoreline configuration developed by a groin or jetty built perpendicular to an initially straight shoreline, blocking the longshore drift of sand. His solutions were partially confirmed by model tests, but did not include the refraction "shadow zone" behind the groin. Bakker, Bretler, and Roos (1971) and Bakker (1969) have further considered the configuration of beaches with groin systems.

Pelnard-Considere (1954), Grijm (1961, 1965), and Bakker and Edelman (1965) have attempted mathematical solutions for the growth of a cuspate river delta on a coast along which sediment is transported by waves. In order to make the problem suitable to mathematical treatment they assume: (1) the river continuously brings a constant supply of sediment to the sea; (2) waves approach the shore at a constant deep-water angle and with a constant energy flux; and (3) wave refraction and diffraction are neglected. Some of the studies use the Pelnard-Considere relationship, equation (10-8), as an approximation to the continuity equation, so that the solutions are necessarily approximate. Some investigations also assume rather unusual forms for the equation relating the beach sediment transport rate to the wave parameters. Grijm (1961, 1965) solves his complicated differential equation by means of a computer to obtain delta shapes such as that shown in Figure 10-4. Even then, two possible solutions are obtained for the delta shape, one in which the angle of incidence is everywhere less than 45 degrees and a second where the angle is more than 45 degrees. Combinations are also possible, as illustrated in Figure 10-4, where solutions are pieced together.

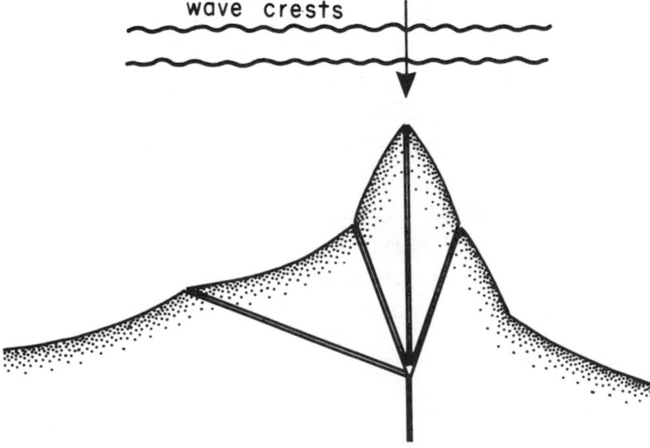

Figure 10-4 Analytical solution for the shape of a delta as obtained by Grijm (1965). In this example, solutions are pieced together to yield a composite delta. [*After* Grijm (1965)]

COMPUTER SIMULATION MODELS

An approach to investigating shoreline configuration that is more versatile than seeking analytical solutions is the method of simulation on a computer. Not only is it more versatile, but the approach is easier to understand and to apply.

In a numerical simulation model on a computer, the equations of sand transport along a beach and the continuity equation can be solved together to any degree of exactness required with any set of boundary conditions. Time variations are easily introduced; for example, the wave conditions could be changed through time, or a river supplying sand to the shoreline could vary its discharge with time. Longshore variations in the wave energy and deep-water angle of approach can also be included, so that wave refraction and diffraction effects may be considered.

The studies of Price, Tomlinson, and Willis (1973) and Komar (1973*a*) first attempted to simulate on a computer beach processes which govern the configuration of the shoreline. Both apply equation (10-1) or its equivalent equation (10-7) to evaluate the longshore movement of sand under an oblique wave approach to the shoreline.

As diagramed in Figure 10-5, the shoreline is divided up into a series of cells of uniform width Δx and with individual lengths $y_1 \ldots y_{i-1}, y_i, y_{i+1}, \ldots, y_n$ beyond some base line. The narrower the cells (the smaller Δx), the more nearly the series of cells approximates the true shoreline. Changes in the shoreline configuration are brought about by the littoral drift S_i (m^3/day) which shifts sand alongshore from one

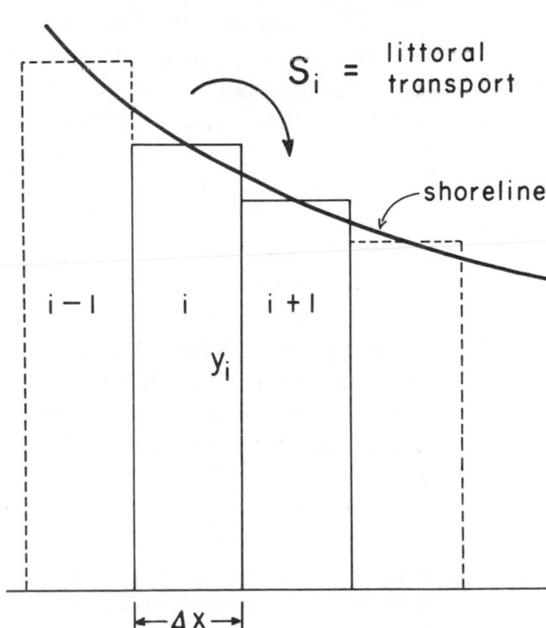

Figure 10-5 The shoreline divided into a series of cells of uniform width Δx, extending outward distances y_i from a base line (x-axis). Each cell would be like that depicted in Figure 10-3. The term S_i denotes the littoral sand drift alongshore from cell i to cell $i + 1$.

cell to the next. From our continuity relationship, equation (10-4), we have

$$\Delta y_i = (S_{i-1} - S_i)\frac{\Delta t}{d\,\Delta x} \tag{10-9}$$

for the shoreline advance or retreat Δy_i in cell i, where S_i is the rate of littoral drift from cell i to cell $i+1$ (S_{out}), and S_{i-1} is the littoral drift from cell $i-1$ into cell i (S_{in}); Δt is the increment of time (days). Note that Δy_i is positive when $S_{i-1} > S_i$, indicating net deposition, while erosion occurs and Δy_i is negative when $S_i > S_{i-1}$, that is, when sand leaves cell i at a faster rate than it enters.

If a particular cell has other sources or losses of sand besides littoral drift, then these can be easily included. For example, if the cell is at the mouth of a river supplying sand at the rate S_r (m^3/day), then equation (10-9) can be expanded to

$$\Delta y_i = (S_r + S_{i-1} + S_i)\frac{\Delta t}{d\,\Delta x} \tag{10-10}$$

In the models it is important that the Δy_i values remain relatively small, so that there are no sudden "jumps" in the shoreline position. This usually requires that the time increment Δt be kept small.

Any simulation model of shoreline changes then simply involves: (1) defining an initial shoreline configuration; (2) establishing the sources of sand to the beach, such as rivers, and possible losses of sand from the beach; (3) giving the offshore wave parameters (height, period, approach angle); (4) indicating how the littoral transport of sand along the beach is to be governed by the wave parameters (equation 10-1 or 10-7); and (5) determining how the shoreline is altered from its initial configuration under these conditions at increments of time, Δt, for some total span of time.

The study of Price, Tomlinson, and Willis (1973) examined changes in a beach brought about by the construction of a long groin blocking the longshore drift of sand. The results of their computer simulation model compared favorably with the actual shoreline changes experienced in a laboratory wave basin (Figure 10-6). This indicates both that the numerical approach is valid and that equation (10-1) can be utilized to evaluate the sand transport rate.

At the beginning of this chapter we saw in a descriptive fashion how a shoreline changes its shape to conform to or become congruent with the shape of the refracted wave crests such that $\alpha_b = 0$ everywhere along the shoreline. We can now demonstrate this better with a simulation model. One example, illustrated in Figure 10-7, shows a pocket beach of 1 km length which reorients itself from its originally straight shoreline until it takes on the shape of the refracted wave crests. In this example the wave curvature is taken as parabolic, but any arbitrary shape could have been used. In this model the time increment was $\Delta t = 0.002$ day, and the longshore width of the cells was $\Delta x = 10$ meters.

The effects of differing wave trains arriving at the pocket beach are shown in Figure 10-8. When the wave pattern changes, the beach responds by altering its curvature and orientation to correspond to the new wave conditions. The shoreline then wobbles about within the bay because of the changing wave directions. As would be expected, a large beach may not fully respond to a certain wave train before the

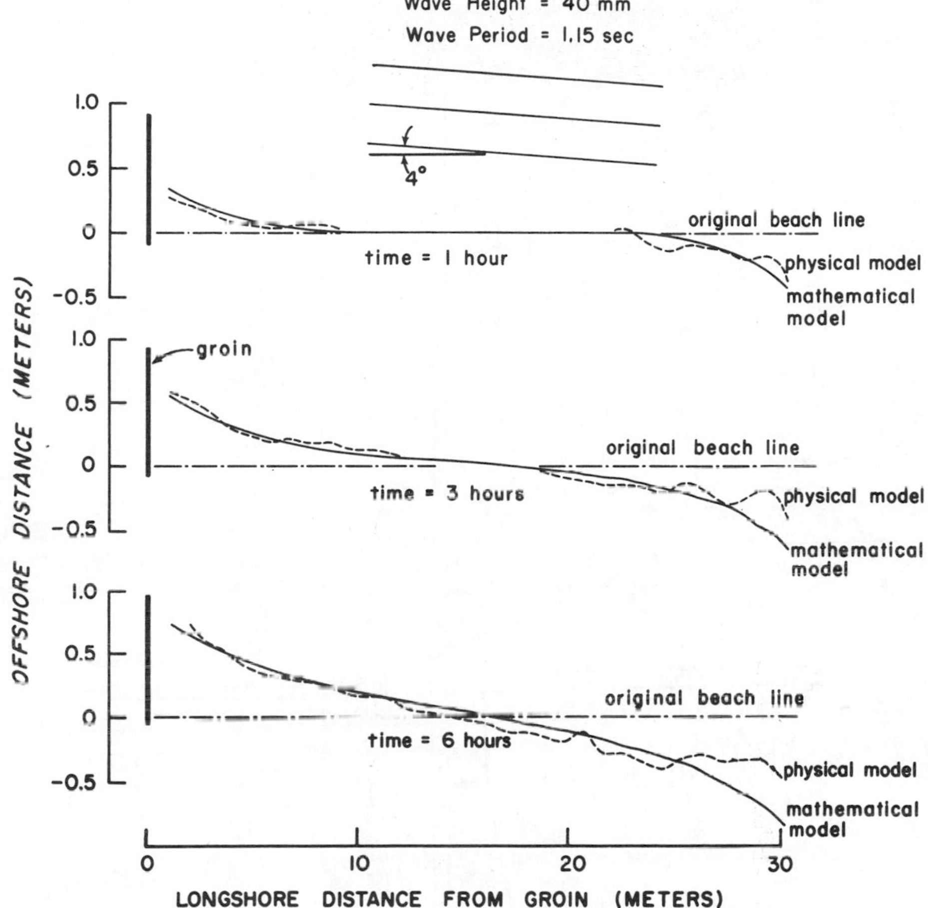

Figure 10-6 Beach reorientation next to a groin due to an oblique wave approach to an initially straight shoreline. The numerical model shorelines calculated with equations (10-7) and (10-9) compare favorably with the results of a wave basin test. [*After* Price, Tomlinson, and Willis (1973)]

wave conditions are changed, so that in general the larger the scale of the beach, the smaller the degree of wobble under the varying waves.

With a source of sand to the beach the equilibrium shoreline would have to make an angle with the refracted wave crests in order for the sand to be moved away from the source area. The conditions of equilibrium shoreline configuration where a river is supplying sand to the beach have been investigated by Komar (1973*a*). Figure 10-9 shows the first 360 days' growth of a delta where the river supplies sand at the rate $S_r = 2 \times 10^4$ m³/day, tending to build out the delta while waves of energy flux 3×10^8 ergs/cm·sec work to flatten the shoreline. It is seen that after some 150 days there is a balance such that the delta continues to grow outward but at a steady rate, maintaining its overall shape in the process. An equilibrium has thus been

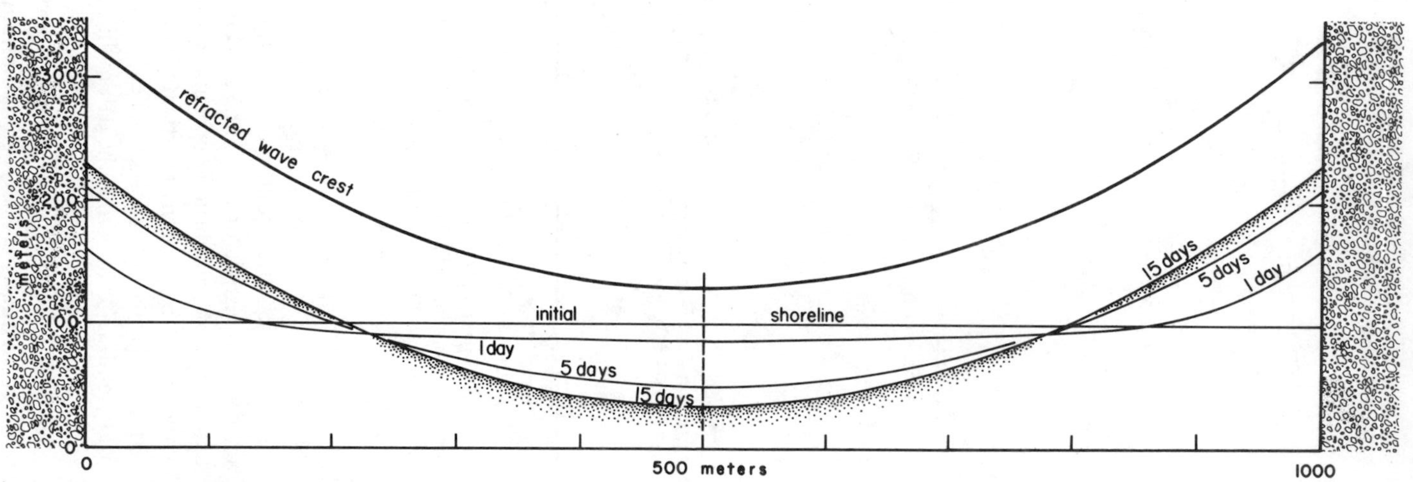

Figure 10-7 An initially straight shoreline between two rocky headlands taking on the same curvature as the refracted wave crests such that after 15 days the sediment drift reduces nearly to zero, since $\alpha_b \simeq 0$ everywhere. In this example, $ECn = 5 \times 10^8$ ergs/cm·sec.

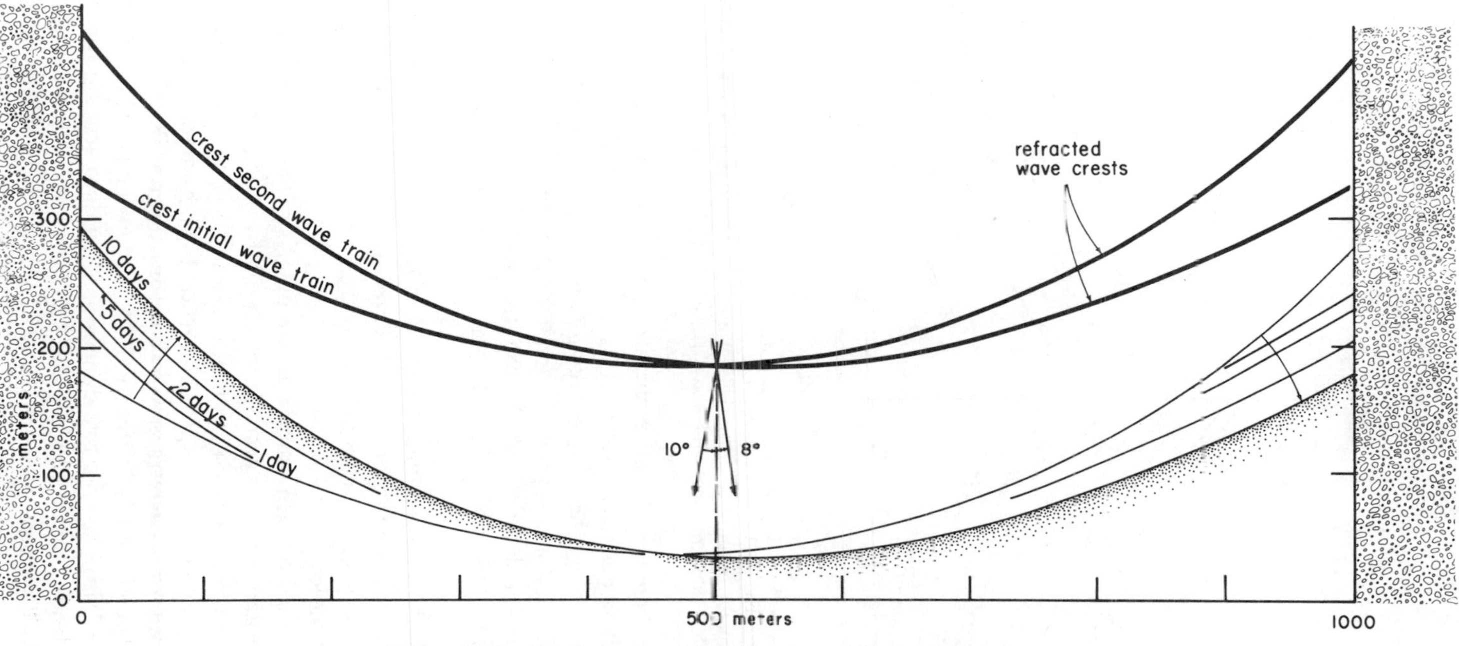

Figure 10-8 Reorientation of beach curvature due to change in wave direction. Such variations in waves would produce a wobbling of the beach between the headlands.

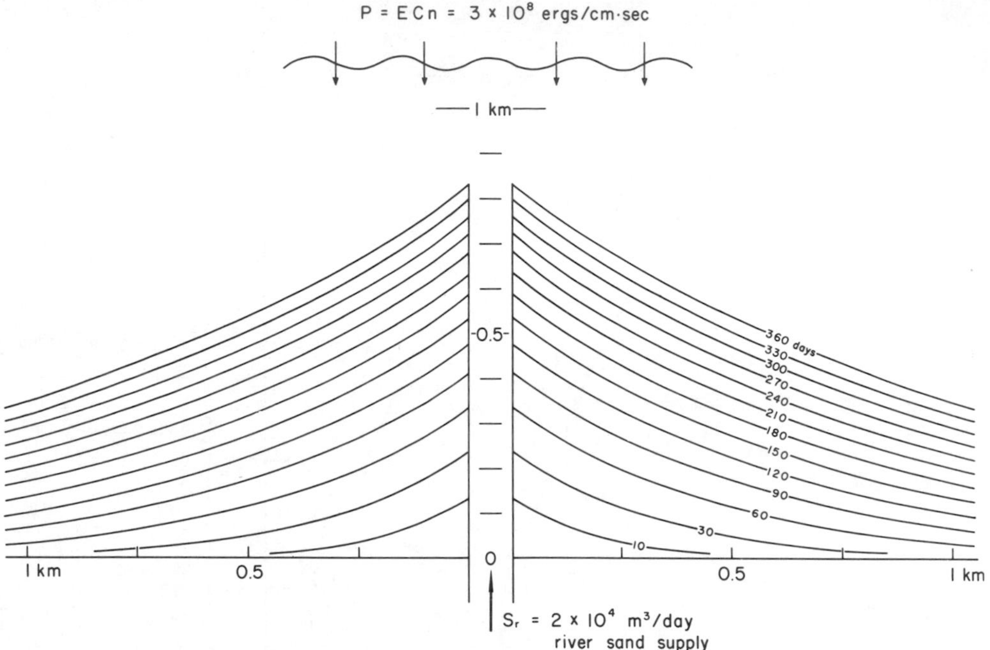

Figure 10-9 The first 360 days growth of a river delta in which waves are arriving straight onshore to an initially straight beach. Values of wave energy flux and river sediment input are shown. [*After* Komar (1973*a*)]

achieved in which there is a shoreline curvature such that the waves are just able to redistribute the sand along the shoreline at the rate it is supplied by the river. In Figure 10-10 the effect of varying the wave energy flux is illustrated; the greater the energy flux for a constant river sand supply, the smaller the wave breaker angles required to redistribute the sand and the flatter the resulting equilibrium shoreline. Decreasing the river sand supply while maintaining the wave energy flux would have had the same effect.

Figure 10-11 shows a combination of the models where a stream supplies sand to a pocket beach. Because of this supply the shoreline builds outward within the bay but maintains a constant equilibrium configuration, since the wave and river conditions are not varied.

The full potential of such numerical computer simulation models is only now being realized. The above models account for wave refraction by prescribing the shape of the wave crests once they reach the beach, already having gone through offshore wave refraction. A more realistic approach, one that will have to be taken in real case studies, is to account for wave refraction by coupling the models of shoreline changes to one of the computer routines [Wilson (1966), for example] which compute wave refaction patterns. This involves a two-dimensional array to account for the offshore topography and hence considerably increases the overall complexity of the model. However, such models would be more realistic, and onshore-offshore shifts of sand

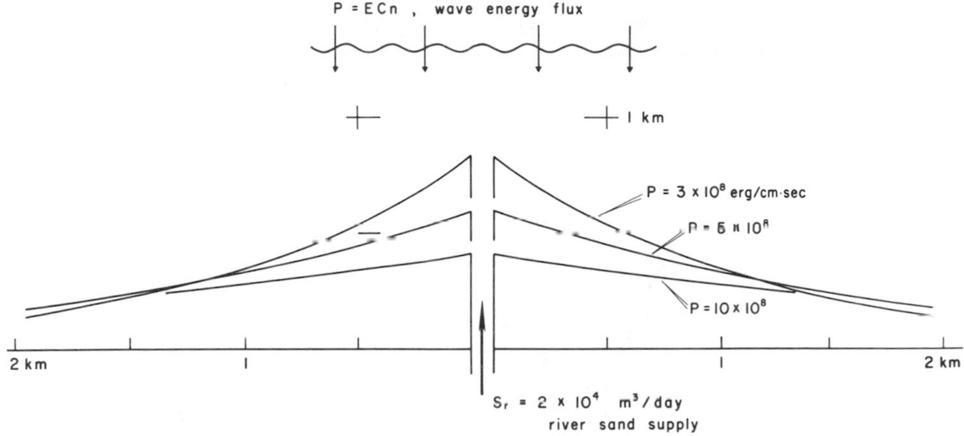

Figure 10-10 Effects on the 360-day delta configuration resulting from different wave energy fluxes while maintaining constant river sediment supply. [*After* Komar (1973*a*)]

could be included as well. Motyka and Willis (1975) have devised one such model, the purpose of which is to investigate the effects on the beach of offshore dredging of sediment. The dredging leaves a hole which affects the wave refraction pattern, which in turn alters the shoreline configuration. Their study, relying entirely on the computer simulation approach, investigated the magnitudes of the beach changes as a function of the dredge hole size and shape. Walton and Dean (1973) also report that they are attempting to model shoreline changes for beach fills in Florida using such techniques.

A somewhat different type of computer simulation model has been developed by King and McCullagh (1971) and applied to the growth of the Hurst Castle Spit in southern England (Figure 10-12). Theirs is a probabilistic model rather than a process model. Storm waves from the southwest produce the main extension of the spit, and northeast waves coming down the Solent between the mainland and the Isle of Wight build the recurves. An increase in the water depth in the direction of growth results in a decrease in the growth rate and an increase in the number of recurves. The method is to select a series of random numbers which determine the order of events in the growth of the spit. The proportion of random numbers allocated to the different wave directions can be adjusted until the simulated spit pattern agrees with the morphology of the real spit. Figure 10-12 represents the best agreement between model and spit found by King and McCullagh (1971). In this way the effects of different proportions of waves coming from different directions can be examined and the effect of the increase in water depth analyzed. It is thus possible to come to some conclusion concerning the variables that are most important in controlling the growth of the spit.

Fox and Davis (1973) have developed an empirical mathematical model to simulate the effects of storm cycles on the beach morphology along the eastern shore of Lake Michigan. The model is empirical in that Fourier analysis of the barometric

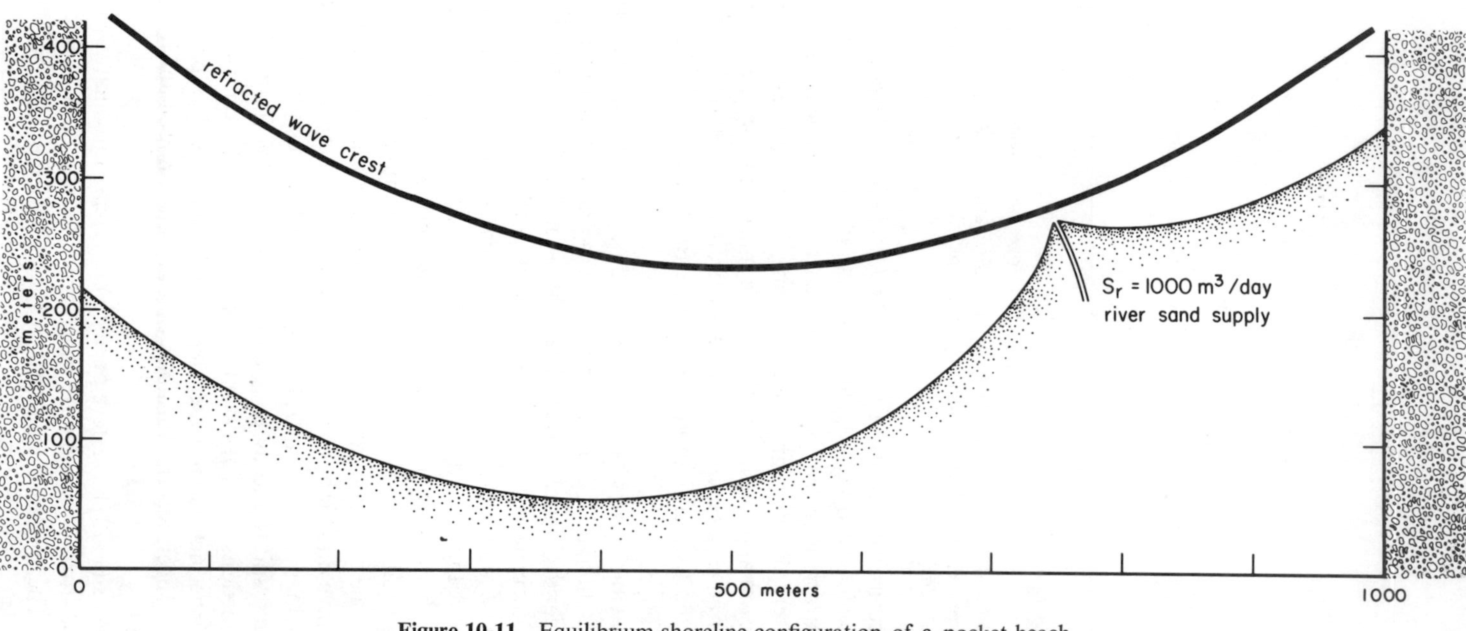

Figure 10-11 Equilibrium shoreline configuration of a pocket beach 1 km in width, where a river supplies sand as shown. In this example, $ECn = 1.5 \times 10^7$ ergs/cm·sec.

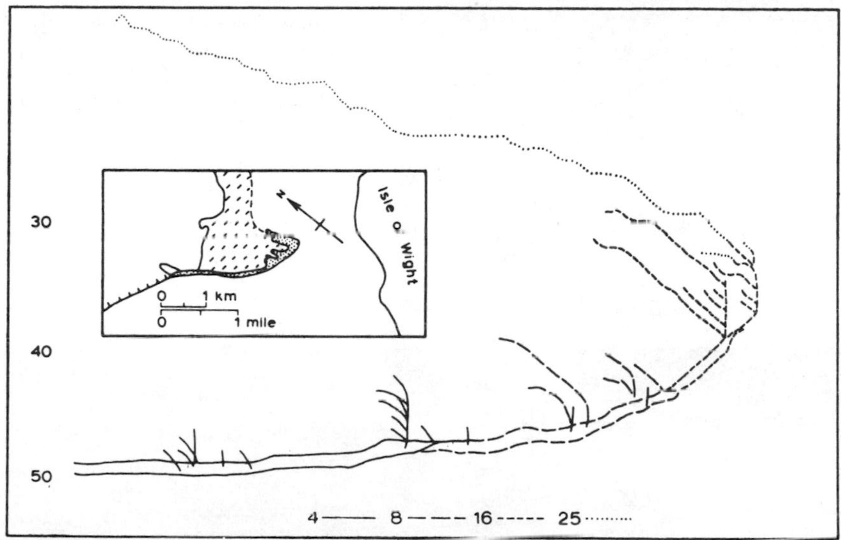

Figure 10-12 The simulated growth of Hurst Castle Spit between the mainland of southern England and the Isle of Wight. The patterns labeled 4, 8, 16, and 25 represent various stages of spit growth, 25 suggesting a possible future stage. [*From* King and McCullagh (1971)]

pressure is used to simulate wave heights and longshore current velocities, which are used in turn to control the morphology patterns. For example, the shoreline sinuosity is empirically determined from the wave and longshore current energies. Cycles of beach changes are thereby related to the cycles of a passing storm center. Although the model appears to work quite well, its use is restricted, since it is empirical rather than being based on actual physical processes which govern longshore current generation and sand transport. Its most severe limitation is that it makes no provision for large swells that are unrelated to the local storm system. Fox and Davis (personal communication) are presently attempting to overcome some of these limitations, adapting the models to beaches in Virginia and Oregon where swell waves become significant.

RHYTHMIC SHORELINE FEATURES

Beaches are seldom entirely straight or smooth in curvature as has been depicted previously in this chapter; rather, they commonly contain crescentic seaward projections or mounds of sediment which trend at right angles to the shoreline (Figure 10-13) and are known as beach cusps, sand waves, shoreline rhythms, or giant cusps (Bruun, 1955; Bakker, 1968; Hom-ma and Sonu, 1963; Zenkovitch, 1967; Dolan, 1971; Shepard, 1952, 1963, p. 195). Although sometimes isolated, they occur more commonly as a series of such forms with a fairly uniform *spacing*, the horizontal distance between successive cusps.

A very wide range of cusp spacings can be found on beaches. On the shores of

(a)

Figure 10-13 A variety of rhythmic shoreline features from (a) small beach cusps at Mono Lake, California (spacing 15 cm); (b) well-developed cusps on a sandy beach in Mexico; (c) large-scale rhythmic topography at Cape Hatteras, North Carolina; (d) classical beach cusps at Alum bay, Isle of Wight, England. [*From* Komar (1973*b*); *Sunset Magazine*, June 1971, p. 31; Dolan (1971); *From* Kuenen (1948)].

(b)

(c)

(d)

ponds and small lakes the spacing may vary from less than 10 cm to 1 meter (Johnson, 1919, p. 467; Evans, 1938; Komar, 1973*b*). Similar small cusps can also be generated in a laboratory wave basin (Longuet-Higgins and Parkin, 1962; Flemming, 1964). On ocean beaches with small waves, the cusp spacing may be less than 2 meters, while those built by large storm waves may be 60 meters apart. Russell and McIntire (1965) give numerous cusp observations and statistics from ocean beaches, with cusp spacings ranging from 6 to 57 meters. Dolan's (1971) measurements of "shoreline rhythms" from the North Carolina coast yielded spacings between successive cusps ranging from 150 to 1,000 meters, with most between 500 and 600 meters. The cusps projected on the average some 15 to 25 meters seaward from the embayments. Shepard's (1952) "giant cusps" ranged up to 1,500 meters in spacing.

Attempts at classifying rhythmic shoreline features generally stress their spacings. Beach cusps are considered to have smaller spacings (less than 25 meters: Dolan and Ferm, 1968; Dolan, Vincent, and Hayden, 1974), while sand waves, rhythmic topography, and giant cusps have larger spacings. These latter terms can be considered to be nearly synonymous, apparently different names for the same or very similar features. Dolan and Ferm (1968) also distinguish *beach cusplets* as having a spacing less than 1.5 meters. Since it appears that cuspate shorelines with a wide range of spacings can be produced by a single mechanism of formation and that there may be more than one mechanism capable of producing a cuspate shoreline, only a genetic classification will be satisfactory. The first attempt at a genetic classification is that of Evans (1938), limited to cusps formed on inland lakes. His theories of origin have not withstood the test of time, so that the classification is not generally useful.

More important than the spacings of the shoreline cusps is their associated offshore morphology. On this basis we can distinguish *beach cusps* and *rhythmic topography* (sand waves, giant cusps). Beach cusps commonly exist as simple ridges or mounds of coarse sediment stretching down the beach face, but where best developed on a steep beach there exist deeper troughs offshore from the cusps and underwater deltas offshore from the embayments (Figure 10-14), so that the offshore topography is a mirror reflection of the shoreline shape (Timmermans, 1935; Kuenen, 1948). In contrast, as distinguished by Hom-ma and Sonu (1963), rhythmic topography consists of a regular series of crescentic and inner bars with a regular spacing in the longshore direction (Figure 10-15). In certain circumstances the rhythmic bars give rise to a rhythmic series of cusps along the shoreline. At other times the rhythmic bars exist with an otherwise straight shoreline. In contrast with beach cusps, which are chiefly a subaerial feature, the underwater morphology is more important to the rhythmic topography and may only secondarily produce a series of cusps along the shoreline. In general, the spacing of the cusps associated with the rhythmic topography is larger than the spacings of beach cusps. Using this distinction, the characteristics of beach cusps will be examined first, followed by rhythmic topography.

Beach Cusps

Probably the most puzzling structures observed on beaches are the cuspate deposits of beach sediment built by wave action and known as *beach cusps*. Because

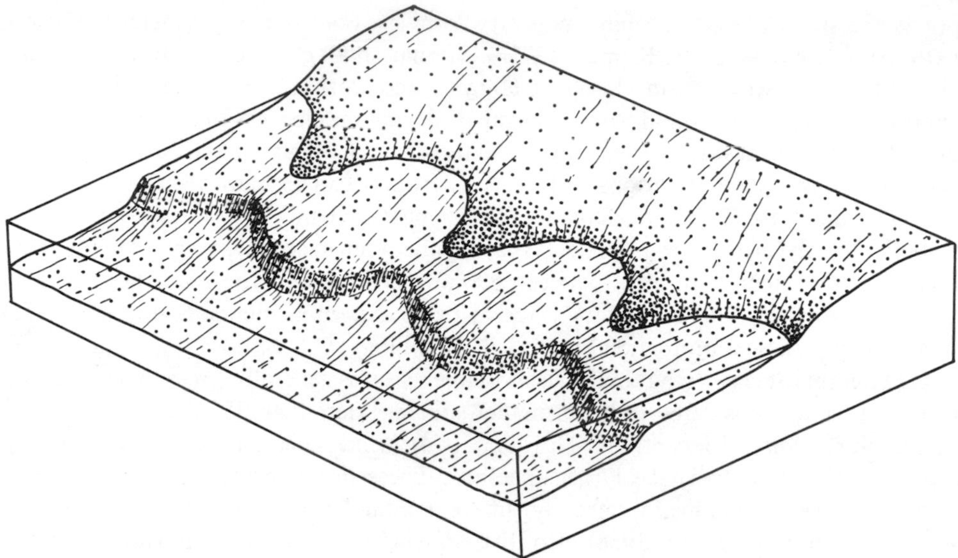

Figure 10-14 Beach cusps and associated underwater "deltas" offshore from the embayments. [*After* Timmermans (1935) and Kuenen (1948)]

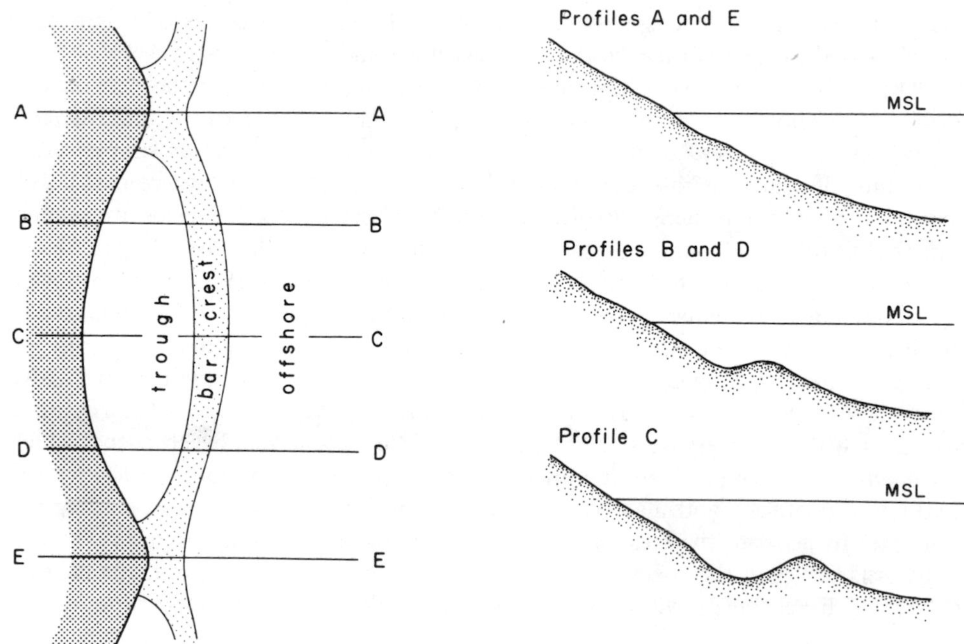

Figure 10-15 Crescentic bars, one type of rhythmic shoreline feature, with associated cusps along the shore. Also shown is the pattern of beach profiles along different sections of the crescentic bars. [*Based on* Hom-ma and Sonu (1963)]

of their regularity, beach cusps have attracted many observers and much speculation as to their origin, Palmer (1834) being the earliest published account. Practically every observation and idea concerning cusps advanced by one author is directly contradicted by that of another. Because of this, we still do not fully understand the basic origin of beach cusps or what controls their rhythmic spacing.

Beach cusps can be formed in any type of beach sediment; Russell and McIntire (1965) have observed cusps in material ranging from basaltic boulders and cobbles, through normal quartz-feldspar sands, to calcareous sands. One of the principal features of beach cusps is their sorting of the beach sediment by grain size. The cusp ridges generally display coarser sediments than do the embayments between the cusps. Even when they appear the same, size analyses often reveal that the cusps are coarser. The simplest forms of beach cusps are as a series of ridges of coarse sediment (commonly gravel) resting on a finer-sand beach face.

The difference in grain size between cusps and embayments results in a difference in their permeability. This has been particularly demonstrated and emphasized by Longuet-Higgins and Parkin (1962). They stressed the importance of a vertical stratification of material in cusp formation, coarse sediment at the surface overlying a mixed-size impermeable layer below. This enables the wave swash to maintain its energy, so that the coarse material can be readily moved about. Once it accumulates into cusps, the coarse-material cusps are less subject to erosion relative to the bays due to their high permeability, which dissipates the swash energy.

The sorting of grain shapes during cusp formation was studied by Flemming (1964) in a laboratory wave basin. Slowly settling grains (disks, plates, blades) tended to be tossed onto the cusps and at the backs of bays, while the more easily rolled spherical grains were dragged down the beach face and concentrated at the foot of the beach.

It is certain that beach cusp formation is most favorable when the waves approach normal to the beach—that is, with their crests parallel to the shoreline (Johnson, 1919; Timmermans, 1935; Longuet-Higgins and Parkin, 1962). This may explain why coastal bays are particularly favorable sites for cusp formation, since oblique waves cannot form so easily. Even this is disputed, however, as Wilson (1904, pp. 106-32) and Gellert (1937) maintain that oblique waves are necessary for cusp development, and Evans (1938, p. 624) and Otvos (1964) maintain that the wave direction is irrelevant. A substantial longshore drift of sand under oblique waves does destroy the beach cusps by first making them highly asymmetric and then washing them away entirely. If the drift is not too great, the cusps may persist in an asymmetrical form. An interesting observation was made by Krumbein (1944), who produced cusps in a wave tank with waves approaching the beach at a 15-degree angle. The cusps produced migrated slowly down the beach, the maximum rate being approximately 30 cm/hr. I do not know of any such field observations of migrating beach cusps; however, the cusps associated with rhythmic topography do commonly migrate alongshore at such rates (page 275). Also, rip currents are known to migrate slowly alongshore under an oblique wave approach (Bowen and Inman, 1969), and it is possible that the cusps observed by Krumbein (1944) were produced by rip currents.

The relationship between rip currents and cuspate shorelines will be examined later when we consider rhythmic topography.

In addition to confirming that cusps form best when the wave crests are parallel to the shoreline, Longuet-Higgins and Parkin (1962) noted that regular waves with long crest lengths are conducive to beach cusp formation. Cusps are not formed by stormy, confused seas.

Tides were once thought to be of importance in the development of beach cusps, but since cusps are formed quite readily in lakes (Evans, 1938; Komar, 1973*b*), wave basins (Longuet-Higgins and Parkin, 1962, Flemming, 1964), and tideless seas (Bagnold, 1940), this is clearly not the case. Cusps are observed to form at all stages of the tide. The presence or absence of tides can, however, influence the morphology of the resulting beach cusps: beach cusps with offshore deltas and troughs (Figure 10-14) form best on tideless or low-tidal-range shorelines, whereas in areas of pronounced tides the cusps tend to become stretched out down the beach face, forming a series of simple ridges.

Johnson (1919) first made the observation, which has generally been reaffirmed by others, that doubling the height of the waves roughly doubles the spacing of the beach cusps. Considering the considerable literature on beach cusps, it is surprising how little data exists whereby this correlation can be tested. Figure 10-16 plots the data obtained by Longuet-Higgins and Parkin (1962) on Chesil Beach, England, which shows this proportional increase in the cusp spacing with increasing wave height. Also shown is that a somewhat better correlation exists between the cusp spacing and the *swash distance*, the width of the beach between the wave break point and the highest point reached by the swash. Williams (1973) found almost no correlation between cusp spacing and the wave height at Stanley Bay, Hong Kong. He did find a good correlation with the swash distance. His observations may be unusual in that he indicated that sargassum growing offshore from the beach may play an important role in controlling beach cusp formation.

At certain times two or more levels of cusps may be found on a particular beach face. Invariably the cusp spacing is greater for those that are highest on the beach, formed presumably by the larger storm waves.

Shepard (1963, p. 201) noted that on Scripps Beach, La Jolla, California (Figure 7-5), the cusp length decreases gradually along the beach from areas of wave convergence (high waves) to areas of wave divergence (low waves). Otvos (1964) reports a similar pattern. This further demonstrates the correlation of the spacing to the wave height or swash distance.

The circulation of the wave swash around the beach cusps and within the embayments is best described by Bagnold (1940) and is depicted in Figure 10-17. The waves break evenly over a straight step line, then the wave surge piles up against the steep promontories of the cusps and is divided by the promontories into two divergent streams, each of which flows into one of the adjoining embayments. These streams head off the surge that has flowed directly up the bay over the shallower underwater deltas. The two side streams from the cusps on either side meet in the center of the bay and together form a return flow down the bay of considerably greater

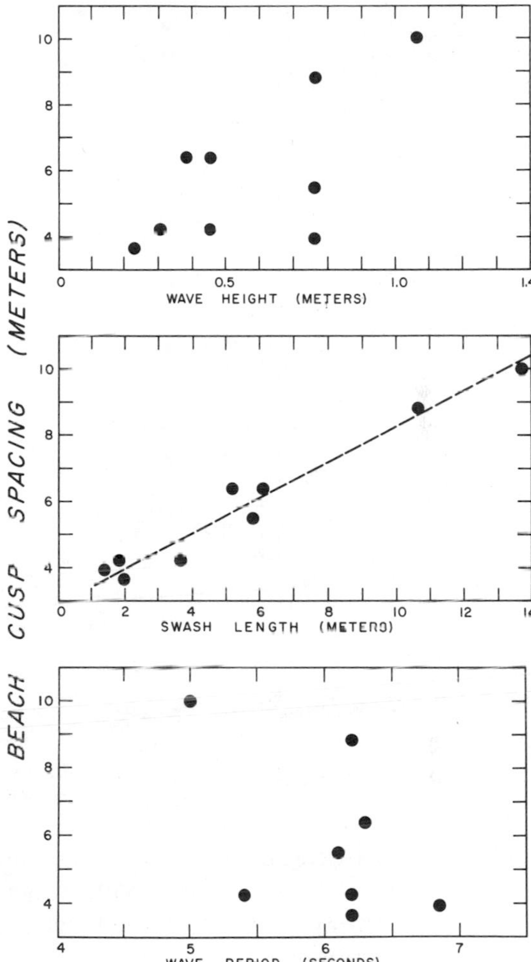

Figure 10-16 Beach cusp spacing related to wave height, swash length, and wave period. Measurements obtained by Longuet-Higgins and Parkin (1962) at Chesil Beach, England.

intensity than the previous upward surge there. This return flow resembles a rip current, but unlike a rip current the flow is discontinuous through time, and the processes of formation are different. Due to the sideways flow of water within the cusp system, there is no dead period (except in front of each cusp promontory) when the water and the sand it carries come momentarily to rest.

The backwash within the bays moves sediment offshore which is deposited to form the deltalike tongues opposite each bay (Figure 10-14). Longuet-Higgins and Parkin (1962) found that dyed pebbles placed in the bay are removed by the backwash, but then are washed back up onto the adjoining cusp promontories, where they come to rest. An equilibrium is eventually reached in which no additional sediment is deposited on the cusps, the prograding cusps having reached a point where the swash energy is strong enough to prevent further deposition. Thus there is a correspondence between the equilibrium beach cusps and the equilibrium shorelines

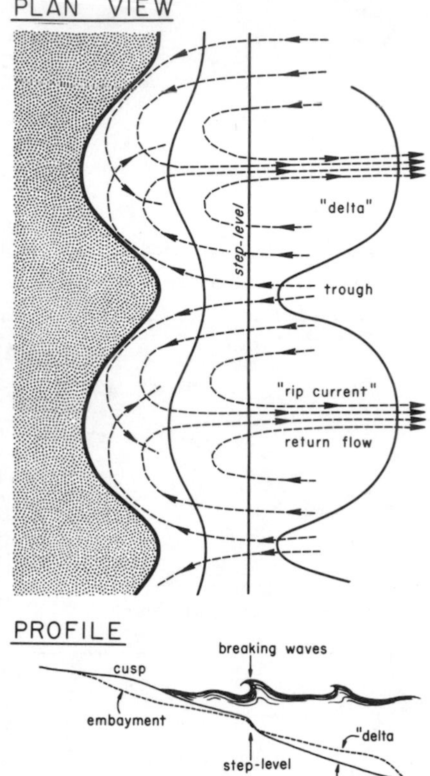

Figure 10-17 Wave swash motions around cusps and within embayments. [*After* Bagnold (1940)]

considered earlier in the chapter. In the case of beach cusps, however, the equilibrium is not one of zero transport, as found for a beach with no sediment source, but rather one in which the sediment is kept in a continual cyclical movement preventing further deposition.

There is some difference of opinion on whether beach cusps form most readily during deposition or erosion of the beach face. Timmermans (1935) observed cusps on the Dutch coast and in his wave tank experiments only when the beach face was oversteep and being eroded. The same was found by Escher (1937). On the other hand, Russell and McIntire (1965) maintain that cusps are developed when deposition on the beach face results in an advancing berm and that they are destroyed by steep storm waves that erode the beach face and cut back the berm. They observed that cusps originate and attain their maximum development during the transition from winter to summer conditions, and are commonly erased when the transition occurs in the opposite direction (Chapter 11).

Several theories have been proposed for the origin of beach cusps. The main requirement of any theory is its ability to explain the very uniform spacings of a given series of cusps and how this spacing relates to the wave parameters and beach sediment material.

Johnson (1910, 1919) first proposed, and Kuenen (1948) later extended, the theory that a regular succession of swash flows on a smooth beach will start to erode any slight depression they encounter on the beach face. The backwash carries the eroded sediment out of the hollows to build deltas opposite them. By continual erosion, the series of hollows enlarge into embayments. As long as the bay water depth is so small that the water passing in and out is able to carry sediment, the enlargement of the bays continues. When the depth in the outer part of the bays approaches a certain limit, erosion gradually slackens. Now refraction of the swash in the bays causes material to be transported out of the bays and onto the cusp promontories. The coarser material is left on the cusps, while the finer is washed back into the bays. The growth of the bays and the prograding of the cusps gradually decreases as the depth of the central area of the bay approaches a maximum and wave swash prevents further deposition on the promontories. It is envisioned that initially there is a great deal of variation in the cusp spacings, but the larger ones grow at the expense of the smaller ones, and smaller ones encroach on larger ones that have already reached their maximum depth. In this way, the theory proposes, a regular rhythmic pattern of cusps is achieved.

Although the theory is appealing, there is doubt whether this approach can account for the equal cusp spacings. For one thing, when the theory was revised by Kuenen (1948), he held a mistaken opinion as to the pattern of the water circulation about the cusps, the circulation he describes being just opposite to that observed by Bagnold (1940) and confirmed by others. The cusps studied by Kuenen (1948) were probably formed by an earlier set of waves and were not in equilibrium with the waves he observed. More important, the process of beach cusp formation does not appear to be nearly so random as depicted by the theory. Experiments in wave tanks and observations of cusp formation on the shores of lakes and the ocean show that beach cusps can be produced with remarkably uniform spacings immediately upon formation (Escher, 1937; Komar, 1973*b*). Finally, the theory does not explain why more often than not cusps are not developed even though conditions seem right according to the hypothesis.

The theory of beach cusp formation proposed by Russell and McIntire (1965) is similar to that of Johnson (1910, 1919) as expanded by Kuenen (1948) except that it is based on a water and sediment circulation as depicted by Bagnold (1940). Cusp development depends mainly on the fact that velocities of flow must be higher to entrain than to carry a sediment load. Because of this, coarse material deposited on the cusps during the brief lull in the water motion remains on the cusps since it cannot be reentrained. Although an improvement over the theory of Kuenen (1948), this hypothesis still does not explain the sudden appearance of uniformly spaced cusps, nor does it develop the relationship between the cusp spacing and the wave parameters.

Two studies have proposed mechanisms by which the wave swash has a rhythmic spacing of intensity which produces a corresponding rhythmic erosion of the beach face. The subsequent cusp development is presumably similar to that outlined by the above theories. Cloud (1966) has suggested that the breaking wave approximates a cylindrical form at the instant of collapse and therefore divides up in the longshore direction according to Plateau's rule: a long liquid cylinder becomes unstable and separates into subequal divisions whose lengths are proportional to the cylinder diameter, the ratio length/diameter being between 15.5 and 16.7. Applying this to the breaking waves, Cloud (1966) hypothesized a similar division of the breaking wave such that the beach cusp spacing should be 15 to 17 times the height of the breakers (or somewhat less, since the largest breakers may be more important than the average breakers). This suggestion has not been pursued, and there is little supporting evidence for it. The measurements of Longuet-Higgins and Parkin (1962), Figure 10-16, indicate an average ratio of cusp spacing to breaker height of approximately 10. On many ocean beaches the beach cusps are separated from the breaking waves by a wide surf zone, and it is difficult to envision how such a mechanism could control the cusp spacing.

Gorycki (1973) has suggested that breaking waves and swash may become structured into a series of "salients" separated by turbulent zones of retarded flow. It is hypothesized that the regular salients give rise to the uniformly spaced beach cusps, the turbulent zones eroding out the bays between the cusps. However, Gorycki's experimental evidence of salients is restricted to patterns with less than 20 cm spacing. He indicates that the tongues of swash that run up the beach face on ocean beaches are large-scale salients which give rise to larger-spaced beach cusps. However, swash tongues are usually very irregular in position along the shoreline (unless cusps are already present), and it is difficult to envision how their changing positions can give rise to rhythmic beach cusps. Komar (1973*b*) ruled out the importance of salients in the formation of small beach cusps along the shores of Mono Lake, California.

Escher (1937) produced beach cusps experimentally in a wave basin and explained their formation by standing waves in the surf zone that are at right angles to the incoming waves and have the same period as the incoming waves; the standing waves could actually be seen. Where the two sets of waves supported one another the resulting higher swash eroded bays, while where they opposed one another cusps were formed. This origin was subsequently ignored because "no records of such standing waves are known" (Kuenen, 1948, p. 35). However, we now know that such standing waves or edge waves may be present on natural beaches (Bowen and Inman, 1969; Huntley and Bowen, 1973). In Chapter 7 we reviewed the study of Bowen and Inman (1969) on how such edge waves may interact with the incoming waves to produce a cell circulation system of longshore currents and rip currents. Bowen and Inman (1969) and Bowen (1973) have also revived the hypothysis that standing edge waves may be responsible for beach cusp development. The rip currents themselves and associated longshore currents may rearrange the beach sediments into a series of beach cusps with the same spacing as the rip currents. The cell circulation is more closely allied with what we are calling rhythmic topography, so our discussion of this will be deferred. In some circumstances closely spaced rip currents do give rise to a series of cusps which would be classified as beach cusps (Komar, 1971). Interacting with the incoming ocean waves, the edge waves may also generate a regular pattern of wave swash on the beach face without developing a full-scale cell circulation (Bowen, 1973). Where the edge waves are in phase with the normal swell waves they support one another and the swash reaches highest on the sloping beach face. Further along the beach where the edge waves and the normal waves are out of phase, the swash would be small and more of the beach would remain dry. The systematic pattern of regular longshore variations in the swash intensity could redistribute the beach sediments into a series of beach cusps.

Unusual beach cusps with spacings of 11 to 59 cm have been studied on the shores of Mono Lake, California (Komar, 1973*b*). They are unusual in that although they possess all the features found in the much larger cusps on ocean beaches, they form when the lake is essentially glassy smooth and only surging, nonbreaking waves arrive at the shoreline. On the basis of this origin, their sudden appearance with a regular spacing, and the magnitudes of their spacings, it was concluded that edge waves probably account for their development.

Where the beach slope is small and the surf zone wide, beach cusps that form on the inner beach face are usually much smaller than the spacings of the rip currents which may also be present. In this case, Bowen and Inman (1969, p. 5,490) suggested that the formation of cusps may be in response to an interaction of the reformed wave bore following breaking with the edge waves which have a smaller modal number n [see equation (10-11), below] than that of the predominant cell circulation system.

I concur with Bowen's (1973) opinion that edge waves are the most probable cause of the rhythmic spacing of beach cusps. However, there is some evidence that argues against the role of edge waves. We saw in Chapter 7 (equation 7-10) that the edge wave length L_e, which will govern the beach cusp spacing, is given by

$$L_e = \frac{g}{2\pi} T_e^2 \sin[(2n + 1)\beta] \tag{10-11}$$

that is, it is a function primarily of the edge wave period T_e and of the beach slope. Due to the presence of the offshore modal number n, there are a number of possible edge wave lengths (and presumably beach cusp spacings) for a given edge wave period and beach slope. Longuet-Higgins and Parkin (1962) examined the possibility of edge waves generating beach cusps, but arrived at a negative conclusion since there was a poor dependence of the cusp spacing on the wave period, assuming the edge waves had the same period as the ocean swell waves (Figure 10-16). This may be due in part to the extreme variability of the beach slope on the cuspate shoreline, the slope also being a factor in the edge wave length equation. Even when it is known beforehand that edge waves are responsible for the cusps little direct correspondence between cusp spacing and edge wave length can be demonstrated because of this slope variability (Komar, 1973*b*). However, this does not explain the total lack of dependence on the wave period found by Longuet-Higgins and Parkin (1962). In addition, Johnson (1919) found that varying the period of waves produced in a laboratory basin had no observable effect on the beach cusps.

In spite of these tentative poor correlations between the beach cusp spacing and the wave period, edge waves remain the only satisfactory explanation for longshore variations in the surf zone properties which can give rise to regularly spaced beach cusps. The evidence for their role in the generation of rhythmic topography is somewhat clearer. It remains possible, however, that other modes of generation may also act in the nearshore to produce beach cusps. Only additional studies can resolve this.*

Rhythmic Topography

As discussed earlier, the offshore morphology within the surf zone and even beyond the breaker zone is of greater importance in rhythmic topography than for beach cusps, which are principally a subaerial feature. *Rhythmic topography* consists of a series of crescentic bars (Figure 10-18), or a regular pattern of longshore bars separated by rip current troughs, or a combination of the two. When cusps accompany the rhythmic topography their spacings are generally larger than those of beach cusps —on the order of 100 meters and more—so that rhythmic topography is a larger-scale feature than beach cusps as we are employing the term.

Rhythmic topography has been reported from widely scattered areas of the world, including Lake Michigan (Evans, 1939; Krumbein and Oshiek, 1950), the Black Sea (Egorov, 1951), the Mediterranean Sea (King and Williams, 1949), the North

(a) Rhythmic Topography on Inner Bar

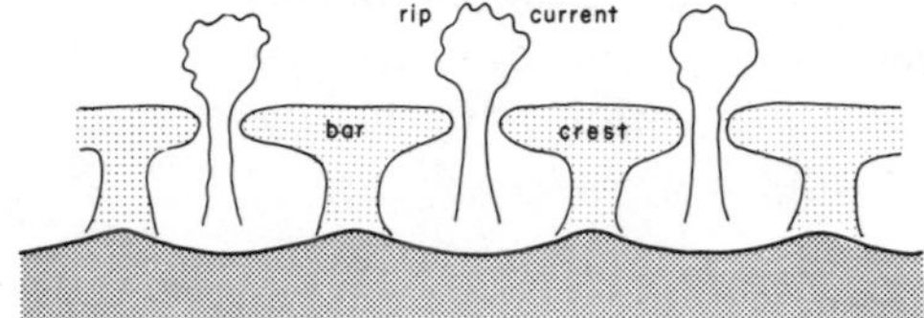

(b) Crescentic Bars

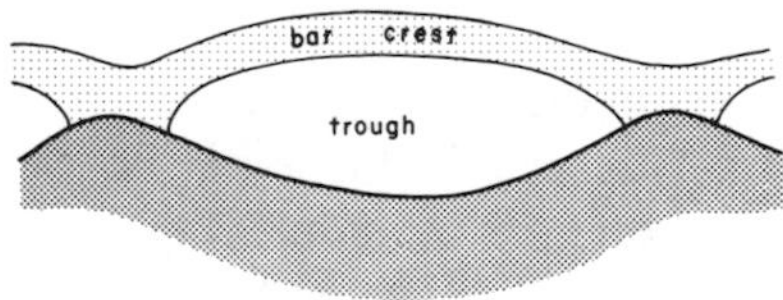

(c) Combination

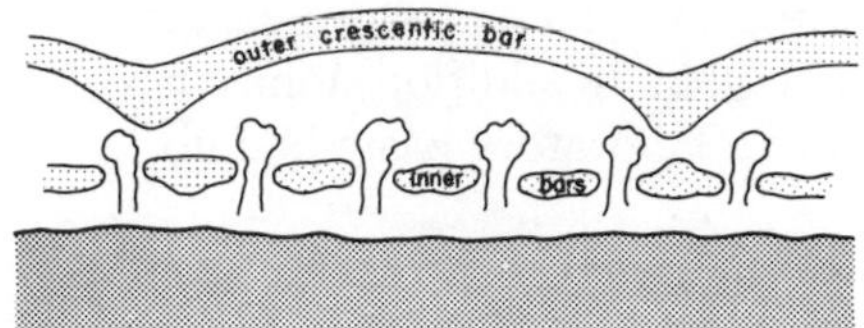

Figure 10-18 The two principal types of rhythmic topography as contrasted with ordinary beach cusps shown in Figure 10-14. The details of the crescentic bars are shown in Figure 10-15.

*See Guza and Inman (1975), who have investigated the generation of subharmonic edge waves and their role in the formation of beach cusps. (Published after this book went to press.)

Sea coast of Denmark (Bruun, 1955), the southeast coast of the United States (Harrison and Wagner, 1964; Dolan, 1971), and the coast of Japan (Hom-ma and Sonu, 1963). Rhythmic topography is not necessarily apparent to an observer confined to the shoreline, since most of its expression is underwater. All the shoreline observer would view is the associated cuspate shoreline, which he might report as large-scale beach cusps. In the other direction, large-scale beach cusps, such as those reported by Trask (1956), are sometimes presumed to be rhythmic topography on the basis of the large cusp spacing.

In some instances the rhythmic topography is known to migrate slowly in the longshore direction, made apparent by the shifting positions of the associated shoreline cusps. The highest rate of migration is that observed by Egorov (1951) in the Black Sea—a rate of 15 to 32 meters in 24 hours. Bruun (1955) reported an average annual displacement of 1,000 meters on the Danish North Sea coast. Other studies report short-term fluctuations in position but with no net long-term migration. Bakker (1968) presents a mathematical theory for sand wave (rhythmic topography) migration in the longshore direction, one in which the migration may be in the opposite direction to the overall sand drift. Van Bendegom (1949) [see discussion in Sonu (1969, p. 394)] demonstrates the effects of migration on local beach erosion and accretion, with an example from the Vlieland coast of the Netherlands. A shoreline cusp with a 200-meter offshore expression migrated an average of 200 m/yr, and as it moved along the shoreline the beach at a particular spot was steadily eroded for about 40 years and then accreted for the next 10 years. Bakker (1968) also discusses this example. The importance of shifting rhythmic topography on local beach erosion is further emphasized by Dolan (1971). He demonstrated that the regular spacing of dune breaching on Bodie Island, North Carolina, during the Ash Wednesday storm of March 7, 1962, matched the rhythmic topography. He also showed that intense beach erosion at Cape Hatteras corresponds to the embayments of rhythmic topography. In addition, groin failure in the area was attributed to a lack of consideration of the changes in profile due to longshore movements of the rhythmic topography. Sonu and Russell (1967) had earlier suggested that abrupt changes in beach profiles repeatedly taken at a stationary traverse could be attributed to such a longshore migration (Chapter 11). Bruun (1955) has hypothesized that a sudden silting of a coastal inlet may result from this longshore migration, bringing a shoal or "sand wave" to the inlet position.

The characteristics of rhythmic topography have been described by Evans (1939), Krumbein and Oshiek (1950), Bruun (1955), Shepard (1952), Hom-ma and Sonu (1963), Sonu (1969, 1972, 1973), Sonu and Russell (1967), Sonu, McCloy, and McArthur (1967), Sonu and van Beek (1971), and others. On the basis of these studies we can distinguish two types of rhythmic topography: (1) rhythmic variations associated with the presence of a cell circulation (rip currents), and (2) crescentic bars. The two types can appear individually on beaches, or they can exist simultaneously on the same beach but remain relatively independent (Figure 10-18). As described by Sonu (1973), the variations in beach morphology associated with the rip currents are apparent on the inner bar, whereas the crescentic bars are an order of magnitude larger and extend into deeper water. The inner-bar rhythms appear to form at a later stage of a storm than the outer crescentic bars (Sonu, 1973, p. 61), independent of the formation of the crescentic bars. The general lack of phase correlation between the inner bars and outer crescentic bars also supports this concept of an independent origin and development. Shoreline topography produced by rip currents will be examined first, and later the origin of the crescentic bars.

The cell circulation of rip currents and associated longshore currents (Chapter 7) may rearrange the beach sediments to produce a regular pattern of rhythmic topography and corresponding cuspate shoreline (Riviere, Arbey, and Vernhet, 1961; Bowen and Inman, 1969; Komar, 1971). Bowen and Inman (1969) suggested that a seaward-flowing rip current would tend to erode a channel for itself, the results being a segmented offshore bar and a system of cusps along the shoreline midway between successive rip currents which occupy the bays. Such a relationship between rip currents and cusps is commonly observed and corresponds to the association within a rhythmic topography.

Sonu (1972) has made a careful study of the relationship of the nearshore topography to the cell circulation on the Gulf of Mexico shoreline. With waves arriving normal to the shoreline, the rip currents occupied troughs with shoals present between the rip currents [Figure 10-19(a)]; no system of shoreline cusps developed, however,

(a) Normal waves

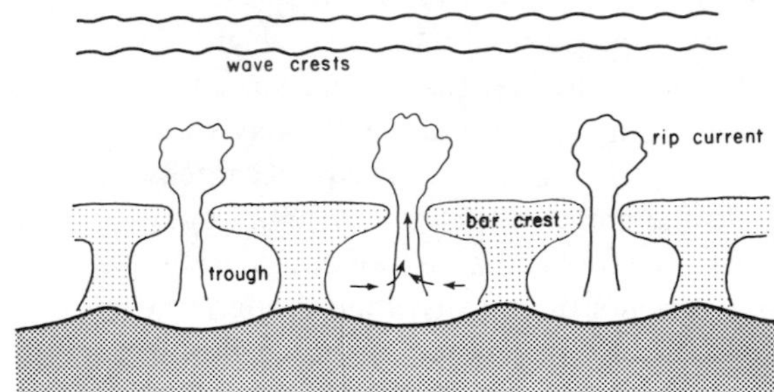

(b) Oblique waves

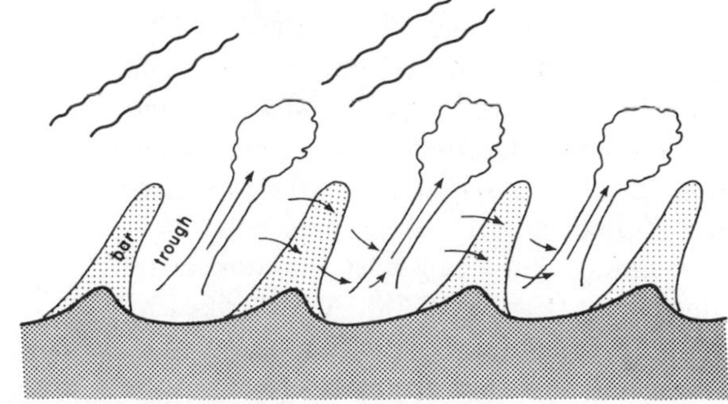

Figure 10-19 Realignment of the bars under oblique wave approach to give a cuspate shoreline rhythm as envisioned by Sonu (1973).

in response to this offshore topography. On the other hand, Davis and Fox (1972) found in Lake Michigan that under similar circumstances cusps built outward in the protected lee of the shoals, giving a pattern of cusps and embayments with rip currents as depicted by Bowen and Inman (1969). Sonu's (1972) view of the association of the cell circulation with this type of rhythmic topography is just opposite to that maintained by Bowen and Inman (1969). Whereas Bowen and Inman (1969) hypothesize that the cell circulation gives rise to the rhythmic topography, the alternating shoals and troughs in the longshore direction, Sonu (1972) maintains that the topography came first and that this topography caused the cell circulation to develop. Most probably Bowen and Inman (1969) are correct in that the initial topography results from a regular series of rip currents; however, once the topography is developed, it plays a strong role in modifying the nearshore cell circulation and the pattern of rip currents.

Sonu (1972) found that when waves arrived at an oblique angle to the beach, the shoals or bars rotated to align with the incoming wave crests, as did the rip currents and their troughs [Figure 10-19(b)]. Where the shoals joined with the shoreline a system of cusps developed, the cusps being the subaerial expression of the shoals, the rip current troughs occupying the embayments between the cusps. Horikawa and Sasaki (1968) [see discussion in Sonu (1973, p. 60)] have similarly produced a system of cusps along the shoreline in a wave basin by the offshore bar rotating and attaching to the shore. Sonu (1973) considers this mechanism one possible means for producing a rhythmic topography. In contrast, Davis and Fox (1972) found that with an oblique wave approach the cuspate shoreline was straightened and that the shoals and rip currents shifted position in the longshore direction.

This skewed rhythmic topography under an oblique wave approach and strong longshore currents has given rise to several theories of origin which stress the analogy with the bed forms produced by a river. Sonu (1969) proposed that this type of rhythmic topography is generated by the instability of the loose surf zone bed perturbated by the longshore current; that is, the oblique ridges of sediment across the surf zone are analogous to dunes produced in a river or by winds. Dolan (1971) suggested that the origin may somehow be similar to the development of meanders in a river. Bruun (1955) indicated that there may be a correspondence between the ridges and the similar migrating bars found in rivers. Since the origin of dunes and bars in rivers remains uncertain, these hypotheses for a similar development under longshore currents are difficult to evaluate. Since this form of rhythmic topography seems to be satisfactorily explained by the cell circulation system under an oblique wave approach, these alternative hypotheses are probably not required.

Once formed, this pattern of *skewed rhythmic topography*, more commonly known as *sand waves* or *transverse bars* (Shepard, 1952), may persist as a relatively stable system. Niedoroda and Tanner (1970) have investigated the effects of such transverse bars on the nearshore circulation system. They found that if the offshore extension of the bar is small, wave energy is focused on the bar position by wave refraction producing higher waves than over the troughs between the bars. This drives a current which flows shoreward along the bar and returns seaward in the troughs.

Although this pattern is similar to that described by Sonu (1972), the currents are much weaker and the return flow does not qualify as a rip current. With a long transverse bar the current is in the opposite direction, flowing seaward along the bar. This is accounted for by the wave energy dissipation through frictional drag over the long bar, causing lower waves over the bar than in the troughs.

The relationship between the cell circulation and the cuspate shoreline has been further investigated by Komar (1971). In addition to forming midway between the rip currents, shoreline cusps at other times develop in the lee of the rip currents. Field examples were cited, some of which could be described as beach cusps and others as rhythmic topography. Of special interest is that in a laboratory wave basin test, cusps were formed in this way and the cusps reached an equilibrium development in which the rip currents that produced the cusps suddenly ceased to exist. The rip currents and all associated longshore currents disappeared once the cusps had formed. It was determined that the equilibrium state consists of a balance between the forces that tend to drive a longshore current from an oblique wave approach to the cusp flanks, on the one hand, and, on the other hand, the forces that normally produce the cell circulation resulting from a longshore variation in wave breaker height (Chapter 7). It is therefore possible for beach cusps to be produced by rip currents, although when the cusps are observed on the beach the rip currents are no longer present. If the balance is not exact, then a system of weakened rip currents could be present, either directly offshore from the cusps or within the embayments.

The *crescentic bar* (*lunate bar*) is essentially a submerged sand bar, concave shoreward, which may or may not have an associated series of cusps along the shoreline (Figures 10-15 and 10-18). Such features are found throughout the world on long straight beaches (Shepard, 1952; Hom-ma and Sonu, 1963), but are particularly well developed in bays such as shown in Figure 10-20. Their development appears to be confined to regions of small tidal range (King and Williams, 1949; Shepard, 1963; Bowen and Inman, 1971), but this is disputed by Sonu (1972), who cites examples on beaches where the spring tidal range is as large as 2.2 meters. However, many of the examples cited by Sonu do not classify as crescentic bars (Bowen and Inman, 1972), being instead beach cusps or rhythmic topography associated with a cell circulation, neither of which requires a small tidal range for its development. Crescentic bars also appear to form best where the beach face slope is small.

The shape of the arcuate crescentic bar is not necessarily symmetrical. With a small longshore drift of sediment they become skewed in the drift direction (Hom-ma and Sonu, 1963, p. 255). With a strong unidirectional drift, the rhythmic bars smooth out into a straight bar. The presence of cuspate bars therefore implies a low longshore sediment drift (Hom-ma and Sonu, 1963, p. 254).

In general, crescentic bars are a larger feature than beach cusps and the rhythmic topography due to rip currents. Their range of wave lengths is difficult to establish, since for many reported occurrences it is not possible to determine whether crescentic bars or some other shoreline rhythm are being described. They appear to range from about 100 to 2,000 meters, with a predominance of 200 to 500 meters. Hom-ma and

Figure 10-20 Well-developed crescentic bars off an enclosed beach near Cape Kalaa, Algeria. [*From* Clos-Arceduc (1962)]

Sonu (1963) describe a system of multiple crescentic bars, the maximum spacing being 1,000 meters; the farther offshore the bars, the larger their spacings.

Like the other rhythmic shoreline forms, there has been a certain amount of controversy concerning the origin of crescentic bars. Sonu (1969, 1973) attributes them to bed perturbation by the longshore current to form dunes such as found in a river. Bowen and Inman (1971) have hypothesized that a role is played by the velocity field associated with edge waves on a sloping beach. They envision that sediments will drift about under the currents of the edge waves until the sand reaches zones where the velocity is below the threshold of sediment motion, the sand depositing in these zones. A schematic diagram of the expected pattern of deposition and crescentic bar formation for edge wave modes $n = 1$ and 2 is shown in Figure 10-21. It is seen that the similarity with observed crescentic bars is encouraging. The bars have a longshore wave length that is one-half that of the edge waves. In the case of $n = 2$, a system of two crescentic bars is predicted. Bowen and Inman (1971) conducted a series of laboratory wave basin experiments that confirmed this basic depositional pattern. They also found that the inner bar of the $n = 2$ case is unstable and is eventually eroded away. Their field examination of the hypothesis centered on the example of crescentic bars shown in Figure 10-20 from the Mediterranean coast of Algiers. This example was examined earlier by Clos-Arceduc (1962, 1964), who attributed their formation to standing oscillations between the headlands in a direction normal to the coast. As pointed out by Bowen and Inman (1972, p. 6,632), Clos-Arceduc's explanation is a good attempt by someone who is not familiar with edge

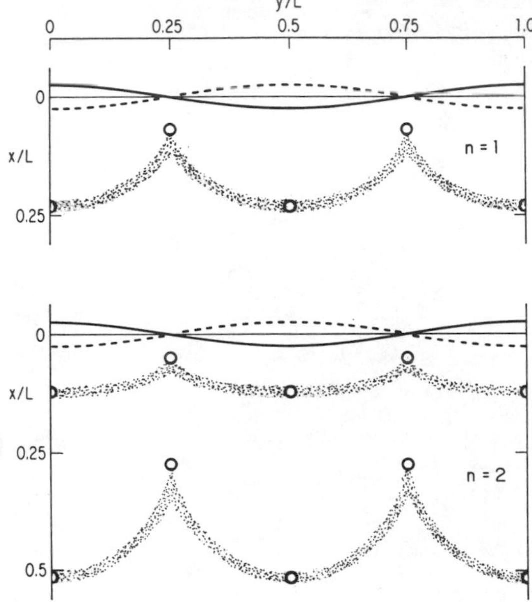

Figure 10-21 Theoretical crescentic bar patterns based on the hypothesis of edge wave generation. [*From* Bowen and Inman (1971)]

waves, but his explanation is not an alternative to theirs. The wave length of the crescentic bars in Figure 10-20 is approximately 500 meters, so that the edge wave length required would be 1,000 meters. The morphology of the bars suggests a mode $n = 2$, and on this basis and using reasonable estimates of the beach slope, Bowen and Inman (1971) utilized equation (10-11) to evaluate the required edge wave period for generation. They obtained $T_e = 40$ to 50 seconds, which of course is much longer than periods of incoming surface waves. Sonu (1972, p. 6,629) has attacked the theory on this basis of requiring long-period edge waves. Bowen and Inman (1972) agree that edge waves associated with crescentic bars are probably not generated from ordinary surface waves, but rather may be driven by surf beat (Munk, 1949; Tucker, 1950) whose periods may be as great as 5 minutes, or possibly by local wind fields (Greenspan, 1956). The more recent measurements of Huntley and Bowen (1973) found edge waves that are a subharmonic of the normal ocean waves. That is, the ocean waves of period 5 seconds were found to give rise to edge waves of period 10 seconds. Since ocean wave periods range up to 15 to 20 seconds, it is conceivable that edge waves of periods up to 40 seconds could be formed by energy fed from ocean waves; these are periods that might account for many observed crescentic bar spacings.

The Carolina Capes

An examination of cuspate shoreline features would not be complete without mention of the large-scale capes, the most famous of which are the Carolina Capes on the southeastern coast of the United States (Figure 10-22): Capes Hatteras, Lookout, Fear, and Romain. Also part of the system is Cape Canaveral in Florida and the less prominent but similar capes developed along the Georgia coast at Tybee Island and Little St. Simons Island. Altogether they have an average cusp spacing of roughly 100 km (Dolan and Ferm, 1968).

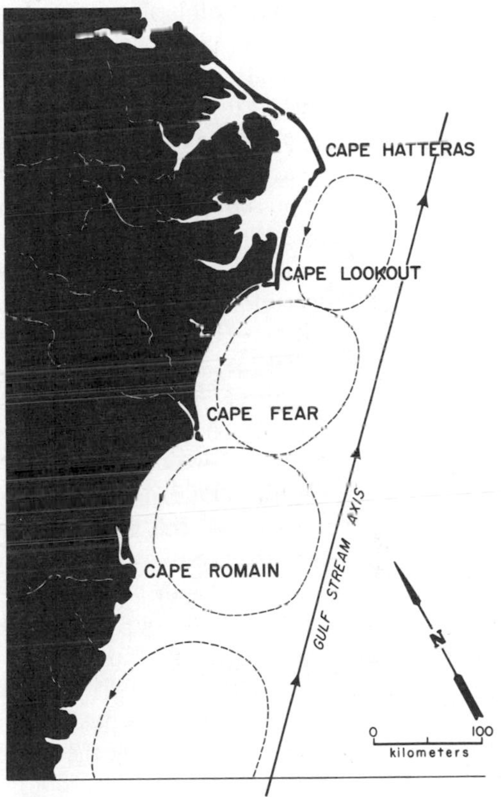

Figure 10-22 The Carolina Capes and suggested eddies from the Gulf Stream, once thought to be the cause of the capes.

Most of the capes are part of the barrier island system which rims the southeast coast. Shoals are present seaward of the capes, extending for tens of kilometers offshore. For example, depths as shallow as 1 meter exist 15 km offshore from Cape Fear on the Frying Pan Shoals. The present barriers are Holocene in age and are the modern counterparts of similar features of Pleistocene age that formed when the sea level was higher than at present (Hoyt and Henry, 1971). Apparently the Carolina Capes have had a long history.

The earliest theory regarding the formation of the Carolina Capes attributed their development to a series of secondary rotational cells or "eddy currents" which developed along the western margin of the Gulf Stream (Tuomey, 1848; Abbe, 1895) as shown in Figure 10-22. Bumpus (1955) investigated the circulation on the shelf in the Cape Hatteras region and found an absence of such eddies. Bumpus indicated that southwesterly winds may pile up water on the south sides of the capes, creating a current in the offshore direction over the projecting shoals. He suggested that such currents might help maintain the capes.

The role of waves in local beach erosion and deposition has been stressed by Cooke (1936, p. 4) and Zenkovitch (1964) in their explanations of cape formation. According to Cooke, the waves scoop out the arcs between the capes, transporting the sediment alongshore and depositing it at the capes to either side, maintaining a smoothly curved shoreline. Zenkovitch (1964, p. 879) explained the capes as "a result of diagonal wave incidence from the north and south." A direct causative role played by waves is difficult to accept, since the waves should tend to concentrate their energy on the projecting capes because of the wave refraction (Chapter 4, Figure 4-22). This should act to erode the capes and straighten the overall shoreline. Indeed, the Carolina Capes presently appear to be retreating under wave attack. The barrier islands are narrowest in the vicinity of the capes and are being driven landward over the onshore lagoon deposits (Hoyt and Henry, 1971). In general, the barrier at each of the capes consists of a single beach ridge, which attests to their lack of progradation and deposition. Historical records plus recent active erosion also demonstrates cape retreat (Rude, 1923).

White (1966) discussed the importance of the shoals seaward of the capes in maintaining the capes even through sea level changes. Because of the shoals, the capes are self-maintaining and relict capes have been able to localize new capes.

White (1966) and Hoyt and Henry (1971) stress the important influence of major rivers on the Carolina Cape positions. Hoyt and Henry (1971) indicate that the concurrence of capes and major rivers is too prevalent to be fortuitous. Each cape coincides with one or more river mouths; some of these rivers, however, presently empty into lagoons behind the barrier islands. They hypothesize that at the beginning of the glacial stage the river gradients were increased due to the lowered sea level. This caused the formation of deltas, which prograded across the shelf as the sea level lowered. Thus a series of delta ridges were deposited perpendicular to the coast and which during the subsequent rise in the sea level became the loci of cape formation.

Ball and Neumann (1968) have indicated that Cape Fear and Cape Canaveral are governed by geologic structure. Cape Fear, for example, is associated with the pronounced Cape Fear Arch.

Dolan and Ferm (1968) also identified along the southeastern U.S. coast a series of secondary capes of 10 km mean spacing, an order of magnitude smaller than the large Carolina Capes. They did not speculate as to their origin.

Cape development along other coasts of the world has been reviewed by Johnson (1919), Steers (1948), King (1972), Shepard (1963), and Zenkovitch (1967). Shepard (1952, 1963) indicates that the Carolina Capes are examples of cuspate forelands. Offshore from many such cuspate forelands are shoals or islands which control their formation. These shoals or islands produce a protected portion of coast in their wave shadow zones, much like that produced by a detached breakwater (Chapter 12, Figure 12-6). Beach sediment is deposited within this protected area to form the cuspate foreland, shaped by the waves that refract and diffract around the shoal or island. If the cuspate foreland eventually connects with the island, it becomes a tombolo (Chapter 2). Shepard (1963, p. 194) has suggested that organic reefs may have developed on the shelf which controlled the positions of the Carolina Capes. Although this mechanism accounts for many cuspate forelands found in other parts of the world, investigations by Gorsline (1963) and Cleary and Pilkey (1968) indicate that this is probably not the cause of the Carolina Capes.

SUMMARY

The large-scale beach configuration tends to approximate an equilibrium in which the wave climate provides precisely the energy and mean wave approach angles required to transport and redistribute the sediments supplied to the beach. The beach configuration, then, is controlled mainly by the curvature and orientation of the refracted wave crests and by the locations and relative importance of the beach sediment sources and losses (sinks).

Superimposed upon this general equilibrium curvature may be a hierarchy of rhythmic shoreline forms (Dolan and Ferm, 1968), ranging from rhythmic topography with cusp spacings up to 2,000 meters down to small, ordinary beach cusps. These rhythmic forms may also be in equilibrium with the prevailing waves and nearshore currents. When all these are present and superimposed upon one another, the resulting shoreline has a very irregular appearance. In spite of this, it still may represent an equilibrium shoreline configuration.

PROBLEMS

10-1 The net annual sand drift at a beach is 200,000 m^3/yr. If the breaking wave height averages $H_b = 2$ m, what "prevailing wave" breaker angle could account for this transport? This example demonstrates that even with a net drift the shoreline need not ordinarily depart markedly from an equilibrium shoreline of zero drift (i.e., one parallel to the wave crests).

10-2 Devise a computer simulation model for the blockage of a littoral drift by a single groin placed transverse to the transport.

10-3 Discuss the relationship between the cell circulation in the nearshore and the resulting shoreline configuration. What sort of equilibrium shoreline configuration could be attained under such a system?

10-4 Discuss the possible role of rhythmic topography in localized erosion of coastal property.

10-5 How would you conduct experiments or a study to investigate the role played by edge waves in the formation of beach cusps?

10-6 A crescentic bar has a wave length of 500 m, and the overall beach slope is approximately 0.015. What edge wave periods would be required for their formation? What might supply such periods?

10-7 Devise a computer simulation model for the development of a cusp in the lee of a rip current where the longshore current is caused by a combined oblique wave approach and longshore variation in wave height as described on page 278. Start with an initially straight shoreline and evaluate the current using equation (7-31).

10-8 Dolan and Ferm (1968) identify a series of secondary capes of 10 km mean spacing between the much larger Carolina Capes. What might be their origin?

REFERENCES

ABBE, C. J. (1895). Remarks on the cuspate capes of the Carolina coast. *Proc. Boston Soc. Nat. His.*, 26: 489–97.

BAGNOLD, R. A. (1940). Beach formation by waves: some model experiments in a wave tank. *J. Inst. Civ. Eng.*, 15: 27–52.

BAKKER, W. T. (1968). A mathematical theory about sand waves and its application on the Dutch Waden Isle of Vlieland. *Shore and Beach*, 36: 4–14.

BAKKER, W. T. (1969). The dynamics of a coast with a groyne system. *Proc. 11th Conf. on Coast. Eng.*, pp. 492–517.

BAKKER, W. T., E. H. BRETELER, and A. ROOS (1971). The dynamics of a coast with a groyne system. *Proc. 12th Conf. on Coast. Eng.*, pp. 1001–20.

BAKKER, W. T., and T. EDELMAN (1965). The coastline of river-deltas. *Proc. 9th Conf. on Coast. Eng.*, pp. 199–218.

BALL, M. M., and A. C. NEUMANN (1968). Crescentic landforms along the Atlantic coast of the United States. *Science*, 161: 710.

BOWEN, A. J. (1973). Edge waves and the littoral environment. *Proc. 13th Conf. on Coast. Eng.*, pp. 1313–20.

BOWEN, A. J., and D. L. INMAN (1969). Rip currents, 2: laboratory and field observations: *J. Geophys. Res.*, 74: 5479–90.

——— (1971). Edge waves and crescentic bars. *J. Geophys. Res.*, 76, no. 36: 8662–71.

——— (1972). Reply. *J. Geophys. Res.*, 77, no. 33: 6632–33.

BRUUN, P. (1953). *Forms of equilibrium of coasts with littoral drift.* Univ. of California Engr. Research Lab. Tech. Report, series 3, Berkeley.

——— (1954). Migrating sand waves and sand humps, with special reference to investigations carried out on the Danish North Sea coast. *Proc. 5th Conf. on Coast. Eng.*, pp. 269–95.

BUMPUS, D. F. (1955). The circulation over the continental shelf south of Cape Hatteras. *Trans. Am. Geophys. Union*, 36: 601–11.

CLEARY, W. J., and O. H. PILKEY (1968). *Sedimentation in Onslow Bay.* Southeast. Geol. Special Publ. no. 1, pp. 1–18.

CLOS-ARCEDUC, A. (1962). Etude sur les vues aeriennes, des alluvions littorales d'allure periodique, cordons littoraux et fostons. *Bull. Soc. Fr. Photogramm*, 4: 13–21.

——— (1964). *La photographie aerienne et l'étude des depots prelittoraux*, Etude de photo-interpretation no. 1. Institut Geographique National, Paris.

CLOUD, P. E. (1966). Beach cusps: response to Plateau's rule? *Science*, 154: 890–91.

COOKE, C. W. (1936). *Geology of the coastal plains of South Carolina.* U.S. Geological Survey Bulletin no. 867, U.S. Government Printing Office, Washington, D.C., 196 pp.

DAVIES, J. L. (1958). Wave refraction and the evolution of shoreline curves. *Geog. Studies*, 5, no. 2: 1–14.

DAVIS, R. A., and W. T. FOX (1972). Coastal processes and nearshore sand bars. *J. Sediment. Petrol.*, 42: 401–12.

DOLAN, R. (1971). Coastal landforms: crescentic and rhythmic. *Geol. Soc. Am. Bull.*, 82: 177–80.

DOLAN, R., and J. C. FERM (1968). Concentric landforms along the Atlantic coast of the United States. *Science*, 159: 627–29.

DOLAN, R., L. VINCENT, and B. HAYDEN (1974). Crescentic coastal landforms: *Zeitschr. für Geomorph.*, 18: 1–12.

EGOROV, E. M. (1951). On the forms of accretive beach related to continuous sediment movement. *Doklady Akademii Nauk*, 80, no. 5 [in Russian].

ESCHER, B. G. (1937). Experiments on the formation of beach cusps. *Leid. Geol. Meded.*, 9: 79–104.

EVANS, O. F. (1938). Classification and origin of beach cusps. J. Geol. 46: 615–627.

——— (1939). Mass transport of sediments on subaqueous terraces. *J. Geol.*, 47: 324–34.

FLEMMING, N. C. (1964). Tank experiments on the sorting of beach material during cusp formation. *J. Sediment. Petrol.*, 34: 112–22.

FOX, W. T., and R. A. DAVIS (1973). Simulation model for storm cycles and beach erosion on Lake Michigan. *Geol. Soc. Am. Bull.*, 84: 1769–90.

GELLERT, J. F. (1937). Strandhorner bei Duhnen-Cuxhaven. *Senckenbergiana Maritima*, 19: 7–12.

GORSLINE, D. S. (1963). Bottom sediments of the Atlantic shelf and slope off the southern United States. *J. Geol.*, 71: 422–40.

GORYCKI, M. A. (1973). Sheetflood structure: mechanism of beach cusp formation and related phenomena. *J. Geol.*, 81: 109–17.

GREENSPAN, H. P. (1956). The generation of edge waves by moving pressure distribution. *J. Fluid Mech.*, 1: 574–92.

GRIJM, W. (1961). Theoretical forms of shorelines. *Proc. 7th Conf. on Coast. Eng.*, pp. 197–202.

——— (1965). Theoretical forms of shorelines. *Proc. 9th Conf. on Coast. Eng.*, pp. 219–35.

GUZA, R. T., and D. L. INMAN (1975). Edge waves and beach cusps. *J. Geophys. Res.*, 80, no. 21: 2997–3012.

HARRISON, W., and K. A. WAGNER (1964). *Beach changes at Virginia Beach.* U.S. Army Coastal Eng. Res. Center Misc. Paper no. 6–64.

HOM-MA, M., and C. J. SONU (1963). Rhythmic patterns of longshore bars related to sediment characteristics. *Proc. 8th Conf. on Coast. Eng.*, pp. 248–78.

HORIKAWA, K., and T. SASAKI (1968). Some considerations on longshore current velocities. *Proc. 15th Coast. Eng. Conf. Japan Soc. Civ. Eng.*, pp. 126–35 [in Japanese].

HOYLE, J. W., and G. T. KING (1958). The origin and stability of beaches. *Proc. 6th Conf. on Coast. Eng.*, pp. 281–301.

HOYT, J. H., and V. J. HENRY (1971). Origin of capes and shoals along the southeastern coast of the United States. *Geol. Soc. Am. Bull.*, 82: 59–66.

HUNTLEY, D. A., and A. J. BOWEN (1973). Field observations of edge waves. *Nature*, 243: 160–61.

JOHNSON, D. W. (1910). Beach cusps. *Geol. Soc. Am. Bull.*, 21: 604–24.

——— (1919). *Shore processes and shoreline development.* Wiley, New York, 584 p. Facsimile edition: Hafner, New York (1965).

KING, C. A. M. (1972). *Beaches and coasts.* 2nd ed. St. Martin's Press, New York, 570 pp.

KING, C. A. M., and M. J. MCCULLAGH (1971). A simulation model of a complex recurved spit. *J. Geol.*, 79: 22–37.

KING, C. A. M., and W. W. WILLIAMS (1949). The formation and movement of sand bars by wave action. *Geog. J.*, 107: 70–84.

KOMAR, P. D. (1971). Nearshore cell circulation and the formation of giant cusps. *Geol. Soc. Am., Bull.* 82: 2643–50.

——— (1973*a*). Computer models of delta growth due to sediment input from rivers and longshore transport. *Geol. Soc. Am. Bull.*, 84: 2217–26.

——— (1973*b*). Observations of beach cusps at Mono Lake, California. *Geol. Soc. Am. Bull.*, 84: 3593–600.

KRUMBEIN, W. C. (1944). *Shore currents and sand movement on a model beach.* U.S. Army Corps of Engrs., Beach Erosion Board Tech. Memo. no. 7, 43 pp.

KRUMBEIN, W. C., and L. E. OSHIEK (1950). Pulsation transport of sand by shore agents. *Trans. Am. Geophys. Union*, 31: 216–20.

KUENEN, PH. H. (1948). The formation of beach cusps. *J. Geol.*, 56: 34–40.

LEWIS, W. V. (1938). The evolution of shoreline curves. *Proc. Geol. Assoc. Eng.*, 49: 107–27.

LONGUET-HIGGINS, M. S., and D. W. PARKIN (1962). Sea waves and beach cusps. *Geog. J.*, 128: 194–201.

MOTYKA, J. M., and D. H. WILLIS (1975). The effect of refraction over dredged holes. *Proc. 14th Conf. on Coast. Eng.*, pp. 615–25.

MUNK, W. H. (1949), Surf beats. *Trans. Am. Geophys. Union*, 30: 849–54.

NIEDORODA, A. W., and W. F. TANNER (1970). Preliminary study of transverse bars. *Mar. Geol.*, 9: 41–62.

OTVOS, E. G. (1964). Observations of beach cusps and pebble ridge formation on the Long Island Sound. *J. Sediment. Petrol.*, 34: 554–60.

PALMER, H. R. (1834). Observations on the motions of shingle beaches. *Philos. Trans. Roy. Soc.* (London), 124: 567–76.

PELNARD-CONSIDERE, R. (1954). *Essai de théorie de l'évolution des formes de rivage en plages de sable et de galets.* Société Hydrotechnique de France, IVes Journées de l'Hydraulique, Les Energies de la Mer, Paris, Question 3.

PRICE, W. A., K. W. TOMLINSON, and D. H. WILLIS (1973). Predicting changes in the plan shape of beaches. *Proc. 13th Conf. on Coast. Eng.*, pp. 1321–29.

RIVIERE, A., F. ARBEY, and S. VERNHET (1961). *Remarque sur l'évolution et l'origine des structures de plage à caractère periodique.* Comptes rendus des seances de l'Academie des Sciences no. 252, Paris.

RUDE, G. T. (1923). Shore changes at Cape Hatteras. *Ann. Assoc. Am. Geog.*, 12: 87–95.

RUSSELL, R. J., and W. G. MCINTIRE (1965). Beach cusps. *Geol. Soc. Am. Bull.*, 76: 307–20.

SHEPARD, F. P. (1952). Revised nomenclature for depositional coastal features. *Bull. Am. Assoc. Petrol. Geol.*, 36: 1902–12.

——— (1963). *Submarine geology*. 2nd ed. Harper & Row, New York, 557 pp.

SILVESTER, R. (1970). Growth of crenulated shaped bays to equilibrium. *J. Waterways and Harbors Div., Am. Soc. Civ. Eng.*, WW2, pp. 275–87.

SONU, C. J. (1969). Collective movement of sediment in littoral environment. *Proc. 11th Conf. on Coast. Eng.*, pp. 373–400.

—— (1972). Comments on paper by A. J. Bowen and D. L. Inman, "Edge wave and crescentic bars." *J. Geophys. Res.*, 77, no. 33: 6629–31.

——— (1973). Three-dimensional beach changes. *J. Geol.*, 81: 42–64.

SONU, C. J., J. M. MCCLOY, and D. S. MCARTHUR (1967). Longshore currents and nearshore topographies. *Proc. 10th Conf. on Coast. Eng.*, pp. 525–49.

SONU, C. J., and R. J. RUSSELL (1967). Topographic changes in the surf zone profile. *Proc. 10th Conf. on Coast. Eng.*, pp. 502–24.

SONU, C. J., and J. L. VAN BEEK (1971). Systematic beach changes on the Outer Banks, North Carolina. *J. Geol.*, 79: 416–25.

STEERS, J. A. (1948). *The coastline of England and Wales.* Cambridge Univ. Press, London, 644 pp.

TANNER, W. F. (1958). The equilibrium beach. *Trans. Am. Geophys. Union*, 39: 889–91.

TIMMERMANS, P. D. (1935). Proeven over den invloed van golven op een strand. *Leidsche Geol. Meded.*, 6: 231–386.

TRASK, P. D. (1956). *Changes in configuration of Point Reyes Beach, California*, 1955–1956. U.S. Army Corps of Engrs., Beach Erosion Board Tech. Memo. no. 91, 61 pp.

TUCKER, M. J. (1950). Surf beats: sea waves of 1 to 5 minutes period. *Proc. Roy. Soc.* (London), series A, 202: 565–73.

TUOMEY, M. (1848). Report on the geology of South Carolina. *Bull. Geol. Surv.* (South Carolina), 293 pp.

VAN BENDEGOM, L. (1949). *Consideration of the fundamentals of coastal protection.* Ph.D. thesis, Technical University, Delft, Netherlands [in Dutch].

WALTON, T. L., and R. G. DEAN (1973). Application of littoral drift roses to coastal engineering problems. *Conf. on Eng. Dyn. in the Surf Zone* (Sidney, Australia), pp. 221–27.

WHITE, W. A. (1966). Drainage asymmetry and the Carolina capes. *Geol. Soc. Am. Bull.*, 77: 223–40.

WILLIAMS, A. T. (1973). The problem of beach cusp development. *J. Sediment. Petrol.*, 43: 857–66.

WILSON, A. W. G. (1904). Cuspate forelands along the Bay of Quinte. *J. Geol.*, 12: 106–32.

WILSON, W. S. (1966). *A method for calculating and plotting surface wave rays.* U.S. Army Coastal Engr. Res. Center Tech. Memo. no. 17, 57 pp.

YASSO, W. E. (1965). Plan geometry of headland bay beaches. *J. Geol.*, 73: 702–14.

ZENKOVITCH, V. P. (1964). Cyclic cuspate sand spits and sediment transport efficiency: discussion. *J. Geol.*, 72: 879–80.

——— (1967). *Processes of coastal development.* Trans. D. G. Fry, ed. J. A. Steers. Oliver and Boyd, Edinburgh, 738 pp.

Chapter 11

BEACH PROFILES AND ONSHORE-OFFSHORE SEDIMENT TRANSPORT

Although nature begins with the cause and ends with the experience, we must follow the opposite course, namely, begin with the experience and by the means of it investigate the cause.

Leonardo da Vinci
Notebooks

One of the more important aspects of a beach is its dynamic personality: the loose granular sediments continuously respond to the ever-changing waves and currents imposed from the adjacent body of water. An equilibrium beach profile may be achieved in a laboratory wave tank, where a constant wave input is maintained. On natural beaches the changing waves give rise to an ever-varying equilibrium which the beach profile attempts to achieve but seldom does. However, the only way in which beach profiles can be understood is in terms of this equilibrium profile and how it is determined by wave and current conditions and the sediments which compose the beach.

This chapter will review what is known about beach profiles: why sand shifts from the exposed beach to the offshore and back again, why coarse-sand beaches are steeper than fine-sand beaches, what governs the number of offshore bars, what are the effects of tides. Models and experimental studies of onshore-offshore sediment transport will also be examined in an attempt to understand the processes involved in beach profile development.

The beach profile is important in that it can be viewed as an effective natural mechanism which causes waves to break and dissipate their energy. The beach serves as a buffer, protecting sea cliffs and coastal property from the intense wave action. If there is a long-term loss of sand from the beach, the beach will be increasingly less capable of serving as a buffer and coastal property erosion will become progressively more probable.

PROFILE CHANGES DUE TO STORMS

If a person takes repeated beach profiles at some fixed location on the coast and uses a reference stake so that they can be compared, he may find long-term systematic changes in the profile as well as short-term fluctuations. The main shift that might be discerned is an annual change commonly referred to as the *summer profile* versus the *winter profile* (Figures 11-1 and 11-2). The summer profile is characterized

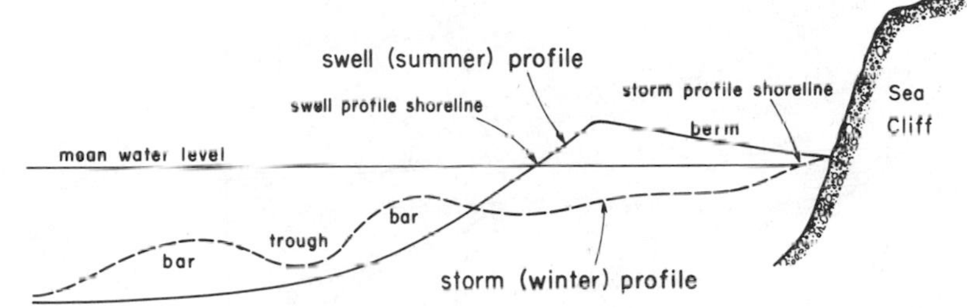

Figure 11-1 The storm beach profile with bars versus the profile with a pronounced berm that occurs under swell wave conditions.

by a wide berm, the flat shoreward portion of the profile (Figure 2-1), and by a smooth offshore profile with no bars except perhaps in relatively deep water. In contrast, the winter profile has almost no berm, the sand having shifted offshore to form a series of bars parallel to the shoreline. The volume of sand involved remains relatively constant; the areas under the winter and summer sections are about the same. The sediment shifts from the berm to bar and back again. The overall profile slope is smaller in the winter than in the summer profile. With repeated profiles, our observer might actually be able to trace the annual destruction and reappearance of the berm (Figure 11-3).

If wave measurements are obtained along with the beach profiles, it will be found that the offshore shift of sand from the berm to the bars takes place during storm conditions of large wave activity. During smaller swell wave conditions the reverse is true: sand shifts back onshore and the berm grows. This accounts for the terminology—*summer* and *winter* profiles. Such shifts in the profile were first observed off the west coast of the United States, where storm waves are typical of the winter and longer-period swell waves occur in the summer (Shepard, 1950*b*; Bascom, 1954). Therefore, the shifts in the profile corresponded with the winter and summer. However, it is peculiar that this terminology is sometimes used elsewhere in the world where the seasonal connotations are not correct. For example, a "winter profile" may develop on a Bermuda beach in mid-August if a storm occurs. Similarly, some of the short-term fluctuations observed in our series of beach profiles could result from a storm in

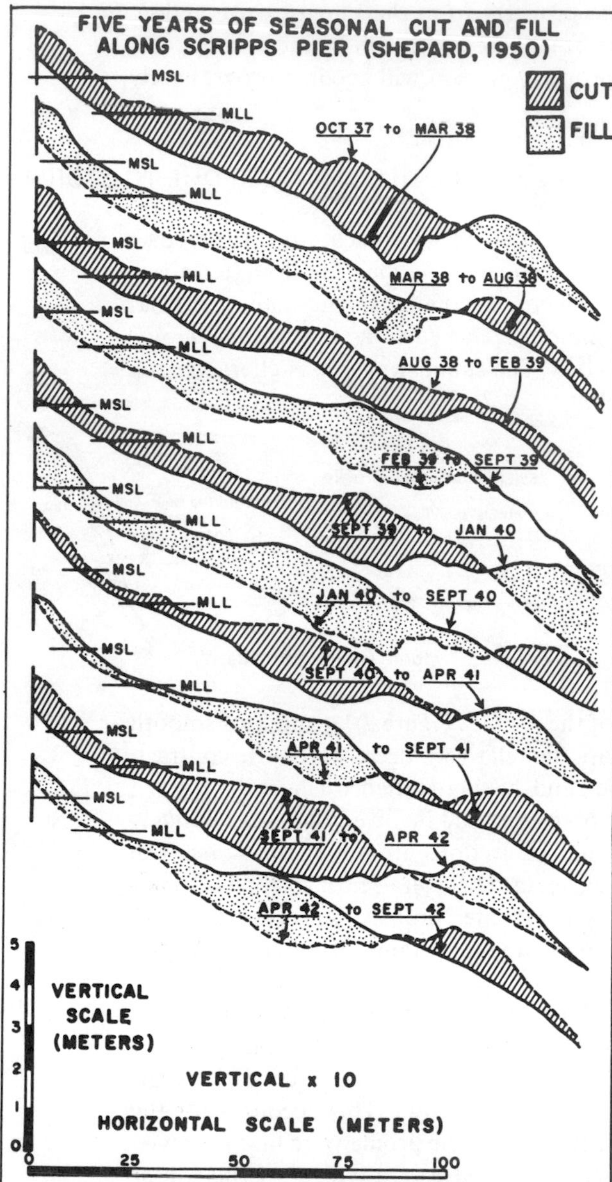

Figure 11-2 Profile changes along Scripps Pier, La Jolla, California, showing the tendency to shift from a more evenly sloping swell (summer) profile to a storm (winter) profile and back again according to the season. [*After* Shepard (1950*b*)]

mid-summer or an extensive quiet period of low waves during the winter months. Johnson (1949) in his wave tank study applied the terms "storm profile" and "normal profile," which eliminates the seasonality. Hayes and Boothroyd (1969) use "storm profile" and "post-storm profile." My preference is for the descriptive terms *storm profile* and *swell profile*, which will be used here.

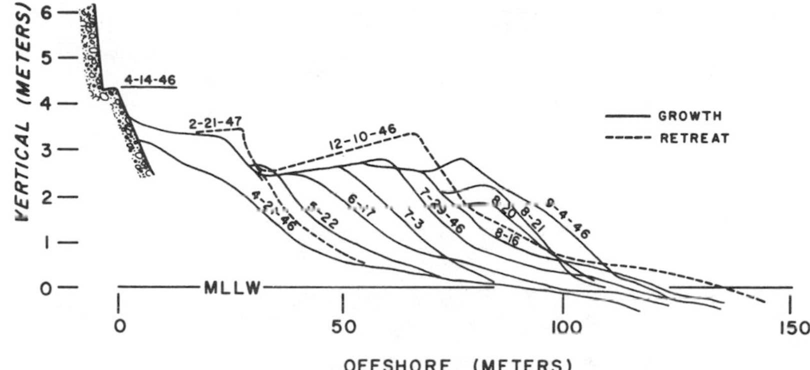

Figure 11-3 The progressive growth and retreat of the berm at Carmel, California, during 1946–47. [*After* Bascom (1954)]

This onshore-offshore shift of sand associated with profile changes from storm to swell conditions is generally correlated with the wave steepness H_∞/L_∞, the ratio of the deep-water wave height H_∞ to the deep-water wave length L_∞, which is related to the wave period T by $L_\infty = (g/2\pi)T^2$. Storm waves have high steepness values, because of both their greater heights and shorter periods, while long swell waves have low steepness values. The wave steepness can be increased either by an increase in the wave height or by a decrease in the wave period.

In his wave channel experiments, Johnson (1949) found profile changes similar to those detected by our hypothetical observer, which he associated with the magnitude of the wave steepness. He determined that with a wave steepness greater than 0.03 an offshore bar always forms (storm profile), whereas if the steepness is less than 0.025 an offshore bar is never formed (swell profile). Scott's (1954) results agree with those of Johnson. Rector (1954) and Watts (1954) both state that a wave steepness of 0.016 is critical in developing a swell profile versus a storm profile with bars. King and Williams (1949) found that a wave steepness of 0.012 is important in governing whether sand moves onshore or offshore in the surf zone and that wave steepness therefore determines the profile type. The differences in the exact value of this critical wave steepness found by the various studies may be due in part to a dependence on the grain size of the beach sediment (Rector, 1954). Some of the variations may also result from the scale of the experiment; results with large waves in a large wave tank or on real ocean beaches might be different than with small waves in a small wave tank. This is indicated by the results of Saville (1957) who, experimenting with waves as large as those on actual beaches, obtained a critical wave steepness of 0.0064, which is much lower than those in the other studies. The grain-size and scale effects on the beach profile type have been further considered by Iwagaki and Noda (1963). Their plot of the beach profile type versus the wave steepness and the ratio of the deep-water wave height to the mean sediment diameter, H_∞/D, Figure 11-4, demonstrates the dependence of critical wave steepness on the wave height and sediment grain size utilized in the experiment. Nayak (1971) has tried to extend the considera-

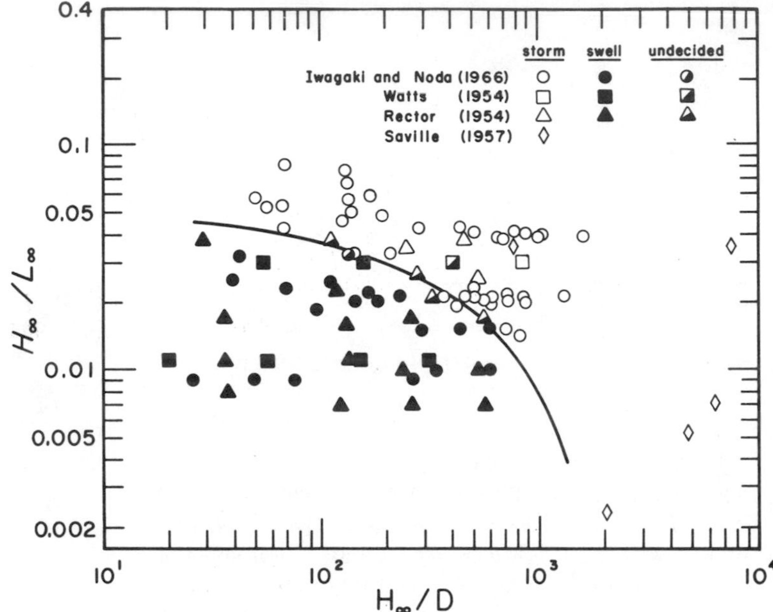

Figure 11-4 The critical wave steepness H_∞/L_∞ at which the beach profile shifts from a storm to a swell profile; D is the mean sediment diameter. [*After* Iwagaki and Noda (1963)]

tion to beach sediments with densities other than that of quartz by using the ratio

$$\frac{H_\infty}{\frac{\rho_s - \rho}{\rho} D}$$

where ρ_s and ρ are respectively the sediment and water densities. Nayak meets with partial success, but his results do not particularly agree with those of Iwagaki and Noda (1963).

Kemp (1961) develops a model for the change in the profile type on the basis of the relationship between the wave period and the time of wave travel between the break point and the limit of swash. When this travel time is greater than or equal to the wave period, the sand shifts offshore to form a bar (storm profile). Using empirical relationships between the distance from breakers to swash limit and the breaker height, Kemp shows that the critical wave steepness is inversely related to the wave period. Because of this, the critical wave steepness on ocean beaches with $T = 10$ seconds would be one-tenth that in a laboratory wave tank with $T = 1$ second. Kemp also demonstrates that the model explains why beaches under laboratory conditions are usually much steeper than those under ocean conditions.

Dean (1973) presents a model for the shift from a storm to a swell profile based on a consideration of the trajectory of a suspended sand particle during its fall to the bottom, acted upon at the same time by the horizontal water particle velocity of the wave. If the fall requires a short time relative to the wave period, then the particle will be acted upon predominantly by onshore velocities. If, on the other hand, the fall velocity is low, then the grains will tend to shift offshore. Based on this model,

Dean finds that the dimensionless ratio $\pi w/gT$, where w is the grain-settling velocity, is important in governing the critical wave steepness. Using the data of Rector (1954), Saville (1957), and Iwagaki and Noda (1963), Dean obtained the equation

$$\text{Critical}\,\frac{H_\infty}{L_\infty} = \frac{1.7\pi w}{gT} \tag{11-1}$$

where the coefficient is empirical. This relationship correctly predicts whether the transport is onshore or offshore in 87.5% of the cases.

Our understanding of the critical wave steepness which governs the shift from the swell profile (summer profile) to the storm profile (winter profile) is still incomplete. This incompleteness is particularly true for field studies. Although there are many field studies that demonstrate an offshore shift of beach sands during storms to form bars and a shoreward return during quiet swell conditions (Shepard and LaFond, 1940; Shepard, 1950*a*, 1950*b*; Bascom, 1954; Strahler, 1966; Gorsline, 1966), none of the studies obtained satisfactory wave measurements that could determine a critical wave steepness. This is also due in part to the great irregularity of profile changes on real beaches. The studies all clearly demonstrate that an increase in the wave height during storm conditions leads to the storm profile with an offshore shift of sand. The dependence on the wave period is less clear. According to the hypothesis of a critical wave steepness, a shift to a storm profile could be accomplished by decreasing the wave period T while maintaining the wave height constant, as well as by an increase in height. This is indicated by pairs of profiles obtained in a wave tank by Iwagaki and Noda (1962), but has not been confirmed on ocean beaches. Shepard and LaFond (1940) indicated that longer-period waves tend to shift sand offshore into bars, opposite to that implied by a critical wave steepness, although the effect of the wave period is not as great as the wave height. Their evidence is not entirely conclusive, but it points out the need for additional field studies which carefully tie the onshore-offshore shifts in beach profiles to the nature of the waves.

Cook and Gorsline (1972) suggest that rip currents are most important in moving sand offshore from the beach during storm conditions when the rips are powerful. They envision that sand is shifted back onto the beach by swell waves when the storm ceases. There is little doubt that rip currents are an important agent in sand removal from the beach during storms. (This was discussed in Chapter 10.) However, the process of sand removal by rip currents cannot be the complete answer to the shift from a swell profile to a storm profile. Such a shift can occur on natural beaches with no apparent rip currents or in a laboratory wave tank where rip currents are not possible.

The development of a storm versus a swell profile can be understood in terms of the directions of sediment transport within the surf zone and beyond the breaker zone. With storm waves the sand seaward of the breaker zone moves shoreward, while sand in the surf zone is transported in an offshore direction (King and Williams, 1949). This convergence of the sand transport directions must result in an accumulation of sand at the breaker position, forming a bar. With flatter swell waves, the sand is moved landward at all depths, within the surf zone as well as beyond the breaker zone, so that it accumulates on the berm. The reasons for this pattern of onshore-offshore transport will be examined later in the chapter.

The beach profile type is of importance to sea cliff and coastal property erosion. With a swell profile the sea cliffs are protected from the wave action by a wide berm and so experience little or no erosion. During storm conditions the sand is shifted offshore and the berm lost, so that the more intense swash is able to reach and erode the sea cliffs. For this reason, on most coasts beach and cliff erosion is most severe during the winter and early spring months. If there is a rapid succession of storms, considerable sand is shifted offshore and erosion of property is particularly acute. In other years the storms are separated by quieter periods, and the beach berm may not be entirely eliminated; property erosion in such years is minimal. Not only the severity but the succession of storms is important.

PROFILE CHANGES DUE TO LONGSHORE SAND TRANSPORT

In addition to onshore-offshore shifts of sand, longshore sand movements may also affect the beach profiles. Within pocket beaches a shift in the wave direction will reorient the shoreline such that the shoreline retreats in the updrift side and advances in the downdrift side of the pocket. This was demonstrated in the computer models of Chapter 10 (Figure 10-8). The best example of this is Boomer Beach, La Jolla, California, where within 24 hours after a change in wave direction the sand shifts to the opposite end of the pocket, up to 3 meters thickness of sand disappearing from the updrift end (Shepard, 1950*b*, p. 5). There may or may not be an onshore-offshore shift of sand as well. At times of high waves the sand will shift offshore into a storm profile at all points along the beach at the same time that the beach is growing outward on the downdrift end of the pocket. Similar shifts can be seen at the terminal ends of long beaches. In southern California, Shepard (1950*b*) found that northwesterly storms of winter cause the southern ends of some beaches to grow seaward, forming a storm profile at the same time, while the southerly wave approach during the summer moves sand to the north, producing a cut but with a swell profile.

When rhythmic topography is present in the nearshore (Chapter 10), the beach profile will depend largely on its position with respect to the topography. This is especially true for crescentic bars, as is seen in Figure 10-15. Sonu and Russell (1967) have pointed out that, depending on the location of the profile, either a storm-type or swell-type profile could be found—the storm type occurring out from the embayment where the bar is present, the swell type found out from the cusps where the beach face shifts seaward and no bar is present. If repeated profiles are obtained from a single transit line, longshore transport under oblique waves may cause the rhythmic topography to migrate alongshore and the successive profiles will show shifts from a swell-type to a storm-type profile and back again even though there are no accompanying changes from storm to swell conditions. Because of this, Sonu and Russell (1966) emphasized that in studies of beach profiles, variations in the longshore direction must be monitored; the problem of beach profiles is three-dimensional rather than two-dimensional.

PROFILE CHANGES DUE TO TIDES

In addition to the response of the beach profiles to storm versus swell conditions, our series of profiles might show alterations which correspond to the tides. This would involve hourly changes resulting from the rising and falling water level of the tides and also longer-term effects due to the differences in the range of spring and neap tides.

Although not pronounced, the profile changes from spring-tide conditions to neap tides are particularly interesting. Thompson and Thompson (1919) and LaFond (1939) demonstrated that on southern California beaches the beach surface a few meters above mean tide level reaches its minimum elevation a few days after spring tide and its maximum elevation following the neap tide. Grunion (the fish *Leuresthes tenuis*) take advantage of this cycle by laying their eggs just after spring tide. The eggs are then buried by sand during the neap-tide period when deposition occurs. At the next spring tide the sand is again removed and the eggs are ready to hatch. This spawning takes place during March to August, when the wave conditions are steadiest, so that the tide-induced changes are most effective. At other times of the year profile changes due to varying wave conditions are much greater than this tidal effect, so that the tidal cycle is masked. This is also true at other beaches: cycles due to spring and neap tides are usually small compared to wave effects.

Several studies have investigated the hourly profile changes resulting from the varying water level of the daily tidal cycles. Strahler (1966) made half-hourly observations of the foreshore changes during one tidal cycle at Sandy Hook, New Jersey. Sand scoured from the swash zone was deposited at the upper limit of the swash and in the step under the breaker zone. As the tide rose, at any set location on the beach face a small amount of deposition occurred, followed by erosion as the site came under the intense swash, in turn followed by deposition as the breaker zone passed over. Just the reverse occurred with the falling tide. A similar pattern of sediment movement was demonstrated by Otvos (1965) and Schwartz (1967) through the use of fluorescent sand tracers. Schwartz (1967) determined that the finer fraction of the sediment was sorted out and deposited at the top of the swash or carried seaward of the breaker zone, while the coarser material accumulated in building the step under the breaker zone.

Duncan (1964) investigated the effects of the water table on this pattern of profile readjustment to a daily tidal cycle. During flood tide the water level rises faster than the water table within the beach, so that the seaward edge of the water table slopes shoreward, as shown in Figure 11-5 (Emery and Foster, 1948; Harrison, Fang, and Wang, 1971; Grant, 1948). During ebb tide the water table slopes seaward. Because of this, during a flood tide water from the wave swash is lost by percolation into the beach and the backwash is weaker than the shoreward swash. Just the opposite is true during the ebb tide, since water is added to the backwash. Duncan (1964)

(a) Flood Tide

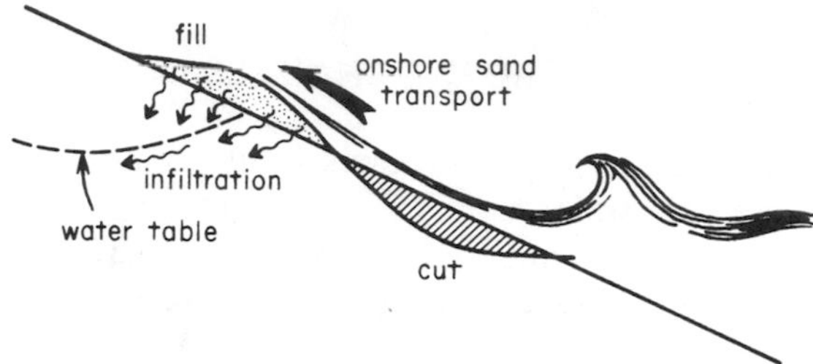

(b) Ebb Tide

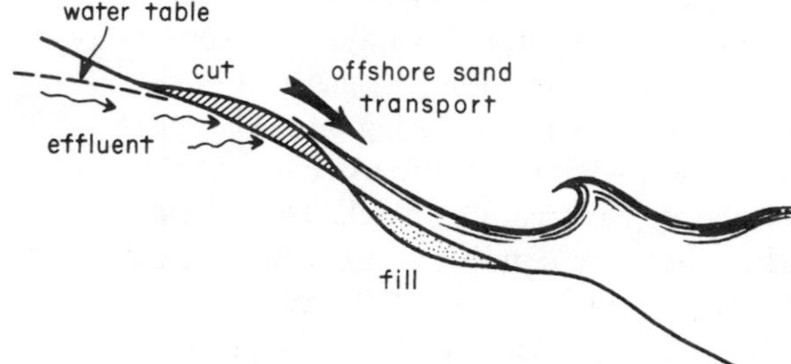

Figure 11-5 Water table effects on the cut and fill of the beach profile during flood and ebb tides. [*After* Duncan (1964)]

found that because of this most of the sediment transported up the beach face by the swash during flood tide is deposited at the top of the swash limit (Figure 11-5). During ebb tide the enhanced backwash removes this sand and deposits it on the shoreward side, where the backwash collides with the incoming surf and loses its transporting capacity.

The importance of the interaction of the swash and the water tables has been further demonstrated by Harrison (1969). Using regression analysis techniques, Harrison found that most of the variability in the quantity of sand eroded from or deposited on the foreshore over the interval from one low water to the next can be explained by (1) the steepness of the breaking waves (storm versus swell profiles), (2) the hydraulic head between the water table within the beach and the swash runup level, and (3) the angle of wave approach to the shoreline. He provides an empirical equation for this net volume change, as well as equations for the resulting advance or retreat of the shoreline and the mean slope of the foreshore.

Intermediate between the daily profile changes associated with the tidal cycle and those of the spring tide-neap tide cycle, Inman and Filloux (1960) found that beaches of the northwest Gulf of California exhibited a fortnightly cycle of erosion and deposition in response to the combined effects of the tides and wave conditions. In this area the higher waters of the spring tide occur in the early afternoon, when the "sea breeze" from the gulf to the desert area is strongest. This correspondence results in a maximum erosion-deposition of the beach profile. At times of neap tides the occurrence of the high water is no longer in phase with the time of maximum sea breeze, so that the energy of the waves is not concentrated at the position of still stand but rather is distributed over a wider stretch of the beach face slope.

In areas such as the northern Gulf of California (Inman and Filloux, 1960), where the tidal range is large compared with the wave height, the beach profiles are characterized by a relatively steep beach face terminated rather abruptly at its base by a wide low-tide terrace. Generally this high-tide beach face is composed of sediments that are coarser than that of the terrace. In the example of Figure 11-6, the

Figure 11-6 Newgale Sands, Pembroke, Wales, showing an example of a "high-tide beach" consisting of shingle, fronted by a fine-sand tidal flat, both exposed at low tide.

steep shingle beach face is distinct from the terrace. In addition, low-tide terraces are characterized by poorly sorted fine-grained sediments, well-formed and complex ripple marks, and more abundant infauna (Chapter 13). Seaward of the terrace may be a "low-tide beach" whose grain size is approximately the same as that of the terrace but with better sorting, and whose slope is somewhat greater than that of the terrace. Because a brief still stand in the water level occurs at high tide and at low tide, the energy of the waves is concentrated at these levels. The water level sweeps rather quickly over the low sloping terrace, and the wave energy tends to become dissipated by the low water depths. The concentration of energy at the high- and low-water levels accounts for the differences between the terrace sediments and the characteristics of the high- and low-tide beaches.

A system of *ridges* and *runnels* may sometimes be found on the low-tide terrace, similar to offshore bars but generally smaller and lower in profile. The ridges are cut by many gaps, formed by water escaping seaward from the runnels as the tide falls. The gaps also allow the water of the rising tide to penetrate quietly into the runnels while the sea is still breaking on the seaward face of the ridge (King and Williams, 1949). With a rising or falling water level from the tides, the breaker position jumps from one ridge to the next with little wave action occurring within the runnels. Such systems have been studied by Gresswell (1937, 1953) on the Lancashire coast of England, between Dieppe, France, and Ostend, Belgium, by Pugh (1953), and at

Blackpool Beach, England, by King and Williams (1949). At Blackpool the ridges and associated runnels tend to increase in size toward the low-tide level. Gresswell (1937) observed that, unlike longshore bars, the ridges build up during calm weather conditions of moderate swell and are reduced in height during storms. King and Williams (1949) indicate that the factors that appear to be of importance to the development of a ridge and runnel system are (1) a large tidal range such that an extensive low-tide terrace is developed with a low offshore slope; (2) low wave energy, such systems forming best in protected areas sheltered from the full ocean fetch; and (3) an abundance of sand.

It should be recognized that the distinction between longshore bars and troughs, systems of ridges and runnels, and bars or dunes generated by the flowing tidal currents is not always clear. For example, Hayes and Boothroyd (1969), Hayes (1972), and Davis et al. (1972) apply the term *ridge and runnel system* to somewhat different features than those described by King and Williams (1949) and King (1959, 1972): they use the term for any bar that eventually extends above the water level. An extensive low-tide terrace is not even required for ridge and runnel formation according to their usage of the term. This is seen by the description of "ridges and runnels" by Davis et al. (1972) in Lake Michigan, where tides are negligible. King and Williams (1949) found no systematic movements of the ridges at Blackpool over several years of observation. This contrasts with the "ridges" of Hayes and Boothroyd (1969) and Davis et al. (1972), which are characterized by extensive migrations (Figure 11-7).

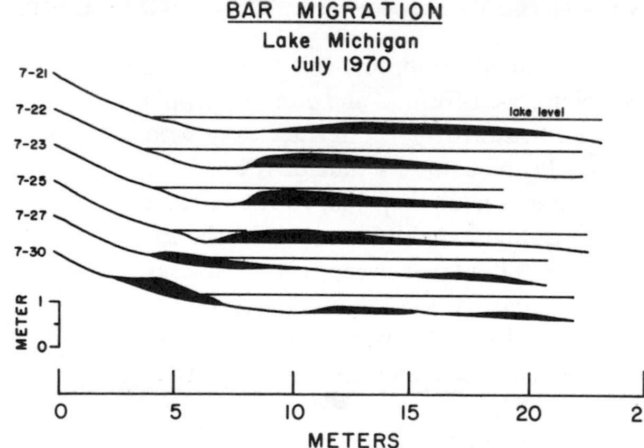

Figure 11-7 Onshore migration of longshore bars or "ridges and runnels" according to Davis et al. (1972).

The landward migration of such "ridges" is the primary mode of onshore sediment transport back to the beach face in the shift from a storm profile to a swell profile. The "ridge" begins its life as a normal longshore bar following a storm, but as it migrates shoreward it eventually reaches above the water level, finally migrating completely onto the beach face. The migrating bars so described appear to be more akin to King and Williams's (1949, pp. 81–82) *swash bars*, which they generated in a laboratory wave tank. The terminology *ridge and runnel* should thus be reserved for the systems of bars on an extensive tidal flat as described by King and Williams (1949) at Blackpool.

PROFILE CHANGES DUE TO WINDS

Onshore winds cause a landward movement of the surface waters which must be compensated by a seaward current at depth. Just the reverse is true with offshore winds, the near-bottom currents being onshore. These currents will be a factor in the onshore-offshore transport of sediments and therefore have a bearing on the response of the beach profile.

King (1959, p. 212) performed experiments in a wave tank that demonstrated the effects of an onshore wind on the beach face configuration. Waves of 4 cm height and deep-water steepness 0.0114 were generated in the tank and allowed to form a profile on an initially smooth gradient of 1: 15. In one case there were no winds; in the second run a wind of 11 m/sec (25 mph) was blown along the tank length in the onshore direction. Without the wind the sand shifted onshore to form a swash bar 5 cm high above the still-water level. With the strong onshore wind the swash bar was completely absent, the combined wind and waves being destructive to the beach profile.

King and Williams (1949) studied the onshore-offshore sand transport in a wave tank both shoreward and seaward of the breakers. One set of experiments involved the superposition of an onshore wind on the wave action. With the onshore wind they found that within the surf zone the sand was transported offshore, even though for the wave steepness used, a swell profile should have developed. Outside the breaker zone the offshore near-bottom current generated by the onshore wind counteracted the landward mass movement of water due to the waves. The former onshore sand transport due to waves alone was converted into a slight seaward transport. Their results will be considered again later in the chapter when the mechanics of onshore-offshore sediment transport are examined.

On natural ocean beaches, separating out the importance of any wind-induced currents from the effects that winds have on the waves and therefore indirectly on the beach profiles is rather difficult. With strong onshore winds steep waves may be locally generated which have a destructive influence on the beach profile in addition to the reaction of the beach to the wind-induced currents that also tend to cut back the beach. With offshore winds the advancing waves tend to be reduced in height by the head winds, so that the waves reaching the shore are of lower steepness. Any constructive effects the near-bottom onshore wind-induced currents might have are enhanced and masked by the reduced wave steepness.

Shepard and LaFond (1940) found in their beach profiles on southern California beaches that a cutting of the beach corresponded with high onshore wind velocities. The maximum amount of cutting for one year occurred at a time when the wind velocity reached a maximum at the same time that the waves were only the fifth largest occurrence for that year.

Observations by King (1953) on Marsden Bay, County Durham, England, also demonstrated such effects of onshore and offshore winds on the profile. With an onshore wind, erosion took place thirteen times on the upper foreshore above mean tide level, while accretion only occurred four times. With an offshore wind, accretion occurred thirteen times and erosion only three times. On the lower foreshore just the reverse took place.

Observations such as these confirm that the strength and direction of coastal winds can be an important factor in the beach profile development. This process is worthy of more study. In any investigation of beach profiles it is necessary that the winds be monitored and an attempt made to assess their effects. Perhaps much of the data variability found in studies of other aspects of beach profiles might be attributable to the variable wind conditions.

LONGSHORE BARS AND TROUGHS

When beach sediment is shifted offshore from the beach face it generally deposits to form a longshore bar with a trough on its shoreward side. Such bars may have a considerable longshore extension; for example, Evans (1940) traced longshore bars for up to 50 km in Lake Michigan. Longshore bars may be periodically broken by gaps due to rip currents. They are generally distinguished from crescentic bars, which have a definite longshore rhythm (Chapter 10).

A good summary of the early literature on the characteristics of longshore bars and troughs can be found in Johnson (1919) and in Evans (1940). Bar-trough systems are often referred to in the early literature as "ball and low," while the term "bar" was used for what we are calling barrier islands and barrier beaches.

Some of the earliest investigations of longshore bars and troughs were made by German scientists along the coasts of the Baltic and North seas. The first description is that given by Hagen (1863), who explained bar formation by a seaward-flowing undertow meeting the shoreward-moving waves, an explanation that is not much different than our present view of origin. Otto (1912) studied a series of bars on the Baltic Sea for a period of five years, obtaining repeated profiles. He demonstrated that most of the changes occur during violent storms. Otto gave the opinion that sand which moved offshore from the trough during storms deposits to form the bar; previous studies had mistakenly suggested that a trough forms seaward of a bar.

The more recent papers by Evans (1940), Keulegan (1948), King and Williams (1949), and Shepard (1950*a*) have contributed most to our understanding of the characteristics and variability of longshore bars and troughs. The study of Keulegan involved the controlled conditions of a laboratory wave channel; the investigations of Evans (Lake Michigan) and King and Williams (Mediterranean Sea) included tideless areas; while the study of Shepard on the California coast reflects the effects of tides. These studies are classics in an understanding of the characteristics and formation of longshore bars and troughs.

All the above studies recognize that breaking waves comprise the most important element in longshore bar formation, because the breakers control the offshore positions of bars, their sizes, and depths of occurrence. These studies all agree that the larger the waves, the deeper the resulting longshore bars and troughs. In his controlled wave tank investigations, Keulegan (1948) found that the bar position is governed by the wave height and wave steepness: an increase in wave height moves the bar seaward to deeper water; holding the height constant and increasing the wave steepness (decreasing the wave period) moves the bar shoreward. This latter dependence on wave steepness has not been demonstrated by field data. Keulegan found that starting with an initially smooth beach face, the bar forms first just shoreward of the breaker position but migrates landward as it grows, the breaker position migrating with it. The fully developed bar is appreciably nearer to the shore than the initial breaker position. In addition to shifting shoreward with the bar, the wave breakers also change character, tending more toward the plunging type than the spilling variety (Chapter 4). The bar is effective in dissipating wave energy. Keulegan found that with bars present, the breaking waves are able to impart a lesser amount of energy to the reformed waves than would be the case if the bars were absent.

Shepard (1950*a*, p. 23) also noted that plunging breakers appear to be more conducive to bar and trough development than are spilling breakers. He found that with spilling breakers that accompany short-period storm waves, the bar-trough system may be entirely eliminated, primarily because of the absence of a definite breaker zone under these conditions.

Keulegan (1948) found that the form of the bar is apparently independent of the size of the generating waves. The ratio of the depth of the trough to the bar depth was practically independent of the wave steepness and beach slope, the ratio having an average value of 1.69. Comparing his laboratory bars with those described by Evans (1940) in Lake Michigan, Keulegan found that the natural bars are somewhat flatter and wider: the trough-depth-to-bar-depth ratio varied between 1.42 and 1.55. The Baltic Sea observations of Otto (1912) and other German observers gave ratios varying from 1.56 to 1.87; the overall average was 1.66, which agrees with Keulegan's laboratory value. The profiles of Shepard (1950*a*) at Scripps Pier and other California piers, his profiles from a University of California wave tank, and profiles reported from Oregon and Washington beaches gave an average ratio of 1.3 using mean sea level and 1.5 using mean lower low water (Shepard, 1950*a*, p. 9). It is difficult to see any systematic pattern in this, demonstrating effects of laboratory versus field studies and the increasing importance of tides.

Most beach profiles show more than a single longshore bar. The number of bars depends mainly on the overall slope of the nearshore—the more gradual the slope the greater the number of bars. Off steep beaches bars are generally absent. At the other extreme, Kindle (1936) noted as many as ten longshore bars in Chesapeake Bay spaced about 25 meters apart, the large number being due to the gentleness of the slope within the bay. More commonly, two or three bars are present. Evans (1940) found that in Lake Michigan there is an average of three bars, rarely more than four, and in a few places where the water deepens rapidly only one is present. He found that the size of the bar increases with distance from the shore, as does the distance between adjacent bars, the depths of the outer bars being greater than bars closer to shore in spite of their larger size. This generally agrees with multiple bars found elsewhere, although Shepard (1950*a*) found examples where a larger outer bar reached upward to a shallower depth than did a smaller inner bar.

Multiple bars can be generated in at least three ways. Each bar individually reflects the average breaking position of waves of a certain size. The deepest bars therefore correspond to the largest waves. Evans (1940, p. 495) demonstrated in a wave tank that an inner and outer bar could be formed by a succession of large waves followed by smaller waves. He first used large waves to generate the outer bar. Subsequent lower waves passed over this outer bar with little effect to themselves or to the bar, and broke closer to the shore, forming an inner bar. King and Williams (1949) found that the outer bar at Sidi Ferruch, North Africa, owed its position to the predominant break point of the storm waves, which are comparatively rare on the Mediterranean Sea but are of great energy, while the inner bar depended on the predominant calm weather waves. It is conceivable that under certain circumstances the two sets of waves forming an inner and outer bar could arrive at the beach at the same time, having been generated by two separate storms.

On a low-sloping beach, once waves break on an outer bar they can often recover and reform as they travel over the deeper shoreward trough. These reformed waves break for a second and perhaps even a third time before they finally reach the shoreline. Inner bars develop where these reformed waves break, giving a multiple-bar system. Since the reformed waves are reduced in size, the topographic relief of the inner bar and trough is smaller.

Still other multiple-bar systems can be explained by the varying water level and breaker position resulting from the tidal cycle. There may be one set of bars for the high-tide water level and a second set for the low-tide position. However, a falling tide will tend to destroy the shallower bars, as shown by King and Williams (1949, p. 81) in their wave tank studies. On ocean beaches the inner bars exposed at low tide are apparently able to persist, although modified and flattened by the varying water level (Shepard, 1950*a*).

During storms, waves break on all bars in a multiple-bar system, the largest waves breaking on the deepest bar and progressively smaller breakers occurring on the inner bars. The waves which finally reach the shoreline are much reduced in size. During times of smaller waves they pass over the deep outer bars without any effect and do not break until they reach the comparatively shallow water over the inner bar. Because of this the inner bar is much more active than the outer bars in deeper water. During the fourteen months of his study, Evans (1940) found little or no movement of the outer two bars in Lake Michigan, while the shallowest inner bar showed considerable migration and variability. He found that the inner bar, built during storm conditions, may be driven inshore by moderate-size waves. These migrations of the inner bar on Lake Michigan, studied by Davis and Fox (1972) and Davis et al. (1972), Figure 11-7, have already been discussed in connection with "ridge and runnel" systems.

Carter, Liu, and Mei (1973) and Lau and Travis (1973) have investigated the formation of multiple-bar sets where several continuous bars are observed parallel to the shoreline. The usual relationship to the breaker zone does not explain these bars, as they exist in water depths too great for nearshore wave breaking. Both studies involve the offshore distribution of the mass transport associated with the waves (Chapter 3). Lau and Travis extend the theoretical development of Carter, Liu, and Mei to account for an offshore sloping bottom. With simple progressive waves approaching the shore the mass transport is everywhere directed onshore. Now, if there is a significant wave reflection from the beach, then a set of standing waves is established, and under these conditions there are reversals in the near-bottom water mass transport. This is shown theoretically in both studies. With reversals in the mass water drift, there are similar reversals in the sediment transport. Sand accumulates where the mass transport converges—that is, at the nodes of the standing wave where the wave amplitude is a minimum. The theoretical development was partially confirmed by Carter, Liu, and Mei in laboratory wave tank experiments. Systems of bars were formed in the flat portion of the tank at the toe of a reflecting beach slope. The results predict that the number of bars is likely to increase when the bottom gradient decreases, and that the spacings between the bars increase in the offshore direction. Both results generally conform with what is observed in nature. According to the theory, for good bar development, wave reflection from the beach should be on the order of 40% (the percentage of wave energy reflected). According to the empirical results on wave reflection, this requires a beach slope of about 17 degrees ($\tan \beta = 0.31$) [Carter, Liu, and Mei (1973)]. This is unusually steep for a natural beach, but even with somewhat lesser reflections (down to about 20% at a beach slope of 10 degrees) there may be sufficient development of a standing wave to produce a system of bars. This intriguing hypothesis of offshore bar development requires field confirmation.

BEACH FACE SLOPE

The slope of the beach face under the wave swash zone is governed by the asymmetry of the intensity of the swash and the resulting asymmetry of the onshore-offshore sand transport. Due to water percolation into the beach face and frictional drag on the swash, the return backwash tends to be weaker than the shoreward uprush. This moves sediments onshore until a slope is built up in which gravity supports the backwash and offshore sand transport. When the same amount of sediment is transported landward as is moved seaward, the beach face slope becomes constant and is in a state of dynamic equilibrium.

The slope of this equilibrium beach face will depend mainly on the amount of water lost through percolation into the beach. This rate of percolation is governed principally by the grain size of the beach sediments: water percolates much more readily into a gravel or shingle beach than into a fine-sand beach. The result is that on the shingle beach the return backwash is much reduced in strength, and therefore its slope is considerably greater than that of beaches composed of fine sand.

A foreshore slope increase with coarse particle sizes has been demonstrated by several field studies (Bascom, 1951; Wiegel, 1964; McLean and Kirk, 1969; Dubois, 1972) and wave tank experiments (Bagnold, 1940; Rector, 1954). Figure 11-8,

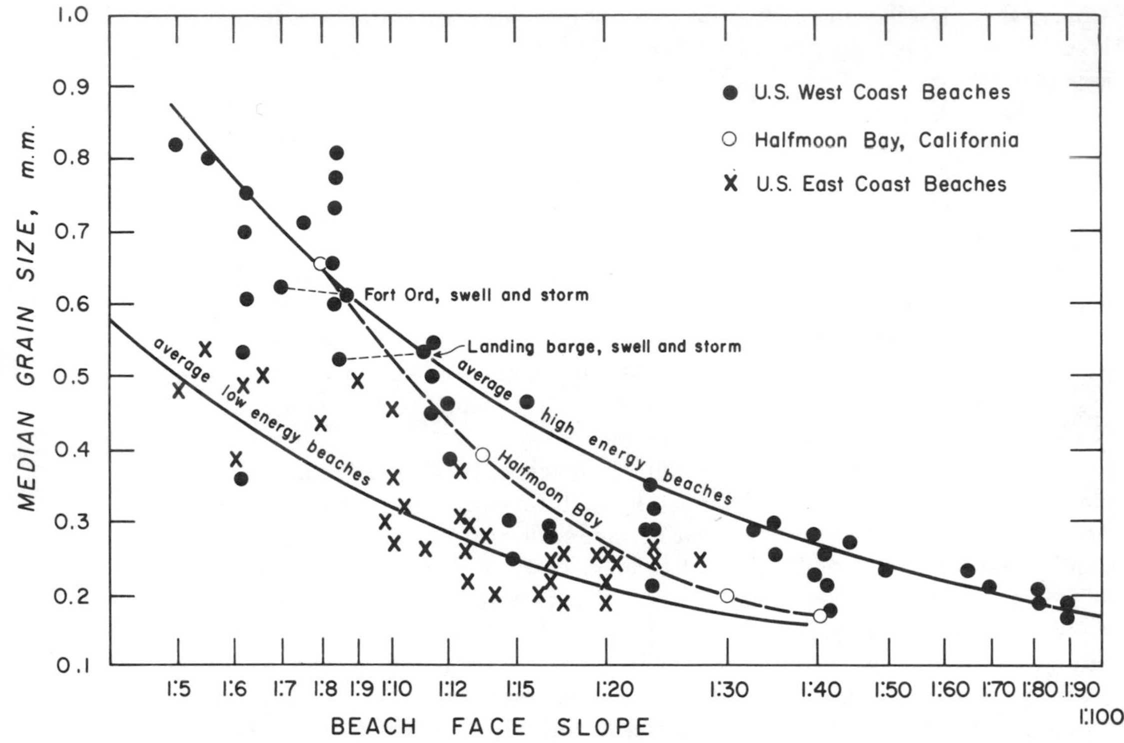

Figure 11-8 The beach face slope as a function of the median grain size of the beach sediments. Also shown is the difference between U.S. west coast and east coast beaches, which reflects the importance of the overall wave energy level on the beach slope. [*After* Bascom (1951) and Wiegel (1964)]

after Bascom (1951) and Wiegel (1964), relates the slope to the grain size at a "reference point" (the part of the beach face subject to wave action at mid-tide level), which standardizes the sampling location and thereby removes some of the variability from such a plot due to offshore variations in grain size. As expected, the coarser the sand, the steeper the beach face slope.

Bascom (1951) and Wiegel (1964) included data only for sand beaches. Figure 11-9, based on the compilation of Shepard (1963, p. 171, table 9), carries the relationship up through cobble beaches. Beach face slopes are seen to be as great as approximately 25 degrees, which is approaching the angle of repose (32 degrees), the limit at which granular material can be piled.

The rate of percolation is affected by the degree of sediment sorting as well as by the grain size itself. Sorting should therefore also have an effect on the beach slope. This was demonstrated by Krumbein and Graybill (1965, pp. 351–53), who found that well-sorted coarse-sand beaches have steeper slopes than poorly sorted coarse-sand beaches. The importance of the degree of sorting has been more thoroughly studied by McLean and Kirk (1969) on mixed sand-shingle beaches in New

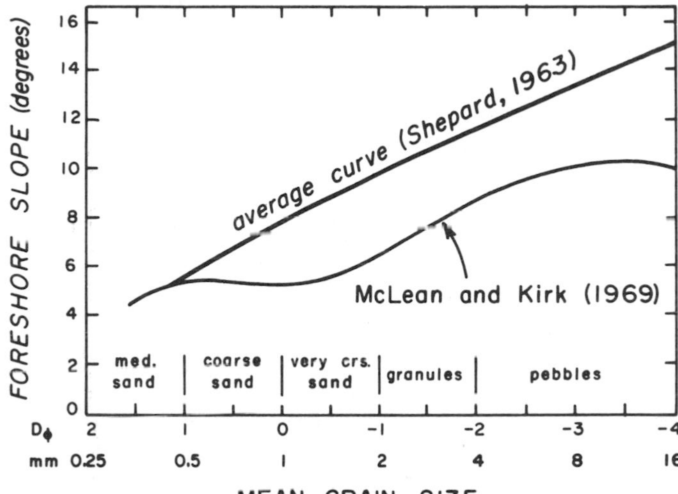

Figure 11-9 Beach face slope versus the grain size according to Shepard (1963) and McLean and Kirk (1969). The wavy nature of the curve of McLean and Kirk demonstrates the importance of the degree of sediment sorting on the beach slope. [*After* McLean and Kirk (1969)]

Zealand. Their curve is compared with the Shepard (1963) curve in Figure 11-9. Due to the overall poorer sorting, the New Zealand beaches have lower slopes than those of Shepard (1963), even for the same medium grain size. Also note that the curve of McLean and Kirk is wavy rather than being smooth. This is due to the nature of the sources of beach sediments. The beaches are made up of modes that are approximately pebbles with diameters of 4 to 16 mm, or sands of diameter 0.5 mm, or mixtures of the two. When the individual modes alone occur, the sediment sorting is good and the resulting beach slope higher; but when the modes are mixed, the sorting is poorer and the percolation and beach slope are reduced. This is why the curve of Figure 11-9 rises (higher slope) for sands of 0.5 mm and for pebbles, but lowers (reduced slope) for intermediate grain sizes.

Bascom's (1951) data is derived from high-energy exposed beaches on the Pacific coast of the United States. Wiegel (1964) added data from lower-energy beaches of New Jersey, North Carolina, and Florida. For a given grain size these low-energy beaches have larger beach face slopes than the exposed Pacific beaches. Included in Figure 11-8 is the average curve of Wiegel for such low-energy beaches. The terms *exposed* and *protected* are also commonly used to describe the large differences in the amount of wave energy reaching a beach.

Also included in Figure 11-8 is a series of points from Halfmoon Bay, California. This bay is partially sheltered by a headland (Figure 11-10) such that the energy is lowest in close proximity to the headland and progressively increases to the south as the sheltering of the headland is lost. In response to the increasing energy level, the beaches farthest from the headland are coarsest: there is a systematic increase in grain size in the longshore direction. The beach face therefore increases its slope, the steepest beaches being the exposed beaches to the south (Figure 11-10). From the trend of the Halfmoon Bay points in Figure 11-8, it is apparent that this increase in slope to the south is not entirely due to the increasing grain size, but also reflects the

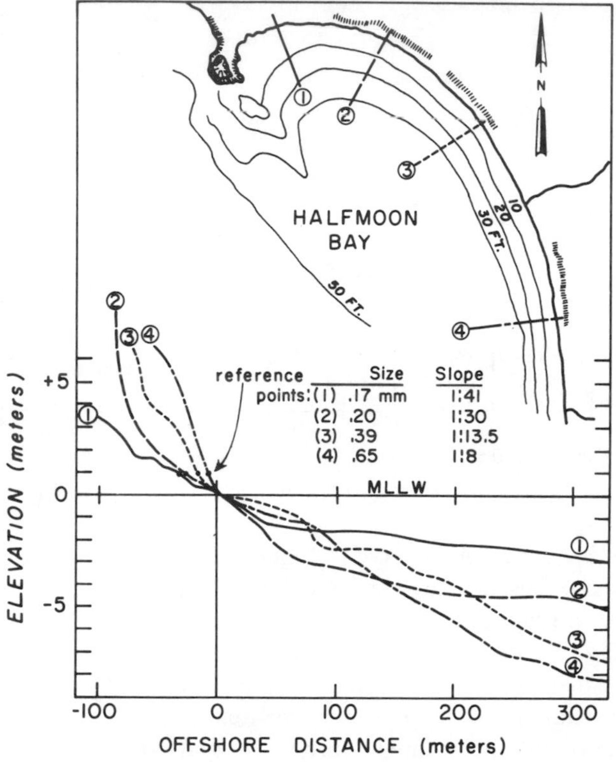

Figure 11-10 Systematic changes in the beach profile slope along the length of Halfmoon Bay, California. [*After* Bascom (1951)]

effects of the energy level itself (the degree of protection), the energy level reaching that of the "low-energy" curve of Wiegel (1964) defined by east coast beaches.

Earlier in this chapter it was seen that when the wave steepness H_∞/L_∞ exceeds a critical value a storm profile results, shifting sand to offshore bars and reducing the beach face slope in the process. Such a dependence on the wave steepness has been demonstrated by Rector (1954) in a laboratory wave channel. With a beach composed of 0.22-mm-diameter sand, he found that the foreshore slope above the still-water level is given by the equation

$$\text{Slope} = 0.30\left(\frac{H_\infty}{L_\infty}\right)^{-0.30} \tag{11-2}$$

which indicates that the greater the wave steepness, the lower the beach slope. Utilizing regression analysis techniques on field data, Harrison (1969) similarly established an inverse relationship between the beach face slope and the wave steepness. Such a dependence on the wave steepness adds scatter to the diagram of Figure 11-8. This is demonstrated by the pairs of points from Fort Ord and Landing Barge. The changes in slope at the two beaches reflect the effects of varying wave steepness—a shift from a swell to a storm profile. Bascom (1951) has used this dependence on the wave steepness to explain the variations in beach slope at Halfmoon Bay. Since the wave period remains everywhere the same along the bay, the wave steepness will be greater at the exposed beaches, where the wave heights are larger. Therefore, in the protected portion of the bay the beach slopes would be higher than the minimum probable slope (storm profile slope) given by the curve of Figure 11-8.

Although the dependence on the wave steepness can help explain differences in slope between exposed and protected beaches with the same grain size, it is not the full answer. The wave steepness may be the same for both ocean waves and waves in a laboratory wave tank or small lake, yet the beach slope will be higher in the low-energy wave tank. In addition to the effects of the wave steepness, there must also be a dependence on the actual energy level or wave height.

To determine the dependence of the beach slope on wave height, independent of wave steepness, laboratory wave tank experiments could be conducted in which wave steepness is held constant while wave height is progressively increased. This of course requires that the wave length (wave period) be raised at the same rate the height is increased if the steepness is to remain the same. Therefore, it is not possible to separate out the effects of the wave height and length (period). King (1972, pp. 325–27) has conducted such experiments and has also obtained field data for comparison. She finds that, holding H_∞/L_∞ constant, an increase in wave height and length results in a decrease in beach slope. This agrees with the differences in the high-energy (exposed) and low-energy (protected) curves of Figure 11-8. King attributed the effects to an increase in wave length, but since she had to augment simultaneously the wave height, the independent effects cannot be assessed. An increase in wave length (period) might be expected to decrease the rate of percolation, since short-period waves would tend to saturate the beach face. This suggests that an increase in wave length (period) would raise the beach slope. However, an enlargement in wave height might be expected to have the opposite effect, enhancing the quantity of backwash and thereby decreasing the beach slope. My interpretation of King's (1972) results is that the decrease in beach slope is brought about by increasing wave heights, the effects of the latter being greater than the incremented wave period. This obviously requires further study.

Using data from twenty-seven beaches in widely varying conditions, King (1972, pp. 330–31) obtained the empirical equation

$$\text{Slope} = 407.71 + 4.20D_\phi - 0.71 \log(E) \tag{11-3}$$

for the beach face slope, where D_ϕ is the sand size in ϕ-units $[D_\phi = -\log_2(D_{mm})]$ and E is the wave energy. This equation accounted for 71.8% of the variability of the beach slope. Besides showing the positive dependence on the grain size, the equation indicates that an increase in the wave energy (height) will result in a decrease in slope. This again agrees with Figure 11-8, showing the importance of the energy level of the beach, and supports our conclusions of the preceding paragraph. Harrison (1969) similarly demonstrated a positive correlation between the slope and grain size and a negative correlation with the energy flux (wave power). In addition, he found a strong correlation with the level of the water table within the beach. The importance of the water table level as compared with the tide level has already been discussed.

Because of the water table level, the beach face tends to have a greater slope during a rising tide level than during a falling tide.

It is apparent that there are several variables which affect the slope of the beach face. The most important is the grain size of the beach sediment, coarser beaches having steeper beach slopes. Added to this is the effect of the wave energy level: for a given grain size, higher energies (wave heights) produce lower beach slopes. Other factors are the wave steepness (storm versus swell profile), the degree of sediment sorting, the level of the water table in the beach, and the stage of the tide. With all these semi-independent factors acting together, a considerable variation in the beach face slope is produced which is generally difficult to sort out and understand. Because of this, quantitative predictions of beach slopes are still remote.

BERM FORMATION

The berm is the nearly horizontal portion of the exposed beach (Figure 2-1), the parcel most familiar to the sunbather. It is formed by sediment brought ashore during swell wave conditions; the berm is a depositional feature. The presence of a berm is not always apparent on a fine-sand beach, since the beach is already "nearly horizontal" everywhere. Distinct berms with sharp berm crests are best developed on medium- to coarse-sand beaches which have a steeper beach face that is more easily discernible from the horizontal berm. Coarse shingle beaches generally do not have as extensive berm development.

In his laboratory wave tank studies, Bagnold (1940) demonstrated that the berm elevation coincides with the wave surge height, the height of the wave runup on the sloping beach face. Bagnold (p. 38) found that

$$\text{Berm elevation} = bH \tag{11-4}$$

where

$$b = \begin{cases} 1.68 \text{ for } D = 0.7 \text{ cm} \\ 1.78 \text{ for } D = 0.3 \text{ cm} \\ 1.8 \text{ for } D = 0.05 \text{ cm} \end{cases}$$

That is, the swash runup and berm elevation is directly proportional to the wave height H. The proportionality factor b is seen to be dependent on the grain size of the beach sediment, being smaller for the coarser sizes, since percolation is greater. Such a correspondence between wave height and berm elevation has been noted by Bascom (1953) on ocean beaches.

Bascom (1954), who made his observations on United States Pacific coast beaches, contains the best description of berm development. After a wave breaks, the water rushes forward up the beach face carrying sand with it, losing velocity as it goes because it is opposed by gravity and friction and because of water loss through percolation. Sand is carried upward as far as the swash extends and is deposited to that extent. As the beach face builds seaward it leaves a nearly horizontal berm, which demonstrates the elevation to which sand was carried by the swash. If the waves were uniformly the same height, a rapidly growing berm would show undulations reflecting the tide level. Strahler (1966), for example, showed that the berm elevation is raised in response to the high waters of the spring tides.

Bascom (1954) noted that the upward growth of the berm depends mainly on the largest waves that are reaching the beach, since their swash passes completely over the crest and deposits the bulk of their sand load atop the berm. The upward growth is greatest at highest tides, and the combination of tides with large waves sometimes produces a shoreward-sloping berm if the crest grows higher as it builds seaward. Figure 11-3 shows an example of this. The swash runs landward down this inclined berm and puddles in a "lagoon." At times enough water gathers to be able to break the berm crest at a low point and flow back out to sea.

High storm waves cut back the berm and tend to destroy it. Bascom (1954, p. 171) has pointed out the interesting paradox that the storm waves also build up a berm of greater elevation due to their greater wave heights. The uprush of the large storm waves adds sand to the top of the berm, even while the beach face is being eroded and the berm extent reduced. A storm may leave a high, narrow berm which can survive until subaerial erosion or a larger storm removes it. A prograding beach may have a series of abandoned storm berms on its landward side.

ONSHORE-OFFSHORE SEDIMENT TRANSPORT

Changes from swell to storm profile and any other variations in the beach profile configuration involve principally onshore-offshore shifts of sediment. Therefore, the profile changes cannot be understood until we comprehend the details of this transport normal to the shoreline. This of course depends in turn on a knowledge of the nearshore currents—wind- and wave-induced unidirectional currents—as well as the pattern of the asymmetrical orbital motions of shallow water waves, including breakers and surf. Such a complete comprehension as is required is still very remote, so that only the broadest understanding is presently possible.

In Chapter 3 it was seen that Stokes waves in shallow water give a forward orbital motion under the wave crests that is short in duration but high in velocity, while the return flow under the troughs is slower but of longer duration. Measurements confirm this pattern. This is due to the steepening of waves in shallow water, isolated crests tending to be separated by wide troughs. Cornish (1898) pointed out that the shoreward component of such waves is much more effective in moving coarser debris, since a shorter-lived current of high velocity will transport material which is too large to be moved at all by the longer-enduring but weaker seaward current. Sand and silt, on the other hand, will readily move a nearly equal distance in both directions. In this way the waves may selectively drive pebbles and cobbles toward the beach.

Johnson (1919, p. 93) summarizes some dramtic examples of this onshore transport of coarser debris. Murray (1853) demonstrated that shingle and chalk ballast dropped into the sea off Sunderland, England, at a distance of 10 to 15 km from land, where the water depth is from 20 to 40 meters, were thrown onshore by storm waves. Gaillard (1904) quotes R. Robinson as authority for pig lead washing up at Madras, India, during a violent storm, the lead having come from a vessel wrecked more than 2 kilometers offshore.

Bagnold (1940, pp. 32–33) found in his wave tank studies that the largest particles are moved only during the most violent part of the orbital motion and progressively creep shoreward, the bigger the particle the more pronounced the creep. The biggest particles therefore collected highest up on the beach.

P. Cornaglia proposed in 1898 an expansion of this selective transport due to stronger onshore orbital motions. Cornaglia's thesis is outlined in a speech by Munch-Peterson (1938). In summary, the approach recognizes the fundamental importance of (1) the influence of waves in trying to move the grains shoreward, and (2) gravity trying to pull the particles offshore in the downslope direction. The pattern of wave orbital motion effects on sediment grains is similar to that described by Cornish (1898), already mentioned. Cornaglia recognized the existence of a minimum water particle velocity u_t necessary to move a given grain size, and established that, depending on the equality or inequality of the areas *b-c-d* and *f-g-h* in Figure 11-11, the particles will either be in equilibrium or tend to have a net motion onshore. As the water depth increases, the difference between trough and crest velocities becomes less until there is not enough difference to move the particle a net distance either onshore or offshore. According to Cornaglia, the influence of gravity becomes increasingly important in tending to pull sediment particles downslope (seaward) at these depths. At some unique depth all of these forces balance and the particle simply moves back and forth. This position of dynamic equilibrium is the so-called *null point*, the location where

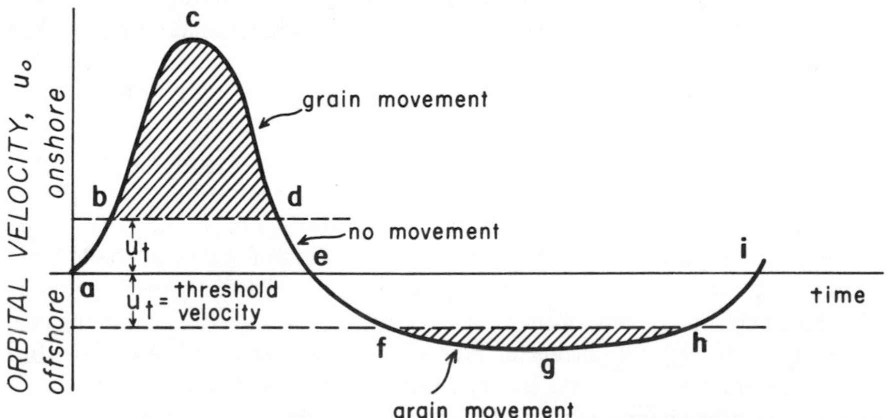

Figure 11-11 The asymmetry of the wave orbital motions with a strong onshore velocity versus a weaker offshore velocity but with a longer duration, causing a net shoreward transport of the coarser sediments. At the null point this onshore transport is offset and balanced by the gravity component acting on the grains, tending to move them offshore.

there is zero net movement for the particular grain size along a line normal to the wave direction. Cornaglia recognized that the null point will vary in position for each grain size on the beach and for different wave conditions. The coarser the particle, the shallower its null-point position. For a given grain size, seaward of its null-point position its movement is offshore, and landward of the point the movement is onshore. If many different sizes are present at a certain location, only one of the grain sizes can be at its null point: the larger grains will move offshore according to the hypothesis, and finer ones will move shoreward.

Ippen and Eagleson (1955), Eagleson, Dean, and Peralta (1958), and Eagleson and Dean (1961) have placed the null-point hypothesis of Cornaglia into mathematical terms and conducted experiments in the laboratory wave tank which at least in part appear to verify the theory. Ippen and Eagleson (1955) used a fixed slope of 1: 15, artificially roughened by sand evenly glued to its surface. Single sediment particles were then introduced and their movements under the waves followed. Figure 11-12 shows typical results. Some grains moved onshore at all depths less than the depth at which threshold was achieved. Other grain sizes and densities demonstrated the existence of null points, with seaward transport offshore and foreward transport shoreward of the null-point position. In the vicinity of the breakers the onshore motion is opposed by the backwash down the slope, so that all grains reach a second point of equilibrium oscillation near the breaker zone. The offshore null-point data of Ippen and Eagleson (1955) is summarized in Figure 11-13, the straight line yielding the relationship

$$\left(\frac{H}{h}\right)^2 \left(\frac{L}{H}\right) \left(\frac{C}{w}\right) = 11.6 \tag{11-5}$$

for the null-point equilibrium, where H, L, and C are respectively the local values of the wave height, length, and phase velocity, and h is the local water depth; w is

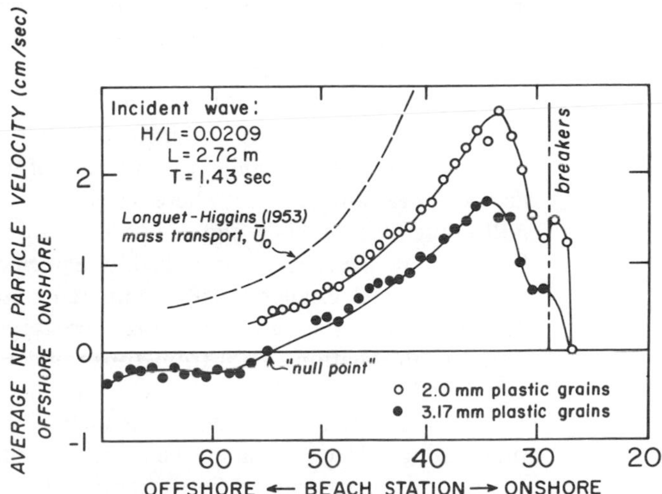

Figure 11-12 Typical results of Ippen and Eagleson (1955) for the onshore-offshore movement of individual grains on a sloping beach. The null point is the position of equilibrium where the grains oscillate under the waves but do not move with a net drift either onshore or offshore. [*After* Ippen and Eagleson (1955)]

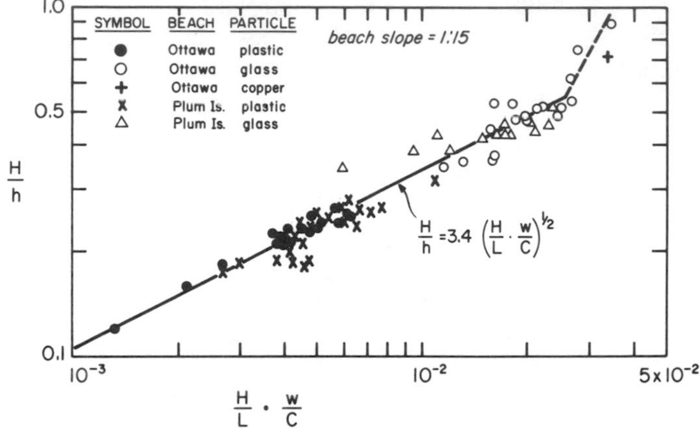

Figure 11-13 Summary of the null-point data for a beach slope of 1: 15. The straight line and equation indicate the null-point depth h for grains of settling velocity w under the wave conditions defined by values for H, L, and C. [*After* Ippen and Eagleson (1955)]

the settling velocity of the grains whose null point exists at that depth h under those wave conditions. The empirical coefficient 11.6 is valid only for the 1: 15 slope tested.

Eagleson and Dean (1961) extended the theoretical considerations and laboratory measurements to obtain

$$D_e^{6/7} \sin \beta = (7.60 \times 10^{-6})\bar{U}_0\left(\frac{\pi}{\nu T}\right)^{4/7} \tag{11-6}$$

for the relationship between the local beach slope β and the null-point grain diameter D_e, where $\bar{U}_0$ is the near-bottom mass transport of water given by Longuet-Higgins (1953) as

$$\bar{U}_0 = \frac{5}{4}\left(\frac{\pi H}{L}\right)^2 C \frac{1}{[\sinh(2\pi h/L)]^2} \tag{11-7}$$

(see Chapter 3, Figure 3-10), where ν is the viscosity of water. Since $\bar{U}_0$ increases onshore as the water shoals, equation (11-6) requires that the bottom slope steepen, or that the grain size increase, or both. In this respect the hypothesis agrees with general observations from nature (Chapter 13).

Setting $\sin \beta = -dh/dx$ in equation (11-6), Eagleson, Glenne, and Dracup (1963) integrated the relationship in the shoreward direction, holding D_e constant, to obtain a theoretical profile for an "equilibrium beach" of uniform grain size where the local beach slope everywhere satisfies the condition of equilibrium oscillation (null point). Figure 11-14 shows one laboratory profile compared with the resulting theoretical profile. The apparent lack of agreement is not promising.

Miller and Zeigler (1958) compared observed onshore variations in median grain size with a theoretical onshore increase based on the null-point hypothesis as expressed in equation (11-5) from Ippen and Eagleson (1955). Although both theoretical and observed distributions agreed insofar as there should be an increase in grain size and an improvement in sorting in the onshore direction as the breaker zone is approached, this cannot be taken as confirmation of the null-point hypothesis, as other mechanisms could be responsible for the sorting. Miller and Zeigler (1964)

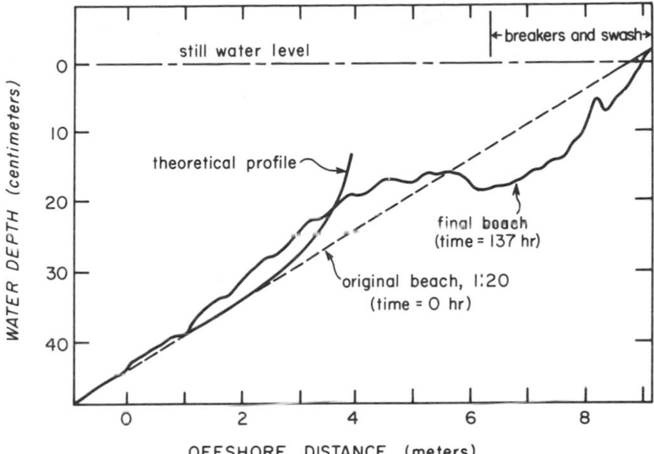

Figure 11-14 The theoretical beach profile determined by the development of Eagleson, Glenne, and Dracup (1963), based on the null-point results of Ippen and Eagleson (1955), compared to an actual profile from a laboratory beach. [*After* Eagleson, Glenne, and Dracup (1963)]

further examined the predictability of sediment patterns, this time in a region of complex nearshore topography including bars attached obliquely to the shoreline. They found that the smooth pattern for median grain diameter showed no indication of the presence of the bars, there being a general increase in grain size onshore. In spite of this, it was concluded that there was little agreement with the theoretical distribution based on the null-point hypothesis. This discrepency appeared to result from (1) the presence of waves arriving obliquely to the shoreline, and (2) mass transport and other wave-induced currents. The studies of Miller and Zeigler (1958, 1964) are considered further in Chapter 13.

According to the null-point hypothysis and the results of Ippen and Eagleson (1955) and Eagleson and Dean (1961), if a sample of sediment with a wide grain-size distribution is placed on the bottom outside the breaker zone, only one diameter will be at its null-point position. In addition, material coarser than this equilibrium diameter should move offshore, while the finer sizes move along the bottom in an onshore direction and collect in the breaker zone. This theoretical prediction conflicts with the results of Zenkovitch (1946) utilizing sand tracers. Zenkovitch tagged different grain sizes with different colors and released the mixture in the offshore zone. He found that the coarsest particles moved onshore and the fine ones offshore, while those with a grain size close to the in situ grains oscillated back and forth in a sort of null-point position of oscillation equilibrium. Although the results of the experiment by Zenkovitch do point to the existence of a null point, they disagree with the theoretical development of Cornaglia, Ippen and Eagleson (1955), and Eagleson and Dean (1961) regarding the onshore-offshore transport of the grains that are out of their equilibrium null-point positions. Their theoretical results also conflict with the previous observations (page 310) that cobbles, shingle, and pig iron dropped into deep water migrate shoreward onto the beach, not seaward as the hypothesis predicts. In discussing Cornaglia's hypothesis, Munch-Peterson (1950) held that it was not valid because the influence of gravity relative to wave forces is too small to be important, no matter what the depth or bottom slope. The laboratory results of Ippen and

Eagleson (1955) and Eagleson and Dean (1961), however, do indicate that gravity may be important, although the results are not conclusive. More probable, mechanisms other than those considered by Cornaglia and by Ippen and Eagleson (1955) dominate in the onshore-offshore transport. They assumed that the grains remain in close proximity to the bottom (bed-load transport predominates), so that the elevation of the moving grains is independent of time. The experiments of Inman and Bowen (1963) and others, which will be discussed in a moment, show that the presence of ripple marks on the sand are important in throwing sand above the bottom and governing the direction of transport. Under such conditions, coarse grains could migrate shoreward while finer sediment drifts offshore. The experiments of Ippen and Eagleson (1955) and Eagleson and Dean (1961) with single-particle motions on a flat, roughened slope probably do not adequately depict the natural movements of grains over a loose sediment bed in the presence of ripple marks.

Bagnold (1963, p. 518) developed a model of sediment transport produced by a coupling of wave action with superimposed linear currents. According to the model, the stress exerted by the wave motion supports and suspends sediment above the bottom, but without causing a net transport, since the wave orbits are closed. Superimposed upon this to-and-fro motion is any unidirectional current which produces a net transport of the sediment, the direction of transport being the same as the current. Since the waves have already supplied the power to put the sand into motion, the unidirectional current can cause a net transport no matter how weak the flow, even if the current taken alone is below the threshold of sediment transport. According to this model, i_θ, the immersed-weight sediment transport rate per unit bed width, is given by the relationship (equation 8-7)

$$i_\theta = K'\omega\frac{\bar{u}_\theta}{u_o} \tag{11-8}$$

where ω is the power the waves expend in placing the sediment in motion near the bottom, and u_o is the horizontal component of the orbital velocity at the bottom such that ω/u_o becomes the stress exerted by the waves; $\bar{u}_\theta$ is the superimposed unidirectional current flowing near the bed which produces the net drift of sediment. The K' is a dimensionless coefficient of proportionality.

As was discussed in Chapter 8, this model has been applied to sand transport on beaches where $\bar{u}_\theta$ became the longshore current, and it was found that the relationship could successfully predict longshore sand drift. This indicates that the model is basically sound. It is apparent that the model may also be applicable to sediment transport outside the breaker zone, where in this case $\bar{u}_\theta$ becomes unidirectional currents such as those driven by winds, tidal currents, normal ocean thermohaline currents, as well as mass transports induced by the waves themselves. Inman and Bowen (1963) made a wave tank study of the applicability of equation (11-8) to this problem and found that matters are complicated in deep water when a current flows over a rippled sediment bottom. They made measurements of the sand transport caused by combined waves and currents traveling over a horizontal sand bed in water 50 cm deep. The waves had heights of 15 cm and periods of 1.4 or 2.0 seconds.

The superimposed currents in all cases flowed in the direction of wave travel (onshore) with steady uniform velocities of 2, 4, and 6 cm/sec. With low superimposed currents the behavior was as expected: the sand transport along the tank increased with increasing velocity. However, at the higher velocities Inman and Bowen obtained the peculiar results in which the sediment transport actually decreased with an increase in current; in one case the sand even moved upchannel, opposite to the direction of current and wave motion. This resulted from the complex periodic movement of the sand above the ripples. At low superimposed current velocities the wave-induced ripple marks remained symmetrical, with eddies forming between the ripples during the onshore and offshore semiorbital motions (Figure 11-15). At times of reversal in the direction of orbital motion the eddies were thrown upward off the bottom, the sand being placed in suspension and therefore most susceptible to the current and a net transport. The sand was thrown somewhat higher off the bottom just

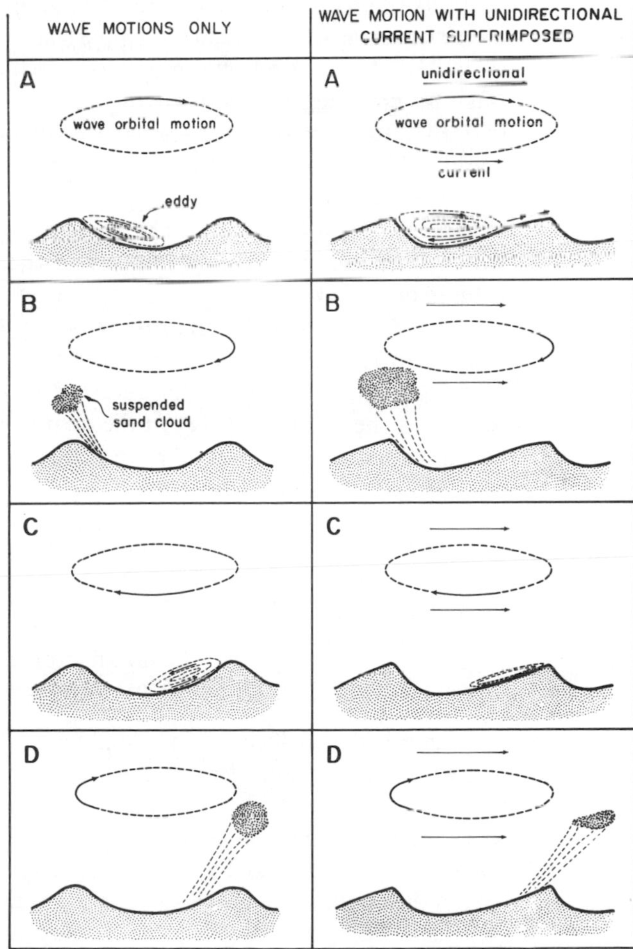

Figure 11-15 The relationship between the sand motions near a rippled sea bed and the orbital motions of the wave action without and with a superimposed unidirectional current. [*Based upon* Inman and Bowen (1963)]

preceding the crest passage, probably because the near-bottom water motions are vertical at that stage. The effect of an increase in the superimposed current was to reduce the symmetry of the ripples by increasing the effective onshore orbital velocity and decreasing the offshore velocity. The resulting asymmetrical ripples had flattened upwave faces and steepened onshore faces. The vortex generated on the downwave face during onshore semiorbits (Figure 11-15) became by far the stronger, and when thrown upward from the bottom was directed in the offshore direction, accounting for the reduction in transport and the one case of an offshore net transport. Such a reversal would be only temporary; with still stronger superimposed onshore currents the wave motion would have become relatively less important and the net transport would again have been strongly in the direction of current and wave travel. Even with just this pilot study of Inman and Bowen (1963) it is apparent that any solution for onshore-offshore transport of sediments under combined wave and current systems is going to be very complex. Only with additional studies of this nature, perhaps with currents opposed to the wave travel direction, and where the wave orbital diameters are several times larger than the ripple spacings such as is the case in ocean conditions (Chapter 13), will we begin to understand the problem. It is conceivable that with an offshore current superimposed on wave motion, fine sediments which are thrown entirely into suspension will be carried offshore, while at the same time coarser sediments which remain near the bottom are carried onshore (Komar, Neudeck, and Kulm, 1972).

Inman and Bagnold (1963) developed a model for the equilibrium beach slope which is governed by the asymmetry of the energy dissipation of the onshore wave orbital motion versus the offshore motion. Because of frictional drag on the wave swash and water percolation into the beach, more sand tends to be transported shoreward than seaward. However, opposing this shoreward movement is the local beach slope such that gravity aids the return flow in moving sand offshore. At equilibrium there must be a balance between the quantity of sand that is carried up and down the beach under the wave action. For this equilibrium, Inman and Bagnold (1963) obtained the relationship

$$\tan \beta = \tan \phi \left(\frac{1 - c}{1 + c}\right) \tag{11-9}$$

for the local beach slope $\tan \beta$, where

$$c = \frac{\text{Local offshore energy dissipation}}{\text{Local onshore energy dissipation}}$$

and $\tan \phi$ is the coefficient of internal friction in the shearing of a granular media and is approximately equivalent to the angle of repose of the sediment (Bagnold, 1954). In deriving this relationship, Inman and Bagnold (1963) applied the energy approach to sediment transport as outlined by Bagnold (1963), the quantity of sediment transported being governed by the local energy dissipation rate (the power). For this reason the equilibrium beach slope is a function of c, the ratio of the local offshore and onshore energy dissipation rates. Inman and Frautschy (1966) further indicated that

$$c \simeq \left(\frac{u_{m\text{-offshore}}}{u_{m\text{-onshore}}}\right)^3 \tag{11-10}$$

the cube of the ratio of the maximum offshore component of orbital velocity to the maximum onshore component.

If there is no difference in the levels of onshore and offshore energy dissipation (or orbital velocities), then $c = 1$ and, according to equation (11-9), $\tan\beta = 0$: the bed will be horizontal. This is because no bottom slope is required to oppose and balance an onshore transport of sand due to the asymmetry of onshore-offshore energy dissipation. At the other extreme, if the asymmetry is large due to pronounced friction and percolation effects, then $c \to 0$ and $\tan\beta \to \tan\phi$: the beach slope will approach the angle of repose for the sediment material. Under these conditions a large equilibrium slope is required to balance the asymmetry of the onshore-offshore dissipation.

The loss in energy between onshore and return flow, which gives rise to the asymmetry of energy dissipation and governs the beach slope, can be due to bed friction or to percolation. *Bed friction*, the drag of the sediment boundary on the water flow, will be greatest for the larger grain sizes. *Percolation* can occur both on the exposed beach face, where it saps the strength of the swash, and underwater, where it occurs horizontally through the bed, associated with pressure gradients between wave crest and trough position (Putnam, 1949); of the two, percolation on the exposed beach face is of course much larger. Like bed friction, percolation effects will be greatest for the coarser grain sizes. The percolation rate increases very rapidly as the grain size is increased from sand to pebbles. Thus while sands are relatively impermeable, beaches composed of pebbles are permeable and cause considerable energy losses. Loss of swash energy approaches a maximum at the upper limit of wave swash. For this reason the value of c in equation (11-9) decreases progressively toward the beach crest, where it may approach zero, there being little or no return flow. Hence, on both model and natural pebble beaches, the beach slope $\tan\beta$ increases progressively upward to the beach crest, at times approaching the angle of repose. Bagnold (1940) has shown that this increase in beach slope toward the beach crest can be prevented by the insertion of an impermeable plate just below the beach surface. Similarly, Longuet-Higgins and Parkin (1962) reduced the slope of a shingle beach by placing an impermeable layer below the beach face.

This model for the equilibrium beach profile, developed by Inman and Bagnold (1963) utilizing the sediment transport principals of Bagnold (1963), appears to be rational and to agree with the overall character of profiles for sand or pebble beaches. However, it remains to be carefully tested; measured values of c must be compared with the resulting beach slope for both laboratory and natural beaches.

King and Williams (1949) obtained wave tank measurements of onshore-offshore sand transport both seaward and shoreward of the breaker zone. Observations were made on three different beach gradients—1: 12, 1: 15, and 1: 20—and under a variety of wave conditions. Seaward of the breaker zone the transport was in all cases found to be onshore, increasing with decreasing water depth, increasing wave height, and increasing wave period. As expected, for a given set of wave conditions, the steeper the slope the smaller the resulting onshore transport. The general onshore transport probably resulted from near-bottom onshore wave-induced currents such as that of Longuet-Higgins (1953), given by equation (11-7). This raised the question of why beaches do not continuously grow seaward if the transport is always toward the beach. King and Williams (1949) concluded that prevailing onshore winds would drive surface water shoreward which would return offshore along the bottom, generating a near-bottom offshore drift of water that would counteract this landward sand movement from the wave action. Repeating the experiments with onshore winds of 13 m/sec [29 mph], they found that the landward movement due to wave action alone is converted into a slight seaward transport.

Inside the breaker zone, King and Williams (1949) found, the direction of transport depends on the wave steepness, the wave-height-to-length ratio. With steep waves the sand moved offshore, while with flat waves the transport was onshore, the critical steepness for the change in direction being approximately 0.012. This of course conforms with the shift from a storm profile to a swell profile. With steep waves, the resulting offshore transport in the surf zone converges with the onshore transport seaward of the breakers; sand therefore deposits in the area of the breakers to form a bar. With the superimposed onshore wind, it was found that even with waves flatter than the critical steepness, the transport in the surf zone, like that seaward of the breakers, tends to be in the offshore direction due to the wind-induced currents.

There have been a number of field studies utilizing sand tracers, which give some indication of patterns of onshore-offshore sediment transport. These will be briefly reviewed. The results of such studies must be interpreted with caution, however, as demonstrated by Price (1969). Using a simple arithmetic model, Price showed that tracer will tend to move toward an area of maximum diffusion rate even if there is no net transport of sand. For example, consider 1,000 particles of tracer introduced at a point outside the breaker zone, as shown in Figure 11-16. The number of particles exchanged between successive ripples increases in the onshore direction, being 50 just shoreward of the tracer and 40 just seaward. With each successive wave stroke more tracer is moved onshore than offshore, even though there is no net transport, since the tracer particles are replaced by untagged grains. After several wave passages it is seen that the "centroid" of the tracer distribution has shifted onshore, even though there has been no net transport of sand in that direction. The results would therefore give a misleading indication of an onshore sand transport. Similarly, in the surf zone there would be an erroneous indication of an offshore transport toward the breaker zone. Although the model of Price is oversimplified, having been introduced only for instructional purposes, it is correct in principle and demonstrates the care that must be taken in the interpretation of tracer distributions.

One probable example of this danger is to be found in some of the results of Ingle (1966), who conducted extensive studies with fluorescent tracers throughout the nearshore region. Ingle (pp. 46–55) found that the tracer dispersion pattern con-

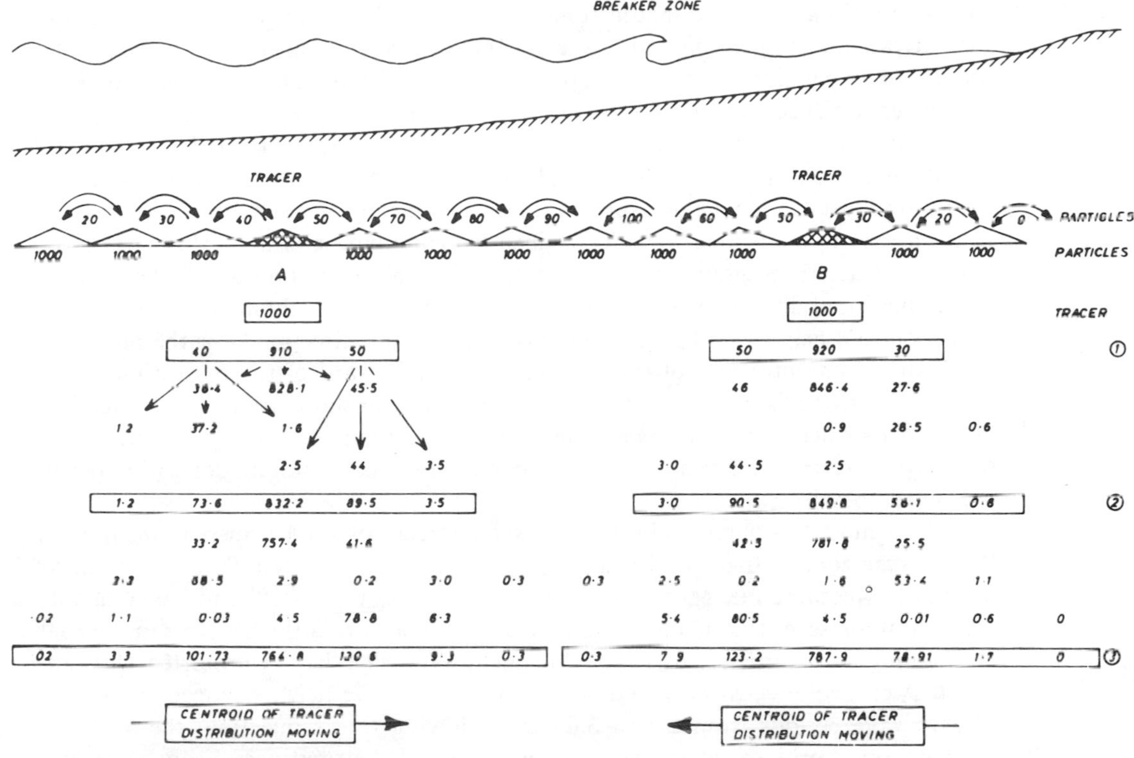

Figure 11-16 Conceptual arithmetic model of Price (1969) on the diffusion of tracer, giving a mistaken impression of a net sediment transport (advection) toward the breaker zone, both within the surf zone and beyond the breakers. [*From* Price (1969)]

sistently indicated a net motion toward the breaker zone of sand both immediately seaward and shoreward of the breakers. This consistent indication of an offshore transport in the surf zone conflicted with the monthly beach profiles, which demonstrated alternating periods of accretion and erosion. This incongruity can be explained by the diffusion model process of Price (1969).

Ingle (1966, pp. 93–100) studied selective onshore-offshore sorting of sand in the surf zone by coating grains finer than 0.25 mm one fluorescent color and coarser grains a second color. Reintroducing these into the surf zone on two occasions, he found that the coarser grains moved diagonally offshore to the breaker zone and then alongshore within the breakers. Grains finer than 0.25 mm moved primarily alongshore under the action of longshore currents, remaining within the surf zone. That the grains were seeking some equilibrium position was further demonstrated by the grain-size distributions of sand across the nearshore: grains coarser than 0.25 mm were essentially restricted to the breaker zone. The tracer grains in the fraction coarser than 0.25 mm moved offshore to their position of equilibrium beneath the breakers, while the finer sizes remained in the mid–surf zone, where the grain size was suitable to that position. It may be that in some cases the offshore shift of sand tracer out of the surf to the breaker zone, observed by Ingle, resulted from the tracer being of a grain size in equilibrium with the breaker zone conditions and not due entirely to the dispersion effects of Price (1969).

Inman and Chamberlain (1959) used natural quartz sand with artificially induced radioactivity as a tracer of offshore sediment movement. The tracer was placed on the bottom just outside the breaker zone in about 3 meters of water, and its subsequent distributions through time were determined by sampling on a grid. The distribution patterns of irradiated grains showed an extremely rapid rate of diffusion, the particles dispersing in all directions and forming an elliptical pattern about the injection site, a pronounced major axis being in the onshore-offshore direction. For the most part the highest concentrations of tracer were found shoreward of the release point, the entire pattern shifting shoreward, indicating a net onshore transport (advection) of sand. This net transport was also indicated by simultaneous measurements of the sand level: erosion occurred in the vicinity of the release point, and deposition occurred on the beach.

The most thorough study utilizing sand tracers of sand transport offshore of the breaker zone is that of Murray (1967). To determine how different sizes move normal to the shoreline, Murray coated the grain sizes 0.25, 0.50, and 1.00 mm with different fluorescent tracer colors and placed them seaward of the breakers in the zone of shoaling waves in Buzzards Bay, Massachusetts. This bay is protected from the ocean swell and affected only by local wind waves. Wave heights ranged from 20 to 36 cm, wave periods from 2.4 to 3.6 sec. Following tracer injection, the area was sampled on a grid to determine subsequent tracer distributions. From these distributions it was found that

1. in all but two of the twenty-one cases the net transport was onshore, the two exceptions being with the 0.25-mm sand size;
2. the coarser the grain size, the higher the rate of onshore transport;
3. the stronger the near-bottom orbital motions under the waves (calculated), the lesser the tendency for onshore movement of all grain sizes; and
4. the finest grain size, 0.25 mm, tends to diffuse outward, forming nearly circular concentration contours, whereas the coarser sizes produce more elongated elliptical patterns with the major axis normal to the shoreline.

The increased onshore transport with increasing grain size agrees with the wave tank observations of Bagnold (1940) with pebbles, and also indicates that the tracer displacements are not due entirely to the effects described by Price (1969). Murray (1967) measured neither near-bottom currents nor winds (although it was indicated that the winds always blew onshore), so that it is not possible to compare their influence on the sediment movement. Murray hypothesized a current distribution like that deduced by Longuet-Higgins (1953) for a wave tank, with an onshore current near the bottom and an offshore current at mid–water depths, some distance above the bottom. The finer-grained 0.25-mm sand is thought to be thrown higher off the bottom, where it becomes more influenced by the offshore current and so tends to be carried offshore (or less onshore). Coarser grains tend to remain near the bottom as bed load, where they are acted upon by the onshore bottom water drift of Longuet-Higgins (1953) and move shoreward. Murray (1967) discounted the importance of wind-induced bottom currents, which would oppose the shoreward mass transport current associated with the waves. However, a field investigation on the Baltic coast by Seibold (1963), utilizing sand tracer grains, produced evidence that an onshore movement of sand took place during strong offshore winds. Thus it would appear that winds can be significant in generating currents that are important in the onshore-offshore transport of sediments.

REFERENCES

BAGNOLD, R. A. (1940). Beach formation by waves: some model experiments in a wave tank. *J. Inst. Civ. Eng.*, 15: 27–52.

——— (1963). Mechanism of marine sedimentation. In *The sea*, ed. M. N. Hill, 3: 507–28. Interscience, New York.

BASCOM, W. H. (1951). The relationship between sand size and beach face slope. *Trans. Am. Geophys. Union*, 32: 866–74.

——— (1954). Characteristics of natural beaches. *Proc. 4th Conf. on Coast. Eng.*, pp. 163–80.

CARTER, T. G., P. L.-F. LIU, and C. C. MEI (1973). Mass transport by waves and offshore sand bedforms. *J. Waterways, Harbors, and Coast. Eng. Div., ASCE*, WW2, pp. 165–84.

COOK, D. O., and D. S. GORSLINE (1972). Field observations of sand transport by shoaling waves. *Marine Geology*, 13: 31–55.

CORNISH, V. (1898). On sea beaches and sand banks. *Geog. J.*, 11: 528–59, 628–47.

DAVIS, R. A., and W. T. FOX (1972). Coastal processes and nearshore sand bars. *J. Sediment. Petrol.*, 42: 401–12.

DAVIS, R. A., W. T. FOX, M. O. HAYES, and J. C. BOOTHROYD (1972). Comparison of ridge and runnel systems in tidal and non-tidal environments. *J. Sediment. Petrol.*, 42: 413–21.

DEAN, R. G. (1973). Heuristic models of sand transport in the surf zone. *Conf. on Eng. Dyn. in the Surf Zone*, Sidney, Australia, 7 pp.

DUBOIS, R. N. (1972). Inverse relation between foreshore slope and mean grain size as a function of the heavy mineral content. *Geol. Soc. Am. Bull.*, 83: 871–76.

DUNCAN, J. R. (1964). The effects of water table and tidal cycle on swash-backwash sediment distribution and beach profile development. *Mar. Geol.*, 2: 186–97.

EAGLESON, P. S., R. G. DEAN, and L. A. PERALTA (1958). *The mechanics of the motion of discrete spherical bottom sediment particles due to shoaling waves.* U.S. Army Corps of Engrs., Beach Erosion Board Tech. Memo. no. 104, 41 pp.

EAGLESON, P. S., and R. G. DEAN (1961). Wave-induced motion of bottom sediment particles. *Trans. Am. Soc. Civ. Eng.*, 126, part 1: 1162–89.

EAGLESON, P.S., B. GLENNE, and J. A. DRACUP (1963). Equilibrium characteristics of sand beaches. *J. Hydraul. Div. ASCE*, 89, no. HY1: 35–57.

EMERY, K. O., and J. F. FOSTER (1948). Water tables in marine beaches. *J. Mar. Res.*, 7, no. 3: 644–54.

EVANS, O. F. (1940). The low and ball of the east shore of Lake Michigan. *J. Geol.*, 48: 476–511.

GAILLARD, D. D. (1904). *Wave action in relation to engineering structures.* U.S. Army Corps of Engineers Prof. Paper no. 31.

GORSLINE, D. S. (1966). Dynamic characteristics of west Florida Gulf Coast beaches. *Mar. Geol.*, 4: 187–206.

GRANT, U. S. (1948). Influence of the water table on beach aggradation and degradation. *J. Mar. Res.*, 7: 655–60.

GRESSWELL, R. K. (1937). The geomorphology of the south-west Lincolnshire coastline. *Geog. J.*, 90: 335–49.

——— (1953). *Sandy shores of South Lincolnshire.* Univ. of Liverpool Press, Liverpool, 194 pp.

HAGEN, G. (1863). *Handbuch der Wasserbaukunst.* 3 Teil-Das Meer, Berlin.

HARRISON, W. (1969). Empirical equations for foreshore changes over a tidal cycle. *Mar. Geol.*, 7: 529–51.

HARRISON, W., C. S. FANG, and S. N. WANG (1971). Groundwater flow in a sandy tidal beach, 1: one-dimensional finite element analysis. *Water Resources Res.*, 7: 1313–22.

HAYES, M. O. (1972). Forms of sediment accumulation in the beach zone. In *Waves on beaches*, ed. R. E. Meyer, pp. 297–356. Academic Press, New York.

HAYES, M. O., and J. C. BOOTHROYD (1969). Storms as modifying agents in the coastal environment. In *Coastal environments: NE Massachusetts*, ed. M. O. Hayes, pp. 290–315. Dept. of Geology, Univ. of Mass., Amherst.

INGLE, J. C. (1966). *The movement of beach sand.* Elsevier, Amsterdam, 221 pp.

INMAN, D. L., and R. A. BAGNOLD (1963). Littoral processes. In *The sea*, ed. M. N. Hill, 3: 529–33. Interscience, New York.

INMAN, D. L., and A. J. BOWEN (1963). Flume experiments on sand transport by waves and currents. *Proc. 8th Conf. on Coast. Eng.*, pp. 137–50.

INMAN, D. L., and T. K. CHAMBERLAIN (1959). Tracing beach sand movement with irradiated quartz. *J. Geophys. Res.*, 64: 41–47.

INMAN, D. L., and J. FILLOUX (1960). Beach cycles related to tide and local wind wave regime. *J. Geol.*, 68: 225–31.

INMAN, D. L., and J. D. FRAUTSCHY (1966). Littoral processes and the development of shorelines. *Proc. Coast. Eng. Spec. Conf., ASCE*, (Santa Barbara, Calif.), pp. 511–36.

IPPEN, A. T., and P. S. EAGLESON (1955). *A study of sediment sorting by waves shoaling on a plane beach.* U.S. Army Corps of Engrs., Beach Erosion Board Tech. Memo. no. 63, 83 pp.

IWAGAKI, Y., and H. NODA (1963). Laboratory study of scale effects in two-dimensional beach processes. *Proc. 8th Conf. on Coast. Eng.*, pp. 194–210.

JOHNSON, D. W. (1919). Shore processes and shoreline development. Wiley, New York, 584 pp. Facsimile edition: Hafner, New York (1965).

JOHNSON, J. W. (1949). Scale effects in hydraulic models involving wave motion. *Trans. Am. Geophys. Union*, 30: 517–25.

KEMP, P. H. (1961). The relationship between wave action and beach profile characteristics: *Proc. 7th Conf. on Coast. Eng.*, pp. 262–77.

KEULEGAN, G. H. (1948). *An experimental study of submarine sand bars.* U.S. Army Corps of Engrs., Beach Erosion Board Tech. Report no. 3, 40 pp.

KINDLE, E. M. (1936). Notes on shallow-water sand structures. *J. Geol.*, 44: 861–69.

KING, C. A. M. (1953). The relationship between wave incidence, wind direction, and beach changes at Marsden Bay, Co. Durham. *Trans. Inst. Brit. Geog.*, 19: 13–23.

——— (1959). *Beaches and coasts.* Edward Arnold, London, 403 pp.

——— (1972). *Beaches and coasts.* 2nd ed. St. Martin's Press, New York, 570 pp.

KING, C. A. M., and W. W. WILLIAMS (1949). The formation and movement of sand bars by wave action. *Geog. J.*, 113: 70–85.

KOMAR, P. D., R. H. NEUDECK, and L. D. KULM (1972). Observations and significance of deep-water oscillatory ripple marks on the Oregon continental shelf. In *Shelf sediment transport*, ed. D. Swift, D. Duane, and O. Pilkey, pp. 601–19. Dowden, Hutchinson & Ross, Stroudsburg, Pa.

KRUMBEIN, W. C., and F. A. GRAYBILL (1965). *An introduction to statistical models in geology.* McGraw-Hill, New York, 574 pp.

LAFOND, E. C. (1939). Sand movement near the beach in relation to tides and waves. *Proc. 6th Pac. Sci. Cong.*, pp. 795–99.

LAU, J., and B. TRAVIS (1973). Slowly varying Stokes waves and submarine longshore bars. *J. Geophys. Res.*, 78, no. 21: 4489–97.

LONGUET-HIGGINS, M. S. (1953). Mass transport in water waves. *Phil. Trans. Roy. Soc.* (London), series A, 245: 535–81.

LONGUET-HIGGINS, M. S., and D. W. PARKIN (1962). Sea waves and beach cusps. *Geog. J.*, 128: 194–201.

MCLEAN, R. F., and R. M. KIRK (1969). Relationship between grain size, size-sorting, and foreshore slope on mixed sand-shingle beaches. *N.Z. J. Geol. and Geophys.*, 12: 138–55.

MILLER, R. L., and J. M. ZEIGLER (1958). A model relating dynamics and sediment pattern in equilibrium in the region of shoaling waves, breaker zone, and foreshore. *J. Geol.*, 66: 417–41.

MILLER, R. L., and J. M. ZEIGLER (1964). A study of sediment distribution in the zone of shoaling waves over complicated bottom topography. In *Papers in marine geology*, ed. R. L. Miller, pp. 133–53. Macmillan, New York.

MUNCH-PETERSON (1938). Littoral drift formula. *U.S. Army Corps of Engrs., Beach Erosion Board Bull.* 4, no. 4 (1950): 1–36.

MURRAY, J. (1853). On movement of shingle in deep water. *Minutes Proc. Inst. Civ. Eng.*, 12: 551.

MURRAY, S. P. (1967). Control of grain dispersion by particle size and wave state. *J. Geol.*, 75: 612–34.

NAYAK, I. V. (1971). Equilibrium profiles of model beaches. *Proc. 12th Conf. on Coast. Eng.*, pp. 1321–39.

OTTO, THEODOR (1912). Der Darss und Zingst. *Jahrb. d. Geo. Gesell. zu Greifswald*, 13: 393–403.

OTVOS, E. G. (1965). Sedimentation-erosion cycles of single tidal periods on Long Island Sound beaches. *J. Sediment. Petrol.*, 35: 604–9.

PRICE, W. A. (1969). Variable dispersion and its effects on the movements of tracers on beaches. *Proc. 11th Conf. on Coast. Eng.*, pp. 329–34.

PUGH, D. C. (1953). Etudes minéralogique des plages Picardes et Flamandes. *Bull. d'Inf. Com. Cent. d'Oceanog. et d'Etudes des Cotes*, 5, no. 6: 245–76.

PUTNAM, J. A. (1949). Loss of wave energy due to percolation in a permeable sea bottom. *Trans. Am. Geophys. Union*, 30: 349–56.

RECTOR, R. L. (1954). *Laboratory study of the equilibrium profiles of beaches.* U.S. Army Corps of Engrs., Beach Erosion Board Tech. Memo. no. 41, 38 pp.

SAVILLE, T. (1957). *Scale effects in two dimensional beach studies.* Int. Assoc. Hydraul. Res., Lisbon, Portugal.

SCHWARTZ, M. L. (1967). Littoral zone tidal cycle sedimentation. *J. Sediment. Petrol.*, 37: 677–83.

SCOTT, THEODORE (1954). *Sand movement by waves.* U.S. Army Corps of Engrs., Beach Erosion Board Tech. Memo. no. 48, 37 pp.

SEIBOLD, E. (1963). Geological investigation of near-shore sand transport. In *Progress in oceanography*, ed. M. Sears, 1: 3–70. Pergamon Press, New York.

SHEPARD, F. P. (1950*a*). *Longshore bars and longshore troughs.* U.S. Army Corps of Engrs., Beach Erosion Board Tech. Memo. no. 15, 31 pp.

——— (1950*b*). *Beach cycles in Southern California.* U.S. Army Corps of Engrs., Beach Erosion Board Tech. Memo. no. 20, 26 pp.

——— (1963). *Submarine geology*. Harper & Row, New York, 557 pp. [Also see 3rd ed. 1973.]

SHEPARD, F. P., and E. C. LAFOND (1940). Sand movements near the beach in relation to tides and waves. *Am. J. Sci.*, 238: 272–85.

SONU, C., and R. J. RUSSELL (1967). Topographic changes in the surf zone profile. *Proc. 10th Conf. on Coast. Eng.*, pp. 502–39.

STRAHLER, A. N. (1966). Tidal cycle of changes on an equilibrium beach. *J. Geol.*, 74: 247–68.

THOMPSON, W. F., and J. B. THOMPSON (1919). *The spawning of the grunion.* Calif. State Fish and Game Comm. Fish Bull. no. 3.

WATTS, G. M. (1954). *Laboratory study on the effect of varying wave periods on the beach profiles.* U.S. Army Corps of Engrs., Beach Erosion Board Tech. Memo. no. 53, 21 pp.

WIEGEL, R. L. (1964). *Oceanographical engineering.* Prentice-Hall, Englewood Cliffs, N.J., 532 pp.

ZEIGLER, J. M., and S. D. TUTTLE (1961). Beach changes based on daily measurements of four Cape Cod beaches. *J. Geol.*, 69: 583–99.

ZENKOVITCH, V. P. (1946). On the study of shore dynamics. *Trans. Inst. Okeanol., Akad. Nauk S.S.S.R.*, 1: 99–112.

Chapter 12

COASTAL ENGINEERING STRUCTURES

The marine engineer has no greater problem to deal with than this. The construction of harbours upon a sandy coast is always risky, resulting in no end of trouble and expense. . . . The interference with the natural sand-travel upon a coast cannot but be injurious; the breaking of any of Nature's laws has a detrimental effect.

Ernest R. Matthews
Coast Erosion and Protection (*1934*)

With established sediment sources and under a certain wave climate, a beach will tend toward a natural equilibrium where the waves are just capable of redistributing the sands supplied from the sources. This equilibrium was examined in Chapter 10. When jetties, breakwaters, or other structures are constructed in the coastal zone, the natural equilibrium will be upset, sometimes with disastrous consequences. Many of the most severe cases of coastal erosion can be attributed to this disturbance of the natural equilibrium.

This chapter will review the types and purposes of structures that are commonly built in the nearshore. Later in the chapter, case studies of the resulting modifications to the coastline will be examined.

TYPES OF STRUCTURES

There are two main purposes for the construction of structures in the nearshore zone: to improve navigation and to diminish coastal erosion. Jetties and breakwaters are built for the protection of boats and to aid navigation, whereas groins and various types of seawalls are constructed to prevent erosion of the coast.

Jetties (Figure 12-1) are built at the mouth of a river or tidal inlet to a bay,

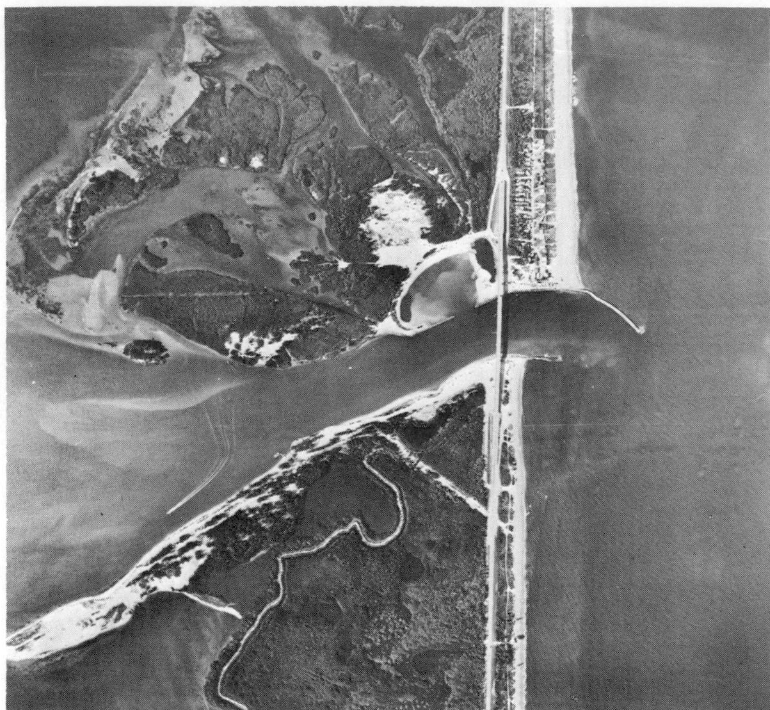

Figure 12-1 Sebastian Inlet with jetties, east coast of Florida. The jetties serve to control the entrance to the lagoon behind the barrier island. Note deposition to north of the north jetty and erosion to the south, indicating a littoral transport from north to south. [*Photograph from* University of Florida archives, *courtesy of* R. G. Dean and T. Walton]

lagoon, or estuary in order to stabilize the channel, to prevent shoaling by littoral drift, and to protect the channel entrance from storm waves. Jetties direct or confine the stream and tidal flow to aid in the channel's self-scouring ability, and help prevent immediate filling if dredging is relied upon to deepen the channel entrance. In order to prevent littoral drift from entering the channel, jetties generally extend through the entire nearshore to beyond the breaker zone. However, in doing so, they also act as a dam to the longshore drift of sand in the nearshore. As the sand moves alongshore under the natural processes of waves breaking obliquely to the shoreline, the drift must stop when it reaches such an obstacle placed across the littoral zone. As a result the sand accumulates on the updrift side of jetties and the shoreline advances. At the same time, on the downdrift side of jetties the sand transport processes continue to operate and so cause sand to drift away from the jetties; erosion and shoreline retreat therefore occur on the downdrift side of jetties. The construction of jetties can therefore interrupt the natural movement of sand along the beach and cause erosion of the adjacent beaches and coastal property. Later in the chapter case studies illustrating this will be examined.

Breakwaters (Figure 12-2) are structures which protect a portion of the shoreline area, thus providing a harbor or anchorage shielded from the waves. Breakwaters are constructed in a variety of shapes, but are generally attached to the coast at one or both ends, with a gap for a boat entrance, and extend outward through the surf

Figure 12-2 The breakwater at Crail, Fife, Scotland, dating back to earlier than the sixteenth century. This harbor provides protection to a small fishing fleet and pleasure boats that operate on the North Sea.

zone. One variety, the *detached breakwater*, is built as a barrier parallel to the shoreline and therefore has no attachment with the coast. Initially it was thought that such a construction would provide a protected area for boats while at the same time allowing the sand to drift alongshore, since there is no direct obstacle across the nearshore. However, breakwaters diminish the wave energy at the shoreline and therefore reduce the capacity for waves to transport the sand alongshore. The result is deposition of littoral sands within the protected lee of the detached breakwater. This is particularly exemplified by the breakwaters constructed at Ceará, Brazil, and Santa Monica, California, which will be examined later.

A variety of seawalls, bulkheads, and revetments are built along the shoreline to prevent property erosion and other damage due to wave action (Figure 12-3).

Figure 12-3 The Beaconsfield seawall, Bridlington, England, circa 1900. [*From* E. R. Mathews, *Coast Erosion and Protection*, 3rd ed. (London: Charles Griffin, 1934)]

They are placed parallel, or nearly parallel, to the shoreline, separating the land from the wave action. A secondary purpose is to diminish slumping of the coastal sea cliffs which they may front. *Seawalls* are constructed of solid or block concrete, steel sheets, timber, or commonly of natural stone known as *riprap*. They are sometimes constructed as a vertical wall; however, this promotes erosion at the toe of the wall, since the wave energy will in part be reflected, leading to increased scour and possible wall failure. More commonly the seawall slopes upward toward the land so as to help diminish the wave energy and decrease reflection, the height of the seawall being sufficient to prevent severe overtopping by wave runup. A seawall affords protection only to the land immediately shoreward and none to adjacent areas along the coast nor to the beach fronting the seawall. When built on an eroding shoreline, the recession continues on adjacent shores.

A *groin* (English: *groyne*), Figure 12-4, is a rib built approximately perpendicular to the shoreline to trap a portion of the littoral drift and thereby build out the beach. This helps prevent further erosion of the existing beach, and since the beach in turn helps to protect the shoreward coastal property, groins also diminish erosion of sea cliffs. Groins are relatively narrow in width and may vary in length from less than 10 meters to over 200 meters. In this regard they appear similar to jetties, although their function is very different. They have the same effect of damming the littoral drift so that the shoreline builds out on the updrift side and erodes on the downdrift side. To protect a large area from erosion, a series of groins, or *groin field*, may be constructed to act together. This enables an extended stretch of beach to be built out and shifts the zone of erosion out of the immediate area to your downcoast neighbor.

Figure 12-4 Groin field south of Tampa Bay, Florida. The orientation of the sand fills between the groins suggests a transport toward the lower right as viewed in the photograph. Note how the sand trapment has increased the beach width. [*Photograph from* University of Florida archives, *courtesy of* R. G. Dean and T. Walton]

Once a groin is filled, it allows littoral drift to pass by its seaward end, so that it traps only a certain quantity of sand. While sand is accumulating between the groins, that sand is prevented from reaching the beaches in the downdrift direction, so that erosion is enhanced, analogous to the erosion that occurs downdrift from jetties. To prevent such damage to adjacent areas of beach, groins may be filled artificially by beach nourishment—that is, by trucking sand out onto the beach. If this is done, the groins will not take their supply of sand from the natural littoral drift, which will continue to reach the downdrift beaches just as it did before groin construction.

The segment of beach between two adjacent groins acts as a small pocket beach. Comparable to our computer-simulated pocket beach oscillations in Chapter 10, the beach between two groins aligns itself with the crests of the incoming waves and may therefore depart from the shore alignment that existed prior to groin construction. Similarly, the beach will oscillate between the groins due to waves arriving from a variety of offshore directions.

There have been many recent suggestions regarding beach protective measures beyond the more traditional approaches already described. At present there is some interest in submerged offshore breakwaters—in some cases for boat moorings, but more often as a means of reducing shoreline erosion. The submerged breakwater stretches approximately parallel to the shore and acts to "trip" the waves and cause premature breaking. This reduces the energy of the incoming waves to a level below that under natural conditions, and thus reduces the rate of erosion of the beach. Such submerged breakwaters may be built of the traditional materials, but long, tubular sacks of polyethylene woven material filled with sand have also been considered. Such tubes, which have been used at the base of eroding sea cliffs in place of riprap or seawalls, have been successful in retarding erosion. Their disadvantage is that they are easily susceptible to damage by vandals. Other suggestions of ways to reduce wave energy include offshore fields of artificial kelp and barriers of bubble-producing devices; however, these have not proved to be very satisfactory.

EFFECTS OF CONSTRUCTION ON ADJACENT BEACHES

When jetties or breakwaters are constructed they upset the natural equilibrium between the sources of beach sediment and the littoral drift pattern. In response, the shoreline must change its configuration in an attempt to reach a new equilibrium.

Jetties and breakwaters act as partial or total dams to the littoral drift, blocking the natural sediment movement along the shoreline. There are numerous examples of this, dating back nearly a century. The initial construction of the harbor for the port of Madras, India (Figure 12-5), took place in 1875, the breakwater extending outward about 1,000 meters from the original shoreline (Vernon-Harcoart, 1881; Spring, 1919; Johnson, 1957; Cornick, 1969). This coast is characterized by a strong littoral sand transport from south to north, so that following the breakwater con-

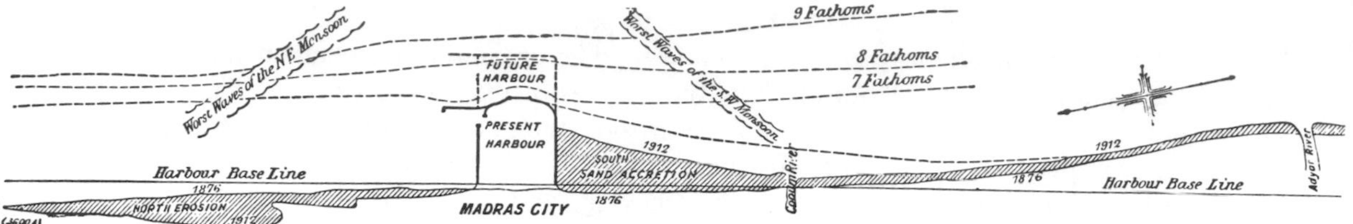

Figure 12-5 Madras Harbor, India, showing accretion on updrift side of harbor and erosion on the downdrift side. [*From* Spring (1919)].

struction, sand accumulated to the south side of the harbor, the shoreline progressively advancing. In the following thirty-six years (1876–1912), more than a million square meters of new land formed. Waves to the north of the harbor continued to transport sand northward; this sand was not replaced, however, because the sand that would have normally been deposited there was trapped behind the breakwater. This resulted in rapid erosion to the shoreline north of the breakwater for a length of 5 km along the shore, and it became necessary to install seawalls to check the destruction. Subsequently, to prevent harbor shoaling, the breakwater was extended seaward and a suction dredge installed to pump sand past the harbor. It is seen that the work done in bypassing the harbor by dredging has replaced the natural transport system due to wave action.

The breakwater at Ceará, Brazil, provides another early example of the resulting deposition-erosion problems (Carey, 1903; Johnson, 1957). A detached breakwater was constructed in 1875, extending for approximately 430 meters more or less parallel to the shoreline. As was already discussed, a detached breakwater intercepts the waves, providing a sheltered area of relatively calm water. The idea in using a detached breakwater is that the littoral drift material will move along the coast uninterrupted by the presence of a structure built across the surf zone, so that no deposition-erosion problem will result. This is a fallacy, since removing the wave action which provides the energy for transporting the littoral sands results in their deposition within the protected area. This is demonstrated by the example of Ceará harbor. The prevailing drift in this section of the Brazilian coast is from east to west. Following construction of the detached breakwater, a tongue of sand grew outward from the original shoreline in the protected lee of the breakwater, finally reaching the breakwater and completely closing off the passage. Sand continued to accumulate on the updrift side of this tombolo, eventualy moving around the seaward side of the breakwater and forming a bar on the downdrift side which joined with the shoreline.

The lessons "learned" at Madras and Ceará were repeated at Santa Monica and Santa Barbara, California. At Santa Monica in 1934 a 600-meter-long detached breakwater was constructed parallel to the shoreline and about 600 meters offshore (Figure 12-6). Immediately following its construction, sand began to deposit in its protected lee (Handin and Ludwick, 1950). Upcoast from the breakwater the shoreline advanced, while on the downdrift side the shoreline eroded. Only dredging prevented attachment of the shoreline to the breakwater with a complete closure of the harbor.

Originally the breakwater at Santa Barbara, California, constructed in 1927–28,

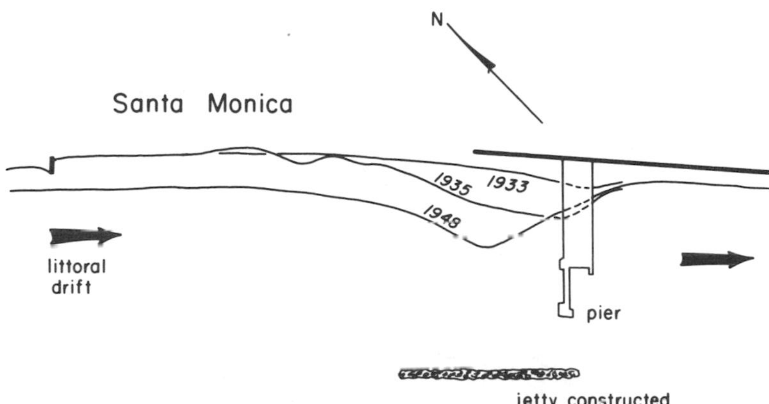

Figure 12-6 Sand deposition in the protected lee of Santa Monica breakwater, California. [*After* Johnson (1957)]

was detached, but in 1930 it was extended and connected to the shoreline to prevent harbor shoaling (Figure 12-7). The predominant waves are from a westerly direction, causing a large littoral transport to the northeast, computed to average about 215,000 cubic meters per year (Johnson, 1953). The breakwater interrupted this littoral drift and, as in the port of Madras, caused deposition on its updrift side (Figures 12-7 and 12-8). Sand accumulated on the west side of the breakwater (Figure 12-8) until the entire area was filled, the sand then moving along the breakwater arm, swinging around its tip, and depositing in the quiet waters of the harbor as a tongue or spit of sand (Figure 12-9). Without dredging, the spit would have eventually grown across the entire harbor mouth, attaching to the opposite shoreline and closing off the harbor. Once this had been accomplished the entire littoral drift would then pass

Figure 12-7 The Santa Barbara breakwater, California, with sand deposition on the updrift side and within the harbor as a tongue of sand. [*Photograph courtesy* R. L. Wiegel]

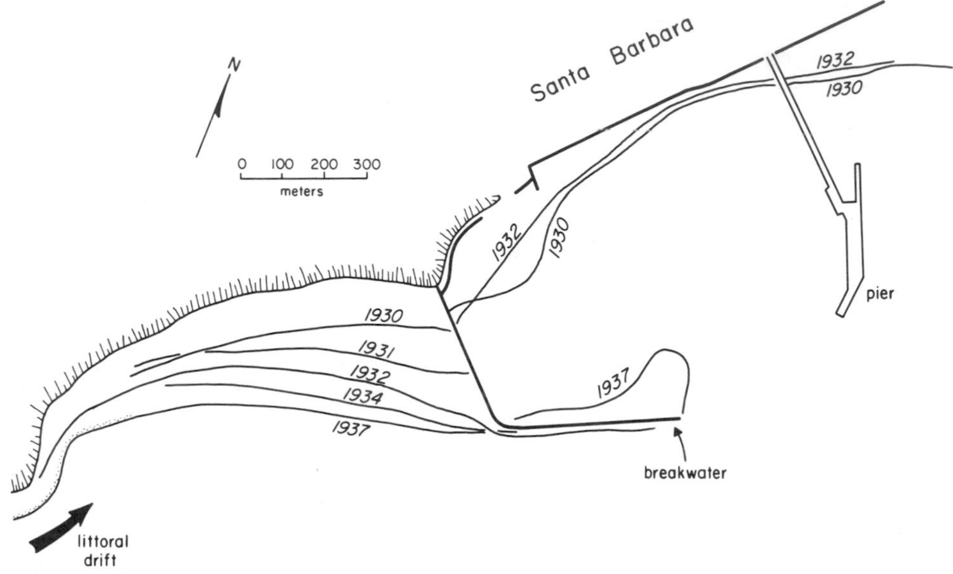

Figure 12-8 Deposition-erosion pattern around the Santa Barbara breakwater. [*After* Johnson (1957)]

around the breakwater, and a new equilibrium shoreline configuration would have been achieved. However, to prevent this closure of the harbor, dredging of the spit was initiated and presently operates continuously. This dredged sand is dumped on the beach to the immediate northeast of the breakwater in order to replenish the sand lost by the blockage of the littoral drift and prevent further erosion which took place following the breakwater construction (Figure 12-10). A more complete case history of the problems at Santa Barbara can be found in Wiegel (1959, 1964).

Jetties have a comparable effect in blocking the littoral sediment transport. Jetties built in 1935 to stabilize the inlet south of Ocean City, Maryland, a seaside resort on the Atlantic coast, trapped the strong southerly littoral drift of sand. The shoreline advanced considerably on the north side of the jetties, opposite Ocean City. Due to the loss of its supply of sand, the beach to the south of the jetties, on the barrier Assateague Island, eroded and the shoreline retreated about 450 meters in the twenty years following jetty construction (Shepard and Wanless, 1971). By 1961 the south beach had eroded to the point where it actually separated from the inner end of the south jetty, leaving a gap of almost 240 meters of open water. Storms in 1962 opened a breach over a kilometer in width, 2.5 km south of the jetties. Subsequent dredging and fill has partially restored the beach and mended the breach, but future storms can be expected to renew the problems.

It is apparent from these examples that whenever a jetty or breakwater is constructed, blocking a net drift of sand along the shoreline, erosion on the downdrift side occurs. It might be inferred from this that if a jetty were constructed in an area of coast where there is essentially zero net drift, then it would not dam any sand

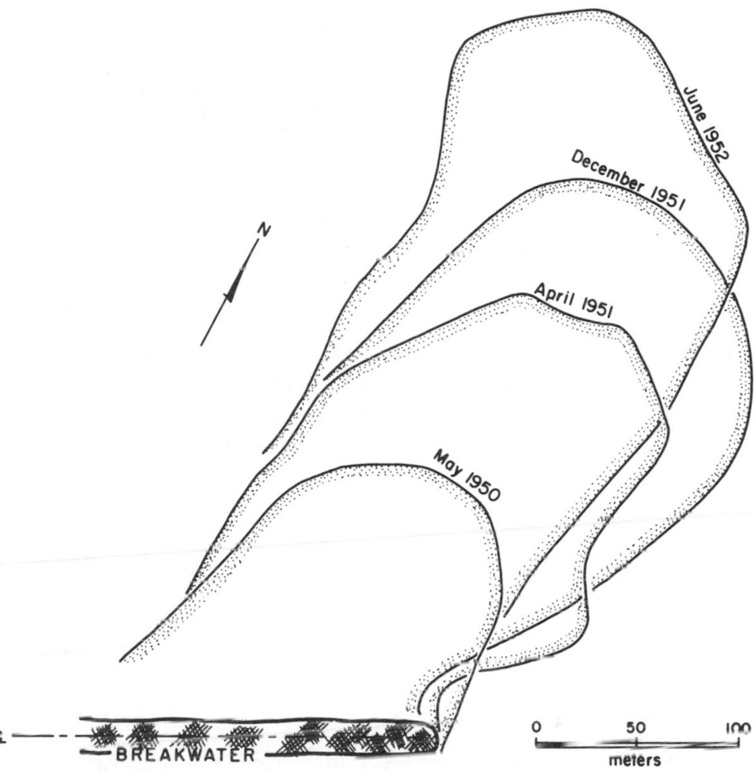

Figure 12-9 The growth of the spit of sand into Santa Barbara harbor at various times after the harbor was dredged in May-June 1949. Curves represent shoreline at mean lower low-tide level. [*After* Wiegel (1964)]

Figure 12-10 Sandyland, 15 km downdrift of the Santa Barbara breakwater, as seen seven years after breakwater construction. [*Photograph by* U. S. Grant IV, *courtesy of* F. P. Shepard]

movement and no deposition-erosion pattern would develop. This is in large part true, and jetties that have proved successful and have not caused erosion are mainly on such coasts. However, under certain circumstances jetties constructed on coasts with a zero net littoral drift have still caused coastal erosion. One example of this is the beach erosion and sand spit destruction at the entrance to Tillamook Bay, Oregon (Terich and Komar, 1973, 1974). A north jetty was constructed at the mouth of the bay in 1914–17 and was later (1933) rebuilt and extended (Figure 12-11). Bayocean Spit to the south of the entrance, separating Tillamook Bay from the ocean, suffered progressive erosion following construction of the jetty, especially in its narrow midportion and southern half. Homes and other buildings of the resort village on the spit were progressively destroyed (see Figure 1-3), and finally in 1952 the spit was breached in its narrow midsection and ocean water poured into the bay, carrying beach sand with it which covered the oyster fisheries there. This breach tended to stabilize as a new bay entrance, the old entrance to the north with the jetty partially closed. In 1956 a dike was built closing off the breach area and restoring the previous bay circulation.

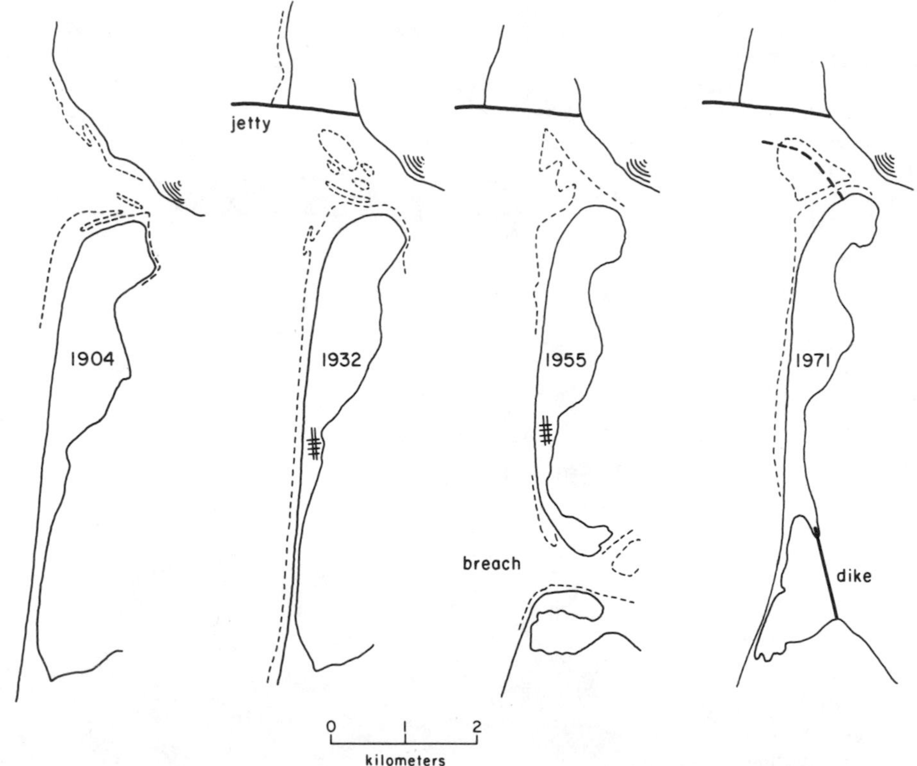

Figure 12-11 The erosion of Bayocean Spit, Tillamook Bay, Oregon, following construction of a north jetty in 1914–17. Heavy dashed line shows position of new south jetty under construction. [*Based on* U.S. Coast and Geodetic Survey charts and aerial photographs]

At the same time that spit erosion was occurring along most of the length of Bayocean Spit, deposition of sand and shoreline advancement was taking place in the immediate vicinity of the jetty. This is especially apparent to the north of the jetty, where the shoreline advanced by some 1,000 meters between 1917 and 1935. In addition, however, the shoreline to the immediate south of the jetty also advanced by some 150 meters; even more important, a large shoal developed at the mouth of the harbor, making the entrance nearly impassable even to small craft. Because of the shoal, a new south jetty was constructed in 1974. A somewhat different pattern of deposition-erosion around the jetty thereby emerged. Near the jetty itself deposition of sand occurred (Figure 12-12), while at greater distances from the jetty, both to the north and south, erosion took place. The reason for this distribution is that the jetty provides small embayments together with the shoreline that existed prior to jetty construction. Sand moves alongshore to fill these embayments until the shoreline

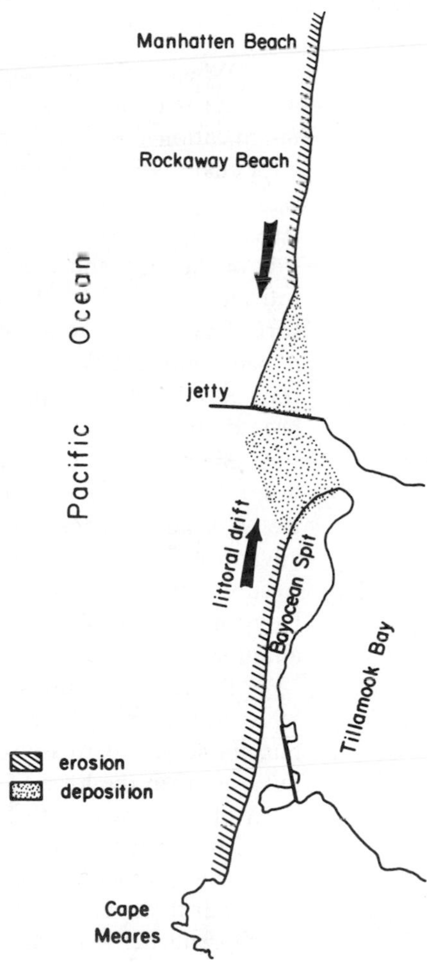

Figure 12-12 Pattern of deposition-erosion around the Tillamook jetty, Oregon. [*After* Terich and Komar (1974)]

is straight and again in equilibrium with the waves. The jetty may also provide areas to either side which are partially protected from the waves. Sand is transported into these areas and accumulates, the weak, diffracted waves being unable to remove the sand. Therefore, the long-term result is an accumulation of sand near the jetty within the "embayments" and sheltered areas created by the jetty construction. To supply this sand to the deposition zone, erosion occurs at greater distances from the jetty. The erosion to Bayocean Spit was especially severe because the length of the beach is relatively small, there being a large rocky headland to the south, so that the erosion per unit length of shoreline had to be large to supply the required sand to the shoals south of the jetty. In contrast, there is a long stretch of beach to the north of the jetty, so that only a small amount of sand need be supplied by each unit length of shoreline.

SUMMARY

When jetties or a breakwater are constructed in the sandy nearshore, they may alter the natural movements of beach sediments and thereby cause erosion. Such a modification upsets the natural equilibrium, and the shoreline configuration undergoes changes in response. The changes are most severe when there is a net drift of sand along the beach under a predominate wave direction, for under such circumstances the jetty or breakwater blocks this transport. Sand accumulates on the updrift side of the jetty and erosion occurs on the downdrift side. In many cases this sand transport blocked by the jetties is artificially bypassed. This involves utilizing some sort of dredge to scoop up the sand that accumulates on the updrift side and dump it back onto the beach on the downdrift side so that it can continue in its natural transport direction. Man then must perform the task that natural forces had previously been doing.

Shoreline modifications are generally smaller if the jetty is located on a coast where there is a very small (nearly zero) net littoral sand transport. Due to the protective shadow zones adjacent to jetties, sand accumulates there and the shoreline advances. Further along the shoreline, away from the jetties, erosion occurs in order to supply the sand close to the jetties. This erosion is generally small—unless the segment of beach is short, as in the example at Tillamook Bay, in which case a large amount of sand must be removed from each unit shoreline length.

Recent suggestions have proposed utilizing partially submerged breakwaters to reduce the nearshore wave energy to a level suitable for safe mooring of boats, but not so low as to stop the sand transport entirely. This may be an improvement, but even then the lowered energy will cause a reduction in the sand transport capability of the waves, so that sand would tend to accumulate on the shoreline in the lee of the breakwater. The only absolute way to prevent the deposition-erosion patterns associated with jetties and breakwaters is simply not to construct such obstacles to the littoral drift. Short of this, one must anticipate what erosion might be caused and take appropriate defensive measures before considerable coastal property is lost. The usual protective measures include the strategic placement of seawalls, riprap, groin fields, beach nourishment, or combinations of these.

REFERENCES

CAREY, A. E. (1903). The sanding-up of tidal harbours. *Minutes Proc. Inst. Civ. Eng.*, 156: 215–302.

CORNICK, H. F. (1969). *Dock and harbour engineering*. 2nd ed. Charles Griffin, London, vol. 2, 352 pp.

HANDIN, J. W., and J. C. LUDWICK (1950). *Accretion of beach sand behind a detached breakwater*. U.S. Army Corps of Engrs., Beach Erosion Board Tech. Memo. no. 16, 13 pp.

JOHNSON, J. W. (1953). Sand transport by littoral currents. *Proc. 5th Hyd. Conf.*, Bulletin 34, State University Iowa Studies in Engineering, pp. 89–109.

——— (1957). The littoral drift problem at shoreline harbors. *J. Waterways and Harbors Div., ASCE*, 83, paper 1211 (April 1957): 1–37.

SHEPARD, F. P., and H. R. WANLESS (1971). *Our changing coastlines*. McGraw-Hill, New York, 579 pp.

SPRING, F. J. E. (1919). Coastal sand travel near Madras Harbor. *Minutes Proc. Inst. Civ. Eng.*, 210: 27–28.

TERICH, T. A., and P. D. KOMAR (1973). The development and destruction of Bayocean Spit, Oregon. *Proc. 3rd Tech. Conf. on Estuaries of the Pac. Northw.*, pp. 113–26.

TERICH, T. A., and P. D. KOMAR (1974). Bayocean Spit, Oregon: history of development and erosional destruction. *Shore and Beach*, 42: 3 10.

VERNON-HARCOART, L. F. (1881). Harbours and estuaries on sandy coasts. *Minutes Proc. Inst. Civ. Eng.*, 70: 1–32.

WIEGEL, R. L. (1959). Sand bypassing at Santa Barbara, California. *J. Waterways and Harbors Div., ASCE*, 85, no. WW2: 1–30.

WIEGEL, R. L. (1964). *Oceanographical engineering*. Prentice-Hall, Englewood Cliffs, N.J., 532 pp.

Chapter 13

LITTORAL SEDIMENTS

I have long discovered that geologists never read each other's works, and that the only object in writing a book is a proof of earnestness, and that you do not form your opinions without undergoing labour of some kind.

Charles Darwin
Letter to J. M. Herbert (c.1843)

Among the many interests of geologists is the identification of ancient beach deposits in the geologic rock record. To aid in such identification, they have conducted studies of modern beaches in the belief that "the present is the key to the past." Geologists have concentrated mainly on the physical properties of the sediments that compose the beach—size distributions of sediment particles, particle shapes (roundness and sphericity)—as well as sedimentary structures typical of beach sediments and organisms that inhabit beaches. By contrasting these properties and structures with those found in modern river sands, desert dune sands, and sediments from deeper water, they expect to be able to identify these various environments in the ancient rock record.

Coastal engineers should also be somewhat familiar with littoral sediments. A common practice today is *beach replenishment*—adding a foreign sand supply to a beach in order to replace sand that has been lost. If the engineer improperly chooses the sand to be ministered to the beach, he may find that all or part of it is quickly shifted to deep water by the waves and his purpose defeated.

COMPOSITION

The sediments of the littoral zone may be composed of any material that is available in significant quantities and is of a suitable grain size. The composition of littoral sediments therefore closely reflects the sources and their relative importance. Most of the beaches of the temperate regions of the earth are composed principally of quartz and feldspar grains, derived ultimately from the weathering of the granitic-type rocks, gneisses, and schists that are abundant on the continents. In addition to quartz and feldspar, beach sediments generally also contain small amounts of heavy minerals such as hornblende, garnet, and magnetite, from which the nature of the source rocks can sometimes be deduced. Examples of the use of heavy minerals to identify sources and to trace sediment movements to and along beaches were discussed in Chapter 9. Micas, chlorite, and minerals of similar platy structure are generally scarce on beaches, as they tend to remain in suspension and are lost to the offshore. The remaining heavy minerals tend to migrate downward into the beach and are concentrated at depth within the beach deposit, while the light quartz and feldspar remain closer to the surface and become enriched there (see page 315 and Figure 13-14). These heavy-mineral concentrates can sometimes approach 100 percent in content. When the beach face is cut back during storm conditions (Chapter 11), the heavy-mineral concentrate may become exposed, and the beach then temporarily becomes a "black-sand beach."

Local pocket beaches with a limited source area may have exotic sediments, even on the continents where quartz-feldspar sand beaches otherwise predominate. For example, Sand Dollar Beach south of Point Sur, California, displays a bright green sand derived from the serpentine of the Franciscan formation. The beautiful pocket beach at Tintagel, Cornwall, England, below the castle of the legendary King Arthur, is made up of flat pebbles and shingle composed of very-fine-grained mica schist from the surrounding metamorphic terrain.

The littoral sediments of volcanic islands often consist almost entirely of fragments of andesite and basalt lavas or individual minerals derived from the lavas. Bennett and Martin-Kaye (1951) described the beach at Black Bay, Grenada, British West Indies, as consisting of nearly 100% augite grains, eroded from volcanic tuffs which form the greater part of the surrounding area. Well known are the green-sand and black-sand beaches of Hawaii, the green sands containing a high percentage of olivine derived from the phenocrysts of the surrounding volcanic rocks and the black-sand beaches consisting of fresh microcrystalline lava and volcanic glass formed where lava flows entered the ocean, the lava exploding when it came into contact with the cold water (Wentworth and Macdonald, 1953; Moberly, Baver, and Morrison, 1965). Other beaches on Hawaii and other oceanic islands range from black volcanic sands to white sands composed entirely of carbonate grains, depending on the relative contributions of the volcanic terrigenous sources and the offshore coral reef sources. The relative proportions depend on how recently there has been volcanism in the area, the intensity of weathering in the hinterland, and the vigor of growth of marine organisms.

Shells and shell fragments are important in many beaches, especially those in the tropics, where biologic productivity is high and the chemical weathering of the land sources tends to be intense. Moberly, Baver, and Morrison (1965) found that for the Hawaiian Islands as a whole foraminiferans form the chief carbonate fraction in most beaches, followed by mollusks, red algae, and echinoid fragments. Coral was a poor fifth—in spite of the fact that the sand is commonly referred to as "coral sands." *Halimeda* (a green algae) occurred only in very small amounts in the Hawaiian beaches, although it is an important constituent in other tropical ocean beaches. Also present were sponge spicules (opal) and crab fragments. In all cases the biologic carbonate remains are predominately broken and worn smooth by wave action.

Shell material may be abundant because the supply of terrigenous sands is either very low or of the wrong grain size for the particular beach. Where other sources are negligible, shells may be the only source of beach material. This explains the unusual shell beach at the far north location of John O'Groats, Scotland (58.5°N latitude), which consists of 97 percent shells (Raymond and Hutchins, 1932). Normally the major shell beaches are found in the tropics, where productivity is high.

The calcium carbonate content of the beaches along the southern Atlantic coast of the United States shows a general increase from north to south because of the increasing productivity of offshore mollusks and the decreasing supply of quartz-feldspar sand to the south (MacCarthy, 1933; Gorsline, 1963; Giles and Pilkey, 1965). To the north of the Carolinas the average beach contains very little calcareous material, usually less than one-half percent. This contrasts with Florida, where there is essentially no continental supply, and broken shells of marine organisms form nearly the entire beach sediment. Giles and Pilkey (1965) found that this pattern is not a simple progressive increase in carbonate content from north to south, however; the content is lower in Georgia than in Florida and increases slightly to the north to a maximum in North Carolina (decreasing markedly further north). Georgia is the deepest recess of the reentrant for this stretch of coast, so that it has the lowest wave energy. Giles and Pilkey (1965) hypothesized that there is less carbonate off Georgia because this lowered wave energy is insufficient to break and grind shells down to the proper size. The grain size of sand is also a minimum on the Georgia coast, increasing both to the north and south; so an alternative possibility is that the decrease is brought about by selective sorting processes. The carbonate grains are generally large in comparison with quartz sands derived from the continent, so that they may be selectively transported and concentrated in the coarser-grained beaches of higher energy both to the north and south of Georgia. Such a sorting by grain size and therefore by composition certainly occurs in the onshore-offshore direction, the shells being concentrated within the high-energy breaker zone while the finer quartz sands are found on the lower-energy portions of the beach profile.

A concentration of shells by selective sorting processes was described by Watson (1971) on the central portion of Padre Island on the Gulf coast of Texas. Central

Padre Island is the site of a convergence of the littoral drift, which causes shells and sand from the entire coast to accumulate there. The shell material is then concentrated on the beach by the winds, which remove and blow inland the finer quartz sand to form dunes. The beaches of central Padre Island are composed of up to 80 percent shells and shell fragments, in spite of the fact that the site is located within a major terrigenous province. Gibbons (1967) described this same type of selective sorting process that concentrates the shell material on the beach at Dee Why, New South Wales, Australia.

Some beaches in the tropics or in arid regions are composed of *oolites*, nearly spherical sand-sized carbonate grains. These are formed by direct precipitation from seawater and so require warm, shallow, agitated waters supersaturated with calcium carbonate. Eardley (1939) has studied beach oolite formation in the Great Salt Lake, Utah, and Rusnak (1960) has described an occurrence in the highly saline Laguna Madre, the lagoon landward of Padre Island, Texas. Oolites are also found on beaches of the Mediterranean Sea coast of Tunisia, but these are apparently not presently forming, but rather have been eroded from fossil oolitic rocks that outcrop in the vicinity (Fabricius, Berdau, and Munnich, 1970). The most famous occurrence of recent oolite sands is in the Bahamas (Newell, Purdy, and Imbrie, 1960), but there they occur on banks near the outer margin of the Bahamian Platform separated from the island beaches by wide lagoons. Oolitic beach sands therefore appear to be rare; beaches are not a prime site for oolite formation.

A few interesting and unusual beaches are the inadvertant by-products of man's behavior. Bascom (1960) has reported the existence of a beach at Fort Bragg, California, composed of old tin cans washed in from the nearby city dump, the cans neatly arranged by the waves (the dump is no longer in use and the beach presently consists of a near-solid mass of cans rusted together). In the coal-mining districts of Britain there are a number of beaches that have a high proportion of sand-sized coal fragments. Bird (1968, p. 87) describes a beach on Brownsea Island in Poole Harbour, Dorset, England, that is composed of broken subangular earthenware fragments derived from a former pipeworks.

GRAIN-SIZE DISTRIBUTION

The mean diameters of grains in beach sediments vary from more than 256 mm for boulders to slightly less than 0.125 mm for very-fine-sand beaches that are well protected. In parts of Labrador, Alaska, Scotland, and Argentina, beaches consisting chiefly of large cobbles and boulders are common. Such coarse material can also be found in many pocket beaches, especially those in the vicinity of rocky headlands. Shingle beaches of flattish pebbles and cobbles are particularly noteworthy in Britain and may even be said to be typical for that area. Indeed, the term *beach* is derived from an Anglo-Saxon word which referred to shingle. More familiar to most beachgoers are the long stretches of sand beaches consisting of grains in the size range 0.25–2.0 mm.

There appear to be three main factors that control the mean grain size of beach sediments; (1) the sediment source, (2) the wave energy level, and (3) the general offshore slope on which the beach is constructed. The importance of the source is obvious. The beach environment will select out the size grains that are appropriate for its particular conditions. If it so happens that the sources provide no appropriate material, then there may be no beach. Assuming the source does provide the proper grain sizes, then there remains a complex relationship between energy level of the processes at the beach, the general offshore slope, and the resulting grain size of the beach deposit. There is a general tendency for the highest-energy beaches with the largest waves to have the coarsest grains. This is particularly apparent if only one small geographical area or only one beach is examined. For example, on an island the more exposed beaches are also generally the coarsest-grained, other factors being constant (Marshall, 1929; Moberly, Baver, and Morrison, 1965; Dobkins and Folk, 1970). One stretch of beach may be coarser at one end than at the other because of the greater wave heights at the coarser-grained end (for example, Halfmoon Bay, California). Such relationships are responsible for many (but not all) examples of longshore selective sorting by grain size. This will be examined later. A simple correlation between grain size and energy level cannot be made, however. This is especially apparent when one realizes that medium-sand beaches may be found in lakes with waves of heights less than 5 cm as well as on ocean beaches of very high energy. Headlands often have small pocket beaches composed of cobbles and boulders, while the long stretches of beach between headlands are of sand. While this may in part be due to higher energy levels on the headlands, also of importance is the offshore slope upon which the beaches are formed, the slope being considerably greater off from the headlands. Bagnold (1963, p. 522) has suggested that *autosuspension* might be the controlling factor—that grains of a certain size and finer become permanently raised above the bottom into suspension and so move offshore without redeposition. For autosuspension the settling velocity w of the grains is related to the beach slope $\tan \beta$ and horizontal current u by

$$w \leqq u \tan \beta$$

According to this relationship, the greater the beach slope the coarser the grains that will remain in permanent suspension and therefore be lost from the beach. For example, taking an average beach slope of 3 degrees ($\tan \beta = 0.05$) and $u = 25$ cm/sec as a fair average of the onshore-offshore currents due to waves, Bagnold obtained $w \leqq 1.25$ cm/sec, which is the settling velocity of 0.15-mm-diameter grains. Finer grains than this should tend to become uniformly dispersed in the turbulent water of the breaker zone and therefore diffuse outward to the offshore. Although protected low-energy beaches contain finer material, open ocean shores rarely include grains less than this 0.15-mm diameter.

Apparently because of the high slope as well as the greater wave energies, even coarse sands can be maintained in autosuspension, so that a headland beach will contain only gravel and cobbles. Initially the bedrock offshore slope would govern what grain size could remain to form a beach. Once the beach is established, the beach slope itself would govern the lower limit of grain size that could remain on the beach.

This complex relationship between grain size, slope, and energy level requires considerably more study, as only general tendencies are apparent, with no quantitative backing other than that discussed in Chapter 11 (Figure 11-8), where the beach slope was empirically related to the mean grain size.

The relationship between source and grain size of sediments available to the beach is more apparent in carbonate beaches than in quartz-feldspar beaches. This is because the carbonate is derived from biologic productivity and in many instances grain-size modes correspond to certain species of plants and animals. An excellent example of this is the study of Folk and Robles (1964) of beach sands on Isla Pérez, an island on the Alacrán Reef complex off the Yucatan peninsula. The beaches there consist of about 60 percent *Halimeda* algae fragments, 25 percent coral, and 15 percent foraminiferans. The structure of the *Halimeda* is such that it breaks up into flat fragments with a mean size of 0ϕ (1-mm)* diameter (Figure 13-1). The coral, principally staghorn coral, either breaks up into lengths governed by the branch length in the parent coral or into a size corresponding to crystal units or wall thickness. Thus the coral provides rather distinct modes at 2ϕ (0.25 mm) and -6ϕ (64 mm). The foraminiferans do not undergo much breakage, and their modal sizes therefore represent essentially the size distributions of the original populations. The grain-size distributions of the Isla Pérez beaches are therefore often multimodal, reflecting the contributions of these several biologic sources as well as physical processes such as energy level. At some locations the physical processes may select out one mode so that, for example, the beach may consist of nearly 100 percent *Halimeda* fragments or 100 percent coral debris.

Generally, beach sands are fairly well sorted; that is, the overall range of grain sizes at any one beach location is small. Beach sediments tend to be better sorted than river sediments, but not as well sorted as dune sands (Mason and Folk, 1958; Friedman, 1967). The sorting values of carbonate beach sands are very similar to those obtained on quartz-sand beaches (Folk and Robles, 1964). Apparently the surf action can accomplish about the same degree of sediment sorting regardless of the composition of the grains.

When the source material contains two or more modal grain sizes, the sorting depends on the mean grain size at the particular location. If the physical processes concentrate the particular grain size that corresponds to a source mode, then the sorting is good. However, if the mean grain size falls between two modal sources, then the sorting is poorer, since two modes are being mixed. For example, mixing a sand mode with a pebble mode yields a poorly sorted deposit. Plots of sorting coefficient versus mean grain size under such beach conditions produce a helical trend, with the best sorting corresponding to the mean grain sizes of the source modes. Examples of this for beach sediments have been found by the studies of Blatt (1958), Folk and Robles (1964), Andrews and van der Lingen (1969), McLean and Kirk (1969), and Sonu (1972). It was seen in Chapter 11 that the study of McLean and Kirk (1969)

*The diameter in ϕ-units is equal to the negative logarithm to the base 2 of the diameter in millimeters.

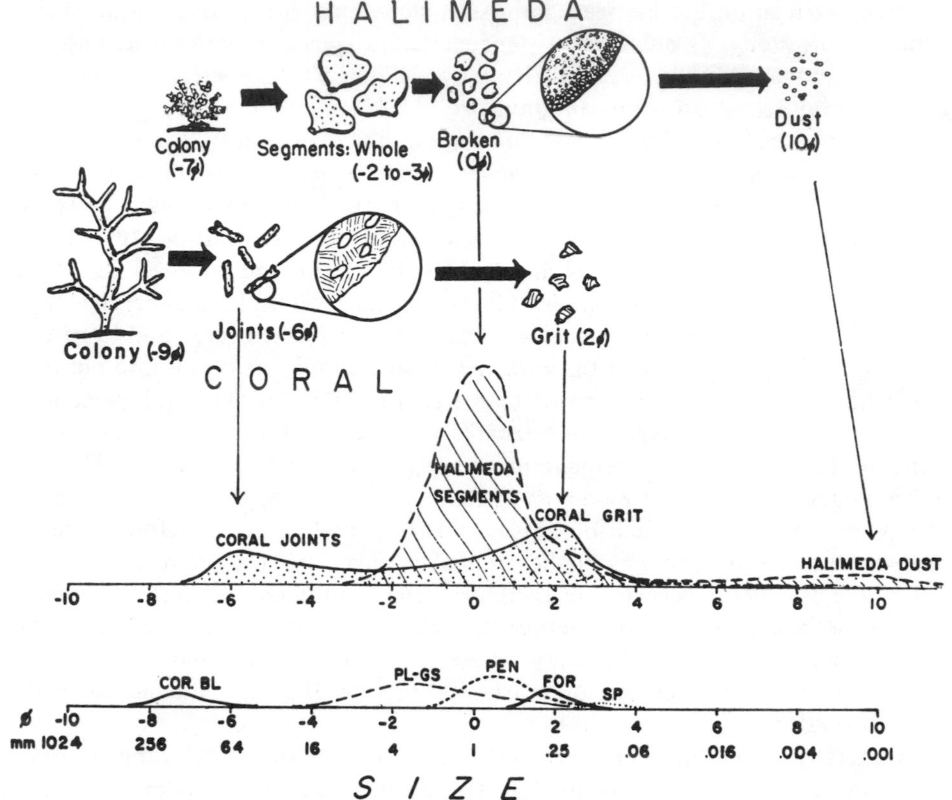

Figure 13-1 The breakup of calcareous algae *Halimeda* and colony coral to form a multimodal beach sand as found on Isla Pérez. The gravel in the beaches is chiefly made of coral, 0ϕ sand is mostly *Halimeda*, and 2ϕ sand is again mainly coral. [*From* Folk and Robles (1964)]

demonstrated a wavy-curve relationship between the beach face slope and the grain size that resulted from this helical trend between sorting (therefore, percolation) and the grain size.

Negative skewness, an overabundance of coarse grains or underabundance of fine grains when compared to the normal Gaussian distribution curve, is a common feature of terrigenous beach sands (Mason and Folk, 1958; Friedman, 1961; Duane, 1964; Hails, 1967; Chappell, 1967). This has also been observed in carbonate beach sands by Stark and Dapples (1941), McKee, Chronic, and Leopold (1959), and by Folk and Robles (1964). As already indicated, there are two possible causes for this—either an addition of a coarse tail to a normal population, or subtraction of a fine tail. The turbulence of the nearshore causes the finest material to be lost offshore, so that the distribution of grain sizes becomes truncated on the fine end. Stark and Dapples (1941) demonstrated that negative skewness can be caused by removal of fines in this way by beach processes. Some of the beach samples of Folk and Robles

(1964) appeared to confirm this further, but in most of the Isla Pérez sediments the negative skewness resulted from the addition of a coarse tail—for example, a few stray coral fragments or gastropods, pelecypods, and so forth. Pebbles or shells scattered within an otherwise sand beach would produce a tendency for negative skewness. This was found by Mason and Folk (1958) on the foreshore of Mustang Island, Texas. On the other hand, fluvial and dune sands tend to have positive skewness because of infiltration of fines (Friedman, 1961; Hails, 1967). Skewness is the most important factor in attempts to use grain-size distributions to distinguish beach, dune, and river sediments (Friedman, 1961). This criterion must be used with caution, however, as some beaches do furnish positively skewed distributions, apparently in some cases inherited from the source area (Andrews and van der Lingen, 1969; Moiola and Weiser, 1968).

Studies such as those of Friedman (1961, 1967) attempt to distinguish beach sands from other environments on the basis of two-dimensional plots of parameters like standard deviation (sorting) and skewness, obtained from the grain-size distribution. Combinations of two parameters thus define fields on the plots that indicate beach sand versus sands from dunes, rivers, and so on. Greenwood (1969) applies multivariate statistics such that all of the parameters can be used simultaneously to distinguish dune, beach backshore, and beach foreshore sands. The product is an equation rather than a two-dimensional plot. The analysis is based on sands from Barnstaple Bay, England, with test checks against "unknowns" from other beaches and dunes.

Visher (1969) recognizes log-normal subpopulations within the total grain-size distribution. Each of these subpopulations is related to a different mode of sediment transport: (1) suspension, (2) saltation, and (3) surface creep or rolling. Each subpopulation has a different mean and standard deviation (sorting). One example from Visher is shown in Figure 13-2, from the foreshore of a beach. The finest material is a suspension population which represents less than 1 percent of the total sample. The main population is that of the moving grain layer or traction carpet of grains in saltation. The two separate saltation populations are presumed to be produced by the swash and backwash, since these represent somewhat different flow conditions. Kolmer (1973) has produced similar subpopulation divisions in a laboratory beach. Figure 13-3 gives an example from Visher of the changes in the proportions of the subpopulations across a beach. In addition to beaches, Visher gives examples from many other environments: dunes, deltas, rivers, and their ancient counterparts, including turbidites. The divisions into subpopulations, their sortings, size ranges, and points of truncation provide insight into the transport processes and modes of deposition of both modern and ancient sediments.

Attempts at utilizing the grain-size statistics (skewness, sorting, and so on) to identify environments in the ancient rock record have not been particularly successful. This is partly because of overlap between the environments: different environments can have the same sediment sorting and so on. In addition, environmental recognition is based on subtle parameters, such as skewness, which may easily be altered by diagenesis.

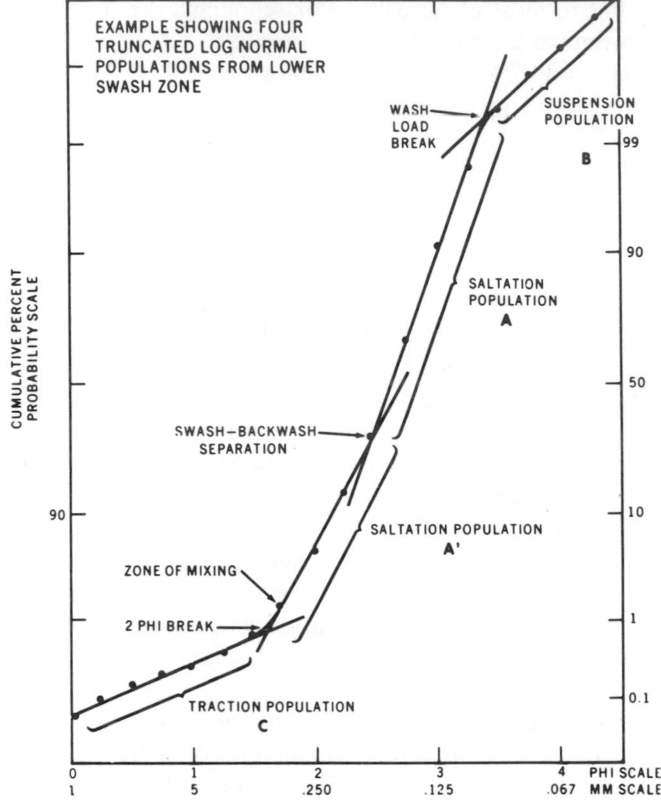

Figure 13-2 The division of the sediment grain-size distribution into subpopulations which are normally distributed, the subpopulations being identified with different modes of sediment transport. [*From* Visher (1969)]

The studies we have already discussed use such statistics as mean grain size, standard deviation (sorting), skewness, and kurtosis to describe the grain-size distributions. Kloven (1966) instead uses the entire grain-size distribution in his analysis, applying factor analysis techniques to the discrete class intervals into which the continuous grain-size spectrum is divided. Imbrie and van Andel (1964) give an excellent description of the factor analysis approach. The technique sorts out from the data end members whose combinations or summations can yield all the observed grain-size distributions. Klovan (1966), for example, applied the approach to samples from Barataria Bay, Louisiana, and determined three end members. One end member was determined to represent beach surf dominance, the second factor was current deposition, and the third was gravitation settling of sediments in protected areas. These three factors were shown to bear a close correspondence to such environments in Barataria Bay. Thus the approach can define the relative effectiveness of each process at the various localities. This is also shown in the study of Solohub and Klovan (1970) in Lake Winnipeg, Manitoba, Canada. None of the previous methods of employing combinations of mean grain size, sorting, and kurtosis were reliable in identifying the depositional environments. Only the factor analysis approach pro-

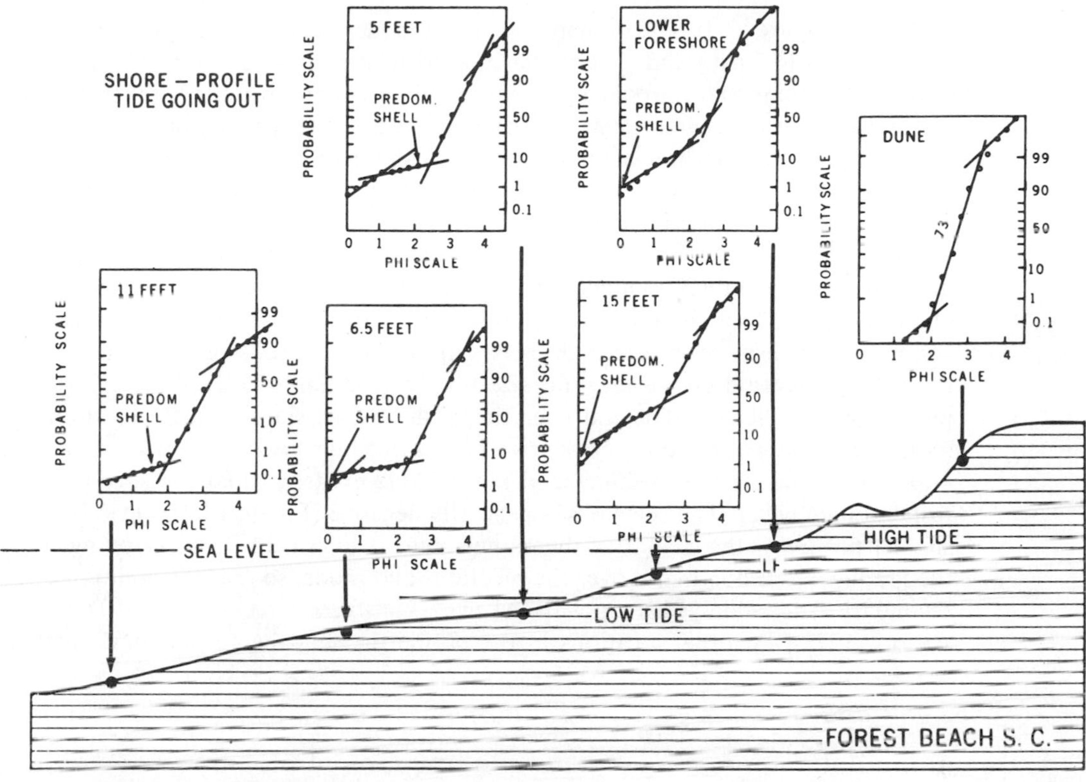

Figure 13-3 Changes in the subpopulations within the grain-size distributions across a beach profile. [*From* Visher (1969)]

duced mappable results consistent with the known depositional environments and energy conditions.

ONSHORE-OFFSHORE GRAIN-SIZE SORTING

A number of studies have investigated the changes in sediment grain size and degree of sorting normal to the shoreline across the beach profile (Krumbein and Griffith, 1938; Krumbein, 1938; Evans, 1939; Bascom, 1951; Inman, 1953; Miller and Zeigler, 1958, 1964; Fox, Ladd, and Martin, 1966; Greenwood and Davidson-Arnott, 1972). The coarseness of the sediment reflects the bottom topography and the local degree of turbulence and wave energy dissipation. The largest sediment particles at any given beach are located at the plunge point of the breaking waves, with a decrease in grain size both toward deeper water and shoreward across the surf and swash zones. This variation is illustrated in the near-tideless Lake Michigan by the study of Fox, Ladd, and Martin (1966), who determined the sediment grain-size parameters from a 100-meter-long traverse across the nearshore at South Haven,

Michigan (Figure 13-4). The principal sediment modes are a medium sand at approximately 1.75ϕ (0.3 mm) and a very coarse sand to granules at -1.0ϕ (2.0 mm), the different locations on the profile represented by different proportions of the two modes. The mean grain size is greatest at the wave plunge point at the base of the beach face, decreasing up the foreshore slope as well as offshore within the trough of the beach profile. There is a second slight coarsening of the sediment over the offshore bar. The incoming waves first break on the offshore bar, but without an intense dissipation and concentration of the wave energy. The waves then reform over the trough and break for a second time, plunging at the base of the beach face and giving up most of their energy. Thus the grain size closely reflects the energy level of the wave processes. The intensity of the swash decreases up the beach face slope, and this is shown in the progressive decrease in grain size up the foreshore. The sediment sorting is poorest at the plunge point and over the offshore bar, since at these locations the sediment consists of mixtures of the two modes, medium sand and very coarse sand-granules. Elsewhere the medium-sand mode alone exists, so that the sorting is correspondingly better. The skewness is generally negative (Figure 13-4), but it becomes slightly positive in the vicinity of the plunge point. This is because at the plunge point the granule mode dominates over the medium-sand mode, so that the addition of the secondary sand mode yields a positive skewness [also see Greenwood and Davidson-Arnott (1972)]. At other locations in the profile the sand mode dominates and the

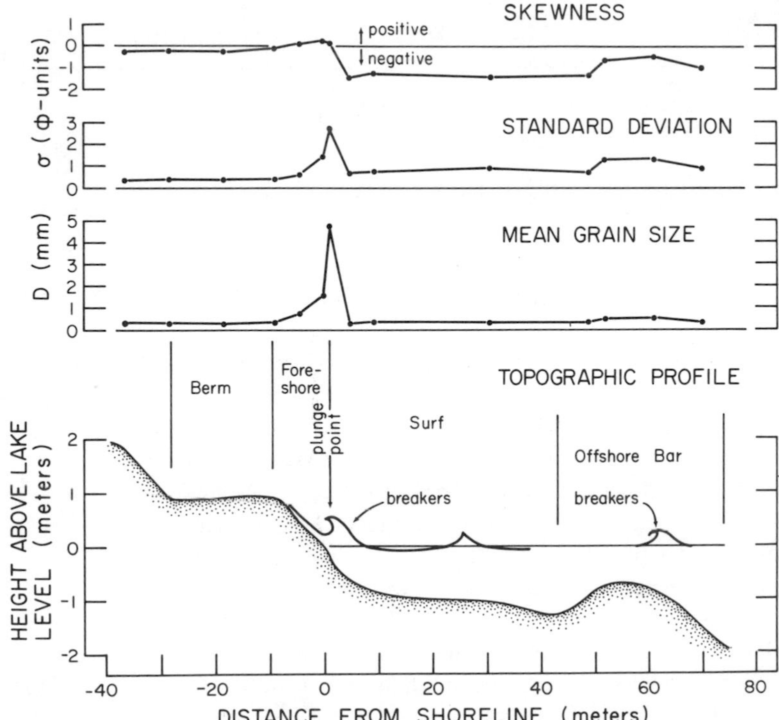

Figure 13-4 Grain-size parameters across a Lake Michigan beach, corresponding to the local wave energy level. [*After* Fox, Ladd, and Martin (1966)]

addition of a small granule mode creates the negative skewness. This illustrates the hazards of using skewness to identify sedimentary environments without giving any thought to the complicating factors of selective sorting processes.

Similar distributions of mean grain size across the beach profile have been found in the other investigations, including studies of gravel and cobble beaches (Krumbein and Griffith, 1938) and of beaches where tides are important (Bascom, 1951; Inman, 1953; Miller and Zeigler, 1958, 1964). Miller and Zeigler (1964) studied an area of irregular bottom topography with spitlike structures extending out from the shoreline at the tip of Cape Cod, Massachusetts. In spite of the irregularity in morphology, there is a fairly regular trend map of median sediment size, with coarsest material in the breaker zone and finer material both onshore and offshore. There was some rhythmic longshore periodicity in the median grain size, but the presence of the spits was not reflected in the offshore grain-size changes. Bascom (1951) found the coarsest mean grain size at the plunge points, but also found the next coarsest material on the exposed summer berm. Bascom hypothesized that this coarse deposit represents material carried up and over the berm crest by maximum wave uprush, whence it cannot return. Alternatively, in some cases it may represent a lag deposit, the finer sand fractions having been selectively removed by winds and carried shoreward to form dunes.

There is not always agreement on the trends followed by the degree of sorting across the beach profile. As already seen, Fox, Ladd, and Martin (1966) found the poorest sorting in the plunge position, due to the mixing of two modes. Similarly, Inman (1953) found the poorest sorting in the breaker and surf zones and the best sorting in the region of the swash zone sands, as well as somewhat improved sorting seaward of the breakers. On the other hand, Miller and Zeigler (1958) found the highest degree of sorting in the breaker zone, with progressively poorer sorting both seaward and shoreward.

Relatively few attempts have been made to quantify or understand the processes responsible for this sorting across the beach profile. Evans (1939) described qualitatively the relationship between the swash energy (velocity) and the resulting pattern of grain size across the foreshore zone. As the swash of the wave ascends the sloping beach face more and more of its energy is lost and the velocity gradually decreases to zero. The energy of the swash has its source in the energy of the incoming wave, but the velocity of the backwash as the flow returns down the beach face is due to gravity, the flow starting from rest and accelerating until it reaches its maximum velocity near the plunge line. Because of these variations in velocity the sediment is sorted as it is moved about on the beach foreshore, the smaller grains accumulating in the slower-moving water near the top of the swash and the coarsest material near the plunge line, where the energy is greatest.

Miller and Zeigler (1958) present a somewhat more advanced model for selective sediment transportation within the breaker and swash zones to explain the resulting grain-size distribution. In the breaker zone they envisioned a net vertical movement of water and sediments near the base of the turbulent breaker such that the finer sizes are lifted farthest above the bottom to be subsequently swept up the beach face or drift offshore. Only the coarsest sizes can remain close to the bottom under this vertical motion and so remain within the breaker zone. This vertical sorting by grain size within the breakers was partially substantiated by measurements of the suspended load. In the foreshore region Miller and Zeigler made an assumption that as the onshore upwash decreases in velocity its sediment load is dropped along the way, the coarsest material first, followed by progressively finer grains up the slope. As a first approximation they assumed that the backwash velocity increases linearly as it starts from zero at the top of the swash runup. Using considerations of the threshold of sediment motion under turbulent flow, they then determined the final sediment distribution across the swash zone following the backwash. Because of the increase in velocity downslope they concluded that there should be a band of graded small sizes near the top of the foreshore, with size increasing progressively down the slope as the breakers are approached. This is of course what is actually found within the swash and breaker zones. Although the study of Miller and Zeigler (1958) represents an advance in attempts to quantify and understand the processes responsible for the sediment sorting, they provide no real quantitative comparisons between their model and the grain-size sorting on some particular beach.

Schiffman (1965) attempted to measure swash and backwash energies and to correlate these with the sediment sorting. He utilized a compression-spring dynamometer to evaluate the water energies in the swash zone on southern California sandy beaches. His results are generalized in Figure 13-5. The measure of energy used is an arbitrary value, defined as the sum of the areas of the incoming force-time curves measured with the dynamometer divided by the time interval over which these forces were acting. Velocities were estimated from the drag equation given the relationship shown in Figure 13-5, where $C_0 = 1.12$ is the drag coefficient. The absolute values of velocity and energy are probably not correct, but can be used for comparison purposes to demonstrate changes across the swash and surf zones. The energy level and grain-size distributions typically show three definite zones inside the breakers: (1) the swash, (2) the surf, and (3) a transition zone separating the two. This transition zone corresponds with the area where the return backwash collides with the base of the next incoming surf bore. Hence the higher energy level. The transition zone is also typified on these California beaches by a bimodal sand-size distribution, a -0.5ϕ (1.4-mm) very-coarse-sand mode being added to the otherwise medium- to fine-sand sediment, again reflecting the higher energy level.

The sand tracer experiments of Ingle (1966) and others in the nearshore have already been discussed in Chapter 11. Tagging different grain sizes various colors, Ingle found that there is a general tendency for the coarser grains to move seaward out of the surf zone to the breaker zone, while at the same time the finer grains remain within the surf zone. The results indicate that a true sorting does take place, each size seeking out the zone in which it is in equilibrium. Other tracer studies likewise support this onshore-offshore sorting by grain size.

Along with the related beach profile morphology changes, the processes responsible for the selective sorting of grain sizes across the profile remain the most poorly documented and understood facets of nearshore processes. Careful and extensive

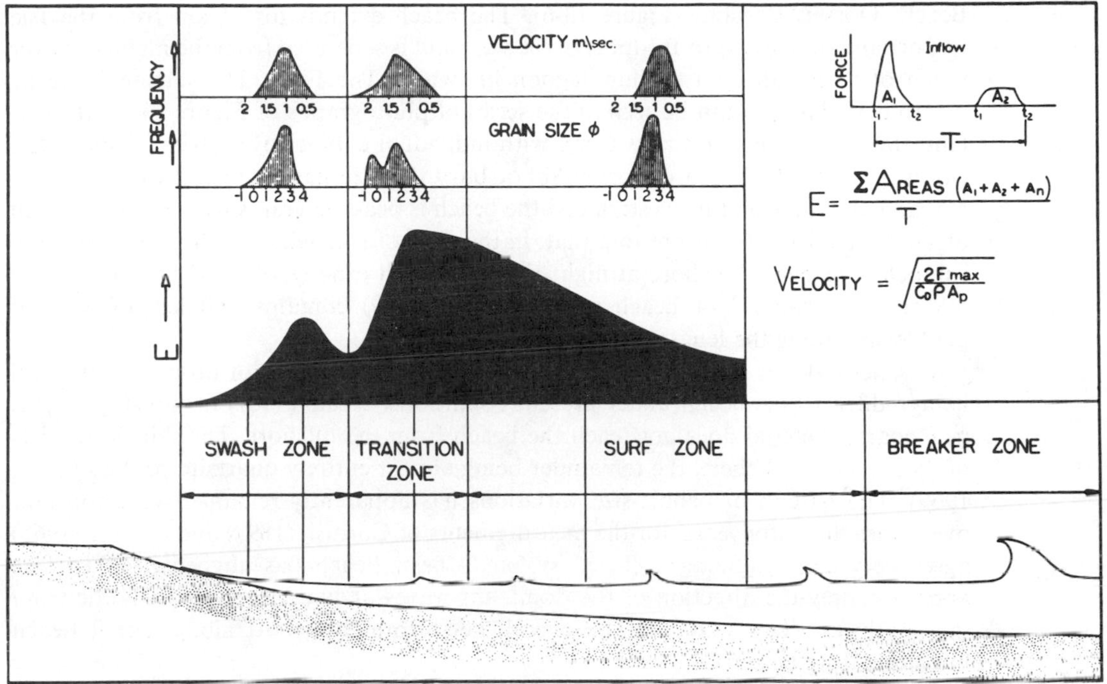

Figure 13-5 Energy level within the inner surf zone and swash zone and its correspondence with the sediment grain size. The energy is high in the transition zone where the backwash collides with the incoming surf. [*From* Schiffman (1965)]

studies are needed along the lines of the groundwork investigation of Schiffman (1965), and possibly utilizing sand tracers.

LONGSHORE SORTING OF BEACH SEDIMENTS

Some stretches of beach demonstrate systematic variations in grain size in the longshore direction as well as across the profile. One good example of this lateral sorting is Halfmoon Bay, California, described by Bascom (1951). This example was discussed in Chapter 11 in connection with changes in beach slope along its length. Halfmoon Bay (Figure 11-10) is protected on its northern end by a rocky headland, while the southern end is fully exposed to the waves. This variation in wave energy is reflected in a progressive change in particle size, the finer sand accumulating in the low-energy sheltered area behind the headland, where refraction reduces the wave energy, and the coarsest sand being found on the exposed southern beach. The grain size is 0.17 mm at its sheltered north end and progressively increases to 0.65 mm at the exposed south end (Figure 11-10). Halfmoon Bay is therefore a clear example of longshore sorting due to longshore variations in the wave energy level.

Probably the best example anywhere of lateral sorting is that found on Chesil

Beach, Dorset, England (Figure 13-6). The beach extends for 28 km from the Isle of Portland on the east to Bridport in the west and is separated from the mainland over much of its length by a shallow lagoon known as The Fleet. The sorting along its length is striking, as can be seen in the series of photographs of Figure 13-6. At Portland the median shingle size is 6 cm, with individual cobbles over 10 cm in diameter. Midway along the beach opposite Abbotsbury the size has decreased to about 1.25 cm, and at Bridport at its western end the beach is pea-size grains with a large portion of sand included. Stories are told that, in the days of smuggling in the area, when the smugglers reached the shore at night they could tell exactly where they were just by taking up a handful of beach pebbles. Carr (1969) contains detailed analyses of pebble size along the length of Chesil Beach.

Chesil Beach appears to be virtually a closed system, with little new material being added to the beach under present conditions. Neate (1967) demonstrated that new coarse material does not reach the beach from the offshore. The shingle consists of 98.5% flint and chert, the remainder being almost entirely quartzite pebbles (Carr, 1969). The pattern of pebble-size variations has apparently remained very constant over more than fifty years, for the measurements of Cornish (1897) and of Carr (1965) agree very closely. Being a closed system, Chesil Beach has aligned itself to face approximately the direction of the dominant waves arriving from the Atlantic from the southwest (Lewis, 1938). The overall net littoral transport along Chesil Beach must therefore be zero.

Such a pronounced example of sorting has of course led to considerable specula-

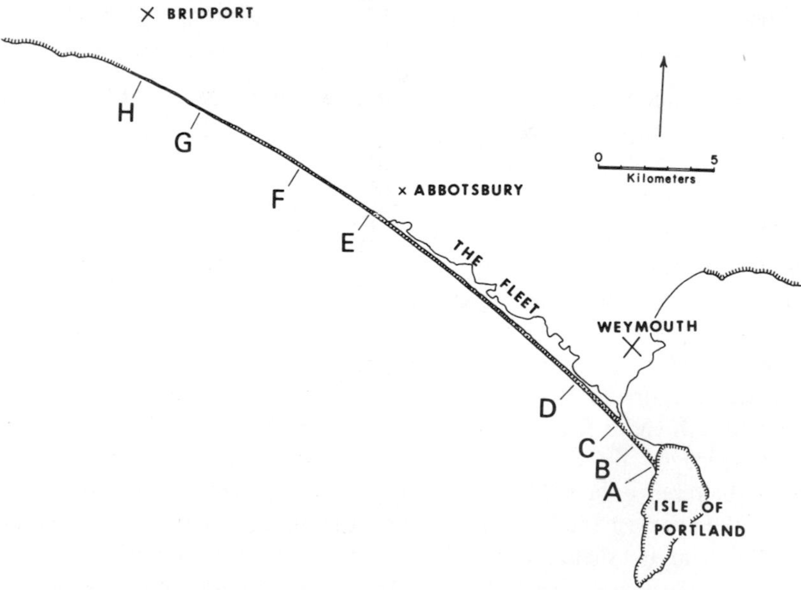

Figure 13-6 Longshore sorting of beach sediments along the length of Chesil Beach, Dorset, England. The map gives locations of beach sediments (ranging from shingle down to pea-size gravel mixed with sand) shown in the photographs on facing page.

A
B
C
D
E
F
G
H

tion as to its origin. Early thoughts on the matter are summarized by the Right Honorable Lord Avebury (*The Scenery of England*, London: Macmillan, 1902, p. 161):

> There has been much difference of opinion whether the shingle travels from east to west or from west to east. The latter view was advocated by Sir John Coode, Sir John Rennie, Sir John Hawkshaw, and Mr. Gregory. Sir J. Prestwich and Sir G. Airy, on the contrary, maintained that the shingle travels from east to west. This view is also prevalent among the fishermen of the locality, and is based on the idea that the stones are gradually worn down, and consequently that the smaller ones are those that have travelled farthest.

The local fishermen seem to have developed the most logical explanation for the sorting. However, their explanation is not correct, since the pebbles are not derived from the Isle of Portland, abrasion appears to be minor, and there is not a westward net transport.

Can the increase in pebble size to the east be accounted for by an increase in the wave energy level, much as at Halfmoon Bay? King (1972, pp. 308–11) appears to support this conclusion, based on the fact that the coarsest material is found where the local offshore slope is the steepest. The reasoning is that the steeper offshore slopes should allow more of the initial wave energy to be concentrated on the beach. The wave measurements obtained by Hardcastle and King (1972) from the beach opposite Wyke Regis and West Bexington, a span of about three-quarters of the total length of the beach, yielded significant wave heights at Bexington that were a factor 1.11 larger than at Wyke Regis, indicating that there is some increase in wave energy toward Portland. On the other hand, Carr (1971) reported that their wave measurements showed no evidence of wave height variations. It appears questionable, then, whether longshore variations in the wave energy along Chesil Beach are sufficient to account for the grain-size sorting.

Lewis (1938) explained the sorting in terms of waves of differing energies reaching the beach from various directions. The beach is approximately perpendicular to the approach of the dominant waves of high energy from the southwest. The next strongest waves come from the west of southwest and would move pebbles of all sizes to the east toward Portland, where they would be blocked. Smaller waves generated in the English Channel arrive from the south of southwest and, according to Lewis, would be capable of moving only the smaller shingle westward. A long-term interplay between large waves from the west and small waves from the south would therefore cause a progressive shift of the coarsest shingle to the east and the finest to the west. The overall net transport could still be zero under this sorting scheme. I like this explanation.

Other examples of longshore variations in grain size do not involve closed systems such as Chesil Beach and apparently cannot be explained in terms of longshore changes in the wave energy level. Instead, they appear to be the product of selective transport processes, with some grain sizes moving alongshore at greater rates than others or with progressive losses of certain grain sizes to the offshore. For example, Pettijohn and Ridge (1932) found a marked but progressive decrease in grain size along the 12-km length of Cedar Point, Ohio, a sand spit which extends out into Lake Erie. The coarsest grain size (approx. 0.5 mm) is found on the beach next to the glacial-deposit sediment source, and the finest sediment (0.15 mm) is located at the terminal end of the spit. Progressive abrasion of the sediment in the direction of transport was ruled out, as was a relationship to a systematic decrease in wave energy along the spit length. Instead, the grain-size decrease was attributed to a selective transport, the smallest sizes moving alongshore at the greatest rate and the coarsest material remaining close to the original source area.

Schalk (1938) found just the opposite trend in grain size away from the source area. On the outer beach of Cape Cod, Massachusetts, Shalk found that the finest sediment is in the mid-cape region, the area of the eroding cliffs which supply the sediments to the beach. Both north and south of this source area (the littoral drift direction diverges from this area) the median grain size on the beach increases. This pattern was shown to result from the offshore loss of the finer grades in the material supplied to the beach by cliff erosion as longshore drift moves the sediment away from the source.

Thus it appears that longshore variations in grain size can be produced in at least three ways: (1) by longshore changes in the wave energy level; (2) by selective rates of transport, the finer grains outdistancing the coarser; and (3) by selective removal of the finer grain sizes from the beach (carried onshore by winds or offshore by waves), leaving the remaining beach sediment coarser in median size. Chesil Beach may provide a fourth mechanism: the interplay of waves from different directions with contrasting energy levels.

ABRASION OF BEACH SEDIMENTS

Considerable effort has gone into studies of the abrasion of sediment grains, especially by geologists who are concerned with the rounding of the grains. What are the rates of grain abrasion? Is it higher in rivers, on beaches, or in sand dunes? What are the effects of grain size and composition?

Twenhofel (1945) provided a useful summary of the then-current thoughts on sand abrasion and rounding. All studies uniformly concluded that particles smaller than about 0.25 mm in diameter do not undergo abrasion and so do not become rounded. This is due to their low inertia in collision with one another. However, such grains can split into smaller fragments by collision with pebbles or larger clasts. Grains of diameter 0.25 to 0.50 mm undergo some abrasion, but at a very low rate, while grains larger than 0.50 mm can be effectively reduced and rounded by abrasion, the abrasion rate increasing with increasing grain size. Such conclusions are based in part on laboratory experiments such as those of Thiel (1940), where sand-water mixtures were tumbled in a steel drum. The conclusions are also based on examinations of grain rounding on natural beaches. For example, MacCarthy (1933) found no evidence of grain rounding during a longshore transport of some 210 km. The actual distance of individual grain travel was of course much greater than this straight-line 210 km. Twenhofel (1945) gave other examples of lack of sand abrasion during littoral transport. The studies also generally concluded that eolian transport is much more effective than either river or littoral transport in abrading and rounding sand (Twenhofel, 1945; Kuenen, 1960).

Surf abrasion of pebbles is considerably greater than that of sand-size grains. Kuenen (1964) performed laboratory experiments on pebble abrasion where the swash action was simulated by a trough containing water and pebbles rocking lengthwise. The pebbles were composed of limestone, quartzite, and chert cut into cubes of sizes ranging from 10 mm to 5.5 cm. Kuenen found that the abrasion rate was largest for medium-size pebbles, since the larger ones rolled smaller distances and the smaller ones lost less per kilometer of travel and did not roll much further. Kuenen pointed out that this result agrees with the fact that on natural pebble beaches the intermediate sizes are the best rounded. The limestone pebbles had the highest abrasion rates, being three times that of the quartzite, chert being by far the most resistant. The initial losses were large, due to the sharp edges of the cubes, but as the clasts became rounded the abrasion losses decreased. Abrasion losses were much higher when the beach consisted entirely of pebbles, being two to three times greater than on a sandy beach. Apparently the main factor in pebble abrasion is the grinding together of the pebbles. On a sandy beach the abrasion results from a sandblasting effect of the sand striking the stationary pebbles and taking out minute chips.

Abrasion of pebbles therefore appears to be significant. For example, Kuenen (1964) estimated that a 10-cm fragment of limestone would require only a week of continuous surf action of 0.5-m-high waves to become a typical well-rounded beach cobble, and one of quartzite would take a month. These rates are for continuous wave action. But because of burial, long-term exposure on the berm, lack of action at low tide, and so forth, it might actually take centuries for rounding.

The effects of abrasion can be seen in fragments of glass on a sandy beach; in time the exposed parts of the glass become rounded and frosted by the miniature chips removed by the sand impact. Such features are also found on natural quartz particles, sand as well as pebbles. A review of sand surface textures can be found in Krinsley and Doornkamp (1973), including many photographic examples obtained with the scanning electron microscope. Many of these abrasion and solution features are indicative of certain environments and have been used to identify their ancient counterparts. For example, Hodgson and Scott (1970) used a combination of grain-size analysis and the grain surface features to identify the presence of beach sands within the Lower Carboniferous formations of Britain.

One material that is easily and quickly abraded and rounded in the surf is shell debris composed of the relatively soft calcium carbonate. Abrasion appears to be far more important than solution of the carbonate. Chave (1960, 1964) experimented on shell destruction in a rolling barrel and found that smaller and lighter shells tend to be broken more rapidly by abrasion than do the more massive forms. On the other hand, in his field studies of shell abrasion, Driscoll (1967) found that the lightest shells are surprisingly durable on sandy beaches because they are carried with the water and so do not suffer the sandblasting action to which the heavier and more stationary shells are subjected. Driscoll found that on beaches of mixed sand and gravel the abrasive effects on the shells are at least five times as great as on the pure-sand beach. Shell surfaces become pitted by impact with pebbles, and breakage is more significant. This shell abrasion is enhanced by selective sorting whereby the shells and shell fragments accumulate naturally together with the pebble concentrations. Driscoll concluded that species possessing greater surface area per unit weight tend to abrade more rapidly because of the sandblasting effects. In all cases the abrasion rates were much less than the rolling-barrel experiments of Chave (1964) would indicate, the abrasion achieved in 2 hours in the barrel taking approximately 100 hours on the natural beach.

GRAIN ROUNDNESS AND SHAPE ON BEACHES

Although such studies are rather tiresome, there has been considerable interest in examining the development and selective transport of sediment grains according to shape and roundness. These grain properties, like grain-size distributions, have also been examined for their potential in distinguishing various sedimentary environments.

Wadell (1932) was the first to show that there is a fundamental difference between roundness and shape of grains and that the two properties are geometrically independent. A grain may be perfectly rounded if its edges are worn away but at the same time be far from spherical in shape. On the other hand, the grain may be quite spherical in shape but still have sharp angular edges—as, for example, in a dodecahedron. Measures of sphericity are still not a completely adequate indication of the grain shape, in that disc-shaped and rodlike particles may have the same sphericity values but still have obviously very different shapes and presumably different characteristics of transport and deposition. Zingg (1935) originally proposed the most widely used method for designating form, whereby grains are classified as spherical, discoid, rodlike, or bladed, based on ratios of the axes of the grains.

This section will review the studies on the development of grain roundness, sphericity, and form, the processes responsible for this development, and how beaches contrast with other environments in this respect. Selective transport by roundness and shape will also be examined, both in the onshore-offshore direction across the profile and alongshore.

Because roundness and sphericity measurements are more easily made on them, pebbles have been more extensively studied than beach sands. The greatest interest and debate has centered on the development of flatter, disclike pebbles on beaches due to abrasion processes. Dobkins and Folk (1970) reviewed the literature on this debate and divided the opinions into three camps: (1) the "agnostics," who believe that there is no systematic difference in the shape of river versus beach pebbles; (2) the "sorters," who believe that discs are thrown high up on the beach and spherical pebbles rolled down to the base underwater, but that abrasion has not changed the overall proportion of discs, and that discs appear more abundant only because they are more easily accessible on the exposed beach; and (3) the "abraders," who believe that surf processes develop discs by abrasion, so that discs are more plentiful than in the source river or sea cliff.

The agnostic and sorter viewpoints both receive support from the study of Sames (1966), who compared beach and river pebbles from Japan composed of resistant chert and quartzite. Both fluvial and littoral pebbles showed a medium stage of roundness, but were different in that the river pebbles still had a certain proportion of rather angular components and showed a wider variety of shapes. In contrast, the beach pebbles showed a uniform degree of high roundness, the angular grades and irregularly shaped particles having been eliminated. However, Sames could find no tendency for beach pebbles to be flatter, the flatness being inherited from the bedded-chert source. The beach environment could not significantly change the flatness by abrasion or selective sorting.

The school of sorting is exemplified by the studies of Landon (1930) and Bluck (1967). Landon showed that spherical pebbles are less stable on a beach than are flat, disclike pebbles, the spherical forms rolling offshore to deeper water while the flat ones remain on the shore. Thus there would be a predominance of flat pebbles on the higher portion of the beach because of selective transport. Excellent examples of this onshore sorting of pebbles by shape are found in the study by Bluck, who investigated the sedimentation of pebble beaches in southern Wales. As shown in Figures 13-7 and 13-8, Bluck found that the beaches could be subdivided into four zones on the basis of pebble shape and size. Furthest shoreward on the gravel ridge is a large-disc zone, typified by cobble-size discs. Next seaward is an imbricated zone composed mainly of imbricated disc-shaped pebbles, the discs dipping seaward, the angle being greater for the pebbles furthest onshore. This offshore-dipping imbrication is enhanced on beaches where the backwash is weak. Where there is strong backwash or an impermeable layer below the discs, the imbrication may be completely destroyed or even dip landward. Seaward of the imbricated zone there is sometimes a sand run, over which the particles move very quickly. These particles accumulate on the seaward side of the sand run to form a band composed of spherical and rod-shaped pebbles. The rod-shaped grains orient their lengths in the onshore direction, normal to the strike of the beach trend. Seaward of this is in turn an area where the spherical and rod-shaped pebbles are infilling a framework of larger cobbles. These cobbles are primarily spherical in shape and form the outer margin of the gravel storm ridge, beyond which is the flat sand berm. Thus the examples studied by Bluck demonstrate a pronounced onshore sorting of pebbles by shape. He attributed the sorting to the backwash rolling of spherical grains as demonstrated by Landon (1930), but also suggested that the sorting is partially produced during storm conditions, when fragments of all shapes are thrown forward by the waves, the discoidal particles being most easily lifted from the sea floor and tending to have a lower settling velocity because of their shape, so that they are thrown further up the beach face than are other particles.

The possibility that disc-shaped fragments are made by special processes of marine abrasion was not supported by the observations of Bluck (1967) in the beach gravels of Wales. There the disclike pebbles are inherited from the source but were concentrated at the top of the beach ridge, as already pointed out. The largest discs are the most disclike and also the most angular. They are the grains least worked upon by the waves, since the disc zone is near the high-tide mark at the top of the ridge. Rather than progressive abrasion causing an accentuating of the disc shape, impact breakage during storm conditions is the favored means of beach particle abrasion for discs.

The best argument for the abrasion of pebbles to a more disclike shape is provided by the study of Dobkins and Folk (1970), who investigated the shape development of pebbles in the rivers and beaches of Tahiti. The primary advantage of conducting the study in Tahiti (among other obvious advantages) is that all pebbles there are of essentially one rock type, basalt, which has no noticeable bedding or crystal orientation and so tends to wear homogeneously. In sampling the beach pebbles, Dobkins and Folk collected across the entire beach profile, thus eliminating any onshore sorting by shape as demonstrated by the Bluck (1967) study. Apparently some beaches did show such onshore sorting patterns, although not as pronounced as those of Bluck. In summary form, the conclusions of Dobkins and Folk (1970) regarding the roundness and form of river pebbles versus beach pebbles in Tahiti are as follows:

1. River pebbles have the lowest roundness and highest sphericity. They are neutral in form; that is, neither discs nor rodlike pebbles predominate. The samples from one river to another are very similar, and river pebbles of all measured sizes have similar properties.
2. Pebbles on high-energy beaches have the highest roundness and lowest sphericity and are decidedly more disclike in shape than river pebbles. There are wide shape variations from one pebble size to the next on any particular beach, but altogether there is a regular decrease in sphericity with increasing grain size.
3. Pebbles on low-energy beaches are intermediate in character, having intermediate roundness and sphericity ranges from extremely low (on sandy beaches) to high, reaching essentially "fluvial" values. There is a wide variation from one pebble size to the next, and from one beach to another. Small pebbles tend to be disclike, while the larger ones are more rodlike.

Dobkins and Folk thus found that abrasion of pebbles on beaches can produce more nearly disclike shapes than found in the river source. They found no correlation between the sphericity of the river pebbles and the sphericity of the corresponding beach pebbles, indicating that the river source had no influence on the shape development of pebbles on the beach. This is probably not true in other areas, since a particular river may initially supply pronounced disclike pebbles because of the source area.

Dobkins and Folk (1970) found that the size of pebbles that achieve the most nearly disclike shape depends on the wave energy and character of the beach surface. For each grain size there is an optimum intensity of wave action that best produces a

Figure 13-7 Sorting by pebble shape across the beach profile, an example from southern Wales. The zones are shown schematically in Figure 13-8. [*From* Bluck (1967)]

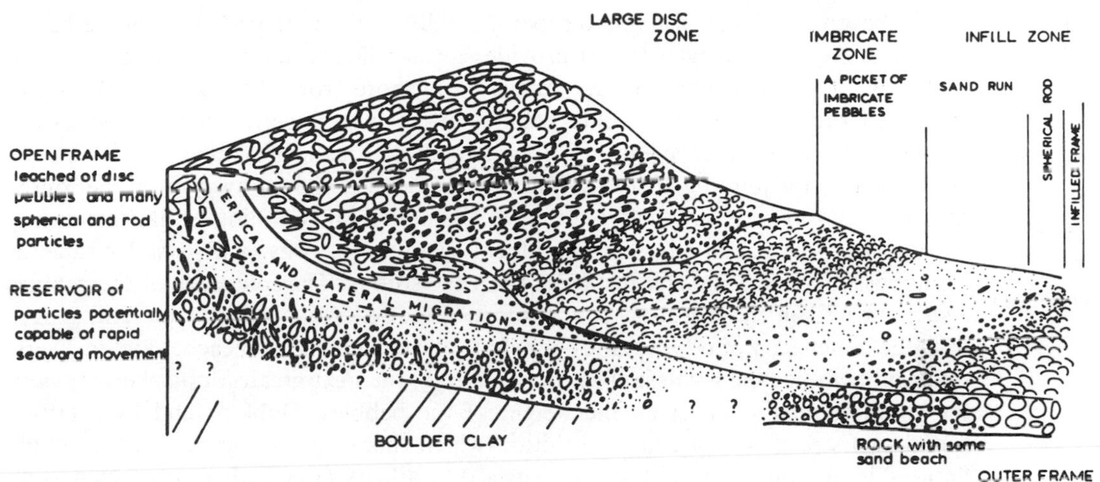

Figure 13-8 Zones of pebble sorting on the beach shown in Figure 13-7, with large disc-shaped pebbles high on the beach and the more spherical and rod shapes at the bottom. [*From* Bluck (1967)]

sliding motion, and this is the wave intensity that develops the best discs. Any pebbles that are larger tend to remain stationary and so are not abraded by sliding to form a disc shape. Smaller pebbles are rolled and tossed randomly by the waves and fall into chinks between larger cobbles, and thus are abraded on all sides and the shape is not altered. This arrangement of pebble abrasion and process agrees with the conclusions of Marshall (1930): the greater the wave energy, the larger the optimum size of the pebbles that can achieve disclike shapes. Dobkins and Folk have even suggested that one could measure shapes as a function of pebble size in ancient conglomerates to determine which size had the most discoidal development and thereby estimate the ancient wave intensity.

Dobkins and Folk (1970) also found that there is a difference in processes and pebble shape achieved on sandy beaches from those achieved on beaches composed entirely of gravel. As already noted, on pure-gravel beaches there is a critical pebble size that achieves maximum disc shape because such pebbles are abraded by sliding across the other pebbles. On sandy beaches, in contrast, there is no critical size; rather, all sizes achieve low sphericity values by sliding abrasion and perhaps by sandblasting of the upper surface of the pebble: all sizes achieve a pronounced disc shape.

One other mechanism of shape sorting of pebbles must be mentioned—that within beach cusp systems. Dobkins and Folk (1970) found one such example in Tahiti. The pebbles on the cusps had remarkable concentrations of rodlike shapes and high sphericity values, even more pronounced than those of river pebbles. On the other hand, strewn on the sand in the bays was an accumulation of especially flat disclike pebbles. Similar to the conclusions of Landon (1930), Dobkins and Folk reasoned that the sand substrate can trap the discs which do not slide well, whereas the rods and more spherical grains roll onto the adjoining cusp horns. Such a shape sorting within beach cusps was also found experimentally by Fleming (1964) in a wave basin.

The study of Grogan (1945) provides an excellent example of the progressive rounding of beach pebbles as they move alongshore from their source. His study examined the rounding of rhyolite pebbles on a Lake Superior beach. The parent rhyolite outcropped at one end of a long stretch of beach, providing angular blocks controlled by the jointing of the rhyolite. As the pebbles moved away from the source they became progressively better rounded, the degree of rounding at first increasing markedly as the rough edges of the blocks were worn away, with smaller rates of roundness increase at greater distances from the source. The sphericity of the pebbles also increased in the direction of transport, although not markedly so. This agrees with the conclusion of other studies of pebble abrasion, both on beaches and in rivers, that abrasion has the greatest effect on rounding, the next greatest on sphericity, and the smallest relative effect on the weight of the pebbles. Dobkins and Folk (1970) found two stretches of beach on Tahiti in which there was an increase in the pebble flatness in the direction of littoral transport. Cailleux (1945) also noted increasing flatness of beach pebbles with transport.

There have been comparatively fewer studies of rounding and shape of beach sands than of pebbles because of the difficulty of making the basic measurements. In the pocket beaches at Cape Arago, Oregon, Rottmann (1973) compared the roundness and sphericity of beach sands with the grain properties within the source bedrocks and terrace sands. That these were in fact the sand sources was demonstrated by supplementary studies of grain size and heavy minerals. It was found that the beach sand grains underwent a statistically significant increase in roundness over that found in the sources. However, there was no accompanying change in mean sphericity between source and beach sands.

Beal and Shepard (1956) examined grain roundness of beach and dune sands on the Gulf of Mexico coast of the United States in order to determine whether the two environments could be distinguished on this basis. They found that there was a consistent difference, the dune sands being rounder than the corresponding sands on the adjacent beach berm and foreshore. Because the roundness value changed abruptly at the boundary between the dunes and the beach, they concluded that the difference must be due to selective sorting, the onshore winds selectively removing the more rounded grains from the beach and transporting them to the dunes. Because of the short transport distances involved, abrasion must be negligible. These conclusions were also supported by the study of Shepard and Young (1961) from worldwide samples.

MacCarthy (1935) also examined adjacent beach and dune sands on the east coast of the United States and found that the dunes had a higher sphericity than the beaches, and he also attributed this to selective sorting by the winds. Beal and Shepard (1956) and Shepard and Young (1961) did not include measurements of sphericity. An increase in sphericity with increasing roundness in selective sorting patterns is typical, as we shall see again in a moment in the discussion of the longshore selective transport of grains. Thus the results of Beal and Shepard (1956) and MacCarthy (1935) taken together probably indicate that there are increases in both roundness and sphericity from beach to dunes at both locations. Krumbein (1942) has suggested that roundness in itself is not important in the selective sorting process, but that because of a positive correlation between grains of higher sphericity and better rounding, the selective sorting by sphericity produces a corresponding sorting by roundness.

The progressive longshore sorting of sands according to shape has been demonstrated by MacCarthy (1933) on Atlantic coast beaches, where he found a progressive decrease in sphericity in the direction of longshore drift. He attributed this to selective sorting wherein the low-sphericity grains outdistance the more spherical ones.

Pettijohn and Lundahl (1943) analyzed samples from the beach on Cedar Point, Lake Erie, for roundness and sphericity and found that the sphericity declined slightly and the roundness considerably along the length of the sand spit in the direction of littoral transport. This is of course opposite to what would be expected if progressive abrasion were significant in causing increased rounding as the sand is transported. Pettijohn and Lundahl, like MacCarthy (1933), explained the changes in terms of selective sorting processes: the less spherical and more poorly rounded grains have lower settling velocities and therefore can more easily remain in suspension and be carried alongshore at higher rates. Similarly, using the same Cedar Point samples, Pettijohn and Ridge (1933) found that heavy minerals of low sphericity were transported selectively along the spit, so that low-sphericity minerals such as hornblende and hypersthene increased in the direction of transport, whereas more equidimensional minerals such as garnet and magnetite lagged behind and decreased in the transport direction.

SEDIMENTARY STRUCTURES

On beaches can be found a great variety of sedimentary structures, including those produced by wave swash action and winds, as well as tracks and burrows made by organisms. Many of these structures can be viewed as diagnostic of the beach environment and so are of interest to geologists in identifying ancient deposits. The purpose of this section is to summarize these structures and review what is known about their origins.

Swash Marks

When the wave swash moves up the beach face its forward edge tends to carry with it sand and debris caught up in surface tension (Evans, 1938). This is deposited at the limit of greatest advance of the swash and is marked by an irregular, thin wavy line of a sand ridge some 1 to 2 mm high, a concentration of mica flakes and flat shell hash, foam, and bits of seaweed and other debris (Figure 13-9). Johnson (1919, p. 514) coined the term *swash mark* for this feature, and this is the name generally used.

Swash marks are typically irregular—first because the initial wave swash is variable in its runup, depending on variations in local wave height and beach topog-

Figure 13-9 Swash marks due to wave runup on beach face, and V-swash marks from the backwash.

raphy. The irregularity is increased because the swash mark of one wave is partly obliterated by a later swash, so that only a portion is preserved. A falling tide produces a series of bifurcating swash marks, the trace from any individual wave swash existing only for a short distance along the beach.

Emery and Gale (1951) have made a careful study of swash and the resulting swash marks. They found that swash marks are observed to form only on the upper part of the beach which is above the water table and therefore fairly dry. Swash marks were rarely found on the lower beach below the level at which the water table outcrops on the beach face, since this zone is covered by a thin, glassy film of water. This film prevents water surface tension from driving the sand forward by the moving water edge, and the downslope flow of the water percolating out of the beach face also tends to wash away any swash mark that may happen to form. Emery and Gale also made extensive measurements on swash mark spacing and found that the average spacing ranges from 3 to 6 meters (10 to 20 feet) for gentle beaches to less than 0.3 meter (1 foot) for steep beaches. There are three reasons for this. First, the water table falls faster in steeper beaches (since they are coarser-grained as well), and as swash marks form only on the dry beach above the water table there is clearly more opportunity for the swash marks to form on steep than on gentle beaches. Second, there is less width of steep beach exposed per unit time by the falling tide, so the swash marks would be compressed together. Finally, Emery and Gale found that the period between successive swashes is greater than the wave period, especially on beaches of low slope. With a low slope the period between the swash maximum was on the order of 15 to 20 seconds where the wave period was approximately 5 seconds. After a large swash has moved up the gentle beach, its considerable volume of water requires a long time to return to the sea. During its extended backwash period it is able to break down several successive advancing swashes. Finally, when most of its water is gone, a later swash can again cross the beach. On steeper beaches most of the water from a given wave swash is able to return to the sea before the next wave arrives. The result is that with a falling tide the spacings between maximum swash runup will be greater on the gentler beach because of the greater elapsed time.

The presence of swash marks in an ancient geologic formation is a sure sign of wave action and a beach environment. Although they are not very common, swash marks are known from several ancient beach deposits [for example, Fairchild, (1901)]. Because the swash tends to be arcuate in shape and to extend only over a limited length of beach, the resulting swash mark is convex shoreward. Twenhofel (1932) has suggested this as a criterion for determining the landward and seaward sides of ancient beaches.

Current Crescents and V-Swash Marks

A shell or pebble placed on the beach face deflects the backwash flow, producing a scour crescentic in shape surrounding it and a characteristic V-shaped structure that widens in the flow direction (Figure 13-9). This feature has not received much attention on beaches other than its use by geologists to identify the direction of the backwash and therefore the offshore side of an ancient beach. Twenhofel (1932) called this beach feature "current marks." The structure is very similar to that produced by unidirectional sheet flow past a pebble in a shallow stream, and in this regard it has been studied by Sengupta (1966) who, following Peabody (1947), used the term *current crescent.* The pebble deflects the water flow so that the velocity is increased around its sides and upstream end. This excavates a semicircular depression or moat on the upstream end of the pebble (Figures 13-10 and 13-11). The enhanced velocity and eddies continue downflow from the object, producing an erosional V, the point being at the pebble. The sand that is eroded redeposits to form a depositional V just interior to this erosional V (Figure 13-10). This is often accompanied by a selective sorting of the light-colored, coarser quartz grains and the finer dark grains, making the structure more apparent (Figure 13-9).

Rhomboid Marks

When present, among the most conspicuous swash zone structures are *rhomboid marks* (Figure 13-11), also commonly referred to as *rhomboid ripple marks* or *rhomboid rill marks* (Otvos, 1964, 1965). These are roughly diamond-shaped or rhombohedral patterns developed on the beach face by the wave backwash; they have been described as resembling the scales of a ganoid fish or the overlapping scale leaves of a cycadean stem. There is some variety among rhomboid marks; Otvos (1965) distinguishes three types.

The rhomboid marks studied by Hoyt and Henry (1963) at Sapelo Island, Georgia, ranged in length from 1 to 46 cm, most commonly being 3.75 to 13 cm long. The ratio of length to width (long diagonal to short diagonal) was from 2: 1 to 5: 1. The long diagonal is always found to be oriented in the direction of flow. Hoyt and Henry also found a good correlation between the length-to-width ratio and

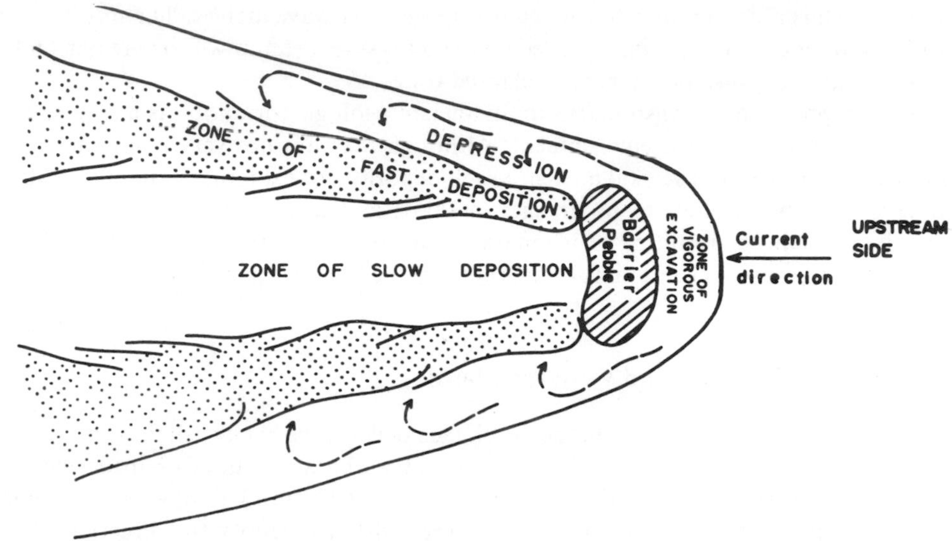

Figure 13-10 Zones of deposition (dotted) and erosion around a pebble caused by sheet flow, whether in a stream or in a backwash, producing a V-swash mark. [*From* Sengupta (1966)]

the beach face slope: the higher the slope, the greater the ratio, the pattern being stretched out more at the higher slope, presumably because of the higher backwash velocities. This finding is comparable to that of Woodford (1935) and Demarest (1947), who found that the angle between the sides of the upstream end of the rhomboid pattern decreases with increasing velocity.

Rhomboid marks are formed under the wave backwash, developing when the water sheet flow depth is less than approximately 2 cm and the velocities still relatively high. The marks develop as the swash recedes, moving down the beach face until left dry by the receding water. They occur on coarse-sand beaches but are more apparent on medium- to fine-sand beaches because they commonly also bring about a selective sorting of the quartz grains and the finer, dark, heavy minerals. The tongues thrown forward are composed mainly of light quartz grains, while the hollows left behind which form the gaps between tongues in the next line are composed principally of the dark, heavy minerals. Therefore, a distinct color pattern is also produced, making the rhomboid marks more noticeable.

The origin of these curious rhomboid marks has not been confirmed. It is not necessary to have crossing or overlapping backwash to establish the pattern. What apparently is required is a mechanism to deflect the sheet flow from its normal course straight down the beach face into a pattern that can produce the rhomboid design. Demarest (1947) called for "dimple" water ripples resulting from water percolating out of the beach face under the receding backwash, the "dimples" deflecting the flow. This explanation has generally been discounted. It has already been seen that large objects such as a shell or pebble placed on the beach face deflect the flow into a charac-

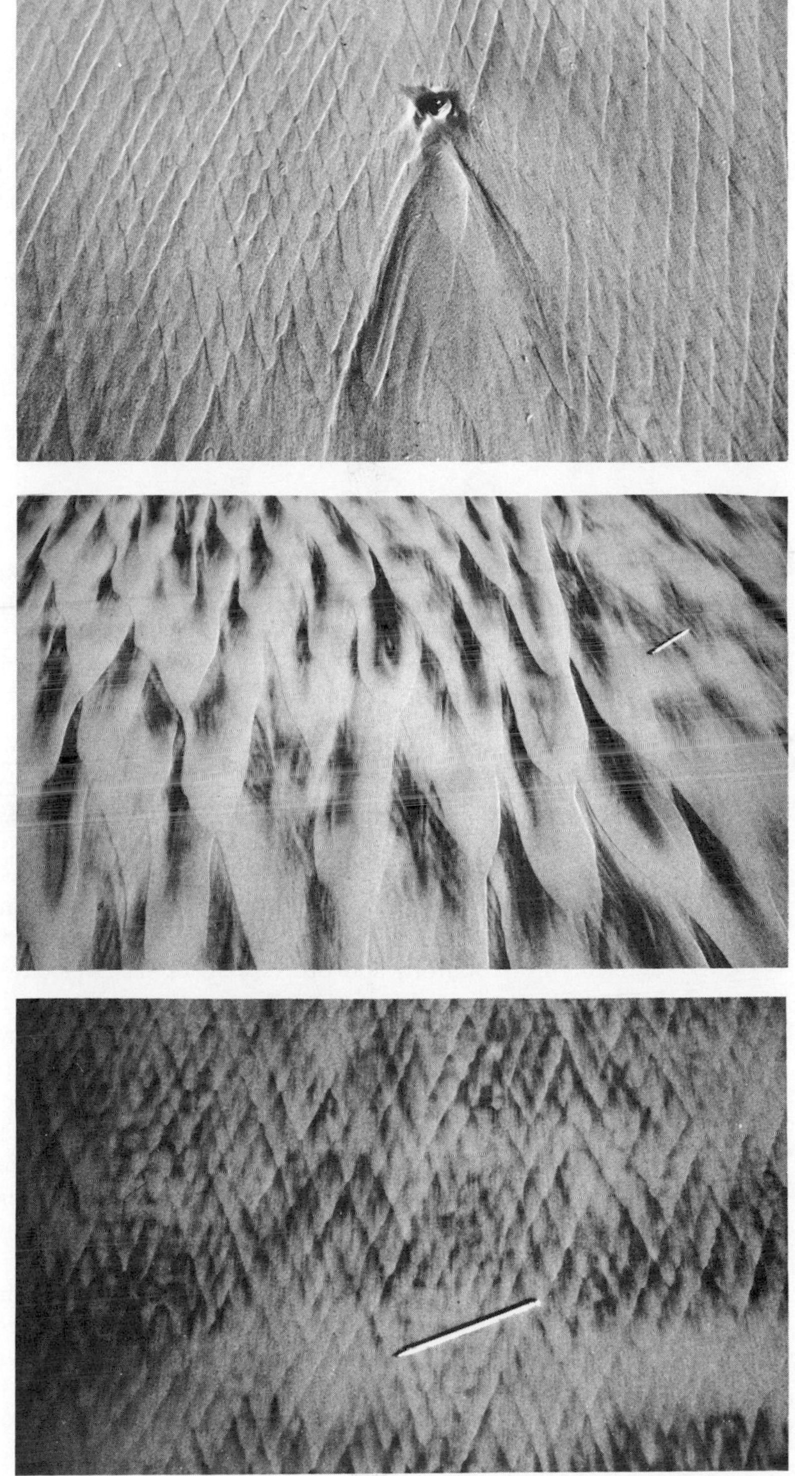

Figure 13-11 Examples of rhomboid marks. The top example also shows a V-swash mark formed downflow from a pebble.

teristic V-shape. If a large number of such objects existed on the beach, their individual V-shape lee flow patterns would interfere with one another and give rise to a rhomboid pattern. However, in the case of most rhomboid marks there are no noticeable objects visible on the beach. It is possible that minor irregularities in the grain surface texture could replace the large visible objects. Johnson (1919, p. 517) indicated that "the thin sheet of water returning down the beach slope appeared to be split into diverging minor currents by every patch of more compact sand or particle of coarser material which impeded its progress, and the crossing of these minor currents resulted in the criss-cross pattern in the sand." Woodford (1935) has suggested that rather than resulting from deflecting currents, the pattern is produced by the interference of the waves that propagate downcurrent from the objects or irregularities when the flow is supercritical—that is, when the Froude number

$$\mathrm{Fr} = \frac{u}{\sqrt{gh}}$$

is greater than unity, where u is the velocity and h is the flow depth ($g = 981$ cm/sec^2 is the acceleration of gravity). With supercritical flow the wave impulses produced by the irregularity on the bottom cause a V-shape pattern whose upstream acute angle α is given by

$$\sin \alpha = \frac{\sqrt{gh}}{u} = \frac{1}{\mathrm{Fr}}$$

A system of irregularities would give the observed rhomboid pattern. I have seen such patterns in small-scale sheet flow in a wash basin, even though the boundary appeared to offer no irregularities to produce the crossing waves. This relationship for the angle would explain why the long axis is always in the flow direction and why the angle decreases with increasing velocity or with increasing beach slope. There is a total lack of the necessary measurements of swash parameters and of the resulting rhomboid swash marks with which to test this hypothesis of Woodford (1935). Measurements by Woodford and others of backwash velocities and depths do indicate that the flow is supercritical.

The type of rhomboid marks discussed thus far is termed "rhomboid rill marks" by Otvos (1965) to distinguish them from "rhomboid ripple marks." This is based on Bucher's (1919) classification of ripple marks formed by uniform flow. Engle (1905) found in his laboratory experiments that the first effect of transportation by a uniform flow was the formation of "small rhomboid, scale-like tongues of sand." With increasing velocity of the current, common current ripples took their place. All of the various rhomboid marks on beaches are clearly closely related in form as well as origin to similar forms produced by unidirectional flow. Only with a careful study of their formation will any classification into different varieties be meaningful.

One point of interest among geologists regarding rhomboid marks is that their presence in ancient rocks would suggest wave processes and a beach environment. In addition, their geometries may possibly be utilized to determine the beach face slope of the ancient beach and perhaps something about the wave properties (Hoyt and Henry, 1963). Twenhofel (1932, p. 574) mentions one ancient example cited in a publication by de Bethune.

Backwash Ripples

There have been no careful observations or extensive studies made of the formation of *backwash ripples* in the swash zone (Figure 13-12). Although they are very common, the only measurements of them reported in the literature are those of Tanner (1965) from the southeast coast of the United States: wave lengths (crest to crest) range typically between 20 and 50 cm. Their heights are very low, so that the *ripple index*, the ratio of length to height, ranges between 30 and 100, much larger than for other water-formed ripples.

Figure 13-12 Low-amplitude backwash ripples, exposed at low tide on the beach face. Ocean is to left.

Backwash ripples are generally observed to migrate slowly downslope and to be asymmetrical. Their leading edges are often bisected by small rills and may show small-scale rhomboid marks superimposed. This concentrates heavy, dark minerals just behind a narrow leading edge of quartz grains, so that the low-height ripples appear as color banding parallel to the length of the beach.

Little study has been made of the origin of backwash ripples. Hayes (1972, p. 344) expressed the opinion that they are formed under supercritical flow conditions ($Fr > 1$) and are comparable to antidunes that form in rivers. By definition, antidunes migrate upstream, opposite to the flow, whereas the backwash ripples migrate with the flow. However, this seaward migration may be a secondary modification following the initial formation at higher Froude number flow. More study of this beach feature is required.

Rill Marks

Rill marks (Figure 13-13) are generally formed by water seeping from the beach face at low tide or following a storm. The water percolates out from the beach and accumulates into small rills which join to form larger rills. The largest rills are much like small-scale braided streams. The result is a miniature dendritic drainage system with the trunk extending downslope, most often in a seaward direction, although the rills are sometimes found on the shoreward side of a bar exposed at low tide or on a shoreward-sloping berm. They also flow into tide pools scoured around large rocks exposed on the beach and hence do not yield any particular onshore or offshore direction. The dendritic patterns of rill marks closely resemble branching plant stems and apparently were commonly mistaken for such when observed in ancient rocks by early geologists (Johnson, 1919, pp. 512–14).

Figure 13-13 Rill marks.

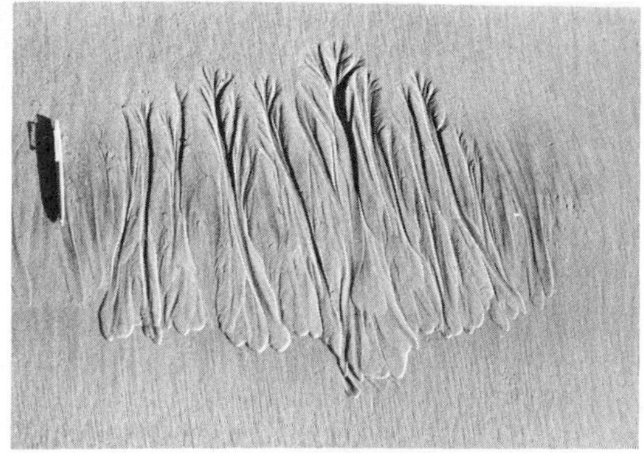

Laminated Beach Sands

One of the principal features of the beach sand deposit when viewed in cross-section are the *laminations* (Figure 13-14) ranging from 1 to 20 mm in thickness

Figure 13-14 Laminations and heavy mineral concentrate shown in a trench cut into beach face. Shovel at right for scale.

(Thompson, 1937; Emery and Stevenson, 1950). They are most apparent on medium- and fine-sand beaches, since there they are enhanced by sorting of the light quartz-feldspar and the dark, heavy minerals into alternating layers. When they are examined closely it is found that each lamination generally consists of coarser quartz sand near its top, with finer-grained heavy minerals concentrated at the base of the layer. A succession of laminations therefore gives alternating light and dark bands. Laminations are also present in coarse-sand beaches, but are not so apparent because of a lack of heavy minerals to enhance their appearance. However, they are generally thicker in coarse-sand beaches.

In sections parallel to the shoreline an individual lamination can typically be traced for 25 meters or more before it disappears (Thompson, 1937). In sections normal to the shoreline a single lamination can seldom be followed for more than 7 meters before it is terminated. The laminations parallel the beach face, so that their original dips correspond to the slope of the foreshore at the time of formation.

Grain-size and composition sorting within an individual lamination have been investigated by Emery and Stevenson (1950) and by Clifton (1969). The measurements of Clifton (Figure 13-15) strikingly show that each lamination consists of a basal fine layer of dark, heavy minerals and an upper coarse, light layer. Within the laminations the contrast between the dark, heavy mineral concentrations and the light quartz zone is gradual, but the interface between adjacent laminations is very sharply delineated.

The studies indicate that laminations are produced in the swash zone, apparently by the backwash flow. The segregation between coarse and fine grains and between light and heavy minerals occurs very quickly within a moving bed-load layer of sand following the maximum erosion during the backwash (Clifton, 1969, p. 557). Thompson (1937) believed that the origin of the laminations is due to variations in the transporting power of the waves. The dark, heavy mineral concentration results from these

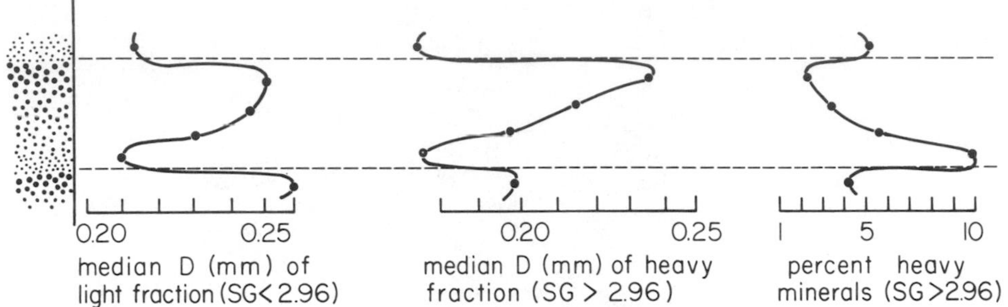

Figure 13-15 Changes in grain size and heavy-mineral proportion through a single thin beach lamination. [*After* Clifton (1969)]

grains being more difficult to transport than those of lower density, so that they settle out first and form the bottom of an individual lamination. This view is not supported by observations of swash processes; but more important, this mechanism could not explain laminations on coarse-sand beaches where there is sorting by grain size, the coarser grains being at the top of the lamination, but no sorting by composition or density since heavy minerals are scarce in these coarse sands. This observation indicates that sorting by grain size is more important than sorting by density.

Bagnold (1954) has demonstrated that when grains of mixed size are sheared together at high rates and at high concentrations, the coarser grains will tend to migrate upward to zones of low shear and the finer grains downward toward the bed, where the shear rates are a maximum. The result would be a sand layer with reverse grading. Inman, Ewing, and Corliss (1966) termed this process *shear sorting* and indicated that it is responsible for the laminations in sand dunes which are comparable to beach sand laminations. This shear-sorting mechanism is most probably the origin of the beach laminations, the backwash shearing the sand as it flows seaward. However, Middleton (1970) has argued against the shear-sorting mechanism of Bagnold (1954). Instead, Middleton believes that the sorting is produced by the smaller particles tending to fall into the spaces between the larger particles and thus displacing the larger particles toward the surface. No definitive experiments have been performed to study the mechanics of the sorting process.

This sorting into laminae gradually tends to drive the smaller-grained heavy minerals downward into the beach face. With time the dark heavies become concentrated at depth within the beach deposit. One example of such a concentration is shown in Figure 13-14. Rao (1957) published a good example of this from the coast of India. When storms cut back the beach these heavy-mineral concentrates become exposed on the beach. This is the case for the example shown in Figure 13-14: during the winter following the photograph, the quartz-rich layer was shifted offshore and the beach face became black from the hornblende-rich sand.

Cross-Laminations in Beach Sands

Cross-laminations in beaches result from changes in the overall beach profile and from migrating sedimentary structures. Thompson (1937) made a careful study of the former, while Clifton, Hunter, and Phillips (1971) investigated the latter.

Studying California beaches, Thompson (1937) noted that the laminations closely parallel the local surface of the beach, so that changes in the beach profile configuration would produce changes in the lamination orientation in those areas where deposition is taking place. He made a systematic study of this, following beach profile variations daily and the resulting patterns of cross-lamination during a storm and the post-storm recovery of the beach. Thompson's investigation was limited to the berm and upper foreshore, since he utilizied trenching. He found that most laminations dip seaward and that the angles between different sets which form the cross-laminations are generally small. He found examples of cross-lamination produced by the burial of formerly eroded berm scarps, and examples of buried channels with rounded floors and coarse, poorly sorted sediments in their axes. Thompson did find some shoreward-dipping laminations, where material from the foreshore was carried up over the berm crest and washed landward down the sloping berm. He also found steeply inclined foreset laminations dipping shoreward (Figure 13-16), produced at

Figure 13-16 Internal cross-stratification, dipping shoreward, produced by the landward migration of a bar. [*From* Hayes (1972)]

the shoreward edge of a bar as it migrated shoreward. Examples of this latter feature are also given in Hoyt (1962), by Hayes (1972) [Figure 13-16], and in Wunderlich (1972).

Based on the conditions of deposition, Thompson (1937) inferred that deposits of the lower foreshore are intricately cross-laminated, dipping in all directions. Being limited to trenching, he could not make actual observations of this area, however. Utilizing Senckenberg boxes and divers, Clifton, Hunter, and Phillips (1971) were able to obtain samples of the structures in this area as well as under the surf, breaker, and offshore zones. On the basis of their observations off Oregon and California, they divided the nearshore into a series of zones or facies trending parallel to the shoreline and reflecting the wave and current processes (Figure 13-17). There were two plane-

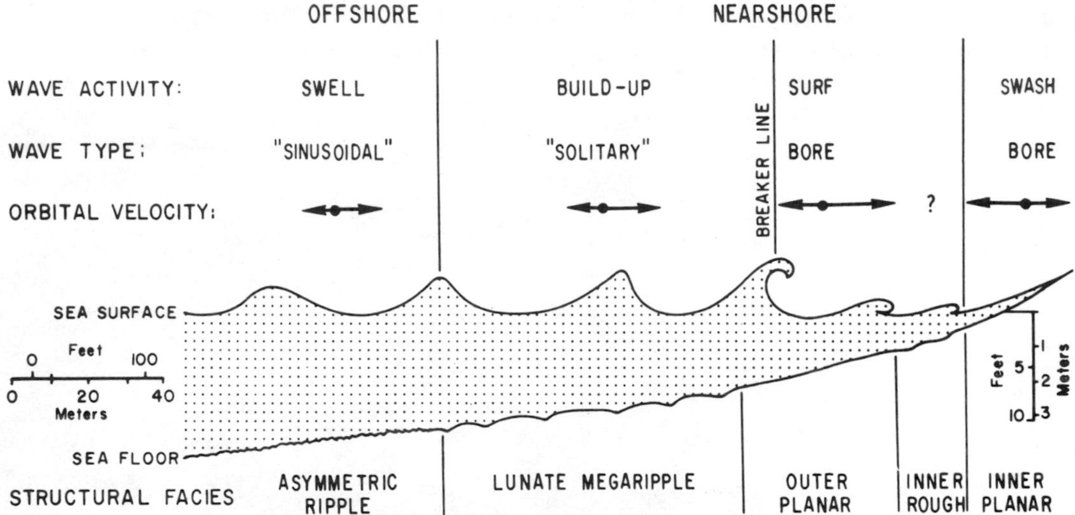

Figure 13-17 Zonation of depositional structures and wave activity within the high-energy nearshore of Oregon. [*From* Clifton, Hunter, and Phillips (1971)]

bed facies characterized in cross-section by parallel laminations. Separating these was an inner rough facies of large, coarse-grained ripples, and to the seaward side was a lunate megaripple facies. The ripples themselves will be discussed later, but here they are important in that Clifton, Hunter, and Phillips found that the migrations of the ripples produced cross-laminations. The large ripples of the inner rough facies formed seaward-dipping trough cross-bedding of medium scale (units 4 to 10 cm thick). Longshore currents were important in this zone, so that there was often a longshore component to the dip of this cross-bedding. The seaward lunate megaripples were found to migrate shoreward, resulting in internal structures of medium-scale cross-stratification with foresets dipping steeply landward. Due to the lunate shape of the megaripples, the geometry of the resulting cross-bedding was of trough shape. There was little other longshore component, since longshore currents were weak outside the breaker zone. There was a distinct size grading up the slipface, which was preserved in the foresets, the base of the foreset lamination being especially coarse in grain size. Clifton, Hunter, and Phillips (1971) found occurrences of asymmetric ripples in the offshore and concluded that they could be responsible for small-scale (units less than 4 cm thick) shoreward-inclined ripple cross-lamination. Such occurrences were also found by Newton (1968) in the North and Baltic seas.

Other studies of laminations and cross-laminations and their distributions through the nearshore zone are those of Wunderlich (1971) and Reineck and Singh (1971) on the west coast of Italy, and the investigations at Sapelo Island, Georgia, by Wunderlich (1972) and Howard and Reineck (1972).

Both Thompson (1937) and Clifton, Hunter, and Phillips (1971) examined outcrops of Late Pleistocene marine terraces along the Oregon-California coast for beach deposits. They found many examples, and cross-lamination was abundant. Clifton, Hunter, and Phillips especially emphasized the sequence of laminations and cross-laminations in identifying the position of the deposit within the nearshore environment.

Sand Domes and Cavernous Beach Sand

Reade (1884) and Johnson (1919, pp. 518–20) described and explained the origin of *sand domes* which appear on the beach when the swash from a large wave sweeps over a portion of beach that was previously dry. As the water seeps down into the beach sand, the miniature domes appear at scattered points on the beach face. They vary in size from 5 to 20 cm in diameter and rise a centimeter or more above the level surface. If it is sectioned, it will be found that the dome is hollow, the wet sand layer being bowed upward over a pocket of air. Reade and Johnson independently explained the formation of sand domes by an accumulation of air that is displaced from the sand pore space as the water seeps down into the beach. Much of this displaced air is seen to bubble up through the wet sand layer to escape the beach. In other places the wet sand layer is quite airtight, so that the expelled air must collect into pockets and so form domes. Emery (1945) also described dome formation as well as black spots or rings produced by the subsequent truncation of domes which raised beach laminations.

Occurring in the same area of the beach reached by the high tide, and closely related in origin to the sand domes, are the *cavernous soft beach sands.* These have been studied by Kindle (1936) and Emery (1945). Kindle's attention was drawn by the fact that sand in this high-tide zone is much softer than elsewhere on the beach. He concluded that the soft sand is produced by the expulsion of the air as the water swash first covers the previously dry sand. The air expulsion itself yields loose, granular packing of the sand, but more important is that air is trapped and remains in patches within the beach, leaving the sand with a honeycombed structure. Emery made a more detailed study of the features of this cavernous sand. He found that the air cavities varied from spherical to roughly ellipsoidal in shape, elongated in the bedding plane. The diameters of the cavities were many times that of the sand grains and usually occurred in layers 20 to 30 cm thick. In one example they constituted 36%

of the volume of the original sand mass, which taken together with the 24% pore space of the original sand gives a total pore space of 56% by volume. Soft, cavernous sand forms best in sand of intermediate grain size, being less common in very fine or very coarse sand. It is less well developed on beaches exposed to large waves than on beaches protected within bays. Sand domes and cavernous sand are absent or rare on lake beaches, apparently because of a lack of water level change from the tides, which is important in their formation. Cavernous sand is not necessarily related to beaches, being very common on tidal sandbars.

Related to the sand domes and cavernous sands are holes 1–4 cm deep and 1–10 mm in diameter, similar to burrows made by sandhoppers. These holes are formed by the stream of air as it escapes upward from the beach, bubbling upward through the covering wave swash (Palmer, 1928, Emery, 1945).

Ripple Marks

In addition to the backwash ripples already discussed, oscillatory and current ripples and combinations thereof can be commonly found on the exposed beach at low tide or observed in the offshore while diving. On the exposed beach the ripples occur most frequently in two locations—in troughs which run parallel to the beach length or strike off across the beach in a former rip current position, and secondly on the extensive tidal-flat areas that are sometimes found between the high-tide and low-tide beaches.

The *ripple marks* found in beach troughs often show a combination of wave oscillatory ripples and current ripples trending roughly at right angles to one another and known as *ladder-back ripples*. Oscillatory ripples are produced by the waves, so their crests parallel the shoreline, while the current ripples are generated by the longshore currents and indicate flow in the longshore direction, unless diverted seaward as a rip current. Depending on the relative strengths of the waves and longshore currents, either the oscillatory or current ripples can dominate in the ladder-back pattern.

The same type of ladder-back ripples can be found on the exposed tidal flat. Here the oscillatory ripples generally dominate, as the longshore currents are much weaker, the cross-current ripples being formed by the runoff of the tide rather than by a true longshore current. When the tidal flat is nearly exposed, the last bit of water drains in the longshore direction, using the troughs of the oscillatory ripples as small channels, finally collecting in larger channels which funnel the water offshore.

Clifton, Hunter, and Phillips (1971) described two large sets of ripples on the high-energy beaches of Oregon and California (Figure 13-17). The cross-laminations produced by these ripples have already been described. The first set occurs in the transition zone of Schiffman (1965), where the backwash collides with the next incoming surf bore, producing a set of standing waves much as would an undular hydraulic jump. These standing waves apparently generate the coarse-sand and gravel ripple marks—or perhaps they should be called *antidunes* in light of their origin. Off steep beaches, Clifton, Hunter, and Phillips (1971) found three to ten such ripples with spacings of from 30 to 60 cm and heights of from 15 to 20 cm. In gently sloping beaches these ripples are commonly totally absent. Seaward of the breaker zone, Clifton, Hunter, and Phillips found a region of lunate megaripples with heights of from 30 to 100 cm and irregular spacing. These had active slipfaces and were migrating shoreward.

There is an extensive literature on the generation of current ripple marks, so this subject will not be reviewed here. Instead we shall examine the formation of oscillatory ripple marks and how they relate to wave motions.

Oscillatory ripple marks offshore of the breaker zone are sometimes found to be asymmetrical in the onshore direction (Evans, 1941; Newton, 1968; Clifton, Hunter, and Phillips, 1971). This may be due to linear currents superimposed on the wave motion so that an intermediate class of ripples between oscillatory and current is produced. Such ripples have been described by Reineck and Wunderlich (1968) and by Harms (1969). The asymmetry of the ripples offshore of the breakers may also be due to the asymmetry of the wave orbits in which a strong onshore motion occurs under the wave crest and a lower-velocity but longer-duration return flow occurs under the trough (Chapters 3 and 11). More important, perhaps, in this region the water wave orbits are many times longer than the ripples generated (Figure 13-18), so in a sense the ripples become somewhat akin to those produced by unidirectional flows. A sloping bottom would also promote the occurrence of asymmetrical ripples. In somewhat deeper water offshore of the breaker zone the ripples are usually more symmetrical and can be clearly identified as wave-induced.

Tanner (1967), Reineck and Wunderlich (1968), and Harms (1969) provide criteria by which one can attempt to distinguish oscillation ripples, current ripples, and combined-flow ripples preserved in ancient strata. The principal criteria are ripple symmetry, ripple pattern, ripple index (spacing-to-height ratio), crest angularity, and slope of faces.

Several laboratory studies of wave-induced ripples have been conducted in wave channels (Hunt, 1882; Evans, 1942; Scott, 1954; Kennedy and Falcon, 1965; Mogridge and Kamphuis, 1973), in pulsating water tunnels (Carstens, Neilsen, and Altinbilek, 1969), and by oscillating a section of bed through still water (Bagnold, 1946; Manohar, 1955). Some of the results of these several studies are mutually contradictory, and the understanding of oscillatory ripple formation is still unsatisfactory.

The study of Inman (1957) is the most comprehensive field investigation of wave-induced oscillatory ripples. His investigations, conducted through the use of underwater breathing apparatus, extended from the surf zone to depths of about 50 meters off the California coast. Figure 13-18, based on his results and supplemented by laboratory measurements, yields an empirical relationship between the ripple spacing λ and the orbital diameter d_o next to the bottom, where d_o is given by

$$d_o = \frac{H}{\sinh\left(2\pi h/L\right)}$$

where H is the wave height, h is the water depth, and L is the wave length (Chapter 3).

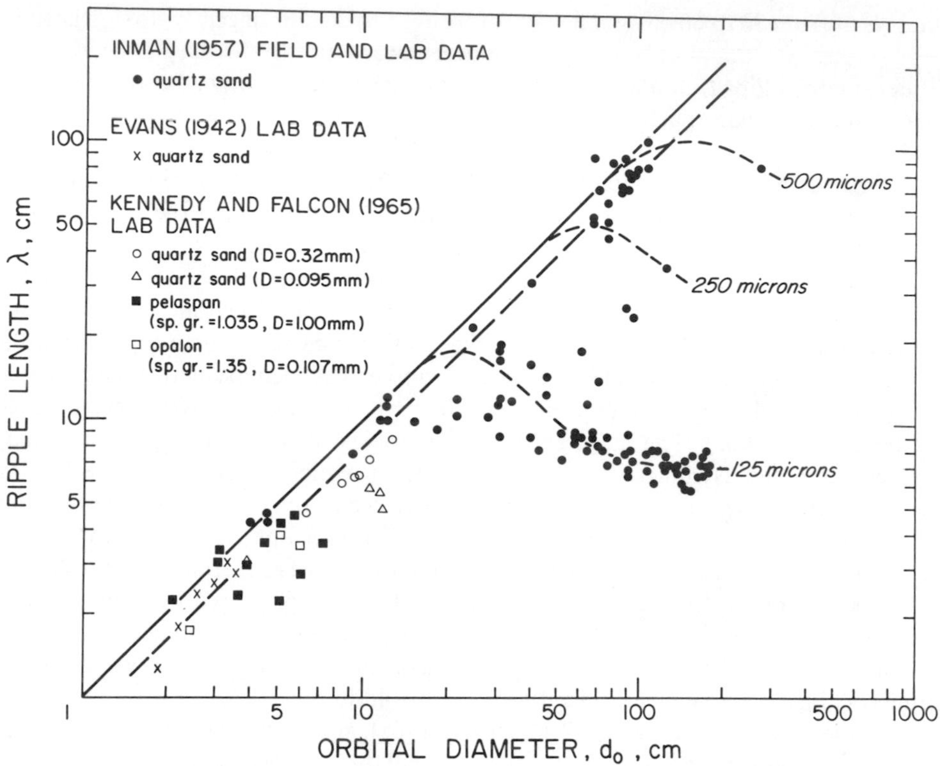

Figure 13-18 Relationship between the ripple spacing λ and the bottom orbital diameter d_0. [*After* Inman (1957) and Komar (1974)]

We see in Figure 13-18 that for sands with median diameters in the range 88–177 microns, as the wave orbital diameter increases the ripple spacing would first increase with approximately $\lambda = d_o$ (more exactly, $\lambda = 0.8d_o$), the ripple length ultimately reaching a maximum of about $\lambda = 21$ cm beyond which λ decreases with a further increase in d_o. Such a relationship between λ and d_o is also demonstrated by the wave channel study of Mogridge and Kamphuis (1973). With very large d_o a stress is finally reached at which ripples cease to exist and to-and-fro sheet sand movement occurs; the λ-versus-d_o curve of Figure 13-18 terminates on the right. This might suggest that in an onshore direction of decreasing water depth and therefore increasing d_o, the ripple spacing would first increase, reach a maximum, and then decrease. This is not generally the case, however, as has been discussed by Komar, Neudeck, and Kulm (1972) and at greater length by Komar (1974). The bottom orbital velocity must of course exceed the threshold velocity necessary for sediment movement. For the conditions in which $\lambda = 0.8d_o$ we must then have for the threshold velocity u_t (equation 8-19)

$$\frac{\rho u_t^2}{(\rho_s - \rho)gD} = 0.21\left(\frac{d_o}{D}\right)^{1/2} = 0.21\left(\frac{\lambda}{0.8D}\right)^{1/2}$$

Taking, for example, $\lambda = 10$ cm and $D = 125$ microns $= 1.25 \times 10^{-2}$ cm for the grain size, this gives $u_t = 11.6$ cm/sec for the required threshold velocity. The actual orbital velocity near the bottom must exceed this threshold velocity so that

$$u_m\,[\text{actual}] = \frac{\pi d_o}{T} = \frac{\pi\lambda}{0.8T} > u_t$$

where T is the wave period. The above example gives

$$T < \frac{\pi\lambda}{0.8u_t} = \frac{10\pi \text{ cm}}{0.8 \times 11.6 \text{ cm/sec}}$$

$$T < 3.4 \text{ sec}$$

Therefore, the $\lambda = 0.8d_o$ line of Figure 13-18 is generally relevant only for short-period waves and hence for very shallow water depths, the type of conditions found only in lakes and confined bays of limited fetch and in the laboratory wave tank. Under ocean conditions the right limb of Figure 13-18 is applicable where $\lambda \ll d_o$. This can be seen in the distribution of the data: the laboratory data forms the $\lambda = 0.8d_o$ line, while the field data of Inman (1957) forms the $\lambda \ll d_o$ curve.

One consequence of this is that in lakes and bays where $\lambda = 0.8d_o$ the ripple lengths will tend to increase onshore, whereas under ocean conditions the opposite is true—the ripples tend to decrease in spacing onshore (Komar, 1974). Due to the complex nature of waves reaching a coast, however, actual patterns of ripple lengths are seldom this systematic. Komar, Neudeck, and Kulm (1972), using an underwater stereo camera system, found oscillatory ripples in an onshore-offshore transect across the Oregon continental shelf at depths up to 125 meters in which the ripple length first increased in the offshore direction, reached a maximum length, and then decreased in length at still greater depths (Figure 13-19). We attributed this variation to waves arriving at the coast from a distant storm in a dispersed state, the longest-period waves arriving first with progressively shorter-period waves following (Chapter 4). The level of competence of sediment movement gradually shifts onshore under such conditions, leaving ripples preserved in deeper water from an earlier, higher-wave-period phase of the storm-produced waves.

The above discussion applies only to sands with median diameter in the range 88–177 microns, sands for which data is available. For coarser sands it is not yet possible to define the dependence of λ on d_o, but it may be similar to that depicted by the curve labeled 250 microns and 500 microns in Figure 13-18. The few available measurements for coarser sands indicate that the sand size is the single most important factor in determining the spacing of the ripples (Cook and Gorsline, 1972). For a given orbital diameter of water wave motion, the ripple lengths produced in 500-micron sands could be more than a factor 10 greater than ripples formed in fine sands. More data is required before the effects of sand size on the ripple spacing can be properly evaluated.

Inman (1957) classified the symmetrical oscillatory ripples into (1) solitary ripples, and (2) trochoidal ripples. *Solitary ripples*, like solitary waves (Chapter 3), have flat troughs separating isolated crests. *Trochoidal ripples* have rounded troughs and

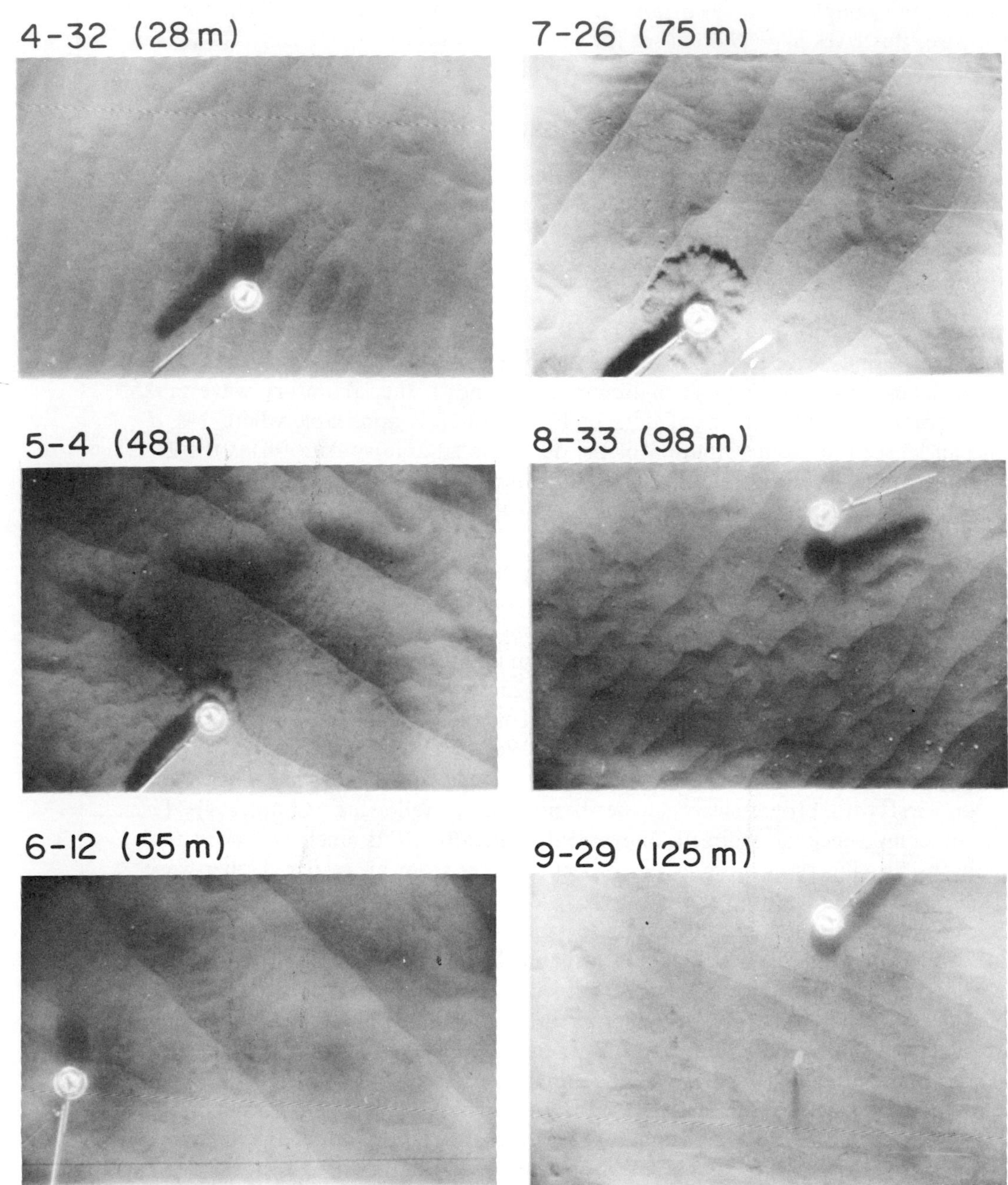

Figure 13-19 Ripple marks observed in bottom photographs in a transit across the Oregon continental shelf. Station numbers and water depths are given for each photo. [*From* Komar, Neudeck, and Kulm (1972)]

somewhat peaked crests that are either sharp or slightly rounded. Solitary ripples are generally restricted to water shallower than some 20 meters, and trochoidal ripples occur in deeper waters.

Inman (1957) has also classified ripple patterns on the basis of the ratio of the average crest length between the points where adjacent ripples bifurcate to the average ripple wave length. Long-crested ripples have ratios of crest length to ripple length greater than 8; intermediate ripples have ratios of 3 to 8; and short-crested ripples have ratios less than 3. Off California, Inman (1957) found that the short-crested ripples occurred furthest offshore and became progressively longer-crested as the beach was approached. However, even at depths greater than 100 meters off Oregon, Komar, Neudeck, and Kulm (1972) observed ripples that were long-crested.

By way of summary, let us examine a hypothetical onshore variation in ripple length and type as we proceed across an oceanic continental shelf from deep water to the beach. The first ripple forms observed in deepest water might be rounded and partially covered with a thin layer of mud; some organism trails tend to smooth and destroy the ripples. These ripples are obviously not in active formation and must have developed earlier under storm conditions of high wave height and long-period waves. In somewhat shallower water we can see some sediment movement, the bottom being swept clean of mud, and primary sharp-crested ripples are formed. At this point we can still be in very deep water. Bottom photographs of the Oregon continental shelf obtained by Komar, Neudeck, and Kulm (1972) revealed symmetrical ripple marks out to depths as great as 200 meters during intense winter storms. During the summer months of low waves, ripples occurred out to depths of 100 meters. Draper (1967) indicated that on the shelf edge west of Britain, wave action should be sufficiently intense to stir fine sand at a depth of 600 feet (183 meters) for more than 20 percent of the year. These deep-water ripples would be of the trochoidal type, and there would be a tendency for them to be short-crested. As we proceed into shallower water the ripple spacing decreases. The ripple pattern becomes more regular, with longer crest lengths. In about 20 meters of water the ripple height begins to decrease and the crests become isolated, so that the ripples are more of the solitary variety than trochoidal type. They continue to be long-crested. As one approaches the breaker zone it is noted that the oscillatory motions of the water are asymmetrical. When the strong onshore flow under the wave crest occurs, asymmetrical ripples are developed, with their steeper faces shoreward. The lower-velocity but longer-duration return flow under the trough may not be powerful enough to modify this form, or if sufficiently strong it would produce an asymmetrical ripple, with its steep face offshore. Just seaward of the breaker zone the onshore orbits under the wave crests are so strong that ripples are entirely wiped out but suddenly reappear under the weaker return flows. In the breaker zone itself both the onshore and return flows are so intense that no ripples are formed, the sand moving to and fro as a carpet within a few centimeters of the bottom. This picture of ripple development conforms to my experience and observations while scuba diving under fairly-high-energy ocean waves. It would differ in some respects in other nearshore environments.

Lebensspuren

There is a conflict in the nearshore zone between the physical processes which generate primary sedimentary structures such as laminations, cross-laminations, ripple marks, and so on, those which have already been considered, and the organisms in the nearshore which tend to obliterate the original stratification and form burrows, tubes, and other *lebensspuren.* The German word *Lebensspur* has been adopted as a collective noun for features that are a record of the life activities imparted by the organism to a substratum (Hertweck, 1972). The remains of the organism itself are excluded. The term approximately corresponds to the English *bioturbation structure.*

There is an extensive literature on lebensspuren and their implications with respect to environment identification in ancient sediment deposits. Only a brief review of a few of the more thorough studies of the nearshore area will be included here.

Reineck and Singh (1971) described the biogenic structures from the Italian coast of the Gulf of Gaeta and investigated the distributions of the structures. Lebensspur is insignificant at water depths less than 2 meters, since the beach physical processes are active at those shallow depths and primary sedimentary structures are preserved. Below 2 meters water depth bioturbation appears, increasing with increasing depth. Between 2 and 6 meters water depth the bioturbated structures are produced mainly by polychaete worms. The lower limit of wave action at that location is 6 meters, so that at greater water depths the grain size becomes finer, bioturbation predominates over the physical processes, and little or no primary structures are found. Although intensely bioturbated, well-shaped burrows are rather rare. The urchin *Echinocardium cordatum* is the main bioturbating animal, but *Turritella communis* (high-spired marine snail) apparently produces the fecal pellets that are abundant in the sediments.

Similar to the study of Reineck and Singh (1971) is the investigation of Howard and Reineck (1972) at Sapelo Island, Georgia. A comparable distribution of primary sedimentary structures and bioturbation was found, the bioturbation increasing at greater water depths as the wave orbital motions decrease in effectiveness and the sediments become finer-grained. On the basis of the structures, Howard and Reineck divided the nearshore into the following zones:

Upper shoreface (low-tide line to 1-m water depth): Low angle seaward-dipping, parallel lamination with bioturbation infrequent

Lower shoreface (1- to 2-m depth): Small-scale ripple lamination with minimal bioturbation

Upper offshore (2- to 5-m depth): Beds of parallel lamination and bioturbated muddy-fine sand; biogenic reworking high

Upper offshore (5- to 10-m depth): Extensive bioturbation, with physical primary structures obscured, in muddy-fine sand

Lower offshore (deeper than 10 m): Clean, medium to coarse sand; trough cross-bedding and bioturbation about equal, with biogenic structures almost exclusively formed by the heart urchin, *Moira atropos*

The studies of Reineck and Singh (1971) and Howard and Reineck (1972) both demonstrate that bioturbation increases with increasing water depth, principally due to the decreasing effectiveness of the wave action which erases the biologic structures and generates primary structures. The locations of the two studies involved similar wave energy levels, and both showed bioturbation beginning in about 2 meters water depth and becoming intense at 5 to 6 meters depth. It can be expected that on higher-energy coasts the region of primary structures will extend to greater depths, bioturbation becoming important only in deeper water at greater distances from the shoreline. The study of Clifton, Hunter, and Phillips (1969) off Oregon extended to some 10 meters depth with no mention of biogenic structures, the intense wave action instead producing megaripples and other primary structures. Kulm et al. (1975) found primary laminations on the Oregon shelf to depths of about 60 meters and sometimes deeper.

The study by Hertweck (1972) provides an excellent investigation of the burrow types and other lebensspuren at Sapelo Island, the same location as the study of Howard and Reineck (1972). Examples of the burrow types of the various animals can be found in the illustrations in Hertweck (1972) as well as in the aquarium study of Frey and Howard (1972). Utilizing the results from the study by Dörjes (1972) on the living populations at the site, Hertweck was able to compare the population with the biogenic structures and speculate on their relative abundance and preservability in the geologic deposits that might be produced. It was found that none of the most common species produce lebensspuren that are particularly suitable for preservation. Those that are most suitable for preservation are:

Ocypode quadrata (ghost crab): burrows in the backshore and dunes
Callianassa major (ghost shrimp): burrows in the foreshore
Callianassa biformis: burrow systems in the upper offshore
Moira atropos (heart urchin): traces in the lower offshore

The first three have preservable lebensspuren, in spite of their low numbers, because of the large sizes of the burrows and the considerable depths to which they extend downward into the sediment. An animal that moves through the sediment and produces locomotion structures may leave more widespread lebensspuren than an animal which forms dwelling structures. This is the case for *Moira atropos*, the heart urchin, who forms a disproportionally large number of lebensspuren considering the small abundance of the organism. Of the total number of organisms living in the area identified by Dörjes (1972), Hertweck (1972) concluded that only about 15% produce distinct lebensspuren and that less than half of those are apt to be preserved: only 7% of the living organisms are likely to form preservable lebensspuren.

Other noteworthy studies of lebensspuren in the nearshore are those of Frey and Mayou (1971) of decapod burrows, Frey (1970), and Hill and Hunter (1973) of the ghost crab *Ocypode quadrata*. Biogenic structures of tidal flats and platforms fronting the beach are discussed in the next section, along with other features typical of that zone.

Other types of lebensspuren besides burrows found on beaches are the tracks and trails of animals walking on the beach face. Tracks of shorebirds are particularly numerous, associated with burrowlike pits made by the birds driving their bills into the sand while feeding. Another "record of life activities" is the circular downbowing of laminations resulting from horse hoofprints, as described by van der Lingen and Andrews (1969). Emery (1944) reported rays of sand of 10 to 25 cm length on the beach face, thrown out by burrowing sandhoppers, a type of amphipod which feeds on kelp washed ashore by the waves, burrowing into the beach during low tide.

Beach Platform and Tidal Flat Sedimentation

We saw in Chapters 5 and 11 that in areas of large tidal range the beach profile consists of a steeper "high-tide beach" fronted by a wide platform, below which in turn is a "low-tide beach." Because of the nearly horizontal character of this platform and the rapid water level changes associated with the ebbing or flooding tide, the shoreline passes relatively rapidly over the platform (Problem 5-3). In addition, the wide expanse of shallow water tends to damp out the waves, further reducing the amount of wave exposure experienced by the platform.

There is not a clear distinction between what we are terming the *beach platform*, which is a part of the beach profile, and *tidal flats*. The primary difference is one of scale, the beach platform being at most a few hundred meters across, while the tidal flat may be several kilometers across. Because of the reduced wave energy the beach platform tends to be finer-grained than either the high-tide or low-tide beach. This is carried to an extreme in true tidal flats, where muds tend to make up the bulk of the deposit; on a beach platform, silts and clays make up only a few percent of the deposit, the bulk being sand. Not many workers make such a distinction between beach platforms and tidal flats, but I do not wish to extend our review to tidal flats that are not very beachlike. In fact, however, the two share many of the same characteristics, and tidal flats have been extensively studied by geologists and biologists. A review of the literature can be found in Reineck and Singh (1973).

The beach platform environment demonstrates the interplay of the waves and the tidal currents that flood and drain the area. The principal primary sedimentary structures one notes on the surface at low tide are the oscillatory ripple marks produced at high tide, modified by the ebbing tide. Having been produced by waves, they tend to be long-crested, with their crests parallel to the shoreline. They are almost always asymmetric, with the steep flanks onshore, appearing much like current ripples produced by an onshore flow. Because of the falling water level, small-spaced ripples are commonly developed on the crests of the earlier-formed larger ripples. The falling water level also rounds off the crests of the larger ripples, sometimes dumping the sand into the adjacent trough, producing a trough that is pointed downward.

Fluctuations in the water level due to the tides cause tidal currents which produce numerous gullies and channels. As the platform becomes exposed the current direction is controlled by the local bottom slope. Water first flows parallel to shore, confined to the troughs of the oscillatory ripple marks. This produces a second set of ripples, this time true current ripples, at right angles to the oscillatory ripples. Taken together the two sets are termed *ladder-back ripple marks.* The water flowing alongshore in the ripple troughs finally collects into gullies, which strike out seaward away from the high-tide beach. The small gullies join to produce larger channels whose flow may reach 1.5 m/sec. Often megaripples and antidunes are produced in the channels (van Straaten, 1950; Klein, 1967; and others). The final product is a system of meandering channels, branching toward their landward ends.

Numerous interruptions in the ripple-marked surface are created by biological activity. These include surface tracks and trails made by snails and crabs crawling over the surface at low tide, various burrow openings, and the fecal castings thrown out of the burrows. Excavations made by stingrays also leave puddles scattered over the flat (Cook, 1971). The biological activity is much more intense on the beach platform than on the high-tide beach, and the wave processes are reduced. Because of this, in cross-section the sediments show only small indications of preserved ripple laminations; for the most part there is considerable reworking by the burrowing organisms, so that lebensspuren are abundant. The study of Howard and Dörjes (1972) on the "beach-related tidal flats" off Sapelo Island, Georgia, especially demonstrates this relationship between sediments and biology. Two flats were investigated, one muddier than the other. However, even the muddy flat showed mud accumulations only as high as 16%, the content normally being 5 to 10% mud, the remainder sand. The mud is concentrated by the organisms into fecal pellets; fecal material actually represents the major fine-grained sedimentary constituent of the muddier flat. Noteworthy differences were found in the fauna of the two platforms (Figure 13-20), even though they are located only a few kilometers apart, apparently due to the differing amounts of mud. On the muddy flat 38% of the total species are polychaetes and 36% are crustaceans, while on the sandy platform 40% of the total species are crustaceans and only 28% are polychaetes. The same tendency is noted in the actual numbers of individuals. As seen in Figure 13-20, there would be a corresponding difference in burrow types at the two locations.

Wind-blown Beach Structures

Sand dunes are commonly formed shoreward of the beach, the onshore winds blowing the sand inland from the beach foreshore and berm. Berry (1973) observed on the western shore of the Gulf of California that within no more than 30 minutes an estimated 30-km/hr wind was effective in eroding away at least the top centimeter of a damp, sandy beach. This demonstrates the effectiveness of coastal winds.

The coastal dune fields formed inland have features typical of all eolian environments; eolian ripple marks, large-scale cross-laminations, and so on. These are described by Bigarella (1972). However, Goldsmith (1973) has shown that there are some distinctive features of the coastal sand dunes that distinguish them from the arid desert dune fields. The presence of vegetation on coastal dunes acts to trap sand, producing low-angle cross-bed dips well below the angle of repose of sand. Taking

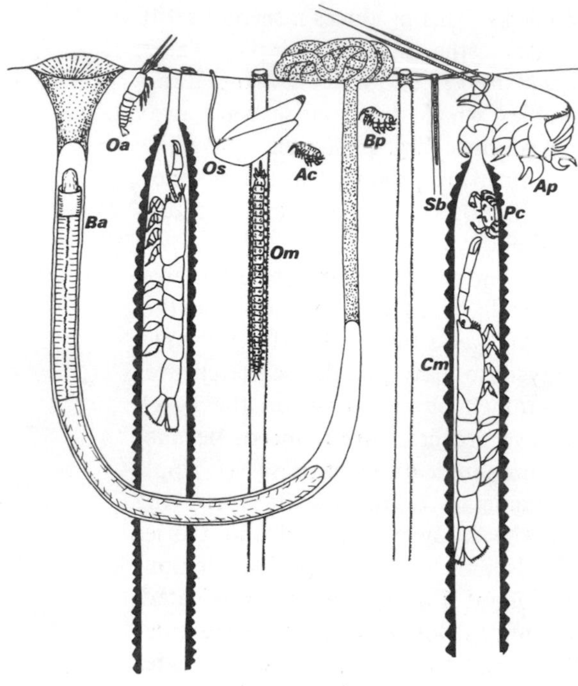

(a)

Figure 13-20 Fauna living in Georgia beach platforms (tidal flats) fronting the high-tide beach foreshore: (a) organisms in a sandy platform; (b) those found living in a muddier platform. Ap = *Albunea paretii*; Ba = *Balanoglossus sp.*; Bp = *Bathyporeia sp.*; Ca and Cm = *Callianassa sp.*; Hf = *Heteromastus filiformis;* Om = *Onuphis microcephala*; Os = *Olivia sayana*; Oa = *Oryrides alphaerostric*; Pc = *Pinnixa chaetopterana*; Sb = *Spiophanes bombyx*; Bp = *Bathypoereia sp.*; Dc = *Diopatra cuprea*; Ma = *Magelone sp.*; Ml = *Mulinia lateralis*; Nv = *Nassarius vibex*. [*From* Howard and Dörjes (1972)]

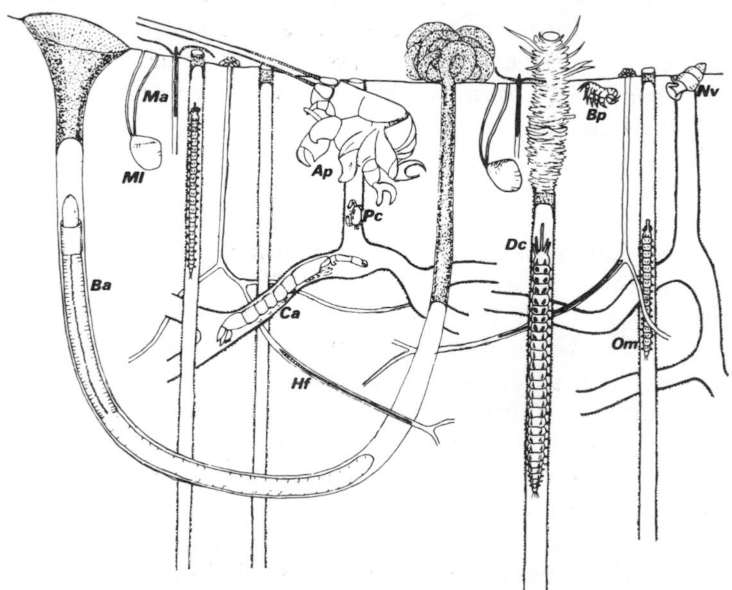

(b)

examples from several locations, Goldsmith found that cross-bedding dips center around two modes, 11–15 deg and 25–35 deg. The low-angle dips were formed by the vertical accumulation of sand behind vegetation hummocks.

In addition to these dune field features, the winds produce a variety of structures on the beach itself, especially on the exposed berm. On Oregon beaches strong winds commonly blow alongshore, parallel to the shoreline, so that the sand is not always driven inland but rather forms patches of low-amplitude miniature barchan dunes or sheets of wind-blown sand on the face of the berm. Any cobble, shell, or piece of driftwood on the beach forms a small wind-shadow dune (Figure 13-21). Berry (1973) described structures 5 to 10 mm high and wide and 10 to 20 mm long left standing on the eroding damp beach by the 30-km/hr winds already mentioned, clearly formed by differential erosion by the wind. The structures were elongated parallel to the direction of the wind, a small quantity of wind-blown sand depositing in the lee of each structure. Because of greater compaction of the sand or differences in strength due to pore-water binding, the little mounds were more resistant to the wind.

Figure 13-21 Wind-blown sand deposits in lee of stones on a beach berm. Ripples are also formed by wind.

Yasso (1966) studied deflation furrows formed by wind erosion of a salt-cemented sand crust on the beach berm. Allen (1967) described a pattern of ridges and hollows produced by foam being driven inland by winds from the surf. On one occasion the pattern extended continuously for almost 4 km along the beach, covering many hundreds of square meters. The ridges were anastomosing with intervening shallow hollows, aligned parallel to the foam patches and transverse to the wind. The spacing of the ridges in the direction of the wind was about 2 to 5 cm, the same as the mean diameter of the bubbles in the patches of foam. Similar in form to the foam ridges of Allen (1967) are eolian microridges identified by Hunter (1969). However, these are formed by sand being transported in saltation by the wind across a wet beach surface, some of the saltating sand being trapped by the surface tension of the water film. The microridges consist of a series of roughly parallel ridges less than 1 cm in wave length, the ridges oriented transversely to the wind direction like ripple marks, but with their upwind slopes steeper than the downwind slope or even vertical and overhanging. Hunter gives examples in the ancient rock record.

Features Characteristic of Polar Beaches

Arctic and Antarctic beaches are affected by wave action for only six months of the year. During the remainder of the year the adjacent sea may be covered with pack ice, which together with shorefast ice prevents waves from reaching the beach. When the air temperature drops below freezing in the fall, salt water begins to freeze on the beach surface, forming a further protective covering. A series of unusual structures are characteristic of these polar beaches, mainly related to the formation of ice on and within the beach as well as ice driven inland from the sea. The studies of Nichols (1961), Rex (1964), Hume and Schalk (1964), Moore (1966), and Greene (1970) describe these features and should be consulted for details.

During the spring breakup of ice, sea ice driven ashore by winds acts much like a bulldozer and drives beach sediments shoreward to form mounds of sand and gravel. Hume and Schalk (1964, p. 267) describe ice-push ridges formed in this way with heights as great as 4.5 meters at Barrow, Alaska. Ice-rafting may also transfer coarse sediment from the surf zone to the back of the beach. When the ice melts it leaves a gravel patch to mark its former location. Spray from the waves freezes to the beach in the fall. This frozen spray coats the beach with a glaze that may extend well inland, forming an ice-and-gravel rampart known as a *kaimoo* (Moore, 1966, p. 593). The melting of the kaimoo leaves a distinctly corrugated ridge of gravel on the beach similar to an ice-push ridge that has been reshaped by wind and surf action (Greene, 1970, p. 421). During the spring breakup, blocks of sea ice and isolated fragments of the kaimoo are left standing on the beach to melt. Surf partly buries these ice blocks with sediment, so that when they finally melt they form pits known as *sea-ice kettles*, commonly as much as 10 to 30 cm deep and 30 to 100 cm in diameter (Greene, 1970, p. 423). During the period of freeze-up, storms may produce a storm *ice foot* consisting of a toe of ice near the high-water mark that results from the freezing of water from the wave swash (Rex, 1964, p. 392). A large quantity of sand and gravel is often washed up the beach and included in the storm ice foot, the gravel-sand-ice being finely bedded, with beds ranging from 5 to 10 cm in thickness. Exceptional cases of ice foot development may rise as much as 2.5 meters above the wave equilibrium surface. If the ice foot thaws in place, the sand and gravel is dropped onto the beach. Joyce (1950) considers the ice foot to be an important agent of erosion because when it breaks up it sometimes rafts away large quantities of sediment to the sea. The ice foot is not restricted to polar beaches; Zumberge and Wilson (1954) have described its formation along the shores of the Great Lakes.

Dillon and Conover (1965) reported an ice-related feature on a Rhode Island beach. They found blocks of laminated, ice-cemented sand exposed on the beach following several days of subfreezing temperatures. Some blocks were tilted, probably due to undercutting by the waves. Other blocks showed contorted laminations similar in appearance to flow structures, apparently the result of melting and refreezing of stressed blocks.

Miscellaneous Structures

A few primary sedimentary structures found on beaches do not fall into any of the previously considered categories. For example, typical raindrop impressions can be found; these are especially common on the beaches of Oregon.

Davis (1965) described *beach pits* which form after the beach sand becomes dry. With the loss of the pore water which helps to bind the grains together, the sand loses some of its strength and a slight disturbance causes the formation of small circular pits in the sand. The floors of the pits are flat and the sides vertical. They range in diameter from 0.33 to 2.5 cm and are about 0.5 to 1.5 cm deep.

BEACHROCK

Beachrock or *beach sandstone* consists of beach sand cemented together by calcium carbonate to form a friable to well-cemented rock. Beachrock forms most commonly on beaches composed of calcareous shell and coral grains, but it can also develop in beaches of quartz sand or other mineral composition. It forms best on sand beaches; shingle or conglomeratic beachrock is less abundant. Modern beachrock occurs in the intertidal zone as a series of thin beds which dip seaward at less than 15 degrees and whose strike is parallel to, or at slight angles with, the trend of the present beach (Ginsburg, 1953).

The formation of beachrock is limited to warm tropical regions. Apparently there is no connection with the rainfall, as beachrock occurs in Arabia with near-zero annual rainfall as well as in Eniwetok or the Caymans with an annual rainfall up to 150 cm (Bricker, 1971, p. 2). The most detailed account is that of Branner (1904), who studied the extensive formation of beachrock along the northeast coast of Brazil. Bricker (1971) contains an up-to-date discussion on the occurrence of beachrock and its origin, with contributions by several authors. Ginsburg (1953) investigated beachrock occurrence in southern Florida, Russell (1962) in Puerto Rico, Stoddart and Cann (1965) in British Honduras, and Emery and Cox (1956) in Hawaii. Beachrock is also found in Samoa (Daly, 1924), the Persian Gulf, the Red Sea, in the Bahamas (Illing, 1954; Scoffin, 1970), Ceylon (Cooray, 1968), and in the Mediterranean (Alexandersson, 1969).

The rate of cementation of beachrock is documented by the incorporation of broken beer bottles (Scoffin, 1970). Either calcite, aragonite, or both may be the cementing material. This may have an important bearing on its origin, since calcite is the form of $CaCO_3$ normally precipitated from fresh water and aragonite from seawater (Illing, 1954). However, as pointed out by Bricker (1971, p. 3), microorganisms and organic coatings may be important in the beachrock cementation process, which might explain the considerable variability in beachrock occurrence and cementing mineralogy. Wholly aragonite cement has been reported in beachrock from Florida (Ginsburg, 1953), the Bahamas (Illing, 1954), and many other locations. Other studies, such as Russell (1962) in Puerto Rico and Branner (1904) in Brazil, identified calcite as the cementing material. Many of the investigations show that cementation occurs in two stages. This is perhaps shown best by the British Honduras beachrock studied by Stoddart and Cann (1965). Intertidal beachrock in an early stage of cementation without exception showed cementation by aragonite. At this stage constituent carbonate grains are unpitted and uneroded, the cement clearly being of external origin. In later stages, the voids become filled and the beachrock well bound. At this stage the grains show considerable solution, so it is apparent that the secondary cement is derived largely from the solution and redeposition of the constituent grains themselves. Careful analysis showed that the primary cement responsible for the initial beachrock cementing is composed of aragonite, while the secondary void-filling cement is probably calcite.

There are two main hypotheses for the origin of beachrock—the action of fresh groundwater, and precipitation from seawater. Field (1919, p. 198) suggested that solution of calcium carbonate from inland sands by groundwater, followed by precipitation as this seeps out through the beach at low tide, would cement the beachrock. Russell (1962) attributes the Puerto Rican beachrock to this origin, since calcite is the cementing agent and because of the well-developed fresh water table in the area. Objections made to this origin are that it does not account for beachrock cemented by aragonite, that it will not explain the rather sporadic occurrence of beachrock, and that it does not account for beachrock in areas that are arid or on sand spits, where no fresh water table could be present; nor does it explain beachrock formation in areas where there is no detrital calcium carbonate for the groundwater to leach for the cementing material.

The localization of beachrock in the intertidal zone is evidence that the cementation is due to the precipitation of calcium carbonate from seawater, primarily as a result of heating and evaporation (Dana, 1851, p. 368; Daly, 1924; Kuenen, 1950). Normal surface seawater is supersaturated with respect to calcium carbonate. Increased temperatures within beach sands would promote precipitation because the solubility product decreases with increasing temperature and because of an increase in the partial pressure of carbon dioxide. Water remaining in the beach at low tide after drainage is in the form of intergranular films, and evaporation would leave aragonite cement. Ginsburg (1953), for example, concluded that the Florida beachrock is formed in this way. This origin for beachrock from precipitation from seawater could not account for calcite cementing as found in some examples, unless the aragonite later converts to calcite, or if the precipitation is controlled by microorganisms or organic coatings. In their careful review regarding beachrock origin, Stoddart and Cann (1965) conclude that beachrock may in fact form in more than one way.

Tanner (1956) identifies possible examples of ancient beachrock from the Permian, Paleocene, Eocene, and Miocene ages.

A curious example of calcareous sandstone that could pass for beachrock was found by Allen et al. (1969), dredged from 75 meters' water depth off the New Jersey coast. It is cemented by aragonite and contains fossils that indicate colder waters and shallower depths than present today. The curious thing is that dating of the shells gave 4,390 years B.P., while the cement gave 15,600 years B.P.: the cement is older than the grains it is cementing. Study of the carbon isotopes of the cement indicated an organic origin for the carbon. This suggested that methane originating from submerged tidal marshes provided the light carbon to cement the sands. The actual deposits (tidal marsh and littoral sands) were formed during the low stand of sea level, the cementing occurring much later, however, when the water depth had reached about 75 meters (determined from the age of the cement and a sea-level-versus-time curve—Chapter 6).

LIFE AT THE SEASHORE

Life in the intertidal zone of the rocky coast and on the sandy beach is particularly difficult, the plants and animals being alternately buffeted by the wave and exposed at low tide to heat prostration and dessication. However, for those organisms that have adapted to such harsh conditions, these zones provide an abundant food supply, so that large populations of individual species can be maintained. The beach acts like a giant sand filter that strains out and concentrates the particulate matter from the water that percolates through the beach. This food can support large populations of species. Dörjes (1971, 1972) found numbers of both species and individuals showing a marked peak in the lower shoreface. Fox (1950) estimated the population of the beach worm *Thoracopphelia mucronata* at La Jolla, California, to be 3,000 per cubic foot; thus a worm bed a mile long by 10 feet and a foot deep (1600 m by 3 m, and 0.3 m deep) would contain some 158 million worms. Large populations of various clams and crustacea can also be found. The beach clam *Donax* may reach 10,000 or more individuals per square meter in beds that extend for several kilometers.

There are many regional books that describe on a popular level the organisms found on the beaches and in tide pools on the rocky coast. Such books aid in indentifying the creatures found as well as telling something of their ecology. *The Sea Shore* by C. M. Yonge (New York: Atheneum, 1963) is a particularly excellent and readable book, covering the subject in general rather than being regional. Good summary papers with numerous references are those of Doty (1957) and Hedgpeth (1957), respectively on the rocky intertidal life and organisms on sandy beaches. Only a brief review is permitted here, and that concentrating on sandy beaches rather than rocky substrate.

The basic food supply consists of microscopic plants and animals and detritus suspended in the water or within the beach sediment itself. Kelp and other seaweeds generally have no place of attachment on the sandy beach, so they do not form a primary food source. The exception is kelp that is washed ashore, upon which beach-hoppers and a few other organisms feed. The animal life of a sandy beach might be classified, on the basis of its feeding habits, into the *active burrowers*, which actually pass the beach sediments through their alimentary tracts (e.g., most of the worms), and the *detritus feeders*, which filter particles of detritus and plankton from the surf waters (e.g., most of the bivalves). The gastropods such as *Olivella* are specialized detritus feeders. Most of the large arthropods are scavengers or predators, although some, like the sand crabs of the genus *Emerita*, have developed a unique method of filter feeding: their featherlike antennae strain the water and catch the detritus carried in the wave swash. All of the organisms that live in the nearshore have special adaptations that allow them to feed but prevent the entry of sand grains. A few forms of echinoderms occur on sandy beaches, usually living just below the tide level. Most familiar are the sand dollars (*Mellita*, *Dendraster*) which wash ashore when dead. They live just beyond the outer reaches of the breaker zone, stacked on edge half-buried in the sand with their mouths just at the sand level. They line up with their diameters parallel to the wave orbital motions so as to present the smallest possible drag. Other echinoderms, including the small opiurans, asteroids of genus *Astropecten*, sea cucumbers, and heart urchins, feed upon detritus or prey upon minute organisms in the sand.

A few small fishes can be found living in the surf. One, the grunion, uses the beach for spawning (Chapter 11). Horseshoe crabs and sea turtles also migrate to the shore to lay their eggs in the sand. The ghost crab *Ocypode* migrates in the opposite direction, returning to the surf to spawn; the return to land is made during the first postlarval (megalopa) stage. Transient shorebirds also are a part of the nearshore ecology, playing the role of scavengers as well as predators.

Zonation of plants and animals with respect to the tide level is readily apparent on the rocky coast. A similar although not as well-defined zonation may be present on the sandy beach. *Ocypode*, the ghost crab, spends most of its life well above the high-tide line, returning to the water occasionally to dampen its gill chamber. It is most abundant at the back of the beach, but may even be found in the shoreward dunes. Beachhoppers and sand fleas are also most abundant at the back of the beach, burrowing into the sand at high water and emerging at low tide to feed on detritus left on the beach. Dahl (1953) has suggested that the ghost crab and large beach-hoppers occupy the same ecological niche, feeding on similar sources and performing the same functions, the two groups having differing geographical distributions, however. Ghost crabs are found in the tropics and the hoppers in temperate latitudes. The hoppers show evidence of an amazing "internal clock" geared to the tides to time their feeding and burrowing activities (Enright, 1963). In the mid-beach zone are the sand crab *Emerita* and the beach clam *Donax*. Both of these groups are found to migrate with the tide up and down the beach. *Donax* reacts to the vibrations caused by the incoming waves at flood tide, emerging from the sand under the swash so as to be carried shoreward and digging into the sand before being swept seaward under the backwash. On the outgoing tide it emerges during the backwash to be carried seaward and digs in during the swash. Further seaward are the various clams and gastropods, while still further seaward in deeper water are the sand dollars, sea cucumbers, and so on. In general, the various sand beach organisms are arranged in zones according to the state of the tide which enables them to obtain the most food while avoiding the most severe physical rigors of the environment.

The intensity of wave action and continually shifting sand make the beach a unique biological environment which requires special adaptation by the organisms that live there. Many of the problems are overcome by simply burrowing deeply into the beach. Filter-feeding bivalves commonly have very long siphons, enabling them to live at depths below the disturbed upper layer of sand. Other bivalves and gastropods adapted for fast burrowing are streamlined in shape such as *Olivella* or have a large muscular foot as in *Polinices* (moon shell). The burrowing ability of bivalves can be attested to by any clam digger. Swash zone organisms such as *Donax* and the sand crab *Emerita* can disappear into the sand with amazing rapidity. They combine bullet-shaped bodies with strong burrowing muscles. The razor clam can bury itself completely in less than 7 seconds. Other clams may be slow diggers but have massive shells to resist the wave shock if they unexpectedly become unprotected from the surf. Another adaptation is with regard to breeding and larval development. Since the larvae are at the mercy of wave motions and nearshore currents, larval life must be of sufficient duration to enable the organisms to find new territory. In breeding, the gametes are usually shed into the water for fertilization; to insure union, large numbers of gametes are emitted.

In addition to their interest in the lebensspuren of nearshore organisms, geologists are also interested in the organisms themselves as potential fossils. The shells found on the beach, although predominantly beach dwellers, are characteristically mixed with many species from other environments. Shells are washed onto the beach from the offshore deeper water and commonly from rocky headlands. Occasionally even shells from rivers and estuaries can be found on the beach. Thompson (1937, p. 740) has suggested that any sandstone or conglomerate containing an appreciable number of shells from various environments is likely to be littoral.

The study of Hertweck (1971), again in the Gulf of Gaeta, Italy, compared the distribution of skeletal remains (potential fossils) with the distributions of living shells investigated by Dörjes (1971). It was found that below the wave base for the region, at depths greater than 6 meters, the shells corresponded to the distributions of the living organisms. At depths shallower than 6 meters the waves were effective and transported the shells of dead animals generally toward the shoreline, although during storms they may be shifted to still deeper water. As a result, the shells found on the beach had little correspondence with the organisms found living there.

Studies have also been made of the selective sorting and orientations of shells on beaches. A particularly good investigation of this sort is that of Nagle (1967). Much of the attention has centered on the orientation of *Donax* shells on the beach, since they are sorted out according to right and left valves. The papers of Muelen (1947), Boucot (1953), Lever (1958), Wobber (1967), and Behrens and Watson (1969) form only a portion of the literature on this subject. There is a suggestion that the sorting can be correlated with the direction of wave approach and hence with the direction of longshore sediment transport. Interpretations of paleodrift directions may be possible from right- and left-valve sorting patterns. In a similar fashion, Reyment (1968) utilized the orientation of orthoconic nautiloids as indicators of shoreline currents in Ordovician deposits.

ANCIENT BEACH DEPOSITS

Ancient beach deposits are of practical importance in that they very often serve as reservoir rocks for oil and gas accumulation. For this reason geologists have been interested in their identification within the stratigraphic rock record. There are many review papers that deal with the identification of beach and related sediments; examples are Martins (1939), Andrews and van der Lingen (1969), Davies, Ethridge, and Berg (1971), and Dickinson, Berryhill, and Holmes (1972).

Identification of beach sands involves studying sedimentary structures and the grain-size distributions, topics that have already been considered in this chapter. Figure 13-22 gives a composite review of the structures found in cross-section within

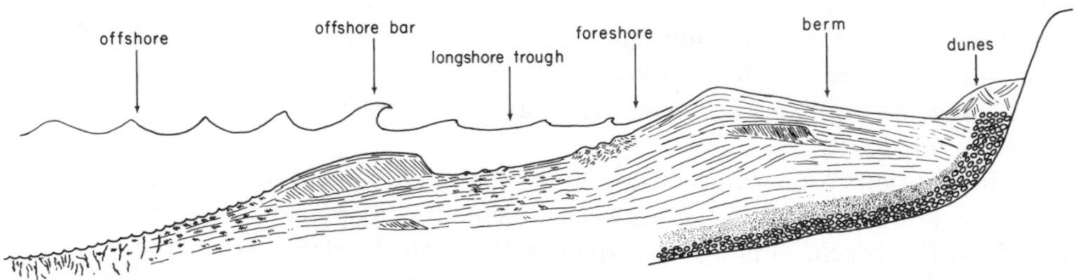

Figure 13-22 Schematic of characteristics within beach subenvironments showing parallel laminations, cross-laminations due to bar migrations, ripple cross-laminations, and burrows in the offshore. Also diagramed is a basal conglomerate layer, overlain by a heavy mineral concentrate, overlain in turn by the normal laminated beach sands.

a beach deposit. Geologists attempt to identify subenvironments within a beach deposit in order to identify the beach as a whole. This is also done by examining the relationship of the sand unit to other facies or sediment units. To aid in the identification of beach deposits, geologists draw heavily on the facies relationships found in "modern" coastal deposits. Four examples of "modern" shoreline deposits will be reviewed here: Galveston Island (a barrier island) on the coast of Texas; the cheniers of the Gulf coast of Texas and Louisiana; the coast of Nayarit, Mexico; and the barrier island deposits on the east coast of the United States.

Galveston Island, Texas, has been studied in detail by Bernard, LeBlanc, and Major (1962), and has become for geologists a prime example of the distribution of facies of a barrier island complex, used to compare with ancient sediments. Galveston Island represents an offlap sequence resulting from the seaward accretion of the barrier during stationary sea level. The barrier island itself is flanked both seaward and landward by silts and clays (Figure 13-23), representing typical offshore and lagoonal deposits, respectively. The vertical sequence of sediments through the barrier is the same as the offshore sequence across the barrier surface and out onto the shelf. The barrier is capped by eolian sands, below which are the beach sands. At still greater depths into the barrier are sands of progressively deepening water. This is shown in

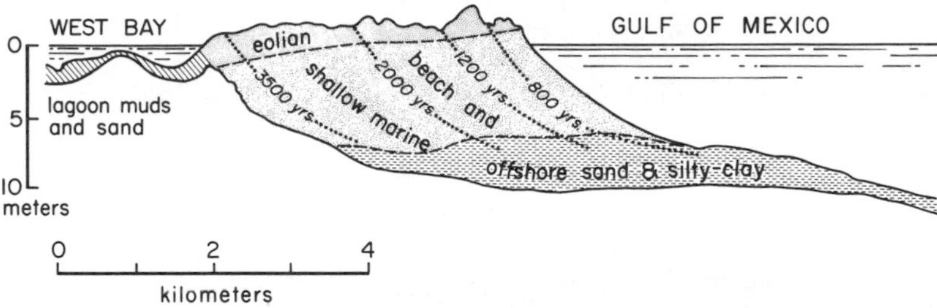

Figure 13-23 Diagramatic cross-section of Galveston Island, Texas, showing the principal environments of sediment deposition and the vertical sequence of facies. [*After* Bernard, LeBlanc, and Major (1962)]

the gradual fining of the sediment grain size at depth within the barrier (Figure 13-23) and also in the sequences of sedimentary structures. The eolian sands that cap the barrier are generally cross-laminated (festoon and planer) and parallel-laminated. Older eolian deposits tend to lose some of these structures and become more massive, apparently from destruction by plants and animals, weathering, and the movement of groundwater. Beach deposits are characterized by well-developed primary sedimentary structures, parallel and cross-laminations being numerous. Burrowing is very scarce and shells are locally concentrated. With increasing depth into the barrier, just as with increasing water depth offshore, the sediments progressively become finer and there is a marked increase in bioturbation, destroying any traces of primary structures. The barrier island sands finally become interbedded with silt and clay layers, again with pronounced bioturbation. Such deposits are represented on the modern shelf at depths greater than about 5 to 6 meters.

Landward from Galveston barrier island are lagoonal sediments consisting of interbedded silts and clays with some fine-sand layers. Animal burrows and plant roots commonly obscure the parallel laminations of the sediments. Laminations are often preserved in the fine sands, which are "washover" sediments carried landward from the barrier island by storms that break through and over the narrow barrier island.

Another sequence of beach ridge deposits that have been extensively studied are in the chenier plain of the Gulf coast of Texas and Louisiana (Russell and Howe, 1935; Fisk, 1955; Byrne, LeRoy, and Riley 1959; Gould and McFarlan, 1959). The *cheniers* themselves are the beaches and are up to 4.5 meters thick, 200 meters in width, and up to 50 km long (Figure 13-24). From the analysis of cores, Byrne, LeRoy, and Riley (1959) were able to identify seven distinct sedimentary facies, including marsh, bay, mudflat, open gulf, and beach. The identifications are based on sediment lithology, the contained faunas, and the facies relationships. The stratigraphy records the postglacial rise in sea level to its present position and the subsequent construction of the chenier plain by coastal progradation. The basal part of the wedge consists of transgressive brackish-water and marine deposits laid down on the underlying

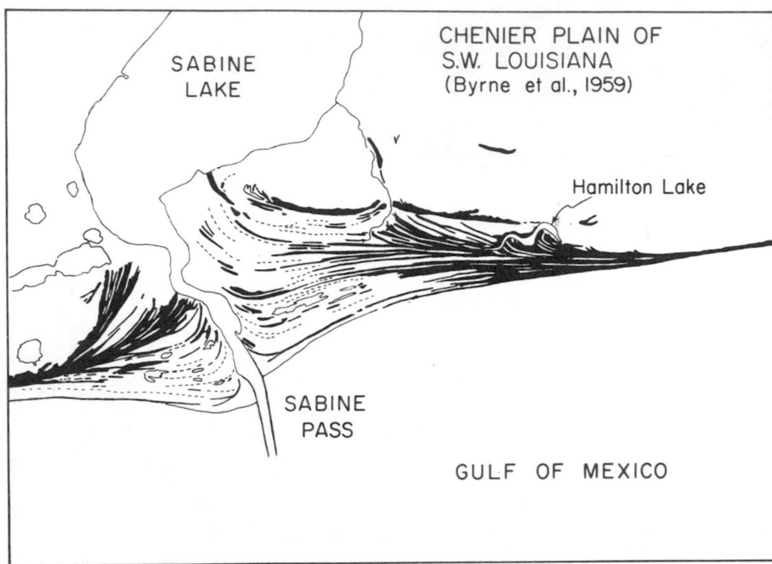

Figure 13-24 Cheniers or beach ridges on the southwest coast of Louisiana. [*After* Byrne, LeRoy, and Riley (1959)]

Pleistocene surface as the sea rose, starting 6,000 years ago and reaching its present level about 3,000 years ago (Chapter 6). Above the transgressive sediments are gulf-bottom sands and silty clays. This facies is identical not only in faunal content, but also in composition, texture, and structure, to deposits which are presently being laid down in the shallow open gulf immediately adjacent to the present-day active beaches. The beach deposits or cheniers consist mainly of sand and shells, with only minor amounts of silt and clay. The shells occur as distinct layers several centimeters thick, concentrated by the wave-sorting processes into lag deposits on the gulf side of the ridge. The sediments on the backslope are characteristically finer-grained. The relict beaches are characterized by a smooth, generally arcuate seaward front and an irregular landward margin. The beach deposits rest on shallow gulf sediments, and are overlapped by organic marsh silts and clays. The ages of the cheniers range from 300 to 2,800 years B.P. (Gould and McFarlan, 1959).

The sequence of beach ridges on the coast of Nayarit, Mexico, studied by Curray, Emmel, and Crampton (1969) have already been discussed in Chapter 2. The lateral and vertical sequences of facies (Figure 13-25) can also serve as a"modern" model to be compared to ancient deposits. The development of the Nayarit beach ridges started on the order of 4,500 years B.P., after the sea level rise associated with the last glacial melting began to slow (Chapter 6). The older alluvium is overlain by a sheet-sand layer several meters thick deposited by the last transgression from the rising sea level. Above that are deposited the regressive coastal sands from the prograding coast. The coastal sand is prograding over the shelf muds, and on the landward side the coastal sand is in turn being covered by lagoonal and fluvial deposits. The littoral sands found in the old beach ridges are well sorted, fine- to medium-grained. Locally there are thick concentrates of shells, mollusks such as *Donax* that still live in the modern beach, the present-day counterpart to the old ridges.

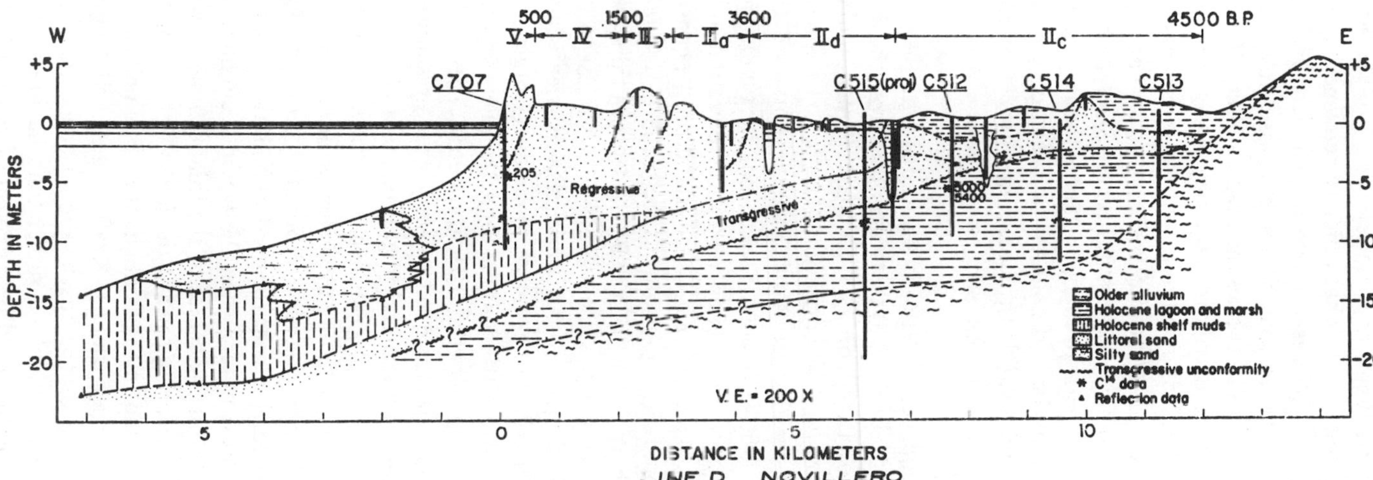

Figure 13-25 Transgressive and regressive sediment facies at the coast of Nayarit, Mexico, forming a series of beach ridges prograding into the sea. [*From* Curray, Emmel, and Crampton (1969)]

Studies by Kraft, Biggs, and Halsey (1973) of the barriers on the east coast of the United States show somewhat different sequences of sediment facies than the studies described above. This is because the barriers and other beaches are slowly migrating landward as the sea level continues to rise, however slowly. As described in Chapters 2 and 6, as the sea level rises the beaches tend to erode during intense storms, the sand washing over the barrier into the bay or lagoon. Such "washover" deposits are also present on the Gulf coast and were described above. However, due to the pronounced landward migrations of the east coast barriers, the resulting vertical sequence of sediments differs. As seen in Figure 13-26, the barrier island sands overlie lagoon and marsh sediments. In many instances the lagoon and marsh sediments are found to outcrop within the surf zone of the beach on the ocean side of the barrier. These are in part eroded away as the system migrates landward. Should the transgression proceed, the shallow marine sands and gravels would eventually override the barrier sequence.

The vertical sequence of deposits associated with the east coast barriers studied by Kraft, Biggs, and Halsey (1973) therefore differ from the coasts that are prograding seaward. At Galveston Island, on the Gulf coast cheniers, and on the coast of Nayarit, all of which are prograding, the barrier sands come to lie above the offshore sediments. In contrast, the east coast barrier sands migrate shoreward and so become deposited over lagoon and marsh sediments. Dickenson, Berryhill, and Holmes (1972), who stressed the importance of barrier migration direction on the resulting vertical sequence of deposits, indicated that Padre Island on the coast of Texas has remained stationary in position, building upward as the sea level rose relative to the land. Padre Island rests with a sharp contact on the Pleistocene. In this case the facies are uniform, or nearly so, in the vertical.

Tidal channels frequently cut through a barrier island and connect with a landward lagoon or marsh. For example, along the Georgia coast inlets are numerous; the barrier islands are from 10 to 30 km long, the inlets separating them up to 3.5 km wide and 15 meters deep (Hoyt and Henry, 1965). These tidal inlets and associated tidal deltas will of course produce sedimentary structures more akin to fluvial channels than to beaches. Appreciable lateral migrations of a channel may obliterate much of the typical beach deposits, replacing them with fluvial-type structures. Cross-bedding and current ripple marks, including megaripples, are common in the inlet sands. Kraft (1971) found a dominance of landward-oriented bedforms due to the high velocities of the flooding tide occurring near the crest of high tide, whereas the highest velocities of ebb occur near lowest water, when the water is confined to only a few channels. Grain size is coarsest near the center of the tidal channel, where lag gravels are common, the grain size decreasing toward the lagoon or marsh. Shells from the bay, the beach, and shelf are found in the tidal channels and deltas.

In addition to barrier islands, Kraft, Biggs, and Halsey (1973) provide what little information there is on the beach sediment deposits and facies where the beach directly abuts an eroding mainland rather than having a separating lagoon. With a transgressing sea the result would be a deposit of beach and dune sands overlying the unconformity from the erosion of the mainland. There is some question whether

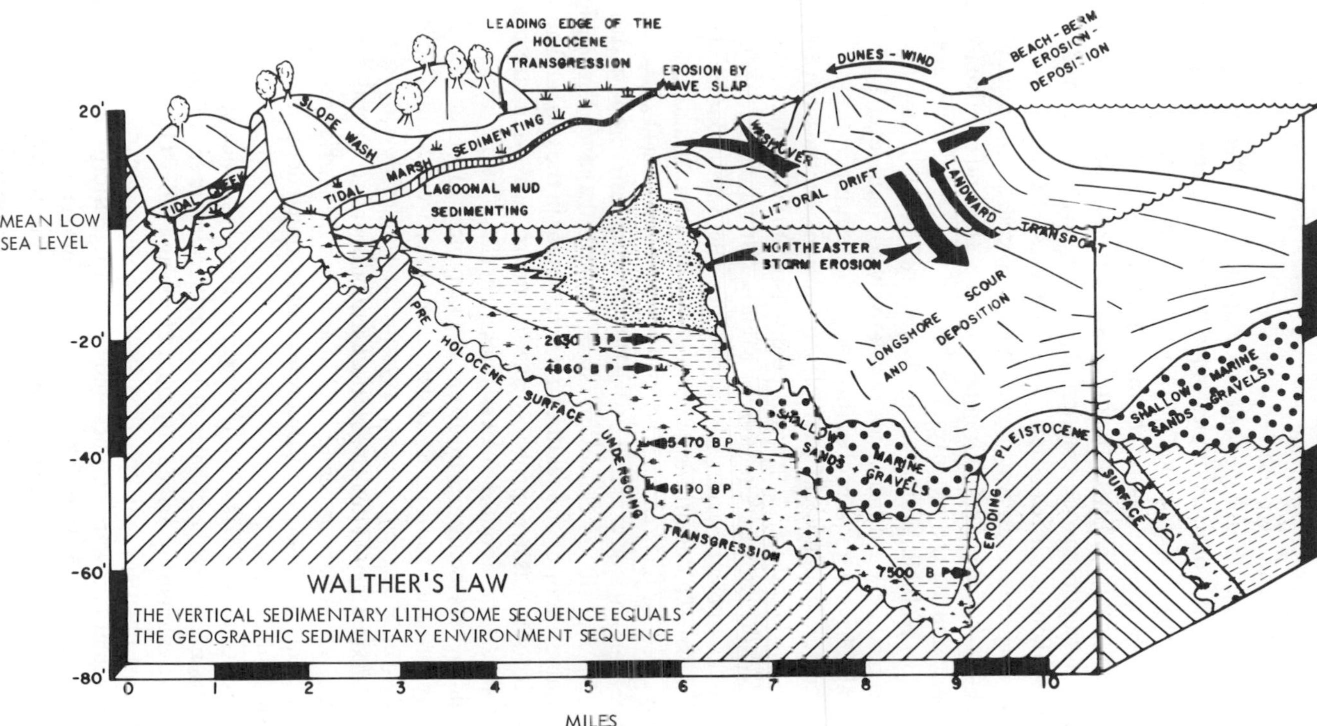

Figure 13-26 Schematic representation of the sedimentary facies in the U.S. east coast barrier islands and of the associated processes of erosion and deposition. [*From* Kraft, Biggs, and Halsey (1973)]

such sedimentary units would be preserved within a stratigraphic sequence. Kraft, Biggs, and Halsey (1973) provide some evidence from drilling that beach deposits formed in this way can be preserved.

Although raised beach deposits of Pleistocene to Recent age are found along many coasts of the world, examples older than Pleistocene are apparently relatively rare. Beach deposits as opposed to other environments are poorly represented in the geologic rock record. One reason for this is that the volume of beach sediment initially formed is small compared with the other environments. Secondly, beach deposits are particularly susceptible to destruction after a sea level change and so might not be preserved. Although raised beach ridges are presently numerous, they have little chance for long-term preservation on our rising coasts, since they are likely to be destroyed by subaerial erosion. Gradual sinking of the land relative to the sea resulting in a transgression of the sea over the land of low relief provides more favorable conditions for geologic preservation. Systems of beach ridges and barrier islands building the land outward into a sea of relatively constant level, such as those already described, also have a good probability for geologic preservation.

Only a few unequivocal examples of ancient beaches have been adequately described in the literature. A few will be discussed briefly here.

Parts of the Lower Cretaceous Muddy Sandstone of Montana have been identified as an ancient barrier island complex (Berg and Davies, 1968; McGregor and Biggs, 1968). This is an important oil-bearing deposit and has been extensively drilled and cored for that reason. Davies, Ethridge, and Berg (1971) discuss the Muddy Sandstone as an ancient example of barrier deposits, as there is a striking similarity to the modern deposits on Galveston Island, described above. The vertical succession of sedimentary structures and textures are very similar in the two. Offshore fine-grained sediments that are massive, probably due to intense burrowing, are overlain by well-laminated sandstone about 4 meters thick. The laminations are planar, low-angle, with individual sets being up to 30 cm thick, typical of beach deposits. Burrowing is sparingly present, rooting is absent, and micro–cross-laminations are present sporadically, probably representing asymmetric ripples developed under conditions of local reduced energy. Thus the vertical sequences of sediments are much the same as on the Galveston barrier island, and in addition the barrier sandstones are flanked by a vertical sequence of deposits that represent lagoonal deposits, washover fan sands, and marsh sediments.

Dillard, Oak, and Bass (1941) traced in the subsurface shoestring sands within the Pennsylvanian Cherokee Shales of Kansas. Again, these are important oil producers and so are extensively drilled. Byrne, LeRoy, and Riley (1959) compare these with the Gulf coast cheniers and find they bear the same linear shape, smooth seaward margins, irregular landward outlines, and bifurcating patterns. Byrne, LeRoy, and Riley (1959) reference other examples of possible ancient chenier sands.

The study of Campbell (1971) of the Upper Cretaceous Gallup Formation in New Mexico is a particularly good example of an identification of beach deposits that compares favorably with investigations of recent sediments. Campbell is able to find a regressive sequence of dune sandstone, coal swamp deposits, backshore, foreshore, and offshore sands, and offshore siltstones and mudstones. The arrangement of the facies is closely analogous to the modern sequence at the coast of Nayarit (Figure 13-25) studied by Curray, Emmel, and Crampton (1969). The identifications are based on stratigraphic relationships, external geometry, internal bedding pattern, sedimentary structures, texture and composition, and fossil content. The dune sandstones consist of a complex of trough-shaped beds, whereas the beaches have even, parallel bedding, the backshore having laminae dipping both onshore and offshore, while the foreshore sands have laminae dipping uniformly seaward. Both wave ripples and current ripples are found, and swash and rill marks are present. There is a gradual seaward decrease in grain size from the beach to the offshore.

Johnson and Friedman (1969) identified nearshore deposits within the Upper Devonian of New York. The identification is based on sediment texture, structures, geometry, and associations with deposits from other related environments (lagoonal and offshore). Their study is another example of contrasting nearshore sediments with other environments, since dune, tidal, and offshore sediments are also present within the formation, which taken together is a deltaic environment.

The Lyons Sandstone of the Colorado Front Range is a well-known deposit that provides an example of the difficulties sometimes involved in identifying ancient environments. Several investigators have concluded that the Lyons is principally an ancient beach deposit. Thompson (1949) investigated this formation, comparing it with his earlier observations (Thompson, 1937) on recent California beaches. Thompson identified what he believed to be swash marks, rhomboid marks, and bubble impressions formed by the escape of air trapped in dry sand by the abrupt flooding by the wave swash (page 375), as well as a variety of animal tracks. The Lyons is well laminated, commonly with the crude parallelism found in beaches. Stratigraphically below the Lyons are deposits of fluvial origin, and above the Lyons are thin, interstratified limestones containing marine gastropods and pelecypods (Thompson, 1949). From these several lines of evidence Thompson concluded that the Lyons is principally a beach deposit with some associated wind-blown dune sands, a typical association found on many present-day coasts. Vail (1917) had earlier concluded that the formation is dominated by sediments of eolian origin, admixed with stream sediments, and Walker and Harms (1973) have also interpreted the deposit as eolian in light of recent studies of dune sands as well as of beach sands. The large-scale steep sets of cross-strata common in the formation are interpreted as accretion faces of dunes. The low parallel ripples are interpreted as wind-formed, and the circular impressions are identified as raindrop imprints. Avalance structures on forest bedding planes of the Lyons are ancient counterparts of avalanche structures on slipfaces of modern dunes. The Lyons is convincingly shown by Walker and Harms (1973) to be chiefly dune deposits; this should serve as a cautionary note in the identification of ancient beach sediments.

CONCLUDING THOUGHTS

The past two decades have seen an intensive effort by geologists to identify the various sedimentary structures of the nearshore environment, the grain size, and the organisms that live there and their effects on the sediments. Such studies will be helpful in the identification of ancient beach deposits. On the whole this phase of beach study is nearing completion. What remains is a need for a better understanding of the physical processes that give rise to the various structures. With such a knowledge we might be better able to interpret the ancient physical conditions associated with the beach deposits—for example, to determine the approximate wave energy level, wave period range, water depth, and beach slope, and from these in turn evaluate the wave fetch distance or beach exposure. Much work is to be done in this direction.

REFERENCES

ALEXANDERSSON, T. (1969). Recent littoral and sublittoral high-Mg calcite lithification in the Mediterranean. *Sedimentology*, 12: 47–61.

ALLEN, J. R. L. (1967). A beach structure due to wind-driven foam. *J. Sediment. Petrol.*, 37: 691–92.

ALLEN, R. C., E. GAVISH, G. M. FRIEDMAN, and J. E. SANDERS (1969). Aragonite-cemented sandstone from outer continental shelf off Delaware Bay: submarine lithification mechanism yields product resembling beachrock. *J. Sediment. Petrol.*, 39: 136–49.

ANDREWS, P. B., and G. J. VAN DER LINGEN (1969). Environmental significant sedimentologic characteristics of beach sands. *N. Z. J. Geol. and Geophys.*, 12: 119–37.

BAGNOLD, R. A. (1946). Motion of waves in shallow water; interactions between waves and sand bottom. *Proc. Roy. Soc.* (London), series A, 187: 1–15.

——— (1954). Experiments on a gravity-free dispersion of large solid spheres in a Newtonian fluid under shear. *Proc. Roy. Soc.* (London), series A, 225: 49–63.

——— (1963). Mechanics of marine sedimentation. In *The sea*, ed. M. N. Hill, 3: 507–28. Interscience, New York.

BASCOM, W. N. (1951). The relationship between sand-size and beach-face slope. *Trans., Am. Geophys. Union*, 32: 866–74.

——— (1960). Beaches. *Sci. Am.*, 203: 80–94.

BEAL, M. A., and F. P. SHEPARD (1956). A use of roundness to determine depositional environments. *J. Sediment. Petrol.*, 26: 49–60.

BEHRENS, E. W., and R. L. WATSON (1969). Differential sorting of pelecypod valves in the swash zone. *J. Sediment. Petrol.*, 39: 159–65.

BENNETT, H. S., and P. MARTIN-KAYE (1951). The occurrence and derivation of an augite-rich beach sand, Grenada, B.W.I. *J. Sediment. Petrol.*, 21: 200–204.

BERG, R. R., and D. K. DAVIES (1968). Origin of Lower Cretaceous muddy sandstone at Bell Creek field, Montana. *Am. Assoc. Petrol Geol. Bull.*, 52: 1888–98.

BERNARD, H. A., R. J. LEBLANC, and C. F. MAJOR (1962). Recent and Pleistocene geology of southwest Texas, field excursion no. 3. In *Geology of the Gulf Coast and central Texas*, Geol. Soc. Am. Ann. Meeting Guidebook, pp. 175–224.

BERRY, R. W. (1973). A note on asymmetrical structures caused by differential wind erosion of a damp, sandy forebeach. *J. Sediment. Petrol.*, 43: 205–6.

BIGARELLA, J. J. (1972). Eolian environments: their characteristics, recognition, and importance. In *Recognition of ancient sedimentary environments*, ed. J. K. Rigby and W. K. Hamblin, pp. 12–62. Soc. Econ. Paleon. and Min. Special Publ. no. 16.

BIRD, E. C. F. (1968). *Coasts.*MIT. Press, Cambridge, 246 pp.

BLATT, H. (1958). *Sedimentation on New Jersey beaches.* Master's thesis, Univ. of Texas, Austin, 159 pp.

BLUCK, B. J. (1967). Sedimentation of beach gravels: examples from south Wales. *J. Sediment. Petrol.*, 37: 128–56.

BOUCOT, A. J. (1953). Life and death assemblages among fossils. *Am. J. Sci.*, 251: 25–40.

BRANNER, J. C. (1905). Stone reefs on the northeast coast of Brazil. *Geol. Soc. Am. Bull.*, 16: 1–12.

BRICKER, O. P. (1971). *Carbonate cements.* John Hopkins Press, Baltimore, 376 pp.

BUCHER, W. H. (1919). On ripples and related sedimentary surface forms and their paleographic interpretation. *Am. J. Sci.*, 197: 149–210.

BYRNE, J. V., D. O. LEROY, and C. M. RILEY (1959). The chenier plain and its stratigraphy, southwestern Louisiana. *Proc. 9th Ann. Meeting Gulf Coast Assoc. Geol. Soc.*, 9: 237–60.

CAILLEUX, A. (1945). Distinction des galets marins et fluviatile. *Soc. Géol. France Bull.*, series 5, 15: 375–404.

CAMPBELL, C. V. (1971). Depositional model—upper cretaceous Gallup Beach shoreline, Ship Rock area, northwestern New Mexico. *J. Sediment. Petrol.*, 41: 395–409.

CARR, A. P. (1969). Size grading along a pebble beach: Chesil Beach, England. *J. Sediment. Petrol.*, 39: 297–311.

——— (1971). Experiments on longshore transport and sorting of pebbles: Chesil Beach, England. *J. Sediment. Petrol.*, 41: 1084–104.

CARSTENS, M. R., F. M. NEILSON, and H. D. ALTINBILEK (1969). *Bed forms generated in the laboratory under an oscillatory flow: analytical and experimental study.* U.S. Army Corps of Engineers Coastal Engr. Res. Center Tech. Memo. no. 28, 93 pp.

CHAPPELL, JOHN (1967). Recognizing fossil strand lines from grain-size analysis. *J. Sediment. Petrol.*, 37: 157–65.

CHAVE, E. K. (1960). Carbonate skeletons to limestone. *Problems, N.Y. Acad. Sci.*, series 2, 23: 14–24.

——— (1964). Skeletal durability and preservation. In *Approaches to paleoecology*, ed. J. Imbrie and N. Newell, pp. 377–87. Wiley, New York.

CLIFTON, H. E. (1969). Beach lamination: nature and origin. *Mar. Geol.*, 7: 553–59.

CLIFTON, H. E., R. E. HUNTER, and R. L. PHILLIPS (1971). Depositional structures and processes in the non-barred high-energy nearshore. *J. Sediment. Petrol.*, 41: 651–70.

COOK, D. O. (1971). Depressions in shallow marine sediment made by benthic fish. *J. Sediment. Petrol.*, 41: 577–602.

COOK, D. O., and D. S. GORSLINE (1972). Field observations of sand transport by shoaling waves. *Marine Geology*, 13: 31–55.

COORAY, P. G. (1968). A note on the occurrence of beachrock along the west coast of Ceylon. *J. Sediment. Petrol.*, 38: 650–54.

CORNISH, V. (1898). On sea beaches and sand banks. *Geog. J.*, 2: 628–47.

CURRAY, J. R., F. J. EMMEL, and P. J. S. CRAMPTON (1969). Holocene history of a strand plain, lagoonal coast, Nayarit, Mexico. In *Lagunas costeras, un simposio*, ed. A.A. Castanares and F. B. Phleger, pp. 63–100. UNAM-UNESCO.

DAHL, E. (1953). Some aspects of the ecology and zonation of the fauna on sandy beaches. *Oikos*, 4, no. 1: 1–27.

DALY, R. A. (1924). *The geology of American Samoa.* Washington Carnegie Inst. Pub. no. 340, pp. 93–143.

DANA, J. D. (1851). On coral reefs and islands. *Am. J. Sci.*, series 2, 11: 357–72, 12: 25–51, 165–86, 329–38.

DAVIES, D. K., F. G. ETHRIDGE, and R. R. BERG (1971). Recognition of barrier environments. *Am. Assoc. Petrol. Geol. Bull.*, 55: 550–65.

DAVIS, R. A. (1965). Beach pitting: an unusual beach sand structure. *J. Sediment. Petrol.*, 35: 495–96.

DEMAREST, D. F. (1947). Rhomboid ripple marks and their relationship to beach slope. *J. Sediment. Petrol.*, 17: 18–22.

DICKINSON, K. A., H. L. BERRYHILL, and C. W. HOLMES (1972). Criteria for recognizing ancient barrier coastlines. In *Recognition of ancient sedimentary environments*, ed. J. K. Rigby and W. K. Hamblin, pp. 192–214. Soc. Econ. Paleon. and Min. Special Publ. no. 16.

DILLARD, W. R., D. P. OAK, and N. W. BASS (1941). Chanute oil pool, Neosho County, Kansas. In *Stratigraphic type oil fields*, ed. A.I. Levorsen, pp. 57–77. Am. Assoc. Petrol. Geol., Tulsa, Okla.

DILLON, W. P., and J. T. CONOVER (1965). Formation of ice-cemented sand blocks on a beach and lithologic implications. *J. Sediment. Petrol.*, 35: 964–67.

DOBKINS, J. E., and R. L. FOLK (1970). Shape development on Tahiti-Nui. *J. Sediment. Petrol.*, 40: 1167–203.

DÖRJES, J. (1971). Der Golf von Gaeta (Tyrrhenisches Meer), IV. das Makrobenthos und seine küstenparallele Zonierung. *Senckenbergiana Maritima*, 3: 203–46.

——— (1972). Georgia coastal region, Sapelo Island, U.S.A.—sedimentology and biology, VII: distribution and zonation of macrobenthic animals. *Senckenbergiana Maritima*, 4: 183–216.

DOTY, M. S. (1957). Rocky intertidal surfaces. In *Treatise on marine ecology and paleoecology*, ed. J. W. Hedgpeth, 1: 535–85. Geol. Society Am.

DRAPER, L. (1967). Wave activity at the sea bed around northwestern Europe. *Mar. Geol.*, 5: 133–40.

DRISCOLL, E. G. (1967). Experimental field study of shell abrasion. *J. Sediment. Petrol.*, 37: 1117–23.

DUANE, D. B. (1964). Significance of skewness in recent sediments, western Pamlico Sound, North Carolina. *J. Sediment. Petrol.*, 34: 864–74.

EARDLEY, A. J. (1939). Sediments of Great Salt Lake, Utah. *Am. Assoc. Petrol. Geol. Bull.*, 22: 1359–87.

EMERY, K. O. (1944). Beach markings made by sand hoppers. *J. Sediment. Petrol.*, 14: 26–28.

——— (1945). Entrainment of air in beach sand. *J. Sediment. Petrol.*, 15: 39–49.

EMERY, K. O., and D. C. COX (1956). Beach rock in the Hawaiian Islands. *Pacif. Sci.*, 10: 382–402.

EMERY, K. O., and J. F. GALE (1951). Swash and swash marks. *Trans. Am. Geophys. Union*, 32: 31–36.

EMERY, K. O., and R. E. STEVENSON (1950). Laminated beach sand. *J. Sediment. Petrol.*, 20: 220–23.

ENGELS, H. (1905). Untersuchungen über die Bettausbildang gerader oder schwach gekrümmter Flussstrecken mit beweglicher Sohle. *Zeitschr. für Bauwesen*, 55: 663–80.

ENRIGHT, J. T. (1963). The tidal rhythm of activity of a sand-beach amphipod: *Zeitschr. für Vergleichende Physiol.*, 46: 276–313.

EVANS, O. F. (1938). Floating sand in the formation of swash marks. *J. Sediment. Petrol.*, 8: 71.

——— (1939). Sorting and transportation of material in the swash and backwash. *J. Sediment. Petrol.*, 9: 28–31.

——— (1941). The classification of wave-formed ripple marks. *J. Sediment. Petrol.*, 11: 37–41.

——— (1942). The relation between the size of wave-formed ripple marks, depth of water, and the size of the generating waves. *J. Sediment. Petrol.*, 12: 31–35.

FABRICIUS, F. H., D. BERDAU, and K. O. MUNNICH (1970). Early Holocene ooids in modern littoral sands reworked from a coastal terrace, southern Tunisia. *Science*, 169: 757–60.

FAIRCHILD, H. L. (1901). Beach structure in Medina sandstone. *Am. Geologist*, 28: 10.

FIELD, R. M. (1919). Remarks on beachrock in Dry Tortugas. *Carnegie Inst. Yearbook*, no. 18, p. 198.

FISK, H. N. (1955). Sand facies of recent Mississippi delta deposits. *Proc. 4th World Petrol. Cong.*, sec. I/C, 3: 377–98.

FLEMING, N. C. (1964). Tank experiments on the sorting of beach material during cusp formation. *J. Sediment. Petrol.*, 34: 112–20.

FOLK, R. L., and R. ROBLES (1964). Carbonate sands of Isla Perez, Alacran Reef Complex, Yucatan. *J. Geol.*, v. 72: 255–92.

FOX, D. L. (1950). Comparative metabolism of organic detritus by inshore animals. *Ecology*, 31: 100–108.

FOX, W. T., J. W. LADD, and M. K. MARTIN (1966). A profile of the four moment measures perpendicular to a shore line, South Haven, Michigan. *J. Sediment. Petrol.*, 36: 1126–30.

FREY, R. W. (1970). Environmental significance of recent marine lebensspuren near Beaufort, North Carolina. *J. Paleontol.*, 44: 507–19.

FREY, R. W., and J. D. HOWARD (1972). Georgia coastal region, Sapelo Island, U.S.A.—sedimentary and biology, VI. radiographic study of sedimentary structures made by beach and offshore animals in aquaria. *Senckenbergiana Maritima*, 4: 169–82.

FREY, R. W., and T. V. MAYOU (1971). Decapod burrows in Holocene barrier island beaches and washover fans, Georgia. *Senckenbergiana Maritima*, 3: 53–77.

FRIEDMAN, G. M. (1961). Distinction between dune, beach, and river sands from their textural characteristics. *J. Sediment.* Petrol., 31: 514–29.

——— (1967). Dynamic processes and statistical parameters compared for size frequency and distribution of beach and river sands. *J. Sediment. Petrol.*, 37: 327–54.

GIBBONS, G. S. (1967). Shell content in quartzose beach and dune sands, Dee Why, New South Wales. *J. Sediment. Petrol.*, 37: 869–78.

GILES, R. T., and O. H. PILKEY (1965). Atlantic beach and dune sediments of the southern United States. *J. Sediment. Petrol.*, 35: 900–910.

Ginsburg, R. N. (1953). Beachrock in south Florida. *J. Sediment. Petrol.*, 23: 85–92.

Goldsmith, V. (1973). Internal geometry and origin of vegetated coastal sand dunes. *J. Sediment. Petrol.*, 43: 1128–43.

Gorsline, D. S. (1963). Bottom sediments of the Atlantic shelf and slope off the southern United States. *J. Geol.*, 71: 422–40.

Gould, H. R., and E. McFarlan (1959). Geologic history of the chenier plain, southwestern Louisiana. *Trans. Gulf Coast Assoc. Geol. Soc. 9th Annual Meeting*, 9: 261–70.

Greene, G. H. (1970). Microrelief of an Arctic beach. *J. Sediment. Petrol.*, 40: 419–27.

Greenwood, B. (1969). Sediment parameters and environmental discrimination: an application of multivariate statistics. *Can. J. Earth Sci.*, 6: 1347–58.

Greenwood, B. and R. G. D. Davidson-Arnott (1972). Textural variation in the subenvironments of the shallow-water wave zone, Kouchibouguas Bay, New Brunswick. *Can. J. Earth Sci.*, 9: 679–88.

Grogan, R. M. (1945). Shape variation of some Lake Superior beach pebbles: *J. Sediment. Petrol.*, 15: 3–10.

Hails, J. R. (1967). Significance of statistical parameters for distinguishing sedimentary environments in New South Wales, Australia. *J. Sediment. Petrol.*, 37; 1059–69.

Hardcastle, R. J., and A. C. King (1972). Chesil Beach sea wave records. *Civ. Eng. and Pub. Works Rev.*, 67, no. 788 (March): 299–300.

Harms, J. C. (1969). Hydraulic significance of some sand ripples. *Geol. Soc. Am. Bull.*, 30: 363–96.

Hayes, M. O. (1972). Forms of sediment accumulation in the beach zone. In *Waves on beaches*, ed. R. E. Meyer, pp. 297–356. Academic Press, New York.

Hedgpeth, J. W. (1957). Sandy beaches. In *Treatise on marine ecology and paleoecology*, ed. J. W. Hedgpeth, 1: 587–608. Geol. Soc. Am.

Hertweck, G. (1971). Der Golf von Gaeta (Tyrrhenisches Meer), V: Abfolge der Biofaziesbereiche in den Vorstand- und Schelfsedimenten. *Senckenbergiana Maritima*, 3: 247–76.

——— (1972). Georgia coastal region, Sapelo Island, U.S.A.—sedimentology and biology, V: distribution and environmental significance of lebensspuren and in-situ skeletal remains. *Senckenbergiana Maritima*, 4: 125–67.

Hill, G. W., and R. E. Hunter (1973). Burrows of the ghost crab *Ocypode quadrata* (Fabricius) on the barrier islands, south-central Texas coast. *J. Sediment. Petrol.*, 43: 24–30.

Hodgson, A. V., and W. B. Scott (1970). The identification of ancient beach sands by the combination of size analysis and electron microscopy: *Sedimentology*, 14: 67–75.

Howard, J. D., and J. Dörjes (1972). Animal-sediment relationships in two beach-related tidal flats, Sapelo Island, Georgia. *J. Sediment. Petrol.*, 42: 608–23.

Howard, J. D., and H.-E. Reineck (1972). Georgia coastal region, Sapelo Island, U.S.A.—sedimentology and biology, IV: physical and biogenic sedimentary structures of the nearshore shelf. *Senckenbergiana Maritima*, 4: 81–123.

Hoyt, J. H. (1962). High-angle beach stratification, Sapelo Island, Georgia. *J. Sediment. Petrol.*, 32: 309–11.

Hoyt, J. H., and V. J. Henry (1963). Rhomboid ripple mark, indicator of current direction and environment. *J. Sediment. Petrol.*, 33: 604–608.

——— (1967). Influence of island migration on barrier-island sedimentation. *Geol. Soc. Am. Bull.*, 78: 77–86.

HUME, J. D., and M. SCHALK (1964). The effects of ice-push on Arctic beaches. *Am. J. Sci.*, 262: 267–73.

HUNT, A. R. (1882). On the formation of ripple-mark. *Proc. Roy. Soc.* (London), series A, 34: 1–19.

HUNTER, R. E. (1969). Eolian microridges on modern beaches and a possible ancient example. *J. Sediment. Petrol.*, 39: 1573–78.

ILLING, L. V. (1954). Bahaman calcareous sands. *Am. Assoc. Petrol. Geol. Bull.*, 38: 1–95.

IMBRIE, J. and TJ. H. VAN ANDEL (1964). Vector analysis of heavy mineral data. *Geol. Soc. Am. Bull.*, 75: 1131–56.

INGLE, J. C. (1966). *The movement of beach sand.* Elsevier, Amsterdam, 221 pp.

INMAN, D. L. (1953). *Areal and seasonal variations in beach and nearshore sediments at La Jolla, California.* U.S. Army Corps of Engineers Beach Erosion Board Tech. Memo. no. 39, 134 pp.

——— (1957). *Wave-generated ripples in nearshore sands.* U.S. Army Corps of Engineers Beach Erosion Board Tech. Memo. no. 100, 67 pp.

INMAN, D. L., D. C. EWING, and J. B. CORLISS (1966). Coastal sand dunes of Guerrero Negro, Baja California, Mexico. *Geol. Soc. Am. Bull.*, 77: 787–802.

JOHNSON, D. W. (1919). *Shore processes and shoreline development.* Wiley, New York, 584 pp. [See 1965 facsimile by Hafner, New York.]

JOHNSON, K. G., and G. M. FRIEDMAN (1969). The Tully clastic correlatives (upper Devonian) of New York State: A model for recognition of alluvial, dune (?), and offshore sedimentary environments in a tectonic delta complex. *J. Sediment. Petrol.*, 39: 451–85.

JOYCE, J. R. F. (1950). Notes on ice-foot development, Neny Fjord, Graham Land, Antarctica. *J. Geol.*, 58: 646–49.

KENNEDY, J. F., and M. FALCON (1965). *Wave-generated sediment ripples.* M.I.T. Hydrodynamics Laboratory Report no. 86, Cambridge, 86 pp.

KINDLE, E. M. (1936). Dominant factors in the formation of firm and soft sand beaches. *J. Sediment. Petrol.*, 6: 16–22.

KING, C. A. M. (1972). *Beaches and coasts.* 2nd ed., St. Martin's Press, New York, 570 pp.

KLEIN, G. DEV. (1967). Paleocurrent analysis in relation to modern marine sediment dispersal patterns. *Am. Assoc. Petrol. Geol. Bull.*, 51: 366–82.

KLOVAN, J. E. (1966). The use of factor analysis in determining depositional environments from grain-size distributions. *J. Sediment. Petrol.*, 36: 115–25.

KOLMER, J. R. (1973). A wave tank analysis of the beach foreshore grain size distribution. *J. Sediment. Petrol.*, 43: 200–204.

KOMAR, P. D. (1974). Oscillatory ripple marks and the evaluation of ancient wave conditions and environments. *J. Sediment. Petrol.*, 44: 169–80.

KOMAR, P. D., R. H. NEUDECK, and L. D. KULM (1972). Observations and significance of deep-water oscillatory ripple marks on the Oregon continental shelf. In *Shelf sediment transport*, ed. by D. Swift, D. Duane and O. Pilkey, pp. 601–19. Dowden, Hutchinson & Ross, Stroudsburg, Pa.

KRAFT, J. C. (1971). *A guide to the geology of Delaware's coastal environments*. Prepared for Geol. Soc. Am. ann. meeting and field trip, Nov. 1971, Boulder, Col., 220 pp.

KRAFT, J. C., R. B. BIGGS, and S. D. HALSEY (1973). Morphology and vertical sedimentary sequence models in Holocene transgressive barrier systems. In *Coastal geomorphology*, ed. D. R. Coates, pp. 321–54. State Univ. of New York, Binghamton.

KRINSLEY, D. H., and J. C. DOORNKAMP (1973). *Atlas of quartz sand surface textures*. Cambridge Univ. Press, Cambridge, 91 pp.

KRUMBEIN, W. C. (1938). Local areal variation of beach sands. *Geol. Soc. Am. Bull.*, 49: 653–58.

——— (1942). Settling-velocity and flume-behavior of non-spherical particles. *Trans. Am. Geophys. Union*, 23: 621–33.

KRUMBEIN, W. C., and J. S. GRIFFITH (1938). Beach environment in Little Sister Bay, Wisconsin. *Geol. Soc. Am. Bull.*, 49: 629–52.

KUENEN, PH. H. (1950). *Marine geology*. Wiley, New York, 568 pp.

——— (1960). Experimental abrasion, 4: eolian action: *J. Geol.*, 68: 427–49.

——— (1964). Experimental abrasion, 6: surf action. *Sedimentology*, 3: 29–43.

KULM, L. D., R. C. ROUSH, J. C. HARLETT, R. H. NEUDECK, D. M. CHAMBERS, and E. J. RUNGE (1975). Oregon continental shelf sedimentation: interrelationships of facies distribution and sedimentary processes. *J. Geol.*, 83: 145–75.

LANDON, R. E. (1930). An analysis of beach pebble abrasion and transportation. *J. Geol.*, 38: 437–46.

LEVER, J. A. (1958). Quantitative beach research, I: the "left-right phenomenon"—sorting of Lamellibranch valves on sandy beaches. *Basteria*, 22: 21–51.

LEWIS, W. V. (1938). Evolution of shoreline curves. *Proc. Geol. Assoc.*, 49: 107–27.

MACCARTHY, G. R. (1933*a*). Calcium carbonate in beach sands. *J. Sediment. Petrol.*, 3: 61–67.

——— (1933*b*). The rounding of beach sands. *Am. J. Sci.*, 25: 205–24.

——— (1935). Eolian sands, a comparison. *Am. J. Sci.*, 30: 81–95.

MCGREGOR, A. A., and C. A. BIGGS (1968). Bell Creek field, Montana; a rich stratigraphic trap. *Am. Assoc. Petrol. Geol. Bull.*, 52: 1869–87.

MCKEE, E. D., J. CHRONIC, and E. B. LEOPOLD (1959). Sedimentary belts in the lagoon of Kapingamarangi Atoll. *Am. Assoc. Petrol. Geol. Bull.*, 43: 501–62.

MCLEAN, R. F., and R. M. KIRK (1969). Relationship between grain size, size-sorting, and foreshore slope on mixed sandy-shingle beaches. *N.Z. J. Geol.* and *Geophys.*, 12: 138–55.

MANOHAR, M. (1955). *Mechanics of bottom sediment movement due to wave action*. U.S. Army Corps of Engineers, Beach Erosion Board Tech. Memo. no. 75, 121 pp.

MARSHALL, P. (1929). Beach gravels and sands. *Trans. N.Z. Inst.*, 60: 324–65.

MARTINS, J. H. C. (1939). Beaches. In *Recent marine sediments*, ed. P. D. Trask, pp. 207–18. Amer. Assoc. Petrol. Geol., [Reprinted in 1968 by Dover, New York].

MASON, C. C., and R. L. FOLK (1958). Differentiation of beach, dune, and aeolian flat environments by size analysis, Mustang Island, Texas. *J. Sediment. Petrol.*, 28: 211–26.

MIDDLETON, G. V. (1970). *Experimental studies related to problems of flysch sedimentation*. Geol. Assoc. of Canada, Special Paper no. 7, pp. 253–72.

MILLER, R. L., and J. M. ZEIGLER (1958). A model relating sediment pattern in the region of shoaling waves, breaker zone, and foreshore. *J. Geol.*, 66: 417–41.

——— (1964). A study of sediment distribution in the zone of shoaling waves over complicated bottom topography. In *Papers in marine geology*, ed. R. L. Miller, pp. 133–53. Macmillan, New York.

MOBERLY, RALPH, L. D. BAVER, and ANNE MORRISON (1965). Source and variation of Hawaiian littoral sand. *J. Sediment. Petrol.*, 35: 589–98.

MOGRIDGE, G. R., and J. W. KAMPHUIS (1973). Experiments on bed form generation by wave action. *Proc. 13th Conf. Coast. Eng.*, pp. 1123–42.

MOIOLA, R. J., and D. WEISER (1968). Textural parameters: an evaluation. *J. Sediment. Petrol.*, 38: 45–53.

MOORE, G. W. (1966). Arctic beach sedimentation. In *Environment of the Cape Thompson region, Alaska*, ed. Wilimovsky and Wolfe, pp. 587–608. U.S. Atomic Energy Comm., Oak Ridge, Tenn.

MUELEN, H. V. D. (1947). Nieuwe gezechtspunten. *Tet Zeepard*, 7: 10–12.

NAGLE, J. S. (1967). Wave and current orientation of shells. *J. Sediment. Petrol.*, 37: 1124–38.

NEATE, D. J. M. (1967). Underwater pebble grading on Chesil Bank. *Proc. Geol. Assoc.*, 78: 419–26.

NEWELL, N. D., E. G. PURDY, and J. IMBRIE (1960). Bahamian oolitic sand. *J. Geol.*, 68: 481–97.

NEWTON, R. S. (1968). Internal structure of wave-formed ripple marks in the nearshore zone. *Sedimentology*, 11: 275–92.

NICOLS, R. L. (1961). Characteristics of beaches formed in polar climates. *Am. J. Sci.*, 259: 694–708.

OTVOS, E. G. (1964). Observations on rhomboid beach marks. *J. Sediment. Petrol.*, 34: 683–87.

——— (1965). Types of rhomboid beach surface patterns. *Am. J. Sci.*, 263: 271–76.

PALMER, R. H. (1928). Sand holes of the strand. *J. Geol.*, 36: 176–80.

PEABODY, F. E. (1947). Current crescents in the Triassic Moenkopi formation. *J. Sediment. Petrol.*, 17: 73–76.

PETTIJOHN, F. J., and A. C. LUNDAHL (1943). Shape and roundness of Lake Erie beach sands. *J. Sediment. Petrol.*, 13: 69–78.

PETTIJOHN, F. J., and J. D. RIDGE (1932). A textural variation series of beach sands from Cedar Point, Ohio. *J. Sediment. Petrol.*, 2: 76–88.

——— (1933). A mineral variation series of beach sands from Cedar Point, Ohio. *J. Sediment. Petrol.*, 3: 92–94.

RAO, C. B. (1957). Beach erosion and concentration of heavy mineral sands. *J. Sediment. Petrol.*, 27: 143–47.

RAYMOND, P. E., and F. HUTCHINS (1932). A calcareous beach at John O'Groats, Scotland. *J. Sediment. Petrol.*, 2: 63–67.

READE, T. M. (1884). Miniature domes in sand. *Geol. Mag.*, 21: 20–22.

REINECK, H.-E., and I. B. SINGH (1971). Der Golf von Gaeta (Tyrrhenisches Meer), III: die Gefüge von Vorstrand und Schelfsedimenten. *Senckenbergiana Maritima*, 3: 185–201.

——— (1973). *Depositional sedimentary environments*. Springer-Verlag, New York, 439 pp.

REINECK, H.-E., and F. WUNDERLICH (1968). Classification and origin of flaser and lenticular bedding. *Sedimentology*, 11: 99–104.

REX, R. W. (1964). Arctic beaches, Barrow, Alaska. In *Papers in marine geology*, ed. R. L. Miller, pp. 384–99. Macmillan, New York.

REYMENT, R. A. (1968). Orthoconic nautiloids as indicators of shoreline surface currents. *J. Sediment. Petrol.*, 38: 1387–89.

ROTTMANN, C. J. F. (1973). Surf zone shape changes in quartz grains on pocket beaches, Cape Arago, Oregon. *J. Sediment. Petrol.*, 43: 188–99.

RUSNAK, G. A. (1960). Some observations of recent oolites: *J. Sediment. Petrol.*, 30: 471–80.

RUSSELL, R. J. (1962). Origin of beach rock. *Zeitschr. für Geomorphol.*, 6: 1–16.

RUSSELL, R. J., and H. V. HOWE (1935). Cheniers of southwestern Louisiana. *Geog. Rev.*, 25: 449–61.

SAMES, C. W. (1966). Morphometric data of some recent pebble associations and their application to ancient deposits. *J. Sediment. Petrol.*, 36: 126–42.

SCHALK, M. (1938). A textural study of the outer beach of Cape Cod, Massachusetts. *J. Sediment. Petrol.*, 8: 41–54.

SCHIFFMAN, A. (1965). Energy measurements in the swash-surf zone. *Limnol. and Oceanog.*, 10: 255–60.

SCOFFIN, T. P. (1970). A conglomeratic beachrock in Bimini, Bahamas. *J. Sediment. Petrol.*, 40: 756–59.

SCOTT, T. (1954). *Sand movement by waves.* U.S. Army Corps of Engineers, Beach Erosion Board Tech. Memo. no. 48, 37 pp.

SENGUPTA, S. (1966). Studies on orientation and imbrication of pebbles with respect to cross-stratification. *J. Sediment. Petrol.*, 36: 362–69.

SHEPARD, F. P., and R. YOUNG (1961). Distinguishing between beach and dune sands. *J. Sediment. Petrol.*, 31: 196–214.

SOLOHUB, J. T., and J. E. KLOVAN (1970). Evaluation of grain-size parameters in lacustrine environments. *J. Sediment. Petrol.*, 40: 81–101.

SONU, C. J. (1972). Bimodal composition and cyclic characteristics of beach sediments in continuously changing profiles. *J. Sediment. Petrol.*, 42: 852–57.

STARK, J. T., and E. C. DAPPLES (1941). Nearshore coral lagoon sediments from Raiatea, Society Islands. *J. Sediment. Petrol.*, 11: 21–27.

STODDART, D. R., and J. R. CANN (1965). Nature and origin of beach rock. *J. Sediment. Petrol.*, 35: 243–73.

TANNER, W. F. (1956). Examples of probable lithified beachrock. *J. Sediment. Petrol.*, 26: 307–12.

——— (1965). High-index ripple marks in the swash zone. *J. Sediment. Petrol.*, 35: 968.

——— (1967). Ripple mark indices and their uses. *Sedimentology*, 9: 89–104.

THIEL, G. A. (1940). The relative resistance to abrasion of mineral grains of sand size. *J. Sediment. Petrol.*, 10: 103–24.

THOMPSON, W. O. (1937). Original structures of beaches, bars, and dunes. *Geol. Soc. Am. Bull.*, 48: 723–52.

——— (1949). Lyons sandstone of Colorado Front Range. *Am. Assoc. Petrol. Geol. Bull.*, 33: 52–72.

TWENHOFEL, W. H. (1932). *Treatise on sedimentation.* 2nd ed. Williams & Wilkin, Baltimore, 2 volumes, 926 pp. [Reprinted in 1961 by Dover, New York.]

——— (1945). The rounding of sand grains. *J. Sediment. Petrol.*, 15: 59–71.

VAIL, C. E. (1917). Lithologic evidence of climatic pulsations. *Science*, 46, no. 1178: 90–93.

VAN DER LINGEN, G. J., and P. B. ANDREWS (1969). Hoof-print structures in beach sand. *J. Sediment. Petrol.*, 39: 350–57.

VAN STRAATEN, L. M. J. U. (1950). Giant ripples in tidal channels. *Koninkl. Ned. Aardrijkskde Genoot.*, 67: 336–41.

VISHER, G. S. (1969). Grain size distribution and depositional processes: *J. Sediment. Petrol.*, 39: 1074–106.

WADELL, H. (1932). Volume, shape, and roundness of rock particles. *J. Geol.*, 40: 443–51.

WALKER, T. R., and J. C. HARMS (1973). Eolian origin of flagstone beds, Lyons sandstone (Permian), type area, Boulder County, Colorado. *Mountain Geologist*, 9: 279–88.

WATSON, R. L. (1971). Origin of shell beaches, Padre Island, Texas. *J. Sediment. Petrol.*, 41: 1105–11.

WENTWORTH, C. K., and G. A. MACDONALD (1953). *Structures and forms of basaltic rocks in Hawaii.* U.S. Geol. Survey Bulletin no. 994, Washington, D.C., 98 pp.

WOBBER, F. J. (1967). The orientation of Donax on an Atlantic coast beach. *J. Sediment. Petrol.*, 37: 1233–51.

WOODFORD, A. O. (1935). Rhomboid ripple marks. *Am. J. Sci.*, 29: 518–25.

WUNDERLICH, F. (1971). Der Golf von Gaeta (Tyrrhenisches Meer), II: Strandaufbau und Stranddynamik. *Senckenbergiana Maritima*, 3: 135–83.

——— (1972). Georgia coastal region, Sapelo Island, U.S.A.—sedimentology and biology, III: beach dynamics and beach development. *Senckenbergiana Maritima*, 4: 47–79.

YASSO, W. E. (1966). Heavy mineral concentrations and sastrugi-like deflation furrows in a beach salcrete at Rockaway Point, New York. *J. Sediment. Petrol.*, 36: 836–38.

ZINGG, T. (1935). Beitrag zur Schotteranalyse. *Schweiz. Min.-Petr. Mitt.*, 15, no. 1: 39–140.

ZUMBERGE, J. H., and J. T. WILSON (1954). Effects of ice on shore development. *Proc. 4th Conf. on Coast. Eng.*, pp. 201–5.

SUBJECT INDEX

L

M

N

O

P

R

S

T

W

AUTHOR INDEX

*References given in italics.

C

D

K

L

M

N

O

P

Q

R

S

T

U

V

W

Y

Z

LOCATION INDEX

O

P

R

S

T

U

V

W

Y